Modern Agronomy

Modern Agronomy

Edited by Roy Tucker

Syrawood Publishing House
New York

Published by Syrawood Publishing House,
750 Third Avenue, 9th Floor,
New York, NY 10017, USA
www.syrawoodpublishinghouse.com

Modern Agronomy
Edited by Roy Tucker

© 2020 Syrawood Publishing House

International Standard Book Number: 978-1-68286-849-2 (Hardback)

This book contains information obtained from authentic and highly regarded sources. Copyright for all individual chapters remain with the respective authors as indicated. All chapters are published with permission under the Creative Commons Attribution License or equivalent. A wide variety of references are listed. Permission and sources are indicated; for detailed attributions, please refer to the permissions page and list of contributors. Reasonable efforts have been made to publish reliable data and information, but the authors, editors and publisher cannot assume any responsibility for the validity of all materials or the consequences of their use.

Trademark Notice: Registered trademark of products or corporate names are used only for explanation and identification without intent to infringe.

Cataloging-in-Publication Data

Modern agronomy / edited by Roy Tucker.
   p. cm.
Includes bibliographical references and index.
ISBN 978-1-68286-849-2
1. Agronomy. 2. Crops. 3. Soil management. I. Tucker, Roy.
SB91 .M63 2020
631--dc23

TABLE OF CONTENTS

Preface IX

Chapter 1 **High-Throughput Screening of Sensory and Nutritional Characteristics for Cultivar Selection in Commercial Hydroponic Greenhouse Crop Production** 1
Atef M. K. Nassar, Stan Kubow and Danielle J. Donnelly

Chapter 2 **Colonization and Spore Richness of Arbuscular Mycorrhizal Fungi in Araucaria Nursery Seedlings** 29
Carlos Vilcatoma-Medina, Glaciela Kaschuk and Flávio Zanette

Chapter 3 **Effect of Irrigation and Mulch on the Yield and Water use of Strawberry** 35
Rahena Parvin Rannu, Razu Ahmed, Alamgir Siddiky, Abu Saleh Md. Yousuf Ali, Khandakar Faisal Ibn Murad and Pijush Kanti Sarkar

Chapter 4 **Phenetic Analysis of Cultivated Black Pepper (*Piper nigrum* L.)** 45
Y. S. Chen, M. Dayod and C. S. Tawan

Chapter 5 **Agronomic Evaluation of Bunching Onion in the Colombian Cundiboyacense High Plateau** 56
Carlos H. Galeano Mendoza, Edison F. Baquero Cubillos, José A. Molina Varón and María del Socorro Cerón Lasso

Chapter 6 **Plant Sources, Extraction Methods, and uses of Squalene** 64
M. Azalia Lozano-Grande, Shela Gorinstein, Eduardo Espitia-Rangel, Gloria Dávila-Ortiz and Alma Leticia Martínez-Ayala

Chapter 7 **Microorganisms in Soils of Bovine Production Systems in Tropical Lowlands and Tropical Highlands** 77
Licet Paola Molina-Guzmán, Paula Andrea Henao-Jaramillo, Lina Andrea Gutiérrez-Builes and Leonardo Alberto Ríos-Osorio

Chapter 8 **Diversity among Modern Tomato Genotypes at Different Levels in Fresh-Market Breeding** 86
Krishna Bhattarai, Sadikshya Sharmas and Dilip R. Panthee

Chapter 9 **Bridging a Gap between Cr(VI)-Induced Oxidative Stress and Genotoxicity in Lettuce Organs after a Long-Term Exposure** 101
Cristina Monteiro, Sara Sario, Rafael Mendes, Nuno Mariz-Ponte, Sónia Silva, Helena Oliveira, Verónica Bastos, Conceição Santos and Maria Celeste Dias

Chapter 10 **Effect of Integrated Inorganic and Organic Fertilizers on Yield and Yield Components of Barley** 109
Tolera Abera, Tolcha Tufa, Tesfaye Midega, Haji Kumbi and Buzuayehu Tola

Chapter 11	**Distribution, Density, and Abundance of Parthenium Weed (*Parthenium hysterophorus* L.)**	116
	C. M. Maszura, S. M. R. Karim, M. Z. Norhafizah, F. Kayat and M. Arifullah	
Chapter 12	**Evaluation of Chickpea Varieties and Fungicides for the Management of Chickpea Fusarium Wilt Disease (*Fusarium oxysporum* f.sp. *ciceris*) at Adet Sick Plot**	124
	Yigrem Mengist, Samuel Sahile, Assefa Sintayehu and Sanjay Singh	
Chapter 13	**Effect of Planting Material Type on Experimental Trial Quality and Performance Ranking of Sugarcane Genotypes**	131
	Michel Choairy de Moraes, Ana Carolina Ribeiro Guimarães, Dilermando Perecin and Manuel Benito Sainz	
Chapter 14	**Genetic Variability and its Implications on Early Generation Sorghum Lines Selection for Yield, Yield Contributing Traits, and Resistance to Sorghum Midge**	139
	Massaoudou Hamidou, Abdoul Kader M. Souley, Issoufou Kapran, Oumarou Souleymane, Eric Yirenkyi Danquah, Kwadwo Ofori, Vernon Gracen and Malick N. Ba	
Chapter 15	**Sesame Yield Response to Deficit Irrigation and Water Application Techniques**	149
	E. K. Hailu, Y. D. Urga, N. A. Sori, F. R. Borona and K. N. Tufa	
Chapter 16	**Assessment of an Invasive Weed "Maimaio" *Commelina foecunda* in the Sesame Fields**	155
	G. Zenawi, T. Goitom and B. Fiseha	
Chapter 17	**Screening Selected *Solanum* Plants as Potential Rootstocks for the Management of Root-Knot Nematodes (*Meloidogyne incognita*)**	162
	Benjamin A. Okorley, Charles Agyeman, Naalamle Amissah and Seloame T. Nyaku	
Chapter 18	**Growth and Yield Performance of Roselle Accessions as Influenced by Intercropping with Maize**	171
	Emmanuel Ayipio, Moomin Abu, Richard Yaw Agyare, Dorothy Ageteba Azewongik and Samuel Kwame Bonsu	
Chapter 19	**Productivity and Water use Efficiency of Sorghum [*Sorghum bicolor* (L.) Moench] Grown under Different Nitrogen Applications**	178
	Hakeem A. Ajeigbe, Folorunso Mathew Akinseye, Kunihya Ayuba and Jerome Jonah	
Chapter 20	**Analysis of Direct and Indirect Selection and Indices in Bread Wheat (*Triticum aestivum* L.) Segregating Progeny**	189
	Zine El Abidine Fellahi, Abderrahmane Hannachi and Hamenna Bouzerzour	
Chapter 21	**Insecticide Seed Treatments Reduced Crop Injury from Flumioxazin, Chlorsulfuron, Saflufenacil, Pyroxasulfone, and Flumioxazin + Pyroxasulfone + Chlorimuron in Soybean**	200
	N. R. Steppig, J. K. Norsworthy, R. C. Scott and G. M. Lorenz	

Chapter 22 **Transforming Triple Cropping System to Four Crops Pattern: An Approach of Enhancing System Productivity through Intensifying Land use System** 207
Md. Aminul Islam, Md. Jahedul Islam, M. Akkas Ali,
A. S. M. Mahbubur Rahman Khan, Md. Faruque Hossain and Md. Moniruzzaman

Permissions

List of Contributors

Index

PREFACE

The science and technology of the production and use of plants for food, fiber, fuel and land reclamation may be termed as agronomy. The field encompasses the areas of plant physiology, plant genetics, soil science and meteorology. Some of the issues dealt under agronomy are the management of the environmental impact of agriculture, creation of healthier food and generation of energy from plants. Plant breeding, soil fertility, weed control, irrigation and drainage, pest control, etc. are important focus areas of agronomic studies. Modern agronomy is concerned with the production of the crops with improved nutritional value and increased crop yields under diverse environmental conditions. This is possible through selective breeding of crops. In addition, the quality of the soil needs to be preserved and the effects of erosion controlled. The field of agronomy addresses all these vital issues. The various sub-fields of agronomy along with technological progress that have future implications are glanced at in this book. It includes some of the vital pieces of work being conducted across the world, on various topics related to agronomy. For someone with an interest and eye for detail, this book covers the most significant topics in this field.

All of the data presented henceforth, was collaborated in the wake of recent advancements in the field. The aim of this book is to present the diversified developments from across the globe in a comprehensible manner. The opinions expressed in each chapter belong solely to the contributing authors. Their interpretations of the topics are the integral part of this book, which I have carefully compiled for a better understanding of the readers.

At the end, I would like to thank all those who dedicated their time and efforts for the successful completion of this book. I also wish to convey my gratitude towards my friends and family who supported me at every step.

Editor

High-Throughput Screening of Sensory and Nutritional Characteristics for Cultivar Selection in Commercial Hydroponic Greenhouse Crop Production

Atef M. K. Nassar,[1,2,3] Stan Kubow,[2] and Danielle J. Donnelly[1]

[1] Plant Science Department, Macdonald Campus of McGill University, 21111 Lakeshore Road, Sainte Anne de Bellevue, QC, Canada H9X 3V9
[2] School of Dietetics and Human Nutrition, Macdonald Campus of McGill University, 21111 Lakeshore Road, Sainte Anne de Bellevue, QC, Canada H9X 3V9
[3] Plant Protection Department, Faculty of Agriculture, Damanhour University, Damanhour, Albeheira 22516, Egypt

Correspondence should be addressed to Danielle J. Donnelly; danielle.donnelly@mcgill.ca

Academic Editor: Kent Burkey

Hydroponic greenhouse-grown and store-bought cultivars of tomato (cherry and beefsteak), cucumbers, bibb lettuce, and arugula were investigated to see if they could be distinguished based on sensory qualities and phytonutrient composition. Only the more dominant sensory criteria were sufficiently robust to distinguish between cultivars and could form the core of a consolidated number of criteria in a more discriminating sensory evaluation test. Strong determinants for cultivar selection within each crop included the following: mineral analysis (particularly Cu, Fe, K, Mg, and P); total carotenoids (particularly β-carotene, lycopene, and lutein); total carbohydrate (except in arugula); organic acids; total phenolics and total anthocyanins (except in cucumber). Hydroponically grown and store-bought produce were of similar quality although individual cultivars varied in quality. Storage at 4°C for up to 6 days did not affect phytonutrient status. From this, we conclude that "freshness," while important, has a longer duration than the 6 days used in our study. Overall, the effect of cultivar was more important than the effect of growing method or short-term storage at 4°C under ideal storage conditions.

1. Introduction

Modern agriculture aims to maximize yield and quality of plants with efficient use of resources and labor. Growing plants in a soilless environment on a supportive substrate (e.g., coir, peat, or rockwool) conserves water, nutrients, and space and is defined as hydroponics [1, 2]. Hydroponics has long been considered a powerful method of commercial vegetable and fruit production where growing conditions (light, temperature, medium composition, etc.) are fully controlled. More recently, advances in biocontrol technology have enabled pesticide-free produce. The acreage of hydroponic greenhouses ranges from 3,000 to 4,000 in the United States and Canada [2]. Canadians consume large amounts of vegetables ~337 g/d, about 123 kg/capita/yr [3].

Important considerations for hydroponically grown greenhouse (GH) produce are microclimate and fertigation components that could potentially affect various plant characteristics. It is our contention that if these are held constant, selection of nutritionally optimal cultivars can be performed, based on a combination of sensory and phytonutrient characteristics. The latter includes biochemical composition and proximate traits. Fruits and vegetables are rich in phytonutrients including anthocyanins, carotenoids, minerals, organic acids, polyphenolics, and vitamins (e.g., vitamins C and E) [4, 5]. Phytonutrients contribute to the general acceptability of any food as they influence taste, color, odor, and nutritional quality. Consumption of fruits and vegetables increases the intake of antioxidant compounds. These dietary antioxidants act as scavengers of reactive oxygen and nitrogen species and prevent lipid oxidation. This decreases DNA damage related to oxidative stress, implicated in causing metabolic diseases that can lead to cancer, as well as cardiovascular and other diseases [6, 7].

Four salad crops were examined in this study including tomato, cucumber, and the leafy greens arugula and bibb lettuce. Tomato consumption in Canada in 2007 was 87 g/capita/d or 32 kg/capita/yr [3]. Canadian consumption of cucumber and lettuce were 4.76 and 9.98 kg/capita/yr, respectively, in 2009 [8] while statistics on Canadian arugula consumption could not be located. Tomato contains phytochemicals including carotenoids (especially β-carotene and lycopene), flavonoids, folate, minerals, polyphenols, and vitamins (C and E) [9]. Epidemiological studies showed that tomato phytochemicals (especially lycopene) reduce cancer (e.g., prostate cancer), act as in vivo antioxidants, enhance cell-to-cell communication, modulate cell-cycle progression, and alter xenobiotic metabolism [9]. Regular consumption of tomato has been related to reduced incidence of both prostate cancer and heart diseases (e.g., myocardial infarction) [10–12], as well as promotion of healthy eye function [13–16]. Cucumbers contain significant amounts of triterpenoid cucurbitacins that possess antioxidant, anti-inflammatory, and anticancer properties [17]. In particular, leafy salad vegetables (arugula and lettuce) provide antioxidant compounds, fiber, polyphenolics, and vitamin C [18–20]. Green leafy vegetables are also recommended as a source of the provitamin A carotenoids [21].

Few studies have been conducted on the phytonutrient composition of hydroponically produced vegetables. As these are grown under defined conditions of light, temperature, and fertigation, the nutritional content of specific cultivars may not vary widely from crop to crop. Although there may be some differences in nutritional composition related to season, differences are primarily related to light effects [22, 23]. Nutritional differences based on cultivar were explored for a range of crops grown in the Macdonald Campus Greenhouse and harvested fresh for comparison with store-bought produce of the same species.

Selection of the most pleasing and nutritious cultivars for each food species would contribute to maximum consumer satisfaction and health. Following sensory characterization and phytonutrient analysis, a method was developed to highlight the most revealing factors for high-throughput cultivar selection within each species (cherry and beefsteak tomato, cucumber, arugula, and bibb lettuce). A range of complex statistical methods was considered. Among the methods examined for data classification and mining were partial least squares or projection to latent structures (PLS), principal component analysis (PCA), variable clustering, classification trees, linear and quadratic discriminant analysis, neural networks, and support vector machines [24]. Following comparative study, a decision was made to use a combination of PLS and PCA, which yield much richer results than other methods [24–26] and combine them with VARCLUS (SAS procedure for division of variables into hierarchical clusters) to verify decision-making.

So the objectives of this research were to (1) utilize taste panels to assess the appearance, taste, and texture acceptability of different cultivars of GH tomato (cherry and beefsteak types), cucumber, and two leafy green crops (arugula and bibb lettuce), (2) compare the phytonutrient composition of different cultivars of these fresh GH crops with store-bought (SB) produce, and (3) identify the most revealing factors (including sensory and phytonutrient data) that could be used in a high-throughput screening of a large number of cultivars, within each crop species, to expediently identify the most appealing and nutritious ones.

2. Materials and Methods

2.1. Plant Materials. Six cherry tomato cultivars including 'Apero', 'Favorita', and 'Juanita' were grown hydroponically in the greenhouse. Seeds of 'Apero' and 'Favorita' were bought from Johnny's Selected Seeds (ME, USA) while seeds of 'Juanita' were bought from De Ruiter Seeds (Monsanto; Montreal, QC, Canada). Fruit of three cherry tomato cultivars were purchased from local stores and tested under their trade names: Jardino (product of EU; from an IGA store on Louis-Menard St.), Fruiterie (product of Mexico; from Inter-Marche on Cote Vertu Blvd.), and Cherries (product of Mexico; from Inter-Marche on Saint Laurence St.).

Four cultivars of beefsteak tomato including 'Arbason', 'Caramba', 'Geronimo,' and 'Trust' were grown hydroponically in the greenhouse. Seeds of 'Arbason,' 'Geronimo,' and 'Trust' were bought from Johnny's Selected Seeds and seeds of 'Caramba' were bought from De Ruiter Seeds. Fruit of three beef steak tomatoes were bought from local grocery stores and tested under their trade names: Kaliroy (produced in Mexico; from IGA on Louis-Menard St.), BionatureL (produced in Mexico; from Loblaws on Jean-Talon St.), and BionatureP (produced in Mexico; from Provigo on St. Urbain St.). 'Diva' Mini Cucumber was hydroponically grown in the greenhouse from seed bought from Johnny's Selected Seed. Three SB cultivars were tested under their trade names: Cool Cukes, Lebanese, and Mini cucumber.

Four cultivars of arugula and three of bibb lettuce were evaluated. The arugula 'Astro' was grown hydroponically from seeds bought from Johnny's Selected Seeds. Three arugula were SB and tested under their trade names: ADO (bought from Adonis) and BW and PRO (bought from Provigo). The bibb lettuce 'RexMT0' was grown in the greenhouse from seeds purchased from Johnny's Selected Seeds. Two store-bought bibb lettuce were tested under their trade names: ADO (bought from Adonis) and IGA (bought from IGA).

Hydroponically grown cherry tomato, cucumber, and leafy vegetables were stored in a household fridge (4°C) for 0, 3, and/or 6 d (based on produce availability). This enabled comparison of "fresh" and cold-stored produce.

2.2. Panelists and Taste Tests. Taste tests were carried out at McGill University in a taste test laboratory (McGill University Research Ethics Board Approval # 953-1110). Fifty persons were recruited and went through two rounds of training and then individuals with consistent ratings were selected. Panelists included 25 males and females from 18 to 52 years of age with no known allergies to any of the tested fruits and vegetables. They tested hydroponic greenhouse grown and store-purchased produce, including tomato, cucumber, arugula, and lettuce. Before each taste test session, panelists were instructed regarding terminology, test procedures, and the nature of the samples. The sensory analysis taste test for each crop was conducted over several sessions during 1 d, with a variable number of 10–20 panelists per session and results were subjected to statistical analysis.

2.2.1. Sample Preparation for Sensory Evaluation.
Hydroponically grown and SB tomato, cucumber, arugula, and bibb lettuce were rinsed well under tap water and air-dried for at least 15 min. Tomatoes were cut into thin round slices about 6 mm thick and then into halves, just before serving. Cucumber samples were cross-sliced (2-3 mm) just before serving. Panelists were served a plate containing, in the following order, 2-3 whole fruit of each cherry tomato cultivar, 2-3 slices of each cucumber cultivar, 2-3 half slices of each beefsteak tomato cultivar, and finally 2-3 leaves of each leafy green cultivar. Panelists were individually supplied with plain soda crackers and directed to have a bite after sampling and then sip some water to rinse their palate between each sample. More samples were given upon request and retasting was conducted when necessary.

2.2.2. Panelist Responses.
Panelists scored their responses onto a print-out paper sheet that was prepared specifically for each crop. After sensory evaluation, panelists were asked to answer demographic questions (multiple choice), which included age, gender, and frequency of fresh vegetable consumption. Tomato ripeness, sweetness/saltiness, juiciness, and general acceptability of cherry or beefsteak varieties were scored on 9-point hedonic scales, where 9: ripe, 5: neutral, 1: unripe. The same scale was used for sweetness/saltiness and responses were scored: 1: salty; 5: not salty or sweet; 9: sweet. Juiciness was scored on a 9-point score where 1: dislike; 5: neutral; and 9: like. General acceptability also was given scores from 1 to 9, where 1: dislike and 9: like.

Cucumber sensory characteristics including flesh color and firmness, aroma of the fruit, and flesh taste sweetness/bitterness traits were scored on a 9-point scale, where 1: dislike and 9: like for flesh color, firmness, and fresh aroma of the fruit. Sweetness/bitterness were scored on a 9-point scale, where 1: bitter, 5: not bitter or sweet, and 9: sweet. Arugula sensory characteristics including green/grassy, intensity of bitterness, astringent, and overall liking were rated on a 9-point hedonic scale, where 1 was the lowest and 9 was the strongest taste and liking. Bibb lettuce was evaluated for color, sogginess/freshness, off-odor, and overall quality. Sensory evaluation of lettuce was done using the same hedonic 9-point scale.

2.3. Mineral Analysis Using ICP-OES.
Greenhouse and SB produce were hand-rinsed under running tap water and then blotted onto paper towels and air-dried for 1-2 h. Samples consisted of 1 g cross-sectional slices from the middle of each of 7–12 whole fruit or up to 20 leaves of leafy vegetables. These were incubated overnight in 3 mL nitric acid (trace metal analysis grade, Fisher Scientific Co., ON, Canada) in a 10 mL Oak Ridge centrifuge tube (Thermo Scientific, NY, USA) placed in a fume hood. On the following day, samples were digested using a heating block (Thermolyne heater type 16500 Dri Bath model DB16525; Thermolyne, Dubuque, IA 52001, USA). The samples were heated to 105°C until no nitrous oxide gases (brown gases) were evolved. Samples were diluted (1 : 4) with Type-1 water (18 Ωcm) and mixed thoroughly (flipping tubes over) prior to injection into the ICP-OES apparatus for analysis. Control elemental stock standard solution (J. T. Baker, St. Louis, MO, USA) was used to calibrate the instrument before sample injection. The inductively coupled argon plasma optical emission spectrometer (ICP-OES) used for mineral analysis in this study was a model VISTA-MPX CCD Simultaneous ICP-OES (Varian Australia PTY Ltd., Australia). The settings were as follows: power 1.2 kW, plasma flow 15 Lmin^{-1}, argon pressure 32 L min^{-1} (600 kPa), nebulizer flow 0.75 Lmin^{-1}, auxiliary flow 1.5 L min^{-1}, pump rate 15 rpm, viewing height 10 mm, replicate reading time 10 s, and instrument stabilization delay 15 s [27]. Mineral contents were expressed as g or mg of macro- or microminerals, respectively, per 100 g DW based on the dry matter content of each variety.

2.4. Sample Preparation for Phytonutrient Analyses

2.4.1. Freeze-Drying and Grinding of Plant Samples.
Plant samples (about 1 kg of beef-steak, cherry tomato, or cucumber and 0.5 kg of arugula and bibb-lettuce leafy vegetables; all randomly selected, 3 replicates/variety) were homogenized and cut into thin slices (0.5–1 cm) and fast-frozen on aluminum plates (diameter of 25 cm) using liquid nitrogen and then freeze-dried at −60 to −70°C for up to 4 d in a freeze-dryer (Christ Freeze-Dryer, Gamma 1-16 LSC, Osterode, Germany). Freeze-dried samples (2 samples/replicate) were ground into a fine powder in liquid nitrogen and stored in a −80 C Freezer (Thermo Electron Corporation, OH, USA). Sample weights were recorded before (fresh weight) and after (dry weight) freeze-drying to calculate the sample dry matter content. Freeze-dried samples were used for phytonutrient analyses.

2.4.2. Preparation of Crude Extracts for Phytonutrient Measurements.
About 100 mg of freeze-dried powder was extracted with 2 mL of 90% methanol (MeOH). Samples were vortexed at maximum speed for 60 s, sonicated (Branson 2200, Branson Ultrasonics Corporation, CT, USA) for 30 min, and centrifuged at 3,500 rpm for 15 min at 4°C. Supernatants were collected into 15 mL Falcon tubes. The remaining pellet was reextracted with 1 mL of 90% MeOH and supernatants were combined. Crude extract was used to measure total soluble phenolics (Folin Ciocalteu or FC test), the antioxidant scavenging capacity using 2,2-diphenyl-1-picrylhydrazyl (DPPH), and 2,2′-azino-bis-3-ethylbenzothiazoline-6-sulphonic acid (ABTS) (hydrophilic phase) [28]. The remaining pellet was reextracted twice, each time with 1 mL hexane, and supernatants were combined. This extract was dried in a Speed-Vac (Thermo Savant, Waltham, MA, USA) and the residues resolubilized in 2 mL of 95% ethanol and used for the evaluation

of antioxidant scavenging capacity using ABTS (lipophilic phase).

2.5. Total Phenolics. Total extractable phenolic contents of GH or SB produce were evaluated with Folin-Ciocalteu (FC) reagent according to [29] using gallic acid (GA) as a standard. A 100 μL sample aliquot of extract or standard dilution was mixed with 2 mL water followed by 200 μL FC reagent (2 N). Tubes were vortexed and incubated at room temperature (RT) for 5 min; then 1 mL of aqueous sodium carbonate solution (20%) was added. Samples were vortexed and kept at RT for 1 h. Absorbance was measured at 765 nm in a Beckman DU 640 spectrophotometer (Beckman Instruments, Fullerton, CA, USA) using 1 cm disposable cells. All measurements were replicated 2 times. Total phenolic content was expressed as milligrams of gallic acid equivalent (mg GAE) per 100 g DW.

2.6. Antioxidant Capacity via ABTS (Hydrophilic and Lipophilic Phases). ABTS (7 mM) stock solution was prepared in 18 Ωcm^{-1} water. Radical cation of ABTS (ABTS$^{\bullet+}$) was produced by reacting ABTS solution with potassium persulfate ($K_2S_2O_8$) (2.45 mM) in the dark at RT for 12–16 h before use for complete oxidation of ABTS [30]. Oxidation of ABTS starts immediately after adding the $K_2S_2O_8$ but absorbance is not maximal or stable until after period more than 6 h. The radical is stable in this form for more than 2 d when stored in the dark at room temperature. The ABTS$^{\bullet+}$ solution was diluted with 95% ethanol to an absorbance of 0.70 ± 0.02 at 734 nm. About 1.2 mL of diluted ABTS$^{\bullet+}$ solution (A734 nm = 0.700 ± 0.020) was added to 100 μL of sample extracts (hydrophilic or lipophilic) or Trolox standards (0–15 μM) in ethanol and absorbance was reported 1 min after initial mixing and up to 5 min. Antioxidant scavenging activity was expressed as μg Trolox equivalent (TE)/100 g DW [28].

2.7. Antioxidant Capacity Using DPPH Radical Scavenging Capacity. Radical scavenging capacity assay was performed as described by [28]. About 100 μL of methanolic crude extract was added to 1.5 mL of DPPH (2.5 μM in methanol) and shaken vigorously. After incubation at RT for 30 min, the absorbance of the remaining DPPH molecules was determined at 517 nm in a Beckman DU 640 spectrophotometer using 1 cm disposable cells. The mean values were obtained from triplicate determinations. Antioxidant activity was expressed as mg GAE/100 g DW.

2.8. Total Anthocyanin Content (TAC). Total anthocyanin content was extracted following the method previously described [31–33]. One hundred micrograms of each freeze-dried sample was mixed with 4 mL acetone and the extracted material was separated from the cake by filtration on a Buchner funnel. The filter cake remnant was reextracted with 70% (v/v) aqueous acetone twice. Filtrates were combined and partitioned with chloroform (1:2 acetone:chloroform, v/v) and stored overnight at 1°C. The aqueous portion containing the TAC was recovered and separated from the residual acetone at 40°C and resolubilized in 0.01% HCl. Extractions were repeated two times. All extracts were stored at −70°C until being analyzed. The TAC in plant produce extracts were determined by using the pH differential method [34]. Absorbance was measured at 510 and 700 nm in a Beckman DU 640 spectrophotometer in potassium chloride (0.025 M, pH 1.0) and sodium acetate (0.4 M, pH 4.5) buffers. A molar extinction coefficient of 26,900 L cm^{-1} mol^{-1} and a molecular weight of 449.2 were used for calculation of total monomeric anthocyanin. Results were expressed as mg of cyanidin-3-glucoside equivalents (mg CGE)/100 g DW.

2.9. Total Carotenoids in Hexane and Ethanol Extracts

2.9.1. Hexane Extraction. Total carotenoids were extracted using hexane as described by [35]. A freeze-dried sample of 0.2 g was weighed into an amber vial and 5 mL of 0.05% (w/v) butylated hydroxyl toluene (BHT) in acetone and then 5 mL of 95% ethanol and 10 mL of hexane were added. Samples were stirred on a magnetic stirring plate, placed on ice, and transferred onto an orbital shaker (Lab-Line Instruments Inc., Illinois, USA) at 180 rpm for 15 min. After shaking, 3 mL of deionized water was added and shaking was resumed for an additional 5 min (still on ice). Sample vials were left at RT for 5 min to allow for phase-separation. Absorbance of the upper hexane layer was recorded at 450 nm in a 1 cm path length quartz cuvette with a hexane blank in a Beckman DU 640 spectrophotometer. Total carotenoid content was calculated for sample weight using the absorbance at 450 nm based on a standard β-carotene curve [36].

2.9.2. Ethanol Extraction. Total carotenoids were assayed following the method of [37] to exclude the chlorophyll content. Approximately 0.2 mg of freeze-dried sample was extracted in 5 mL of 95% ethanol. Extracts were vortexed for 60 s and sonicated for 30 min and then centrifuged at 3,700 rpm for 10 min at 4°C before transfer to 1 cm disposable cells. Absorbance of supernatants was read at 470, 648.6, and 664.1 nm in a Beckman DU 640 spectrophotometer. Concentrations of total carotenoids (including xanthophylls) were determined by the following equations:

$$\text{Chl}_a\,(\mu g/mL) = (13.36 * A_{664.1} - 5.19 * A_{648.6}),$$
$$\text{Chl}_b\,(\mu g/mL) = (27.43 * A_{648.6} - 8.12 * A_{664.1}),$$
$$\text{Total Carotenoid}\,(\mu g/mL)$$
$$= \frac{(1000 * A_{470} - 2.13 * \text{Chl}_a - 97.64 * \text{Chl}_b)}{209}. \quad (1)$$

2.10. β-Carotene, Lutein, and Lycopene Measurements. The carotenoid compounds in freeze-dried produce samples were extracted using a modified method from [38]. A freeze-dried sample of 0.1 g was extracted with 5 mL of extraction solvent (hexane/acetone/ethanol: 50:25:25 v/v/v) in dark glass vials. The extracts were vortexed for 30 min and 1 mL of water was added. The upper layer was collected and the extract was evaporated to dryness in a Speed-Vac. The residue was dissolved to a final volume of 0.5 mL of ethanol/acetone 65:35 (v/v). The extract was passed through a 0.20 μm nylon membrane filter and 20 μL was injected for HPLC analysis.

Solvents and the HPLC running method under isocratic conditions were as described by [38]. Analysis was performed using a reverse phase Varian 9012 HPLC Gemini-NX (5 μm, 100 mm × 4.6 mm) C-18 column equipped with a tertiary pump, refrigerated autosampler, and single variable UV wavelength detector. The mobile phase consisted of methanol (solvent A) and acetonitrile with triethylamine (9 μM) (solvent B) at 90:10 at a flow rate of 0.9 mL min^{-1}. The column temperature was 30°C and the absorbance was read at 475 nm.

2.11. Measurement of Organic Acids. The organic acids, ascorbic, citric, fumaric, and malic, were determined following [39] using a HPLC (Varian 9012, Varian Chromatography Systems, CA, USA) equipped with a tertiary pump, refrigerated autosampler, and single variable UV wavelength detector. A 100 mg freeze-dried sample was extracted with 5 mL of 0.008 N H_2SO_4 in type-1 water (18 Ωcm). Samples were shaken for 1 min on an orbital shaker (Lab-Line Instruments Inc., Illinois, USA) and centrifuged at 3,000 rpm for 10 min at 4°C. Supernatant was collected and filtered through a 20 μm nylon syringe filter (Fisher brand, Fisher Scientific, ON, Canada). Extracts were analyzed using a reverse phase HPLC Gemini-NX (5 μm, 100 mm × 4.6 mm) C-18 column (Phenomenex Inc., CA, USA) and a 4.6 mm × 2.0 mm guard column. The mobile phase was 0.008 N H_2SO_4 in water at 1.0 mL min^{-1} under isocratic conditions at 245 nm for ascorbic acid and at 210 nm for malic, citric, and fumaric acids. Extraction was repeated twice for each sample. The data obtained was expressed as mg of organic acid per 100 g DW.

2.12. Determination of Total Carbohydrates. About 20 mg of freeze-dried sample was added to 5 mL of 1 M HCl and placed into a preheated shaker water bath at 100°C for 2 h [40]. Digested supernatant was collected and used for total carbohydrate measurements. Total carbohydrates were determined using the phenol-sulfuric acid colorimetric method described by Nelsen [41]. One mL of 5% w/w phenol and 5 mL of concentrated sulfuric acid were added to 1 mL of acid-digested sample in a 15 mL glass tube. Tubes were allowed to stand for 30 min in a shaker (180 rpm) water bath at 30°C. The absorption was read at 485 nm in a 1 cm quartz cell against a blank consisting of 1 mL of 5% w/w phenol, 5 mL of concentrated sulfuric acid, and 1 mL of deionized water. Standard solutions of glucose from 10–100 mg L^{-1} were used for preparation of a standard curve.

2.13. Total Dietary Fiber. Freeze-dried samples were analyzed for soluble and insoluble fiber using the method of AOAC (1995) [42, 43]. The method includes enzymatic hydrolysis with α-amylase (heat stable, Sigma A3306), protease (from *Bacillus licheniformis*, Sigma P3910), and amyloglucosidase (from *Aspergillus niger*, Sigma A9913). A 1 g quantity of freeze-dried sample was suspended in 50 mL phosphate buffer and a series of enzymes were added in the following sequence in a water bath (Reciprocal Shaking Bath Model 25, Precision Scientific, Il, USA): (1) 50 μL of thermo-resistant α-amylase at 95–100°C for 35 min, (2) 100 μL of protease at 60°C for 30 min, and then following pH correction to 4.0–4.7, (3) 300 μL of amyloglucosidase at 60°C for 30 min. After soluble fiber precipitation with ethanol (95% v/v) at 60°C, the sample was filtered. The crucibles containing the residues were dried at 105°C, then cooled in a desiccator, and weighed. Residue was divided into two portions. The first part was used for estimation of protein content using a LECO nitrogen analyzer and a %N × 6.25 conversion factor to estimate the total protein content. The second part was incinerated at 525°C overnight to measure the ash content. Total dietary fiber was calculated using the following equation:

$$\text{Total Dietary Fiber\%} = \left[\frac{(\text{weight of residue} - \text{protein} - \text{ash} - \text{blank})}{\text{sample weight}} \right] \times 100. \quad (2)$$

2.14. Statistical Analyses. Statistical analysis was performed using the statistical analysis systems (SAS) software (version 9.2, 2009). Analysis of variance (ANOVA) was used to detect significant differences between varieties for individual attributes. Sensory evaluations were analyzed using the general linear model (GLM) procedure of SAS (Version 9.2, Cary, NC, USA). The data were first analyzed to determine panelist's age and gender effects in a randomized complete block design (RCBD). Least square means of panelists' responses were separated when effects were significant in the ANOVA table ($P \leq 0.05$). Panelist, panelist's age, gender, and crop varieties were analyzed as main factors and were studied for their interactions. When effects of panelists were not significant they were pooled. The means were compared using Scheffe's Multiple Comparison Procedure ($P \leq 0.05$).

Phytonutrients were analyzed using the GLM procedure of SAS (Version 9.2, Cary, NC, USA, 2011). Analysis of variance (ANOVA) was used to differentiate between main factors for cultivars within each crop and between crops for individual phytonutrient characteristics. The data were first analyzed for normality using PROC UNIVARIATE. When effects were significant in the ANOVA table ($P \leq 0.05$) least square means of measurements were separated and compared using Scheffe's Multiple Comparison Test ($P \leq 0.05$).

2.14.1. Identification of the Dominant (Useful) Variables. Identification of the dominant variables was done after running the variable cluster analysis (VARCLUS), principal component analysis (PCA), and partial least square (PLS) statistical analysis. Based on the hierarchical clustering patterns, classification of variables in nonoverlapping clusters was estimated by analyzing the correlation or the covariance matrices subsequently.

Cluster analysis is a separation method that divides the data set into a number of groups or clusters (tree dendrogram; tree leaves and leaflets) based on similarity. The data points within each group (cluster; one major leaf with one or several leaflets) are more similar to one another than they are to the data points within other clusters. The distance between clusters on the x-axis (horizontal axis) indicates how similar to one another the elements of a cluster are. The more distinct a cluster is, the longer the branch along the x-axis that connects it to a larger cluster is.

Then PCA and PLS were run to confirm and visualize in space the cluster (group) classification of variables based on the weight of each variable in relation to the variance (PLS) and covariance (PLS and PCA) matrices [24, 26, 44]. Clustering of variables based on their contribution to the overall variance was done using the VARCLUS procedure. VARCLUS output mainly classified variables based on their R^2 values. Three R^2 values were compared for better selection of variables: R^2 with own (the greater the better clustering), R^2 with nearest cluster (the lower the better), and $1 - R^2$ ratio (the lower the better).

Correlation loading plot is a circular graph showing the variation accounted for by each extracted factor (of one or more variable(s)) and is generated by partial least square (PLS) analysis. The amount of variation in the data points is relative to their distance from the origin. So, the outside circle shows the data points that are most similar and grouped together (highly correlated). Each successive circle working from the outside to the inside indicates decreasing levels of explained variation. The correlation between any two variables is relative to the length of the projection of the point corresponding to one variable on a line through the origin passing through the other variable. The sign of the correlation (negative or positive) corresponds to which side of the origin the projected point falls on.

Multidimensional preference analysis is another mapping technique for the two principal components that show most of the variance and is generated by the principal component analysis (PCA). A biplot displays the independent and dependent variables in a single plot by projecting them onto the plane that accounts for the most variance. Points that are tightly clustered and point in the same direction share the most attributes in common. Pointing in the opposite direction suggests negative correlation. The longer the projection, the relatively more important the variable.

For data mining and discriminant analysis, statistical analysis was done on two steps: (1) significant variation among the 34 variables was identified (based on MANOVA table) and the nondiscriminant variables were identified for a drop-list. (2) Variable clustering analysis was done according to the contribution of each variable to the overall variance using the VARCLUS procedure. Then, their possible interrelationships were studied through visualization of a network structure using both principal component analysis (PCA) and partial least squares regression (PLS).

3. Results and Discussion

3.1. Sensory Traits of Fruits and Vegetables

3.1.1. Sensory Traits of Cherry and Beef Steak Tomatoes. Data presented in Table 1 show sensory traits of cherry and beefsteak tomato from the greenhouse and Montreal area grocery stores. Tomato ripeness was expressed by fruit firmness and color. Unripe tomatoes are firmer. As they ripen, they soften until they reach the fully ripe state where their color (red or orange) is most intense and the fruit are slightly soft to the touch [45].

There was no effect of panelist gender or age on cherry tomato ripening classification. No differences occurred in ripeness score between the GH cherry tomatoes 'Apero,' 'Favorita,' 'Juanita,' and the SB Fruiterie while SB Jardino was scored as less ripe than the others. Sweetness or saltiness of tomato is a trait related to soluble sugars and mineral content of tomato [39, 46]. Juiciness represents the amount of juice/moisture perceived in the mouth. Mouth feeling is a measure of granularity of tomato texture. GH 'Apero' and 'Favorita' and SB Fruiterie had significantly sweeter taste, were juicier, and had a better mouth feel and greater general acceptance compared with both GH 'Juanita' and SB Jardino (Table 1). There was no effect of age or gender on panelist response.

No differences in sensory traits were found between GH beefsteak tomato cultivars (Table 1). The GH beefsteak tomatoes all scored similarly for juiciness and mouth feel but were significantly firmer than both the SB varieties (CHS and Plain Jane). Firmness also was reported to be highly correlated with general acceptability of field-grown tomato [45, 47]. GH 'Arbason,' 'Caramba,' and SB 'Plain Jane' were considered riper and more generally acceptable than the other cultivars. No differences were discerned in sweetness/saltiness and all of them were considered more sweet than salty. Sex, but not age, was important to determination of sensory traits of beefsteak tomato. Male panelists gave greater scores than females for ripeness, sweetness, juiciness, mouth feel, and general acceptability.

3.1.2. Sensory Traits of Cucumber. Results of sensory characterization of cucumber cultivars were presented in Table 2. Hydroponic greenhouse-grown 'Diva' was ranked above the two SB cucumber cultivars. Flesh color and fruit firmness of 'Diva' were more accepted compared with SB cucumbers. 'Diva' and Lebanese were sweeter than Mini Cucumber but all three were similar for fruit aroma. Panelists scored sensory characteristics of cucumber similarly without apparent age or gender bias.

3.1.3. Sensory Traits of Arugula. Results of sensory characterization of arugula (Rocket salad) cultivars are presented in Table 3. All arugula cultivars scored similarly for overall quality and greeny/grassy trait (a measure of freshness). The GH arugula cultivar had an intensely bitter and astringent taste compared with the SB cultivars. Although all participants in this test had never eaten arugula before, their gender and age significantly affected their responses. Male panelists had greater scores for greeny/grassy compared with female participants. Age of participants affected sensory traits of arugula with no obvious trend. For example, panelists, 21–41 years old, had greater scores for overall quality than panelists of the age of 19-20 (3 panelists) and ≥ 43 (2 participants) years. For the intensity of bitterness, only one panelist (a 29 years old) scored less than all other participants. Also, only 2 panelists (49 years old) had greater scores of greeny/grassy trait than 2 panelists (43 years old). Similarly, four panelists

TABLE 1: LS mean score values[1] ± SE of panelist responses to different sensory traits and general acceptability of greenhouse-grown ('Apero,' 'Favorita,' and 'Juanita') and store-bought (Fruterie and Jardino) cherry tomato and greenhouse-grown ('Arbason,' 'Caramba,' 'Geronimo,' and 'Trust') and store-bought (CHS, del Campo, and Plain Jane) beefsteak tomato.

Cultivar	Ripeness*	Sweetness/saltiness	Juiciness	Mouth feeling	General acceptability
Cherry tomato					
Apero	8.36 ± 0.183^a	8.08 ± 0.197^a	8.27 ± 0.185^a	8.02 ± 0.209^a	8.22 ± 0.204^a
Favorita	8.15 ± 0.183^a	7.50 ± 0.192^{ab}	7.81 ± 0.185^a	7.35 ± 0.209^{ab}	7.83 ± 0.204^{ab}
Juanita	8.11 ± 0.183^a	6.94 ± 0.193^b	6.64 ± 0.196^b	6.96 ± 0.222^b	6.94 ± 0.215^c
Fruterie	7.99 ± 0.183^a	7.49 ± 0.198^{ab}	8.31 ± 0.185^a	7.12 ± 0.225^b	7.42 ± 0.220^{bc}
Jardino	6.81 ± 0.197^b	5.80 ± 0.192^c	6.07 ± 0.218^b	5.59 ± 0.226^c	5.82 ± 0.215^d
Beefsteak tomato					
Arbason	8.27 ± 0.287^a	5.37 ± 0.292^a	6.27 ± 0.280^{ab}	5.82 ± 0.297^{ab}	5.96 ± 0.303^b
Caramba	6.72 ± 0.287^{bc}	5.72 ± 0.292^a	6.87 ± 0.280^{ab}	6.07 ± 0.297^{ab}	6.26 ± 0.303^a
Geronimo	7.77 ± 0.287^{ab}	5.77 ± 0.292^a	6.32 ± 0.280^b	6.12 ± 0.297^a	6.41 ± 0.303^a
Trust	7.77 ± 0.287^{ab}	5.82 ± 0.292^a	7.07 ± 0.280^a	6.62 ± 0.297^a	6.66 ± 0.303^a
CHS	7.92 ± 0.287^{ab}	6.57 ± 0.292^a	7.37 ± 0.280^a	6.42 ± 0.297^a	6.71 ± 0.303^a
del Campo	6.17 ± 0.287^c	6.02 ± 0.292^a	6.47 ± 0.280^{ab}	6.22 ± 0.297^a	6.01 ± 0.303^{ab}
Plain Jane	4.97 ± 0.287^c	5.17 ± 0.292^a	5.32 ± 0.280^c	5.32 ± 0.297^b	4.91 ± 0.303^c

[1]LS means were compared using Scheffe's Multiple Comparison Procedure. Superscripts not sharing the same letter within each column are significantly different ($P \leq 0.05$). *Tomato ripeness was scored on 9-point hedonic scales, where 9: ripe, 5: neutral, 1: unripe. Sweetness/saltiness responses were scored as 1: salty, 5: not salty or sweet, and 9: sweet. Juiciness was scored on a 9-point scale, where 1: dislike, 5: neutral, and 9: like. General acceptability also was given scores from 1 to 9, where 1: dislike and 9: like.

TABLE 2: LS mean score values[1] ± SE of panelist responses to different sensory traits of greenhouse-grown ('Diva') and store-bought (Lebanese and Mini Cucumber) cucumbers.

Cultivar	Flesh color*	Firmness	Aroma	Sweetness/bitterness
Diva	7.50 ± 0.253^a	7.81 ± 0.260^a	6.77 ± 0.319^a	6.80 ± 0.276^a
Lebanese	5.99 ± 0.267^b	6.69 ± 0.304^b	6.47 ± 0.308^a	6.14 ± 0.304^a
Mini Cucumber	5.61 ± 0.257^b	5.54 ± 0.276^c	6.33 ± 0.324^a	5.51 ± 0.305^b

[1]LS means were compared using Scheffe's Multiple Comparison Procedure. Superscripts not sharing the same letter within each column are significantly different ($P \leq 0.05$). *Cucumber flesh color and firmness, aroma of the fruit, and flesh taste sweetness/bitterness traits were scored on 9-point scales, where 1: dislike and 9: like for flesh color, firmness, and fresh aroma of the fruit. Sweetness/bitterness were scored on a 9-point scale, where 1: bitter, 5: not bitter or sweet, and 9: sweet.

TABLE 3: LS mean score values[1] ± SE of panelist responses to different sensory traits of greenhouse-grown ('Astro') and store-bought (ADO, BW, and PRO) arugula.

Cultivar	Overall quality*	Intensity of bitterness	Green/grassy	Astringency
Astro	6.07 ± 0.262^a	6.40 ± 0.378^a	7.47 ± 0.225^a	6.40 ± 0.309^a
ADO	6.92 ± 0.255^a	4.27 ± 0.370^b	7.42 ± 0.229^{ab}	4.35 ± 0.308^b
BW	6.91 ± 0.258^a	4.09 ± 0.402^b	6.74 ± 0.212^{ab}	3.61 ± 0.315^b
PRO	6.62 ± 0.274^a	5.10 ± 0.369^b	6.63 ± 0.231^b	4.07 ± 0.288^b

[1]LS means were compared using Scheffe's Multiple Comparison Procedure. Superscripts not sharing the same letter within each column are significantly different ($P \leq 0.05$). *Arugula sensory characteristics including green/grassy, intensity of bitterness, astringency, and overall liking, were rated on 9-point hedonic scales, where 1 was the lowest and 9 was the strongest taste and liking.

(18 years old) scored the astringent characteristic greater than 1 participant, 25 years old. Mixed responses of panelists to the flavor of arugula could be due to the presence of glucosinolates/isothiocyanates, which give both positive and negative sensory properties [48].

3.1.4. *Sensory Traits of Bibb Lettuce.* Overall quality and freshness were not different between GH 'RexMT0' and the SB bibb lettuce cultivars and all had acceptable odor (Table 4). Panelists liked the color of 'RexMT0' and the IGA bibb lettuce but not ADO. No age or sex bias was apparent for panelist response to overall quality of bibb lettuce. However, for lettuce color only, males had greater scores compared with female participants.

3.2. *Mineral Content of Fruits and Vegetables*

3.2.1. *Mineral Content of Cherry Tomato.* Least squares means results for the macro mineral (Ca, Mg, K, and P; g/100 g FW) contents of three hydroponically grown GH and

TABLE 4: LS mean score values[1] ± SE of panelist responses to different sensory traits of greenhouse-grown ('RexMT0') and store-bought (ADO and IGA) bibb lettuce.

Cultivar	Overall quality*	Color	Sogginess/freshness	Off-odor
RexMT0	6.75 ± 0.237[a]	7.19 ± 0.212[a]	3.23 ± 0.348[b]	1.59 ± 0.63[a]
ADO	6.51 ± 0.235[a]	5.86 ± 0.226[b]	4.32 ± 0.325[a]	1.67 ± 0.70[a]
IGA	7.23 ± 0.212[a]	7.11 ± 0.198[a]	3.67 ± 0.311[ab]	1.38 ± 0.59[a]

[1]LS means were compared using Scheffe's Multiple Comparison Procedure. Superscripts not sharing the same letter within each column are significantly different ($P \leq 0.05$). *Bibb lettuce was evaluated for color, sogginess/freshness, off-odor, and overall quality. Sensory evaluation of lettuce was done using the same hedonic 9-point scales where 1 was the lowest and 9 was the strongest taste and liking.

three SB cherry tomato varieties are presented in Table 5. There were no clear distinctions in overall macro mineral content between GH and SB cherry tomato. However, the GH 'Apero' had the greatest overall content of macrominerals. In particular, it had greater K content but similar Mg content compared with 'Juanita.' Also, GH 'Apero' and 'Juanita' (but not 'Favorita') had greater Mg and P content compared to all of the SB varieties. The SB varieties 'Cherries' and 'Fruiterie' (but not 'Jardino') had greater Ca levels compared with the GH cultivars.

Levels of microminerals Cu, Fe, and Na (Table 5) and Al, As, Cr, Se, and Zn (mg/100 g FW) (Table 6) varied widely in GH and SB cherry tomato. As with the macrominerals, there were no clear distinctions in overall micro mineral content between GH and SB cherry tomato. However, 'Apero' had the greatest Fe and Na contents compared with other GH and SB varieties but similar Cu levels to 'Cherries.' Store-bought cherry tomato also showed less Na content compared with GH varieties. Cd was detected only in the GH 'Apero' and 'Juanita' (but not 'Favorita') with no differences between them. Hydroponically grown GH cherry tomato showed consistently less Al content compared with SB varieties while the SB variety cherries were uniquely high in Al content among all of the cherry and beefsteak tomato cultivars. Overall mineral content was not affected by fridge-storage of cherry tomato cultivars for 6 d (Tables 5 and 6). However, storage resulted in increased Fe levels in 'Juanita' and As content in 'Apero' while levels of Se were dramatically reduced in 'Favorita' and 'Juanita.'

3.2.2. Mineral Content of Beefsteak Tomato. Concentrations of macrominerals (Ca, K, Mg, and P; g/100 g FW) of the four GH beefsteak tomato cultivars (Arbason, Caramba, Geronimo, and Trust) and the three SB varieties (Kaliroy, BionatureL, and BionatureP) are presented in Table 5. Overall, the GH cultivars were superior in macromineral content to SB varieties. However, clear cultivar differences were apparent among both the GH and the SB beefsteak tomato groups. The cultivars with the greatest tissue concentrations of all macrominerals were 'Caramba' (greatest in all except Ca and P) and 'Geronimo' and 'Trust' (greatest in Ca, and less than 'Caramba', but greater than the other cultivars for other macrominerals).

Microminerals (mg/100 g FW), Cu, Fe, and Na (Table 5) and Al, As, Cd, Cr, Pb, Se, and Zn (Table 6), varied between GH and SB beefsteak tomato varieties. 'Caramba' had the greatest Cu, Fe, and Na compared with other beefsteak tomato cultivars. The GH group had the least Al concentrations and As and Cd were detected only in 'Arbason.' 'Geronimo' showed the least Cr content but this was similar in the other GH cultivars. Also, 'Geronimo' had the greatest Pb level but this was similar in several other cultivars, both GH and SB. Selenium was detected, at similar concentrations, only in two cultivars, Caramba and Geronimo. Very high Zn content occurred in three of the four GH beefsteak cultivars. These Zn levels were much greater than in the other crops tested. 'Caramba' had the greatest Zn content followed by 'Geronimo' and 'Trust.' The other beefsteak cultivars (Arbason, Kaliroy, BionatureL, and BionatureP) had relatively low tissue levels of Zn.

3.2.3. Mineral Content of Cucumber. Results of least square means of Ca, K, Mg, and P (g/100 g FW) of one GH cultivar (Diva) and three SB cultivars of cucumber are presented in Table 5. No differences were found for Ca and Mg levels among GH and SB cucumber. Similar K concentrations were reported for 'Diva,' 'Lebanese,' and 'Mini Cucumber.' 'Diva' had greater P content than 'Mini Cucumber' and 'Cool Cukes' but was not different from 'Lebanese.' Hydroponically grown GH 'Diva' led the SB varieties in Cu content but had similar Fe content to 'Cool Cukes' (Tables 5 and 6). Na level was greater in 'Diva' compared with cultivars Mini Cucumber and Cool Cukes but was not different from 'Lebanese.' 'Diva' had greater concentrations of Zn and Al compared to the SB varieties and Cr and Pb concentrations were similar for GH and SB cucumbers. Overall, it was apparent that fridge-storage of the cucumber 'Diva' for 3 or 6 d somewhat affected its mineral content (Table 5). While storage did not affect Ca, Cu, K, and Fe, it increased Mg, Na, P, and Zn and reduced Al levels, compared with fresh cucumber (Table 6). Also, both As and Cd were detected, at similar concentrations for both storage intervals, only after 3 and 6 d storage but were not detected in the fresh sample. The rationale for differences in mineral content before and after storage is unclear, but the results may reflect reduced moisture content after storage, difficulty in complete extraction of minerals from fresh cucumber tissue, and/or potential matrix interferences.

3.2.4. Mineral Content of Arugula. Arugula from the GH (1 cultivar) and SB (3 cultivars) were similar in both macro- and micromineral content, except for Fe, Na, and Al, where all of the SB cultivars had consistently greater levels (Tables 5 and 6). Fridge-storage for 6 d of GH 'Astro' did not have a great impact on mineral content (Tables 5 and 6). Fresh 'Astro' was

TABLE 5: LS mean values[1] of macrominerals (Ca, K, Mg, and P; g/100 g DW) and microminerals (Cu, Fe, and Na; mg/100 g DW) in hydroponically grown (GH) and store-bought (SB) cherry and beefsteak tomato, cucumber, arugula, and lettuce cultivars.

Crops	Varieties	Macrominerals				Microminerals		
		Ca	K	Mg	P	Cu	Fe	Na
Cherry tomato								
GH	Apero	0.08^b	5.51^{ab}	0.40^a	1.14^a	1.93^a	13.81^a	65.90^{ab}
GH	Apero-6*	0.09^b	6.10^a	0.41^a	1.11^a	1.72^{ab}	12.35^{ab}	73.86^a
GH	Favorita	0.03^c	2.92^c	0.15^d	0.50^b	0.71^c	5.48^d	23.98^f
GH	Favorita-6	0.03^c	2.95^c	0.16^d	0.50^b	0.72^c	5.12^{de}	26.23^{ef}
GH	Juanita	0.07^b	4.88^b	0.31^{abc}	1.03^a	1.37^b	10.08^{bc}	38.13^{de}
GH	Juanita-6	0.08^b	5.25^{ab}	0.35^{ab}	1.14^a	1.61^{ab}	13.86^a	47.54^{cd}
SB	Jardino	0.08^b	1.20^d	0.11^d	0.16^b	0.51^c	1.35^e	20.96^f
SB	Cherries	0.17^a	2.33^{cd}	0.24^{bcd}	0.54^b	1.45^{ab}	7.20^{cd}	52.24^{bc}
SB	Fruiterie	0.15^a	3.21^c	0.18^{cd}	0.42^b	0.17^c	4.27^{de}	19.83^f
	Mean	0.09^D	3.82^A	0.26^C	0.73^B	1.13^B	8.17^A	40.96^{CD}
Beefsteak tomato								
GH	Arbason	0.11^b	3.50^c	0.28^c	0.57^a	1.16^c	6.97^c	23.81^{cd}
GH	Caramba	0.14^b	7.94^a	1.07^a	0.22^b	6.67^a	22.04^a	144.00^a
GH	Geranimo	0.23^a	5.49^b	0.67^b	0.22^b	4.92^b	15.34^b	90.58^b
GH	Trust	0.21^a	5.80^b	0.64^b	0.30^b	4.66^b	12.68^b	75.23^{bc}
SB	Kaliroy	0.10^b	1.64^{cd}	0.07^c	0.24^b	0.27^c	1.69^c	19.26^{cd}
SB	BionatureL	0.11^b	1.39^d	0.07^c	0.22^b	0.02^c	1.63^c	18.04^d
SB	BionatureP	0.11^b	1.79^{cd}	0.06^c	0.16^b	0.04^c	1.34^c	11.97^d
	Mean	0.14^{CD}	3.94^A	0.41^A	0.28^D	2.53^A	8.81^A	54.70^C
Cucumber								
GH	Diva	0.20^a	2.70^a	0.14^b	0.53^b	0.65^a	3.27^{ab}	24.51^b
GH	Diva-3	0.23^a	2.52^a	0.24^a	0.75^a	0.61^a	3.51^{ab}	40.46^a
GH	Diva-6	0.22^a	2.70^a	0.24^a	0.85^a	0.75^a	4.12^a	38.99^a
SB	Cool Cukes	0.18^a	1.56^c	0.11^b	0.25^d	0.11^b	1.28^c	15.60^c
SB	Lebanese	0.25^a	2.09^{ab}	0.14^b	0.39^{bc}	0.04^b	2.68^b	20.55^b
SB	Mini Cucumber	0.24^a	1.96^{ab}	0.11^b	0.32^{cd}	0.12^b	2.48^b	16.01^c
	Mean	0.22^C	2.25^B	0.17^D	0.52^C	0.38^D	2.89^C	26.02^D
Arugula								
GH	Astro	2.21^a	3.61^a	0.28^a	0.62^a	0.47^{bc}	6.23^b	70.25^c
GH	Astro-6	1.45^b	3.80^a	0.34^a	0.54^{ab}	0.31^c	4.06^c	49.35^c
SB	ADO	2.18^a	2.67^b	0.42^a	0.64^a	0.69^b	10.89^a	573.67^{ab}
SB	BW	1.57^{ab}	2.67^b	0.29^a	0.61^a	1.22^a	10.63^a	637.72^a
SB	PRO	1.62^{ab}	2.45^b	0.30^a	0.38^b	0.47^{bc}	10.00^a	459.03^b
	Mean	1.81^A	3.04^B	0.33^B	0.56^C	0.63^{CD}	8.36^A	358.00^A
Bibb lettuce								
GH	RexMT0	0.42^{ab}	3.90^a	0.24^b	0.96^b	0.71^b	5.34^b	31.45^b
GH	RexMT0-6	0.35^b	3.26^a	0.16^c	0.76^b	0.52^b	3.16^b	36.41^b
SB	ADO	0.46^a	3.71^a	0.31^a	1.40^a	1.85^a	13.14^a	243.60^a
SB	IGA	0.42^{ab}	4.08^a	0.18^{bc}	0.84^b	0.39^b	3.79^b	37.66^b
	Mean	0.41^B	3.74^A	0.22^{CD}	0.99^A	0.87^C	6.36^B	87.28^B

[1] LS means were compared using Scheffe's Multiple Comparison Procedure. Small case superscripts within species and large case superscripts between species not sharing the same letter are significantly different ($P \leq 0.05$). *Produce (based on availability) was fridge-stored (4°C) for 3 and/or 6 days.

similar in mineral content to Astro after 6 d storage except for greater Ca and Fe (levels reduced by 34.39 and 34.83%, resp.; Table 5) and Cr and Pb (not detected in stored produce) (Table 6).

3.2.5. Mineral Content of Bibb Lettuce. Data presented in Tables 5 and 6 showed no differences between GH and SB bibb lettuce in K (g/100 g FW) and As (mg/100 g FW) mineral concentrations. However, the SB variety ADO had greater Mg

TABLE 6: LS mean values[1] of microminerals, including Al, As, Cd, Cr, Pb, Se, and Zn (mg/100 g DW) of greenhouse-grown (GH) and store-bought (SB) cherry tomato, beefsteak tomato, cucumber, arugula, and bibb lettuce.

Crops	Varieties	Al	As	Cd	Cr	Pb	Se	Zn
Cherry tomato								
GH	Apero	2.51^d	0.24^b	0.02^{ab}	0.05^a	0.29^a	11.32^{bc}	4.90^a
GH	Apero-6*	3.57^d	1.29^a	0.04^a	0.08^a	0.34^a	0.79^c	4.35^{ab}
GH	Favorita	2.30^d	0.32^b	—	0.05^a	0.25^a	41.64^a	1.96^c
GH	Favorita-6	3.46^d	0.03^b	—	0.04^a	0.27^a	0.94^c	1.87^c
GH	Juanita	3.70^d	0.86^{ab}	0.02^{ab}	0.06^a	0.23^a	36.73^a	4.26^{ab}
GH	Juanita-6	3.05^b	0.67^{ab}	0.01^b	0.05^a	0.21^a	0.91^c	5.03^a
SB	Jardino	10.22^c	—$	—	0.03^a	0.20^a	24.59^{ab}	1.16^c
SB	Cherries	34.91^a	—	—	0.06^a	0.16^a	—	2.73^{bc}
SB	Fruiterie	24.67^b	—	—	0.03^a	0.32^a	23.10^{ab}	1.18^c
	Mean	9.82^B	0.57^A	0.02^B	0.05^B	0.25^B	17.50^B	3.05^C
Beefsteak tomato								
GH	Arbason	4.31^c	0.41^a	0.01^a	0.12^{ab}	0.17^{ab}	—	2.70^c
GH	Caramba	3.10^c	—	—	0.08^{ab}	0.23^{ab}	19.93^a	43.29^a
GH	Geronimo	2.32^c	—	—	0.04^b	0.37^a	23.68^a	29.74^b
GH	Trust	2.23^c	—	—	0.12^{ab}	0.13^b	—	25.39^b
SB	Kaliroy	13.96^{ab}	—	—	0.16^a	0.17^{ab}	—	0.71^c
SB	BionatureL	11.33^b	—	—	0.14^a	0.22^{ab}	—	0.68^c
SB	BionatureP	16.95^a	—	—	0.18^a	0.11^b	—	0.50^c
	Mean	7.74^C	0.41^{AB}	0.01^B	0.12^A	0.20^B	21.81^{AB}	15.15^A
Cucumber								
GH	Diva	28.55^a	—	—	0.04^a	0.19^a	—	2.83^b
GH	Diva-3	2.72^d	0.54^a	0.03^a	0.06^a	0.14^a	32.99^a	3.41^{ab}
GH	Diva-6	2.11^d	0.38^a	—	0.03^a	0.09^a	28.60^{ab}	4.02^a
SB	Cool Cukes	10.24^c	—	—	0.03^a	0.15^a	21.66^b	1.21^c
SB	Lebanese	19.82^b	—	—	0.03^a	0.16^a	24.53^{ab}	1.73^c
SB	Mini Cucumber	16.95^b	—	—	0.02^a	0.19^a	13.80^b	1.08^c
	Mean	13.40^A	0.50^{AB}	0.03^B	0.04^B	0.15^B	24.32^A	2.38^C
Arugula								
GH	Astro	5.76^b	0.25^{ab}	—	0.02^a	0.34^a	28.36^a	3.44^c
GH	Astro-6	5.54^b	0.15^b	—	—	—	—	2.56^c
SB	ADO	13.38^a	0.56^a	—	0.03^a	0.42^a	34.77^a	6.74^b
SB	BW	11.26^a	0.14^b	—	0.04^a	0.47^a	40.09^a	13.07^a
SB	PRO	9.80^a	—	0.04^a	—	0.47^a	22.92^a	3.13^c
	Mean	9.15^{BC}	0.27^B	0.04^{AB}	0.03^B	0.43^A	31.54^A	5.75^B
Bibb lettuce								
GH	RexMT0	2.87^{bc}	0.81^a	0.03^{bc}	—	—	—	6.73^b
GH	RexMT0-6	2.24^c	0.42^a	0.01^c	—	—	—	4.63^c
SB	ADO	8.58^a	0.57^a	0.16^a	—	—	—	9.08^a
SB	IGA	4.90^b	0.88^a	0.07^b	—	—	—	3.41^c
	Mean	4.64^D	0.67^A	0.07^A	—	—	—	6.03^B

[1] LS means were compared using Scheffe's Multiple Comparison Procedure. Small case superscripts within species and large case superscripts between species not sharing the same letter are significantly different ($P \leq 0.05$). *Produce (based on availability) was fridge-stored (4°C) for 3 and/or 6 days. —$: not detected.

and P contents compared with both GH 'RexMT0' and the SB variety IGA. Similar Ca content was found in GH 'RexMT0' and the SB 'ADO' and IGA. Store-bought variety ADO had greater Al, Cd, Cu, Fe, and Na levels compared with both GH 'RexMT0' and the SB variety IGA. Few differences were observed in mineral content between fresh and fridge-stored (6 d) GH 'RexMT0.' Storage did not alter the concentrations of Ca, Cu, K, Fe, Na, and P but reduced Al, Cd, Mg, and Zn levels.

3.3. Dry Matter (%) of Fruit and Vegetables. Cherry tomato cultivars had the greatest dry matter percentages (9.22%)

TABLE 7: LS mean values[1] of dry matter content (%), total dietary fiber (%), and total carbohydrate content mg glucose/100 g DW equivalents of greenhouse-grown (GH) and store-bought (SB) cherry tomato, beefsteak tomato, cucumber, arugula, and bibb lettuce.

Crops	Varieties	Dry matter (%)	Total carbohydrates	Total dietary fiber (%)
Cherry tomato				
GH	Apero	11.02^a	55.48^{bc}	1.37^a
GH	Apero-6*	11.06^a	52.27^c	1.79^a
GH	Favorita	9.57^b	53.81^{bc}	1.35^a
GH	Favorita-6	10.14^{ab}	60.87^{abc}	1.46^a
GH	Juanita	10.28^{ab}	64.10^{abc}	1.80^a
GH	Juanita-6	10.16^{ab}	49.76^c	1.88^a
SB	Jardino	6.00^d	78.80^a	1.70^a
SB	Cherries	6.28^d	57.06^{abc}	1.82^a
SB	Fruiterie	8.50^c	76.92^{ab}	1.65^a
	Mean	9.22^A	61.00^B	1.65^C
Beefsteak tomato				
GH	Arbason	6.45^a	49.02^{ab}	2.14^a
GH	Caramba	5.79^a	49.01^a	2.08^a
GH	Geronimo	4.91^a	62.71^a	1.72^a
GH	Trust	4.65^a	41.68^b	1.65^a
SB	Kaliroy	5.18^a	50.24^{ab}	1.66^a
SB	BionatureL	2.81^a	55.14^{ab}	1.90^a
SB	BionatureP	5.95^a	55.76^{ab}	1.84^a
	Mean	5.11^C	51.94^C	1.86^{BC}
Cucumber				
GH	Diva	5.20^a	75.03^{ab}	2.68^a
GH	Diva-3	4.69^a	70.18^{ab}	1.87^a
GH	Diva-6	5.10^a	66.72^b	1.64^a
SB	Cool Cukes	4.81^a	92.07^a	1.65^a
SB	Lebanese	5.00^a	81.89^{ab}	1.70^a
SB	Mini Cucumber	4.64^a	60.75^b	1.67^a
	Mean	4.91^C	74.44^A	1.87^{BC}
Arugula				
GH	Astro	8.47^a	24.95^a	3.12^a
GH	Astro-6	7.64^a	13.54^b	2.69^a
SB	ADO	8.32^a	10.65^b	2.92^a
SB	BW	8.55^a	9.80^b	3.43^a
SB	PRO	5.79^b	9.38^{ab}	2.07^a
	Mean	7.79^B	13.67^E	2.85^A
Bibb lettuce				
GH	RexMT0	4.57^b	32.53^{ab}	2.03^a
GH	RexMT0-6	4.65^b	42.21^{ab}	2.14^a
SB	ADO	5.25^a	44.53^a	2.19^a
SB	IGA	4.39^b	27.49^b	2.16^a
	Mean	4.71^C	36.69^D	2.23^B

[1]LS means were compared using Scheffe's Multiple Comparison Test. Small case superscripts within species and large case superscripts between species not sharing the same letter are significantly different ($P \leq 0.05$). *Produce (based on availability) was fridge-stored (4°C) for 3 and/or 6 days.

followed by arugula (7.79%), then beefsteak tomato (5.11%), cucumber (4.91%), and bibb lettuce (4.71%) (Table 7). For cherry tomato, the GH 'Apero' and 'Juanita' but not 'Favorita' had greater dry matter percentages than the three SB cultivars. There were no differences in dry matter content between GH and SB cultivars for the other crops. There was no effect of storage duration (3 or 6 d) on the dry matter content of any crop.

3.4. Total Carbohydrates. Total carbohydrates of GH and SB crops were expressed as mg glucose equivalent/100 g DW (Table 7). Total carbohydrate levels varied with crop.

Cucumber (74.44) had the greatest total carbohydrate content followed by cherry tomato (61.00) and then beefsteak tomato (51.94) and bibb lettuce (36.69), while arugula (13.67) had the least content. Hydroponically grown GH and SB crops were similar in their carbohydrate content. No differences in total carbohydrate contents occurred between fresh GH cherry tomato, bibb lettuce, and cucumber compared with produce that was fridge-stored for 6 d. However, storage of arugula 'Astro' for 6 d decreased the total carbohydrate levels compared with fresh 'Astro.'

3.5. Total Dietary Fiber (%). Percentages of total dietary fiber (TDF) of GH and SB produce are presented in Table 7. TDF% ranged from 2.07 to 3.43 (arugula), 2.03 to 2.19 (bibb lettuce), 1.64 to 2.68 (cucumber), 1.65 to 2.14 (beefsteak tomato), and 1.37 to 1.88 (cherry tomato), with no differences between GH or SB produce. Fridge-storage (6 d) had no effect on produce TDF%.

3.6. Antioxidant Assays for Fruits and Vegetables

3.6.1. Antioxidant Scavenging Activity (ABTS (Hydrophilic and Lipophilic) and DPPH). Antioxidant scavenging capacity of GH and SB vegetables was evaluated using two oxidizing agents: 2,2′-azino-bis(3-ethylbenzothiazoline-6-sulphonic acid) (ABTS) and 2,2-diphenyl-1-picrylhydrazyl (DPPH) (Table 8). Hydroponically grown GH and SB cherry tomato cultivars had similar ABTS and DPPH antioxidant activity. Similarly, no differences were found between GH and SB beefsteak varieties in the hydrophilic and total ABTS and the DPPH scavenging activities except for GH 'Geronimo' that showed less ABTS lipophilic antioxidant activity than the SB variety BionatureL. However, 'Geronimo' showed similar antioxidant capacity to all other varieties. The GH cucumber 'Diva' had the greatest hydrophilic and total ABTS activity compared with the three SB varieties. No differences were found between Mini Cucumber, Cool Cukes, and Lebanese cucumber in antioxidant scavenging capacity. There were no differences in antioxidant capacity observed between GH arugula 'Astro' and the three SB varieties. For bibb lettuce, the SB variety ADO had greater hydrophilic ABTS antioxidant activity (TE/100 g DW) compared with all other lettuce varieties. Also, GH and SB bibb lettuce showed similar antioxidant activity based on lipophilic and total ABTS and DPPH assays (Table 8).

Antioxidant activity based on ABTS (hydrophilic) (mg TE/100 g DW) showed that, overall, arugula, beefsteak tomato, and cherry tomato had greater activity than bibb lettuce and cucumber. Bibb lettuce showed greater antioxidant activity than cucumber (Table 8). Cherry tomato was reported to contain higher antioxidant scavenging activity compared to beefsteak tomato [49]. ABTS (lipophilic) scavenging activity was greatest for arugula followed by bibb lettuce and beef-steak tomato, cherry tomato, and cucumber varieties. In direct contrast, the DPPH free radical test (mg AAE/100 g DW) showed that, overall, cucumber had greater antioxidant activity than all other crops, while beefsteak tomato, bibb lettuce, and arugula had greater activity than cherry tomato but arugula and cherry tomato were similar in their antioxidant scavenging capacity. Fridge-storage of three GH cherry tomato cultivars, one arugula, and one bibb lettuce cultivar for 6 d did not affect their antioxidant scavenging activity based on either method used (Table 4). For cucumber, fresh samples of 'Diva' had greater hydrophilic and total ABTS activity but similar lipophilic ABTS and DPPH antioxidant scavenging activity compared with fridge-stored (3 or 6 d) produce. Using the ORAC technique, Song et al. [7] reported no differences in the antioxidant capacity among tomato, cucumber, and lettuce.

3.6.2. Total Phenolic Content. Total phenolic content of GH and SB vegetables was expressed as mg of gallic acid equivalent (GAE) per 100 g DW (Table 9). Most cultivars of cherry tomatoes (Apero, Favorita, Juanita, and Jardino), whether GH or SB, had similar total phenolic levels. However, the SB variety Cherries had the least content compared with all other cherry tomato cultivars. Total phenolic content of the GH beefsteak tomatoes was similar to that of the SB varieties.

The only GH cucumber, 'Diva,' had greater total phenolic content compared with the three SB varieties. Similar content of total phenolics was found in GH and SB arugula and bibb lettuce cultivars. Vegetable crops, arranged in descending order based on their total phenolic contents (mg/100 g GAE), were arugula, cherry tomato, beefsteak tomato and cucumber, and bibb lettuce and cucumber, respectively. Lettuce was also reported to have greater content of total phenolics than cucumber by Chu et al. [6]. There was no effect of storage on total phenolic levels of any of the tested vegetable crops (Table 9). Generally, tomato cherry or beefsteak tomato showed similar total phenolics content to cucumber, arugula, and bibb lettuce. This was not in agreement with Song et al. [7]; they reported that tomato cultivars had higher total phenolics than both cucumber and lettuce that had similar content.

3.7. Total Anthocyanin Content (TAC). Comparing between crops revealed that arugula had the greatest TAC content followed by cherry tomato and cucumber, which had TAC similar to beefsteak tomato while bibb lettuce had the least TAC content (Table 9). For cherry tomato, TAC content of GH 'Favorita' was greater than 'Apero' and 'Cherries' but similar to 'Juanita,' 'Jardino,' and 'Fruiterie.' For beefsteak tomato, the GH cultivars generally had greater TAC compared with the SB ones. However, 'Arbason' and 'Trust' had similar TAC content to 'Kaliroy' and 'BionatureL.' Store-bought cucumber 'Lebanese' had greater TAC content than the GH 'Diva' and the other two SB cultivars. Arugula 'Astro' had greater TAC content compared to SB varieties. For bibb lettuce, GH 'RexMT0' had greater TAC content compared with the SB 'ADO' but was similar to 'IGA' (Table 9).

Fridge-storage affected TAC content of the crops in different ways. For example, storage of GH cherry tomato did not affect TAC content of 'Favorita' and 'Juanita' but increased the TAC of 'Apero.' Storage (3 d) increased the TAC of cucumber 'Diva' but it was decreased again after 6 d storage (Table 9). Storage of GH arugula for 6 d decreased the TAC content. There was no difference in TAC content between fresh and stored bibb lettuce 'RexMT0.' The mechanism for

TABLE 8: LS mean values[1] of antioxidant scavenging capacity of hydrophilic and lipophilic phases using oxidizing agents ABTS (mg trolox equivalent (TE)/100 g DW) and DPPH (hydrophilic phase) (mg ascorbic acid equivalent (AAE)/100 g DW) of hydroponically grown (GH) and store-bought (SB) cherry tomato, beefsteak tomato, cucumber, arugula, and bibb lettuce.

Crops	Varieties	ABTS Hydrophilic	ABTS Lipophilic	ABTS Total	DPPH
Cherry tomato					
GH	Apero	0.67^a	0.29^a	0.96^a	11.71^a
GH	Apero-6*	0.69^a	0.30^a	0.99^a	14.20^a
GH	Favorita	0.71^a	0.32^a	1.04^a	13.33^a
GH	Favorita-6	0.69^a	0.30^a	0.99^a	11.57^a
GH	Juanita	0.68^a	0.31^a	0.99^a	11.75^a
GH	Juanita-6	0.69^a	0.29^a	0.99^a	12.12^a
SB	Jardino	0.71^a	0.32^a	1.03^a	12.82^a
SB	Cherries	0.69^a	0.31^a	0.99^a	12.21^a
SB	Fruiterie	0.70^a	0.31^a	1.01^a	12.53^a
	Mean	0.69^A	0.31^C	1.00^B	12.47^C
Beefsteak tomato					
GH	Arbason	0.69^a	0.30^{ab}	0.99^a	15.55^a
GH	Caramba	0.68^a	0.31^{ab}	0.99^a	14.81^a
GH	Geronimo	0.69^a	0.27^b	0.96^a	14.82^a
GH	Trust	0.69^a	0.31^{ab}	1.00^a	14.57^a
SB	Kaliroy	0.63^a	0.30^{ab}	0.93^a	14.86^a
SB	BionatureL	0.68^a	0.33^a	1.01^a	15.50^a
SB	BionatureP	0.69^a	0.29^{ab}	0.97^a	12.03^a
	Mean	0.68^A	0.30^C	0.98^B	14.59^B
Cucumber					
GH	Diva	0.47^a	0.30^a	0.77^a	17.22^a
GH	Diva-3	0.36^b	0.32^a	0.68^{ab}	17.83^a
GH	Diva-6	0.30^{bc}	0.29^a	0.59^{bc}	17.22^a
SB	Cool Cukes	0.30^{bc}	0.30^a	0.60^{bc}	18.15^a
SB	Lebanese	0.25^c	0.30^a	0.55^c	17.07^a
SB	Mini Cucumber	0.33^{bc}	0.30^a	0.63^{bc}	17.48^a
	Mean	0.34^C	0.30^C	0.64^D	17.50^A
Arugula					
GH	Astro	0.70^a	0.49^a	1.19^a	15.52^a
GH	Astro-6	0.67^a	0.47^a	1.14^a	14.51^{ab}
SB	ADO	0.67^a	0.47^a	1.14^a	9.94^b
SB	BW	0.72^a	0.43^a	1.14^a	11.46^{ab}
SB	PRO	0.74^a	0.52^a	1.26^a	14.81^{ab}
	Mean	0.70^A	0.48^A	1.18^A	13.25^{BC}
Bibb lettuce					
GH	RexMT0	0.41^b	0.43^a	0.84^a	15.06^a
GH	RexMT0-6	0.39^b	0.28^a	0.68^a	14.70^a
SB	ADO	0.67^a	0.29^a	0.95^a	12.70^a
SB	IGA	0.48^{ab}	0.44^a	0.92^a	15.35^a
	Mean	0.49^B	0.36^B	0.85^C	14.45^B

[1] LS means were compared using Scheffe's Multiple Comparison Test. Small case superscripts within species and large case superscripts between species not sharing the same letter are significantly different ($P \leq 0.05$). *Produce (based on availability) was fridge-stored (4°C) for 3 and/or 6 days.

variations in TAC between fresh and stored material or the TAC fluctuations during storage needs further study; however, such changes may be related to synthesis and degradation of anthocyanin pigments during storage.

3.8. *Total Carotenoid Levels.* Total carotenoid levels of GH and SB varieties were estimated in two different extracts: hexane or ethanol (Table 9). Overall, for the hexane extracts, there were no differences between GH and SB vegetables

TABLE 9: LS mean values[1] of total phenolics (mg/100 g DW gallic acid (GAE) equivalent), total carotenoids (mg/100 g DW β-carotene), and total anthocyanins (mg/100 g DW cyanidin-3-glucoside) of hydroponically grown (GH) and store-bought (SB) cherry tomato, beefsteak tomato, cucumber, arugula, and bibb lettuce.

Crops	Varieties	Total phenolics	Total anthocyanins	Total carotenoids	
				Hexane	Ethanol
Cherry tomato					
GH	Apero	10.60^{ab}	4.91^{bc}	1170.88^a	8.17^b
GH	Apero-6*	11.00^{ab}	16.04^{ab}	1420.39^a	8.92^b
GH	Favorita	11.06^{ab}	18.42^a	1332.90^a	9.72^b
GH	Favorita-6	12.11^{ab}	12.24^{abc}	1157.47^a	7.44^b
GH	Juanita	10.43^{ab}	10.37^{abc}	1175.16^a	8.51^b
GH	Juanita-6	13.60^a	3.81^c	1212.27^a	8.71^b
SB	Jardino	11.10^{ab}	10.96^{abc}	1282.42^a	14.82^a
SB	Cherries	8.59^b	6.43^{bc}	1221.45^a	14.87^a
SB	Fruiterie	9.91^{ab}	10.52^{abc}	1252.70^a	11.27^a
	Mean	10.99^B	10.41^B	1247.30^C	10.27^A
Beefsteak tomato					
GH	Arbason	8.31^a	4.93^{abc}	1554.79^a	11.83^a
GH	Caramba	7.23^a	11.47^a	1480.51^a	14.35^a
GH	Geronimo	8.26^a	11.03^a	1481.94^a	11.73^{ab}
GH	Trust	9.29^a	10.48^{ab}	1457.40^a	12.51^{ab}
SB	Kaliroy	9.55^a	4.97^{abc}	1486.47^a	8.73^a
SB	BionatureL	8.90^a	3.29^{bc}	1549.62^a	9.11^a
SB	BionatureP	9.26^a	2.17^c	1203.18^a	8.49^a
	Mean	8.68^C	6.91^C	1459.13^B	10.96^A
Cucumber					
GH	Diva	9.65^a	3.90^{cd}	1721.66^a	3.84^{ac}
GH	Diva-3	8.72^{ab}	11.13^{ab}	1782.82^a	3.46^c
GH	Diva-6	9.87^a	2.22^d	1722.46^a	4.73^{abc}
SB	Cool Cukes	6.69^{bc}	9.95^{bc}	1815.45^a	7.90^a
SB	Lebanese	6.99^{bc}	16.66^a	1707.20^a	6.95^{ab}
SB	Mini Cucumber	6.38^c	7.24^{bcd}	1748.34^a	6.05^{bc}
	Mean	8.05^{CD}	8.52^{BC}	1749.66^A	5.45^C
Arugula					
GH	Astro	12.33^{ab}	59.56^a	1552.25^a	1.23^{ab}
GH	Astro-6	11.01^b	31.63^{bc}	1450.70^{ab}	1.46^{ab}
SB	ADO	13.06^{ab}	38.26^{bc}	994.12^b	0.97^b
SB	BW	15.92^a	27.40^c	1145.93^{ab}	1.96^{ab}
SB	PRO	15.68^a	40.84^b	1480.76^{ab}	3.16^a
	Mean	13.60^A	39.54^A	1324.75^{BC}	1.7^D
Bibb lettuce					
GH	RexMT0	6.59^a	2.72^a	1505.92^a	2.82^b
GH	RexMT0-6	7.06^a	1.60^{ab}	1469.75^a	5.80^b
SB	ADO	7.78^a	0.69^b	1270.19^a	13.79^a
SB	IGA	7.47^a	1.63^{ab}	1534.99^a	5.23^b
	Mean	7.22^D	1.66^D	1445.21^B	6.92^B

[1] LS means were compared using Scheffe's Multiple Comparison Test. Small case superscripts within species and large case superscripts between species not sharing the same letter are significantly different ($P \leq 0.05$). *Produce (based on availability) was fridge-stored (4°C) for 3 and/or 6 days.

(cherry tomato, beefsteak tomato, cucumber, and bibb lettuce), except for arugula 'Astro' that had greater total carotenoid content than one (ADO) but not the other two SB varieties. No differences were found among the GH 'Astro,' SB 'BW,' and SB 'PRO' varieties and no differences were observed between the three SB varieties.

For the ethanol extract, GH cherry tomato had lesser concentrations of carotenoids compared with the SB cultivars

(Table 9). All GH varieties had similar total carotenoids. The opposite results occurred for beefsteak tomatoes. Cultivar Caramba had greater carotenoid content compared with the SB cultivars but was not different from the other GH cultivars, Arbason, Geronimo, and Trust. For arugula, the SB 'PRO' had greater total carotenoids than 'ADO' but was not different from 'Astro' and 'BW.' For bibb lettuce, the SB 'ADO' had greater carotenoid levels compared with 'RexMT0' and 'IGA.' There were no distinctions between carotenoid content in fresh or stored cultivars. Fresh and stored cultivars of all vegetables showed similar total carotenoid levels from both hexane and ethanol extracts. Lettuce had lactucaxanthin (7.5 mg/g FM) as the main carotenoid compound while tomato was rich in the lycopene (35.4 mg/g FM), lutein (1.0 mg/g FM), and β-carotene (3.2 mg/g FM) carotenoids [50].

3.9. HPLC Analysis of Major Carotenoids.
The HPLC analysis of the major carotenoid compounds (β-carotene, lutein (xanthophyll), and lycopene) in vegetable produce is summarized in Table 10.

3.9.1. β-Carotene.
β-Carotene levels among cherry tomato cultivars ranged from 107.43 to 390.34 mg/100 g DW. The SB 'Cherries' had the greatest content of β-carotene followed by 'Fruiterie.' GH cherry tomato cultivars showed less β-carotene contents than the SB cultivars but similar to 'Jardino.' 'Favorita' showed greater content compared with 'Apero' and 'Juanita,' which had similar levels of β-carotene. The average content of β-carotene for cherry tomato varieties was greater than that of the beefsteak tomatoes. Beefsteak tomato cultivars had β-carotene levels that ranged from 54.25 to 153.32 mg/100 g DW. GH 'Caramba' and the SB 'BionatureP' and 'BionatureL' showed the greatest β-carotene levels compared with all other cultivars. 'Arbason' and 'Trust' showed the least concentrations of β-carotene. No differences were noted among 'Caramba,' 'Geronimo,' and 'BionatureL.' Also, 'Kaliroy,' 'Trust,' and 'Arbason' had similar contents of β-carotene.

Store-bought cucumber variety Mini Cucumber showed greater β-carotene content (139.05 mg/100 g DW) compared with the other two SB varieties and the GH 'Diva' (Table 10). Cultivar Diva had the least β-carotene level compared with the SB varieties but had similar content to 'Cool Cukes.' Arugula varieties contained great amounts of β-carotene that ranged from 445.22 to 1,365.03 mg/100 g DW. However, there was a huge variation in β-carotene content among arugula varieties where 'ADO' had greater content compared with 'Astro' but was not different from the other SB varieties BW and PRO. No differences were found between the GH 'Astro' and the SB varieties BW and PRO. For bibb lettuce, the SB 'IGA' had greater content of β-carotene compared with 'RexMT0' and 'ADO.' The GH 'RexMT0' had greater β-carotene level than 'ADO.'

Fridge-storage of GH vegetables depressed the content of β-carotene in cherry tomato and bibb lettuce varieties but did not significantly affect cucumber and arugula varieties (Table 10). Storage of cherry tomato cultivars for 6 d decreased the β-carotene contents of 'Apero' (35.35%) and 'Favorita' (35.28%) but not 'Juanita.' Storage of cucumber 'Diva' for 3 and 6 d at 4°C did not change the β-carotene content. Storage of arugula 'Astro' for 6 d did not affect β-carotene compared with the fresh sample. Storage of greenhouse bibb lettuce 'RexMT0' for 6 d significantly decreased β-carotene levels compared with fresh produce.

3.9.2. Lutein.
Lutein concentration in arugula was the greatest among tested crops and ranged from 33.76 to 95.22 mg/100 g DW: manyfold greater than in the other vegetables (Table 10). Cherry tomato and beefsteak tomato cultivars were similar in lutein concentrations. The SB 'Cherries' had greater lutein content compared with other cherry tomato cultivars but was not different from 'Fruiterie' (SB) and 'Favorita' (GH). Cultivars Apero, Favorita, Juanita, Jardino, and Fruiterie had similar lutein levels. Beefsteak tomato that was either GH or SB had similar lutein levels (2.85–5.79 mg/100 g DW). For cucumber, no differences were found in lutein levels between the GH 'Diva' and the SB 'Cool Cukes,' 'Mini Cucumber,' and 'Lebanese.' Store-bought arugula cultivars were similar in lutein content but greater than that of the GH 'Astro' (Table 10). Arugula was reported to be a rich source of lutein when eaten fresh or processed [21, 51]. Also, the SB bibb lettuce 'IGA' had greater lutein content compared with 'ADO' and 'RexMT0,' which had similar lutein levels. There was no effect of fridge-storage on the lutein content of any of the GH vegetables. Similarly, hydroponic lettuce has previously been shown to contain significantly lower lutein, α-carotene, violaxanthin, and neoxanthin content than the conventionally produced lettuce [22].

3.9.3. Lycopene.
Lycopene is the most predominant carotenoid pigment in tomatoes (about 83%) [49]. Lycopene was detected only in tomato varieties and not in cucumber, arugula, or bibb lettuce varieties (Table 10). A wide range of lycopene concentrations was detected in cherry tomato varieties (60.74–139.74 mg/100 g DW). Store-bought 'Cherries' showed the greatest lycopene level but this was similar to GH 'Favorita.' Cultivars Favorita and Fruiterie were similar in lycopene content. No differences occurred in lycopene levels of 'Apero,' 'Juanita,' and 'Jardino.' Beefsteak tomato cultivars had levels of lycopene that ranged from 32.03 to 67.00 mg/100 g DW: about half the levels of cherry tomatoes. Hydroponically GH grown 'Trust' and 'Caramba' had the greatest lycopene levels compared with other cultivars. No differences occurred in lycopene levels between 'Arbason,' 'Geronimo,' 'Kaliroy,' 'BionatureL,' and 'BionatureP.' Fridge-storage of cherry tomato varieties for 6 d decreased lycopene content of 'Favorita' but not 'Apero' or 'Juanita.'

3.10. HPLC Analysis of Organic Acids

3.10.1. Ascorbic Acid (Vitamin C).
Concentrations of ascorbic acid were similar for GH and SB cherry tomato (Table 11). A different trend was noticed for beefsteak tomato, where SB cultivars had greater vitamin C content compared with GH cultivars. The SB 'Kaliroy' had the greatest vitamin C content and the GH 'Geronimo' had the least concentration. Hydroponically grown GH cucumber 'Diva' had greater ascorbic

TABLE 10: LS mean values[1] of β-carotene, lutein, and lycopene (mg/100 g DW) of hydroponically grown (GH) and store-bought (SB) cherry tomato, beefsteak tomato, cucumber, arugula, and bibb lettuce evaluated using high pressure liquid chromatography (HPLC).

Crops	Varieties	β-Carotene	Lutein	Lycopene
Cherry tomato				
GH	Apero	166.16^d	4.03^b	73.55^c
GH	Apero-6*	107.43^e	2.56^b	60.74^c
GH	Favorita	235.53^c	5.00^{ab}	119.75^{ab}
GH	Favorita-6	152.43^d	5.49^{ab}	75.00^c
GH	Juanita	171.30^d	4.18^b	76.60^c
GH	Juanita-6	147.70^d	4.07^b	83.80^c
SB	Jardino	146.55^d	3.95^b	72.36^c
SB	Cherries	390.34^a	9.12^a	139.74^a
SB	Fruiterie	311.53^b	6.52^{ab}	89.56^{bc}
	Mean	203.22^B	4.99^C	87.90^A
Beefsteak tomato				
GH	Arbason	54.25^e	3.65^a	39.00^b
GH	Caramba	146.88^{ab}	5.79^a	66.07^a
GH	Geronimo	111.99^{bcd}	4.69^a	45.92^b
GH	Trust	86.56^{de}	3.35^a	67.00^a
SB	Kaliroy	107.23^{cd}	2.85^a	32.17^b
SB	BionatureL	139.89^{abc}	2.98^a	40.74^b
SB	BionatureP	153.32^a	3.62^a	32.03^b
	Mean	114.30^C	3.88^C	46.13^B
Cucumber				
GH	Diva	37.09^{cd}	10.74^{bc}	—$
GH	Diva-3	22.25^d	5.26^c	—
GH	Diva-6	33.11^d	13.63^{bc}	—
SB	Cool Cukes	56.29^c	23.38^a	—
SB	Lebanese	93.19^b	19.11^{ab}	—
SB	Mini Cucumber	139.05^a	22.95^a	—
	Mean	63.50^C	15.89^B	—
Arugula				
GH	Astro	445.22^b	33.76^c	—
GH	Astro-6	615.75^b	58.22^{bc}	—
SB	ADO	1365.03^a	75.47^{ab}	—
SB	BW	842.91^{ab}	72.14^{ab}	—
SB	PRO	1080.80^{ab}	95.22^a	—
	Mean	869.94^A	66.96^A	—
Bibb lettuce				
GH	RexMT0	223.77^b	19.97^b	—
GH	RexMT0-6	85.70^c	9.89^c	—
SB	ADO	55.00^d	14.78^{bc}	—
SB	IGA	290.27^a	28.38^a	—
	Mean	163.69^{BC}	18.26^B	—

[1] LS means were compared using Scheffe's Multiple Comparison Test. Small case superscripts within species and large case superscripts between species not sharing the same letter are significantly different ($P \leq 0.05$). *Produce (based on availability) was fridge-stored (4°C) for 3 and/or 6 days. —$: not detected.

acid levels compared with SB 'Cool Cukes' and 'Mini Cucumber' but was not different from 'Lebanese.' No difference occurred in ascorbic acid content between the GH arugula 'Astro' and the SB 'BW' and 'PRO.' Cultivar ADO had the least vitamin C content. For bibb lettuce, no differences were found in vitamin C content between 'ADO,' 'RexMT0,' and 'IGA.'

Based on average vitamin C content of cultivars within each crop, beefsteak tomato had the greatest vitamin C content (198.76 mg/100 g DW) followed by cherry tomato (129.45), arugula (73.72), bibb lettuce (67.17), and finally cucumber (37.82). Curiously, fridge-storage of GH 'Apero' and 'Juanita' for 6 d increased vitamin C concentrations but

TABLE 11: LS mean values[1] of organic acids (mg/100 g DW) of hydroponically grown (GH) and store-bought (SB) cherry tomato, beefsteak tomato, cucumber, arugula, and bibb lettuce evaluated using high pressure liquid chromatography (HPLC).

Crops	Varieties	Ascorbic acid	Malic acid	Citric acid	Fumaric acid
Cherry tomato					
GH	Apero	108.15^c	25.48^a	2.04^{bc}	0.14^{ab}
GH	Apero-6*	317.52^a	5.30^{ef}	5.60^a	0.24^a
GH	Favorita	85.01^{cd}	11.93^{def}	4.10^{ab}	0.16^{ab}
GH	Favorita-6	44.94^d	4.45^{ef}	2.01^{bc}	0.13^{ab}
GH	Juanita	81.74^{cd}	18.99^{cde}	4.82^{ab}	0.19^{ab}
GH	Juanita-6	241.30^b	3.21^f	1.60^{bc}	0.15^{ab}
SB	Jardino	82.82^{cd}	37.44^b	4.20^{ab}	0.24^a
SB	Cherries	119.29^c	58.92^a	1.59^{bc}	0.09^b
SB	Fruiterie	84.28^{cd}	32.98^{bc}	0.50^c	0.16^{ab}
	Mean	129.45^B	22.08^B	2.97^B	0.17^B
Beefsteak tomato					
GH	Arbason	117.72^{cd}	4.33^b	2.59^{abc}	0.26^a
GH	Caramba	125.16^{cd}	6.38^b	2.71^{ab}	0.26^a
GH	Geronimo	46.46^d	109.65^a	0.09^d	0.16^{abc}
GH	Trust	178.78^c	4.11^b	1.04^{cd}	0.21^{ab}
SB	Kaliroy	497.97^a	6.34^b	3.54^{ab}	0.10^{bc}
SB	BionatureL	306.23^b	10.14^b	3.92^a	0.18^{abc}
SB	BionatureP	119.00^{cd}	1.21^b	2.10^{bc}	0.08^c
	Mean	198.76^A	20.31^B	2.29^B	0.18^{AB}
Cucumber					
GH	Diva	55.41^b	85.43^{ab}	14.15^{ab}	0.50^a
GH	Diva-3	4.92^d	102.27^a	16.22^a	0.02^c
GH	Diva-6	87.50^a	19.67^b	4.72^c	0.14^c
SB	Cool Cukes	20.32^{cd}	104.46^a	18.11^a	0.30^{abc}
SB	Lebanese	43.18^{bc}	93.04^a	11.04^{abc}	0.34^{ab}
SB	Mini Cucumber	15.60^d	108.31^a	7.67^{bc}	0.01^c
	Mean	37.82^D	85.53^B	11.98^A	0.22^A
Arugula					
GH	Astro	87.07^{ab}	67.20^b	1.21^a	0.06^b
GH	Astro-6	56.91^{bc}	183.18^a	0.15^b	0.01^c
SB	ADO	31.45^c	12.18^c	1.21^b	0.02^c
SB	BW	112.97^a	4.81^c	0.25^b	0.004^c
SB	PRO	80.21^{ab}	82.89^b	0.08^b	0.14^a
	Mean	73.72^C	70.05^B	0.36^C	0.05^C
Bibb lettuce					
GH	RexMT0	65.72^{ab}	221.80^b	0.04^b	0.02^a
GH	RexMT0-6	85.29^a	814.06^{ab}	0.08^{ab}	0.06^a
SB	ADO	54.30^b	1028.70^a	0.11^{ab}	0.12^a
SB	IGA	63.38^{ab}	1323.56^a	0.14^a	0.07^a
	Mean	67.17^C	847.03^A	0.09^C	0.07^C

[1]LS means were compared using Scheffe's Multiple Comparison Test. Small case superscripts within species and large case superscripts between species not sharing the same letter are significantly different ($P \leq 0.05$). *Produce (based on availability) was fridge-stored (4°C) for 3 and/or 6 days.

these were decreased in 'Favorita.' Similarly, storing 'Diva' for 6 d increased the ascorbic acid content compared with fresh cucumber. The observed increased vitamin C content with storage could be due to an increase in ascorbic acid synthesis. No effect of 6 d storage occurred on the vitamin C content of arugula 'Astro' or bibb lettuce 'RexMT0' compared with fresh produce.

3.10.2. Malic Acid. Malic acid contributes a sour taste to vegetables and tends to be present in greater concentrations in unripe fruit. Organic acids, especially malic and citric acids, were reported as the major organic acids in tomato [52]. For cherry tomato, the GH 'Apero' and the SB 'Cherries' had similar malic acid contents that were greater than all other cherry tomato cultivars (Table 11). For beefsteak tomato, 'Geronimo' had the greatest malic acid content while no differences occurred between the other GH and SB cultivars. For cucumber, no differences were found between GH and SB varieties. The GH arugula 'Astro-6' had greater malic acid content compared with SB 'ADO' and 'BW' but was similar to PRO. The reverse trend was found in bibb lettuce where the GH cultivars had less malic acid than the SB cultivars. Malic acid concentration was greatest in bibb lettuce (847.03 mg/100 g DW) while it was 10% of this level and similar for cucumber (85.53) and arugula (70.05) and only 2% of this level in beef-steak tomato (20.31) and cherry tomato (22.08).

Surprisingly, fridge-storage (6 d) of GH arugula 'Astro' and bibb lettuce 'RexMT0' increased the concentration of malic acid by 172.60 and 267.02%, respectively. In contrast, fridge-storage of GH cherry tomato (6 d) did not make any difference in terms of malic acid content for 'Favorita' (Table 11). However, for 'Apero' and 'Juanita,' storage dramatically decreased the concentration of malic acid by 79.20 and 83.10%, respectively. Similarly, storage of the cucumber 'Diva' for 6 d decreased the content of malic acid by 66.64%.

3.10.3. Citric Acid. Citric acid is a natural preservative and often used as a food additive for its flavor. It contributes to vegetables' acidity, color, and taste [48]. Citric acid contents of vegetable crops and fruits arranged in descending order were cucumber (11.98 mg/100 DW), cherry tomato (2.97) and beefsteak tomato (2.29), and arugula (0.36), and bibb lettuce (0.09) (Table 11). Cherry tomatoes 'Favorita,' 'Jardino,' and 'Juanita' had greater citric acid content compared with 'Apero,' 'Cherries,' and 'Fruiterie.' There were no differences in the citric acid concentration of GH cultivars of cherry tomato. For beefsteak tomato, 'BionatureL,' 'Kaliroy,' 'Caramba,' and 'Arbason' had similar citric acid content, which was greater than 'Geronimo.' Cucumber 'Cool Cukes' had more citric acid than 'Mini Cucumber' that was not different from 'Diva' and 'Lebanese.' Citric acid content of hydroponically grown arugula 'Astro' was greater than SB cultivars, which were similar in citric acid content. For bibb lettuce, 'ADO,' 'IGA,' and 'RexMT0-6' had similar citric acid levels.

As with malic acid, different trends occurred in the citric acid content of hydroponically grown vegetables that were stored for 6 d (Table 11). Stored cherry tomato 'Apero' had increased citric acid content (174.5%) but 'Favorita' and 'Juanita' were unaffected. Also, storage of cucumber 'Diva' and arugula 'Astro' showed decreased citric acid content of 66.6% and 87.6%, respectively. Fridge-storage doubled citric acid content in bibb lettuce 'RexMT0.'

3.10.4. Fumaric Acid. Fumaric acid (trans-butenedioic acid) is an acidulent agent that contributes a sour taste to food. Vegetable crops had different fumaric acid content (mg/100 g DW) (Table 11). Cucumber (0.22) had relatively greater fumaric acid concentrations compared with beef-steak (0.18) and cherry tomato (0.17) which were greater than the two leafy greens arugula (0.05) and bibb lettuce (0.07) (Table 11). Cherry tomato 'Jardino' had greater fumaric acid than 'Cherries' but was not different from 'Apero,' 'Favorita,' 'Juanita,' and 'Fruiterie.' Greenhouse-grown beef-steak tomato varieties Arbason, Caramba, Geronimo, and Trust and SB 'BionatureL' had similar fumaric acid contents that were greater than the SB 'BionatureP.' The fumaric acid content of 'Trust,' 'BionatureL,' and 'Kaliroy' was similar.

For cucumber, GH 'Diva' had greater fumaric acid content than 'Mini Cucumber' but similar content to 'Cool Cukes' and 'Lebanese' (Table 11). The SB arugula 'PRO' had the greatest fumaric acid content. 'Astro' had greater fumaric acid than 'ADO' and 'BW,' which were similar. Fumaric acid content of all bibb lettuce cultivars was the same (0.02–0.12 mg/100 g DW).

The effect of fridge-storage on fumaric acid content was inconsistent between GH vegetables as seen with other organic acids (Table 11). There was no effect of storage on fumaric acid content of cherry tomatoes. However, storage of cucumber for 3 or 6 d decreased its fumaric acid content. Storage for 6 d increased the fumaric acid content of the arugula 'Astro' but did not affect the fumaric acid content of the bibb lettuce 'RexMT0.'

3.11. Effect of Storage on Nutritional Characteristics of Fresh Produce. Fridge-storage of cherry tomato for 6 d resulted in increased content of Fe and vitamin C in 'Juanita' and As, TAC, vitamin C, and citric acid content in 'Apero' while levels of Se, vitamin C, and lycopene were reduced in 'Favorita.' Generally storage of GH cherry tomato depressed the content of β-carotene. Malic acid was decreased by 79.20 and 83.10% in 'Apero' and 'Juanita,' respectively.

Also, fridge-storage of the cucumber 'Diva' for 3 or 6 d increased its mineral content. It increased the TAC of cucumber 'Diva' after 3 d in storage but decreased it again after 6 d storage. Six-day storage of 'Diva' increased its the ascorbic acid content and decreased its hydrophilic and total ABTS scavenging activity, the content of malic acid (by 66.64%), citric acid (by 66.6%), and fumaric acid content.

Fridge-storage for 6 d of GH arugula 'Astro' resulted in greater Ca and Fe but Cr and Pb were not detected in stored produce but decreased the total carbohydrate and TAC levels. Storage of GH arugula 'Astro' for 6 d increased the concentration of malic and fumaric acids but dramatically decreased malic (by 79.20%) and citric acids (by 87.6%). Storage of GH bibb lettuce 'RexMT0' for 6 d had reduced Al, Cd, Mg, Zn, and β-carotene levels. On the other hand, it increased the concentration of malic acid and doubled citric acid content.

The observed variations in plant metabolites and antioxidant capacity following storage can be attributed to a variety of factors that include biosynthesis during storage, as well as losses due to degradation caused by enzymatic (i.e., polyphenoloxidase, ascorbate oxidase) and nonenzymatic oxidation reactions. Additionally, a slight drop in tissue dry weight during storage could account for slightly more efficient extraction, particularly minerals.

TABLE 12: Drop-list. Nonsignificant variables (sensory and phytonutrient) that were excluded after the first statistical selection step because these variables do not differentiate among tested varieties of each select crop.

Crop	Variables (ANOVA P value)
Cherry tomato	*Sensory*: ripeness (0.4340), sweetness (0.1785), and mouth feeling (0.1652). *Phytonutrient*: chromium (0.4172), total phenolics (0.4853), antioxidant scavenging activity using ABTS (hydrophilic) (TEACH 0.4706), lipophilic (TEACL 0.2781), total (TEACT 0.4095), DPPH scavenging assay (0.5468), total carotenoids (lipophilic extract) (0.5467), ascorbic acid (0.0651), and total dietary fiber (0.1593) and lutein (0.0712).
Beefsteak tomato	*Sensory*: sweetness (0.5524), juiciness (0.2993), mouth feeling (0.4957), fruit ripeness (0.0641), and general acceptability (0.5228). *Phytonutrient*: total antioxidant capacity using ABTS (hydrophilic) (TEACH (0.7314)), lipophilic (TEACL (0.0782)), total (TEACT (0.7914)), DPPH scavenging assay (0.9069), total carotenoids (lipophilic extract) (0.9069), and total dietary fiber (0.1219).
Cucumber	*Sensory*: Flesh aroma (0.2351). *Phytonutrient*: calcium (0.4110), iron (0.3572), magnesium (0.1290), chromium (0.2893), ABTS antioxidant assay (lipophilic) (TEACL 0.9522), DPPH scavenging assay (0.9488), total carotenoids (lipophilic extract) (0.9489), dry matter (0.1990), malic acid (0.2847), and total dietary fiber (0.1446).
Arugula	*Sensory*: overall quality (0.1453). *Phytonutrient*: calcium (0.1130), magnesium (0.2069), selenium (0.6852), total antioxidant capacity (ABTS) (hydrophilic) (TEACH (0.4296)), lipophilic (TEACL (0.0938)), total (TEACT (0.2823)), and total Dietary Fiber (0.1790).
Bibb lettuce	*Sensory*: sogginess and freshness (0.2596), odor (0.6512), and overall quality (0.1781). *Phytonutrient*: calcium (0.4926), potassium (0.8910), phosphorous (0.0850), antioxidant capacity using ABTS (Total) (TEACT (0.5871)), DPPH scavenging assay (0.0684), total carotenoids (lipophilic extract) (0.0684), ascorbic acid (0.4451), and total dietary fiber (0.5905).

3.12. Ruling Out Nondiscriminant Variables (Compiling a Drop-List). Assessing traits that contribute to define the most nutritious variety is of great importance for long-term breeding programs for vegetables. The graphical representation of the interrelationships among various sensory and phytonutrient parameters is useful to visualize essential relationships [39]. It is important to rule out nondifferentiating traits. It was interesting that many of the sensory evaluation criteria were on the drop-list (Table 12), with a few important exceptions. These included criteria that called on panelists to make an overall assessment of relative crop quality (general acceptability). Most sensory evaluation criteria that are usually used for these crops (pooled from the literature) do not directly discriminate between cultivars. However, they in some cases, such as for cherry tomato, are necessary to enable reflective answers to overall ranking of general acceptability. Also, a few dominant criteria, such as astringency, or degree of bitterness in arugula, or juiciness in cherry tomato, were relatively more useful to discriminate between cultivars than the other criteria used, which could perhaps be eliminated in favor of new ones (Table 12).

Based on the output after running ANOVA for each of the tested populations separately (cherry tomato, beefsteak tomato cucumber, arugula, and bibb lettuce), a first selection process among assayed variables was done (Table 12). It was clear that the estimation of total dietary fiber did not differentiate among varieties of tested crops. Also, measurement of the antioxidant capacity of vegetables and fruit produce using the ABTS (either hydrophilic of lipophilic phases) was useful for discrimination among cucumber varieties but did not distinguish among varieties of cherry tomato, beef-steak tomato, arugula, or bibb lettuce. Similar to ABTS, the DPPH and total carotenoid (lipophilic extract) assays were not suitable to distinguish between varieties of cherry tomato, beefsteak tomato, cucumber, or bibb lettuce (Table 12).

Similarly, some mineral elements were not different in varieties of SB and GH produce (Table 12). For example, chromium (Cr) was the same in all cherry tomato varieties. In cucumber, calcium (Ca), iron (Fe), magnesium (Mg), and Cr were not discriminating mineral elements among assayed varieties. In arugula, Ca, Mg, and selenium (Se) were the same in SB and GH varieties. For bibb lettuce, Ca, potassium (K), and phosphorous (P) were the same in SB and GH produce. Other assays generated similar results for specific vegetable or fruit produce varieties including total phenolics, ascorbic acid, and lutein content and fruit ripeness, sweetness, and the mouth feeling of cherry tomato (Table 12).

Taste test components including fruit ripeness, sweetness, juiciness, mouth feeling, and general acceptability of beefsteak tomato were similar. Dry matter, malic acid, and flesh aroma were not able to distinguish varieties of cucumber. The overall quality term could not distinguish arugula varieties. Ascorbic acid and the sensory variables, odor, sogginess and freshness, and overall quality, were not discriminating variables among bibb lettuce varieties (Table 12). The large-scale similarity among tested varieties might be due to the fact that they were selected carefully to have similar characteristics for better comparison between the hydroponic production system and store-bought produce.

3.13. Dominant Variables for Fruit and Vegetables

3.13.1. Dominant Variables for Cherry Tomato. Figure 1 shows that 100% of the total variation was explained using 9 clusters.

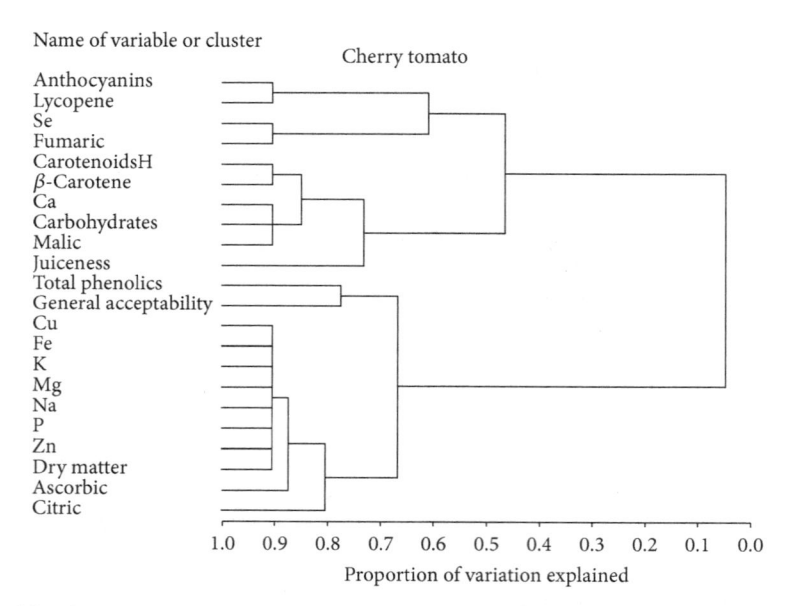

FIGURE 1: Tree dendrogram of the clustering patterns and their proportion of explained variance (9 clusters explained 100% of the variation) using VARCLUS procedure for phytonutrient variables that were used to differentiate among cherry tomato store-bought and greenhouse-grown varieties. CarotenoidsH: carotenoids, hydrophilic fraction.

The VARCLUS method showed that separation of dependent variables into 9 clusters resulted in the best classification. Based on results illustrated in Figures 1, 6, and 11, we identified two sensory criteria (fruit juiciness and general acceptability) and 11 phytonutrient factors (Ca, Cu, Fe, K, Mg, total carotenoids, lycopene, total phenolics, citric acid, fumaric acid, and total carbohydrates) as the most influential for discrimination among cherry tomato varieties.

The PLS and PCA output showed that the eigenvalues of the first 2 components (PLS factors 1 and 2 and PCA components 1 and 2) were responsible for 76.4% of the variance and 69.5% of the covariance, so plotting PLS 1 versus PLS 2 (Figure 11) and PCA 1 versus PCA 2 (86.4%) (Figure 6) identified variables that contributed more than others to the variance. Correlation loading plot (Figure 11) shows that plotting factors PLS 1 and 2 of X and Y scores reveal 76.4% and 69.5% of X and Y variability, respectively, which illustrates the best dimensional rotation and visualization of the variance components.

Figures 6 and 11 confirm that the variables with the most weight in the variance were fruit juiciness, general acceptability, some minerals (Ca, Cu, Fe, K, and Mg), total carotenoids, lycopene, total phenolics, citric acid, fumaric acid, and total carbohydrates. It is also interesting to note that certain variables were negatively correlated with general acceptability and malic acid, including anthocyanins, total phenolics, lycopene, citric acid, and fumaric acid.

3.13.2. Dominant Variables for Beefsteak Tomato. Results in Figure 2 show that 100% of the variation was explained using 9 clusters. The VARCLUS method indicated that the best classification was obtained through separation of dependent variables into 9 clusters. Based on results illustrated in Figures 2, 7, and 12, we identified 15 phytonutrient factors (Ca, Cu, Fe, K, Mg, P, Zn, β-carotene, lutein, lycopene, total phenolics, total anthocyanins, citric acid, fumaric acid, and total carbohydrates) as the most influential for discrimination among beefsteak tomato varieties. Biological network analysis was used to interrelate sensory and phytonutrient traits as discriminating factors of tomato and strong associations were reported between citric acid with tomato aroma, glycine with tomato aroma, and granulosity with dry matter [39]. Similar to our results, sugar and organic acids contributed the most to the flavor of tomato [53–55].

The PLS and PCA output showed that the eigenvalues of the first 2 components (PLS factors 1 and 2 and PCA factors 1 and 2) are responsible for 87.92% of the variance, so plotting PLS 1 versus PLS 2 (Figure 12) and PCA 1 versus PCA 2 (Figure 7) identified variables that contribute more than others to the variance. Correlation loading plot (Figure 12) shows that plotting factors PLS 1 and PLS 2 of X and Y scores reveal 51.2% and 63.7% of X and Y variability, respectively, which illustrate the best dimensional rotation and visualization of the variance components. Figures 7 and 12 confirm the most important variables to distinguish between greenhouse-grown and store-purchased produce were minerals (Ca, Cu, Fe, K, Mg, P, and Zn), carotenoids (β-carotene, lutein, and lycopene), total phenolics, total anthocyanins, citric acid, fumaric acid, and total carbohydrates. Playing a major role in the general acceptability of beef-steak tomato and similar to observations with cherry tomato were the variables total phenolics and anthocyanins; both were negatively correlated with fumaric acid and total carbohydrate contents.

3.13.3. Dominant Variables for Cucumber. Data presented in Figure 3 show that 100% of the total variation was explained using 11 clusters. Based on the VARCLUS method, separation of dependent variables into 11 clusters was the

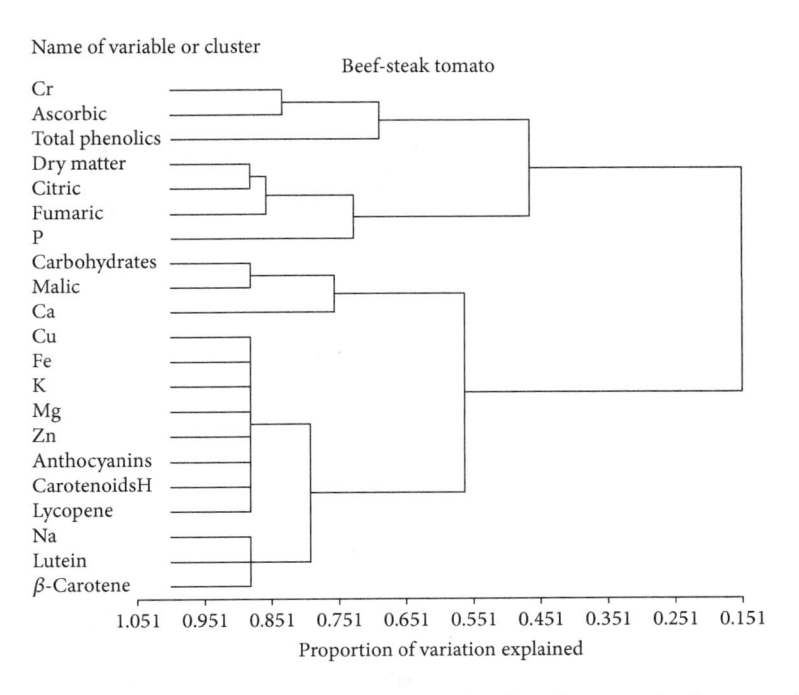

FIGURE 2: Tree dendrogram of the clustering patterns and their proportion of explained variance (9 clusters explained 100% of the variation) using VARCLUS procedure for phytonutrient variables that were used to differentiate among beefsteak tomato store-bought and greenhouse-grown varieties. CarotenoidsH: carotenoids, hydrophilic fraction.

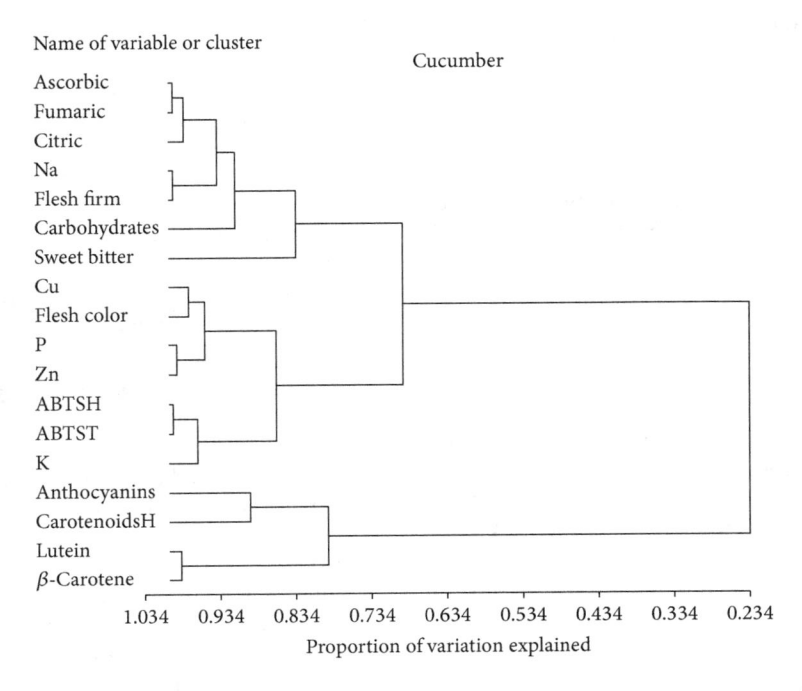

FIGURE 3: Tree dendrogram of the clustering patterns and their proportion of explained variance (11 clusters explained 100% of the variation) using VARCLUS procedure for phytonutrient variables that were used to differentiate among cucumber store-bought and greenhouse-grown varieties. CarotenoidsH: carotenoids, hydrophilic fraction; ABTSH: ABTS, hydrophilic fraction; ABTST: ABTS, total hydrophilic and lipophilic fractions.

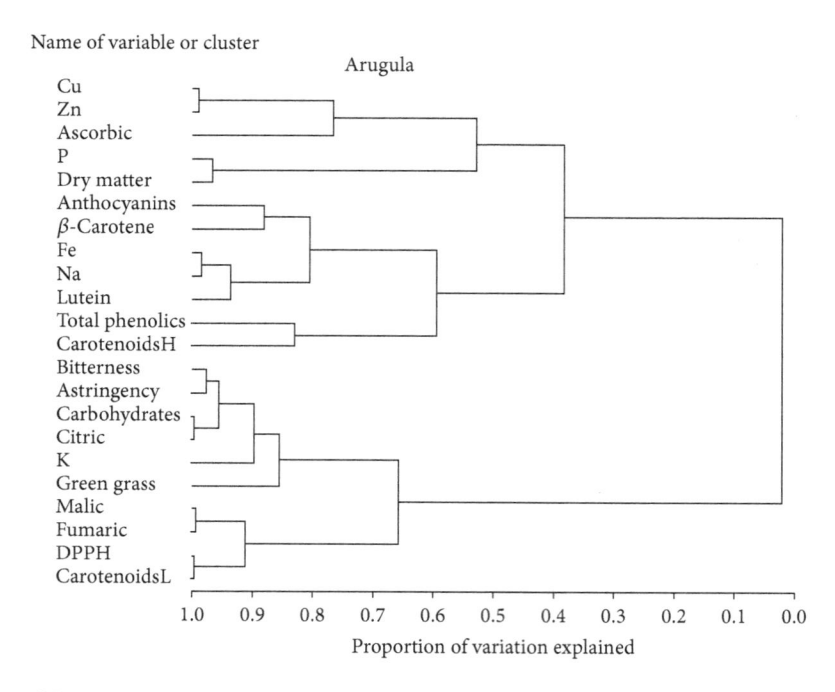

FIGURE 4: Tree dendrogram of the clustering patterns and their proportion of explained variance (13 clusters explained 100% of the variation) using VARCLUS procedure for phytonutrient variables that were used to differentiate among arugula store-bought and greenhouse-grown varieties. CarotenoidsH: Carotenoids, hydrophilic fraction; CarotenoidsL: carotenoids, lipophilic fraction.

best classification (Figure 3). Based on results illustrated in Figures 3, 8, and 13, we identified two sensory criteria (fruit sweetness and flesh firmness) and 10 phytonutrient factors (Cu, K, P, total carotenoids (hydrophilic extract), ABTS, lutein, ascorbic acid, fumaric acid, total carbohydrates, and total anthocyanins) as the most influential for discrimination among cucumber varieties. Ascorbic acid content was an important trait where cucumber had similar levels to carrot [56].

The PLS and PCA output showed that the eigenvalues of the first 2 components (PLS factors 1 and 2 and PCA factors 1 and 2) are responsible for 100% of the variance and 85.4% of the covariance, so plotting PLS 1 versus PLS 2 (Figure 13) and PCA 1 versus PCA 2 (100.0%) (Figure 8) would identify variables that contribute more than others to the variance. Correlation loading plot (Figure 13) shows that plotting factors PLS 1 and PLS 2 of X and Y scores reveal 100% and 85.4% of X and Y variability, respectively, which illustrates the best dimensional rotation and visualization of the variance components. It was clear from data presented in Figures 8 and 13 that firmness and sweetness were positively correlated with minerals (Cu, Fe, P, and Zn) and fumaric and ascorbic acids. These variables collectively were negatively correlated with total carbohydrate, total anthocyanins, and total carotenoids. Those 12 variables express almost all of the variance reported.

3.13.4. Dominant Variables for Arugula. Approximately 13 clusters explained close to 100% of variation among arugula varieties (Figure 4). These clusters were visualized in a tree dendrogram obtained using the VARCLUS and PCA procedures. Plots of the correlation loading (out of PLS analysis) and the multidimensional preference analysis plots showed similar groupings of the variables (Figures 9 and 14) used in the variable selection. Plotting X-scores (scales) and Y-scores explained 66.6 and 46.8%, respectively, of the variance and covariance (Figure 14) and PC1 versus PC2 illustrates 91.66% of the covariance (Figure 9). Two sensory components (astringent taste and green/grassy color) and 11 phytonutrient factors (Fe, Cu, P, K, citric acid, malic acid, ascorbic acid, total carotenoids, β-carotene, total phenolics, and total anthocyanins) were the most influential for discrimination among arugula varieties. It was reported in the literature that arugula is a rich source of β-carotene either fresh or processed [21, 51]. It was reported that main compounds that affect the aroma of arugula were isothiocyanates and derivatives of butane, hexane, octane, and nonane. Specifically, 4-methylthiobutyl isothiocyanate (14.2%), cis-3-hexen-1-ol (11.0%), cis-3-hexenyl butanoate (10.8%), 5-methylthiopentyl isothiocyanate (9.3%), cis-3-hexenyl 2-methylbutanoate (5.4%), and 5-methylthiopentanenitrile (5.0%) were found in concentrations higher than 5.0% [57]. From Figures 9 and 14, a strong positive association was noticed between the astringent taste and green/grassy color, K, citric acid, and total carbohydrates, while negative correlation occurred between astringent taste, green/grassy color, P, K, citric acid, carbohydrates, malic acid, fumaric acid, carotenoids, β-carotene, and total phenolics. Also, contributions from ascorbic acid and total anthocyanins into total variance were limited.

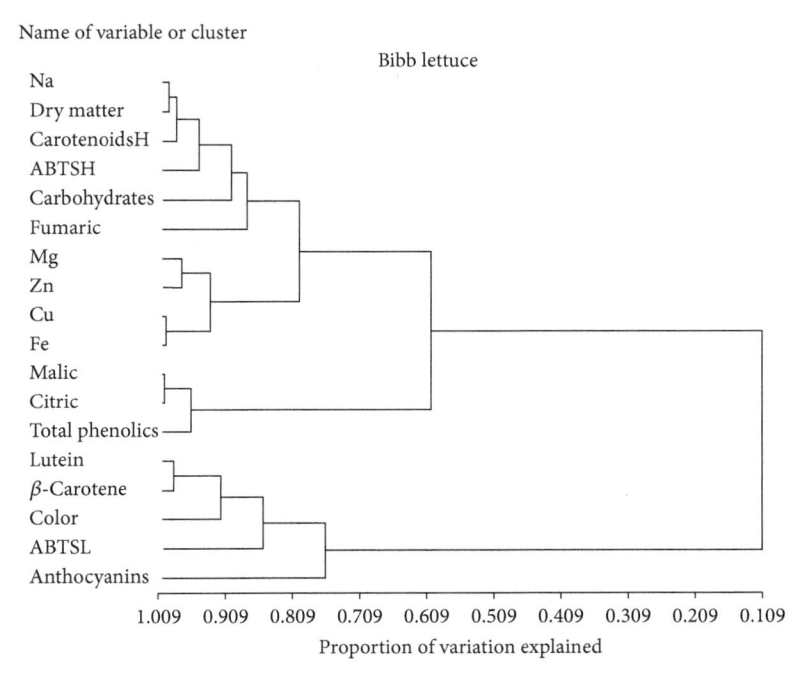

FIGURE 5: Tree dendrogram of the clustering patterns and their proportion of explained variance (13 clusters explained 100% of the variation) using VARCLUS procedure for phytonutrient variables that were used to differentiate among bibb lettuces store-bought and greenhouse-grown varieties. CarotenoidsH: carotenoids, hydrophilic fraction; ABTSL: ABTS, lipophilic fraction.

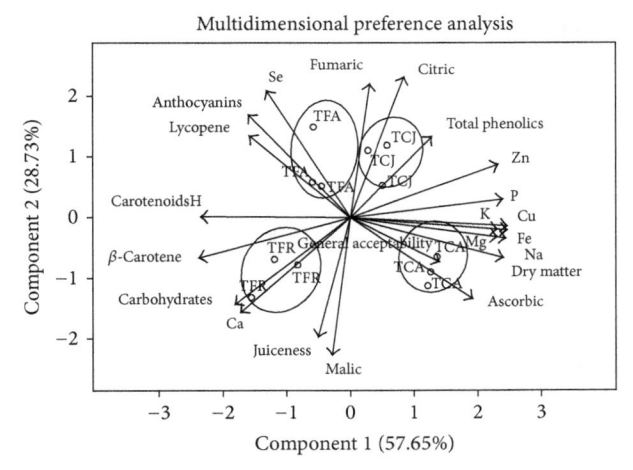

FIGURE 6: Multidimensional preference analysis plot of the PCA procedure. A visual illustration tool that shows similarity classification of the dependent and independent variables but in different space dimension for variables used to differentiate cherry tomato varieties: greenhouse cvs TFA: Favorita, TCJ: Juanita, and TCA: Apero, and the store-bought cv TFR: Fruiterie.

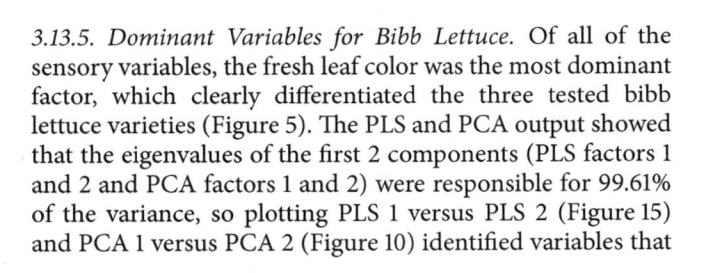

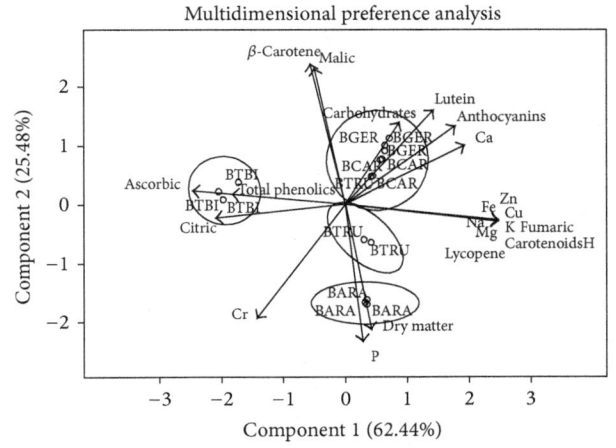

FIGURE 7: Multidimensional preference analysis plot of the PCA procedure. A visual illustration tool that shows similarity classification of the dependent and independent variables but in different space dimension for variables used to differentiate beefsteak tomato varieties: greenhouse cvs BGER: Geronimo, BCAR: Caramba, BTRU: Trust, and BARA: Arbason, and the store-bought cv BTBI: BionatureL.

3.13.5. Dominant Variables for Bibb Lettuce. Of all of the sensory variables, the fresh leaf color was the most dominant factor, which clearly differentiated the three tested bibb lettuce varieties (Figure 5). The PLS and PCA output showed that the eigenvalues of the first 2 components (PLS factors 1 and 2 and PCA factors 1 and 2) were responsible for 99.61% of the variance, so plotting PLS 1 versus PLS 2 (Figure 15) and PCA 1 versus PCA 2 (Figure 10) identified variables that contributed more than others to the variance. The correlation loading plot (PLS) and the multidimensional preference analysis plot (PCA) showed basically similar classification of the dependent and independent variables. Correlation loading plot (Figure 15) showed that plotting factors PLS 1 and PLS 2 of X and Y scores revealed 100% and 82% of X and Y variability, respectively, which illustrates the best dimensional rotation and visualization of the variance components.

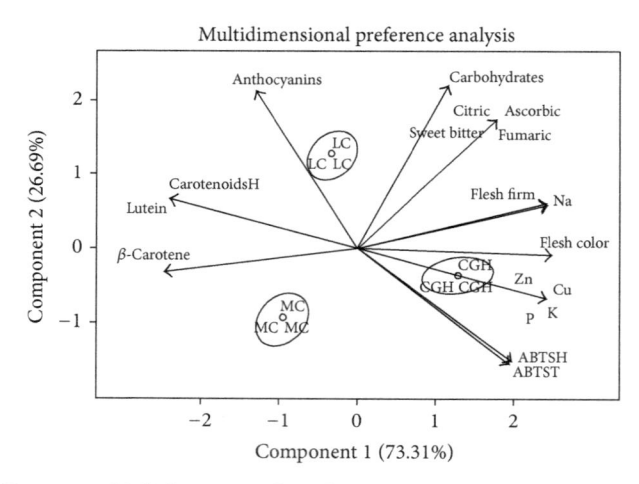

FIGURE 8: Multidimensional preference analysis plot of the PCA procedure. A visual illustration tool that shows similarity classification of the dependent and independent variables but in different space dimension for variables used to differentiate cucumber varieties greenhouse cv CGH: Diva and store-bought cvs LC: Labenese, and MC: Mini Cucumber.

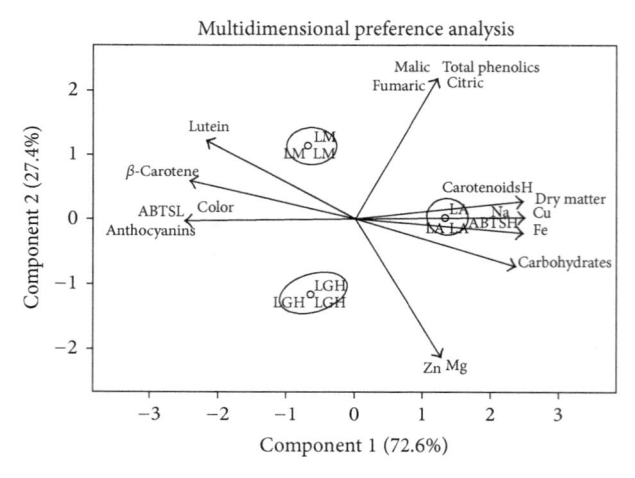

FIGURE 10: Multidimensional preference analysis plot of the PCA procedure. A visual illustration tool that shows similarity classification of the dependent and independent variables but in different space dimension for variables used to differentiate bibb lettuce varieties: greenhouse cv LGH: RexMT and store-bought cvs LM: IGA and LA: ADO.

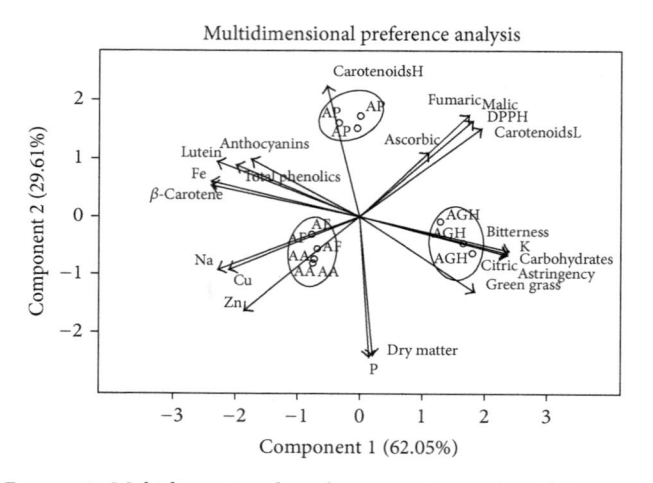

FIGURE 9: Multidimensional preference analysis plot of the PCA procedure. A visual illustration tool that shows similarity classification of the dependent and independent variables but in different space dimension for variables used to differentiate arugula varieties: greenhouse cv AGH: Astro and store-bought cvs AP: PRO, AF: BW, and AA: ADO.

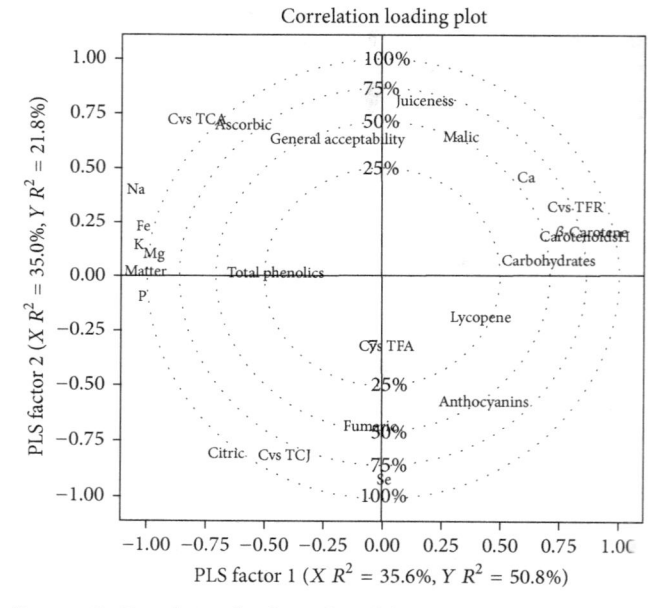

FIGURE 11: Correlation loading plot of the PLS procedure. A visual illustration tool that shows similarity classification of the dependent and independent variables but in different space dimension for variables used to differentiate cherry tomato varieties: greenhouse cvs TFA: Favorita, TCJ: Juanita, and TCA: Apero and the store-bought cv TFR: Fruiterie.

Figure 5 shows that 100% of the total variation was explained using 13 clusters. The VARCLUS method indicated that separation of dependent variables into 13 clusters resulted in the best classification (Figure 5). Based on results illustrated in Figure 5 and Table 6 it was clear that 11 phytonutrient factors (Cu, Zn, Mg, ABTS, total carotenoids (hydrophilic extract), β-carotene, total phenolics, total anthocyanins, citric acid, fumaric acid, and total carbohydrates) were the most influential for discrimination among bibb lettuce varieties.

For bibb lettuce, total anthocyanins, total phenolics, total carbohydrates, citric acid, malic acids, fumaric acid, lutein, and β-carotene were the dominant variables in differentiating between greenhouse-grown and store-bought lettuce. Also, these were the variables that contributed the most to taste and acceptability (Figures 10 and 15).

4. Conclusions

In this study, we examined the possibility of discriminating cultivars of several crop species, both hydroponically grown and store-bought, based on taste panels' sensory analysis,

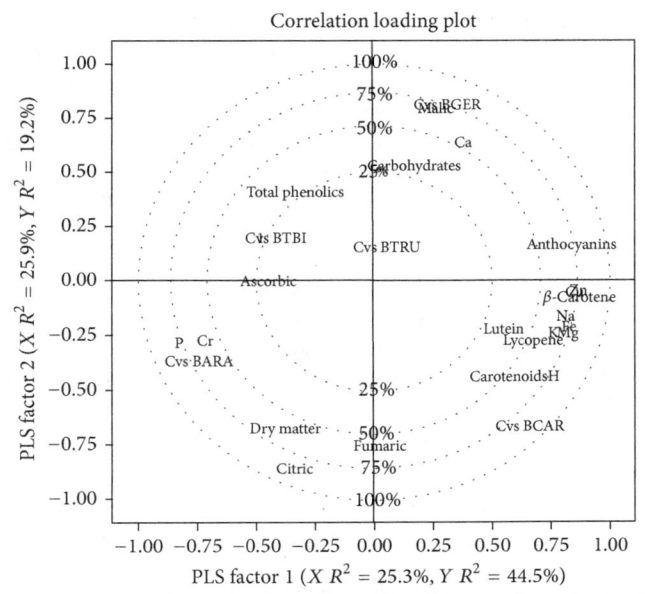

FIGURE 12: Correlation loading plot of the PLS procedure. A visual illustration tool that shows similarity classification of the dependent and independent variables but in different space dimension for variables used to differentiate beefsteak tomato varieties: greenhouse cvs BGER: Geronimo, BCAR: Caramba, BTRU: Trust, and BARA: Arbason and the store-bought cv BTBI: BionatureL.

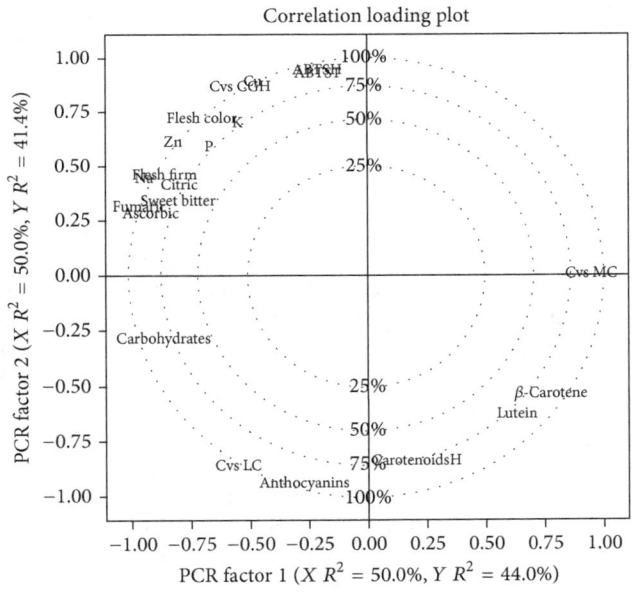

FIGURE 13: Correlation loading plot of the PLS procedure. A visual illustration tool that shows similarity classification of the dependent and independent variables but in different space dimension for variables used to differentiate cucumber varieties greenhouse cv CGH: Diva and store-bought cvs LC: Lebanese and MC: Mini Cucumber.

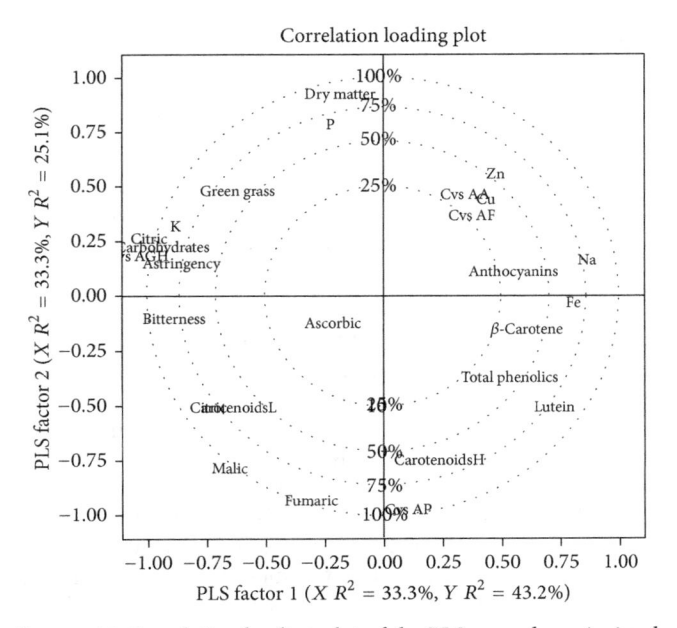

FIGURE 14: Correlation loading plot of the PLS procedure. A visual illustration tool, which shows similarity classification of the dependent and independent variables but in different space dimension for variables used to differentiate arugula varieties: greenhouse cv AGH: Astro and store-bought cvs AP: PRO, AF: BW, and AA: ADO.

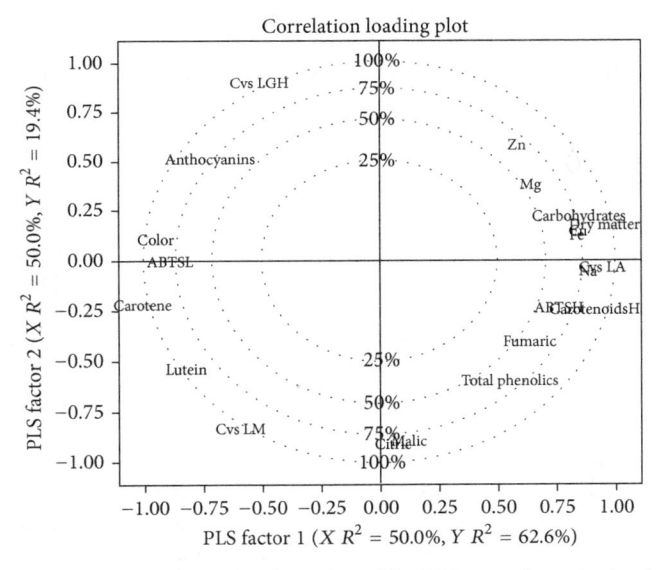

FIGURE 15: Correlation loading plot of the PLS procedure. A visual illustration tool, which shows similarity classification of the dependent and independent variables but in different space dimension for variables used to differentiate bibb lettuce varieties: greenhouse cv LGH: RexMT and store-bought cvs LM: IGA and LA: ADO.

phytonutrient analysis of a long list of phytonutrients, and complex statistical analysis. Our hope was that this strategy could provide plant breeders with sufficient information to both choose the best varieties and design long-term strategies to improve crop breeding programs towards improved human health.

Based on our panelists' responses, the variables we chose for cultivar selection among greenhouse-grown crops (also between these and store-bought crops) were in most cases insufficiently robust to help distinguish between cultivars. The exceptions were the more dominant sensory characteristics such as the "astringent" and "grassy" traits of arugula cultivars, the "juiciness" and "general acceptability" of cherry tomato cultivars, and the "freshness" and "sweetness" of cucumber cultivars. These dominant characteristics could form the core of a consolidated number of criteria in a more discriminating sensory evaluation test.

Among the phytonutrient tests, mineral analysis was powerful for crop cultivar differentiation; in particular, the minerals Cu, Fe, K, Mg, and P were more useful than others to discriminate between cultivars within each crop. Total carotenoids, particularly β-carotene, lycopene, and lutein, were also strong determinants for cultivar selection. Previous literature reported that hydrophilic antioxidant activity (ABTS), ascorbic acid, β-carotene, and lutein were major phytonutrient components of tomato varieties and ascorbic acid was the main antioxidant component [58]. We found that total carbohydrate was a strong variable for cultivar selection within each crop except arugula. Organic acids, including ascorbic, citric, and fumaric acids, were good traits for cultivar selection within each crop. Total phenolics and total anthocyanins were strong factors to differentiate between cultivars within each crop except cucumber.

Our data supports the use of hierarchical cluster analysis, used efficiently in past studies to compare between different types of foods through the analysis of data sets [39, 59]. This statistical approach enabled overall conclusions regarding hydroponically grown in comparison to store-bought produce and helped determine the effect of "freshness" on phytonutrient parameters. Overall, some hydroponic produce was superior in phytonutrient parameters to SB produce, but this was not necessarily apparent with all hydroponic produce or all of these parameters. For example, the cherry tomato cultivar Apero exceeded SB varieties in content of most minerals (Mg, Na, P, K, Fe, and Zn), percent dry matter, and malic acid content. Cultivar Geronimo exceeded SB cultivars in Ca, Cu, K, Fe, Mg, Na, Se, Zn, total anthocyanins, and malic acid content. In addition, cultivar Trust had greater content of Ca, Cu, K, Fe, Mg, Na, Zn, total anthocyanins, and lycopene compared with SB varieties.

The effect of fridge-storage under ideal circumstances (picked fresh, packed into zip-lock plastic bags, and placed at 4°C for 3 or 6 d) did not have a dramatic effect on phytonutrient status. As long as produce is handled well after harvest, it seems that it does not deteriorate too much during this time frame. There are various implications of this to consider. Foremost is that while "fresh-picked" is perceived by consumers to be very important, "freshness" under ideal storage conditions, such as the 6 d of fridge-storage used in our study, did not affect phytonutrient quality. It is apparent that the effect of cultivar is more important than the effect of short-term storage under ideal storage conditions. This is reassuring for suppliers and customers who may be concerned about relative freshness of their produce. For example, the effects of various phenolic and organic acids that were affected by storage in some cultivars are taste-related components that could have an impact on sensory quality. There are many unknowns related to the effects of storage on sensory parameters (this was not done) and phytonutrient characteristics, which could now be explored. Apart from GH arugula that had relatively more intense bitter flavor, GH produce generally had rankings of sensory traits that were either superior or equal to SB produce. The importance of cultivar selection towards improved sensory traits is evident for cherry tomatoes, where 'Apero' showed consistently better sensory characteristics relative to the other GH cherry tomatoes in addition to the SB produce. The advantage of GH cucumbers relative to the two other tested SB produce is apparent for flesh color and flesh firmness indicating a superior product in terms of sensory characteristics.

The current study provides new information concerning key relationships and correlation among sensory analysis, phytonutrient content, and data mining statistics of arugula, bibb lettuce, cucumber, and tomato (beef-steak and cherry) to enable better visualization of essential relationships among vegetable and fruit traits. Plant breeders can use this study to select the more discriminating sensory criteria and strategize to find others and to better select among phytochemical assays and eliminate the often costly analyses with lesser utility for discrimination between genotypes. Ultimately, consumers will benefit by identification and promotion of cultivars with superior phytonutrient content and antioxidant capacity.

Acknowledgments

Thanks are due to Ms. Lauren Rathmell of Lufa Farms for help growing the hydroponic greenhouse produce in the Macdonald campus research greenhouse, purchasing the store-bought produce and other materials for the taste tests, and assisting with plate preparation and other matters during the taste tests. The authors also thank Drs. Z. Farook and M. Baig (Arid Agriculture University, Rawalpindi, Pakistan) for their assistance with the taste and phytonutrient tests and paper review. Thanks are due to Mr. Simon Hebert and his staff for their assistance with greenhouse help during the crop growth and harvesting phases. Thanks are also due to Dr. G. Marquis, Chair, and the other members of McGill's Ethics Board, for their helpful suggestions towards the detailed planning necessary to conduct the Taste Tests (McGill Ethics certificate REB File #: 953-1110). The authors also thank the School of Dietetics & Human Nutrition for use of their sensory analysis facility for conducting the taste tests. Thanks are also due to Drs. M. Lefsrud and Dr. S. Prasher (Bioresource Engineering Dept.), A. Mustafa (Animal Sciences Dept.), and J. Singh (Plant Science Dept.) for the use of the freeze-dryer, ICP OES, LECO, and −80°C freezer, respectively. Authors D. J. Donnelly and S. Kubow are also grateful to the NSERC Discovery Grant program.

References

[1] H. M. Resh, *Hydroponic Food Production: A Definitive Guidebook of Soilless Food-Growing Methods*, Newconcept Press, Princeton, NJ, USA, 6th edition, 2004.

[2] J. B. Jones Jr., *Hydroponics: A Practical Guide for the Soilless Growers*, CRC Press, 2nd edition, 2005.

[3] FAO, FAO STAT, 2011, http://faostat.fao.org/faostat/servlet.

[4] I. T. Johnson, "Antioxidant and antitumor properties," in *Antioxidants in Food*, J. Pokomy, J. N. Yanishlieva, and M. Gordon, Eds., pp. 100–123, Woodhead Publishing, Cambridge, UK, 2001.

[5] O. M. Atrooz, "The antioxidant activity and polyphenolic contents of different plant seeds extracts," *Pakistan Journal of Biological Sciences*, vol. 12, no. 15, pp. 1063–1068, 2009.

[6] Y.-F. Chu, J. Sun, X. Wu, and R. H. Liu, "Antioxidant and antiproliferative activities of common vegetables," *Journal of Agricultural and Food Chemistry*, vol. 50, no. 23, pp. 6910–6916, 2002.

[7] W. Song, C. M. Derito, M. K. Liu, X. He, M. Dong, and R. H. Liu, "Cellular antioxidant activity of common vegetables," *Journal of Agricultural and Food Chemistry*, vol. 58, no. 11, pp. 6621–6629, 2010.

[8] Statistics Canada, *Food Statistics*, SAS, Ed., Statisitics Canada, 2009.

[9] J. K. Campbell, K. Canene-Adams, B. L. Lindshield, T. W.-M. Boileau, S. K. Clinton, and J. W. Erdman Jr., "Tomato phytochemicals and prostate cancer risk," *Journal of Nutrition*, vol. 134, no. 12, pp. 3486S–3492S, 2004.

[10] H. Gerster, "The potential role of lycopene for human health," *Journal of the American College of Nutrition*, vol. 16, no. 2, pp. 109–126, 1997.

[11] L. Arab, S. Steck, and A. E. Harper, "Lycopene and cardiovascular disease," *The American Journal of Clinical Nutrition*, vol. 71, pp. 1691S–1695S, 2000.

[12] A. V. Rao and S. Agarwal, "Role of antioxidant lycopene in cancer and heart disease," *Journal of the American College of Nutrition*, vol. 19, no. 5, pp. 563–569, 2000.

[13] W. Stahl and H. Sies, "Lycopene: a biologically important carotenoid for humans?" *Archives of Biochemistry and Biophysics*, vol. 336, no. 1, pp. 1–9, 1996.

[14] P. W. Simon, "Plant pigments for color and nutrition," *HortScience*, vol. 32, pp. 12–13, 1997.

[15] E. Giovannucci, "Tomatoes, tomato-based products, lycopene, and cancer: Review of the epidemiologic literature," *Journal of the National Cancer Institute*, vol. 91, no. 4, pp. 317–331, 1999.

[16] M. Rodriguez-Carmona, J. Kvansakul, J. Alister Harlow, W. Köpcke, W. Schalch, and J. L. Barbur, "The effects of supplementation with lutein and/or zeaxanthin on human macular pigment density and colour vision," *Ophthalmic and Physiological Optics*, vol. 26, no. 2, pp. 137–147, 2006.

[17] P. K. Mukherjee, N. K. Nema, N. Maity, and B. K. Sarkar, "Phytochemical and therapeutic potential of cucumber," *Fitoterapia*, vol. 84, no. 1, pp. 227–236, 2013.

[18] M. S. DuPont, Z. Mondin, G. Williamson, and K. R. Price, "Effect of variety, processing, and storage on the flavonoid glycoside content and composition of lettuce endive," *Journal of Agricultural and Food Chemistry*, vol. 48, no. 9, pp. 3957–3964, 2000.

[19] C. Nicolle, N. Cardinault, E. Gueux et al., "Health effect of vegetable-based diet: lettuce consumption improves cholesterol metabolism and antioxidant status in the rat," *Clinical Nutrition*, vol. 23, no. 4, pp. 605–614, 2004.

[20] R. Llorach, A. Martínez-Sánchez, F. A. Tomás-Barberán, M. I. Gil, and F. Ferreres, "Characterisation of polyphenols and antioxidant properties of five lettuce varieties and escarole," *Food Chemistry*, vol. 108, no. 3, pp. 1028–1038, 2008.

[21] E. M. M. Tassi and J. Amaya-Farfan, "Carotenoid uptake by human triacylglycerol-rich lipoproteins from the green leafy vegetable *Eruca sativa*," *Ecology of Food & Nutrition*, vol. 47, no. 1, pp. 77–94, 2008.

[22] M. Kimura and D. B. Rodriguez-Amaya, "Carotenoid composition of hydroponic leafy vegetables," *Journal of Agricultural and Food Chemistry*, vol. 51, no. 9, pp. 2603–2607, 2003.

[23] J. Benton Jr., *Hydroponics: A Practical Guide for the Soilless Grower*, CRC Press, Boca Raton, Fla, USA, 2nd edition, 2005.

[24] S. Wold, N. Kettaneh, and L. Eriksson, "PLS in data mining and data integration," in *Handbook of Partial Least Squares: Concepts, Methods and Applications*, V. E. Vinzi, W. W. Chin, J. Henseler, and H. Wang, Eds., Springer Handbooks of Computational Statistics, Springer, Berlin, Germany, 2010.

[25] L. Munck, *The Revolutionary Aspect of Exploratory Chemometric Technology. The Universe and the Biological Cell as Computers. A Plea for cognitive flexibility in mathematical modelling*, Narayana Press, Gylling, Denmark, 2005.

[26] S. Wold, M. Sjöström, and L. Eriksson, "PLS-regression: a basic tool of chemometrics," *Chemometrics and Intelligent Laboratory Systems*, vol. 58, no. 2, pp. 109–130, 2001.

[27] A. M. K. Nassar, K. Sabally, S. Kubow, Y. N. Leclerc, and D. J. Donnelly, "Some canadian-grown potato cultivars contribute to a substantial content of essential dietary minerals," *Journal of Agricultural and Food Chemistry*, vol. 60, no. 18, pp. 4688–4696, 2012.

[28] A. M. K. Nassar, S. Kubow, Y. N. Leclerc, and D. J. Donnelly, "Somatic mining for phytonutrient improvement of 'Russet Burbank' potato," *American Journal of Potato Research*, vol. 91, no. 1, pp. 89–100, 2014.

[29] R. Chirinos, H. Rogez, D. Campos, R. Pedreschi, and Y. Larondelle, "Optimization of extraction conditions of antioxidant phenolic compounds from mashua (*Tropaeolum tuberosum* Ruíz & Pavón) tubers," *Separation and Purification Technology*, vol. 55, no. 2, pp. 217–225, 2007.

[30] R. Re, N. Pellegrini, A. Proteggente, A. Pannala, M. Yang, and C. Rice-Evans, "Antioxidant activity applying an improved ABTS radical cation decolorization assay," *Free Radical Biology & Medicine*, vol. 26, no. 9-10, pp. 1231–1237, 1999.

[31] L. E. Rodriguez-Saona and R. E. Wrolstad, "Extraction, isolation, and purification of anthocyanins," in *Current Protocols in Food Analytical Chemistry*, R. E. Wrolstad, T. E. Acree, H. An et al., Eds., John Wiley & Sons, New York, NY, USA, 2001.

[32] C. R. Brown, R. Wrolstad, R. Durst, C.-P. Yang, and B. Clevidence, "Breeding studies in potatoes containing high concentrations of anthocyanins," *American Journal of Potato Research*, vol. 80, no. 4, pp. 241–250, 2003.

[33] G. A. Garzón, K. M. Riedl, and S. J. Schwartz, "Determination of anthocyanins, total phenolic content, and antioxidant activity in Andes berry (*Rubus glaucus* Benth)," *Journal of Food Science*, vol. 74, no. 3, pp. C227–C232, 2009.

[34] M. M. Giusti and R. E. Wrolstad, "Characterization and measurement of anthocyanins by UV-visible spectroscopy," in *Current Protocols in Food Analytical Chemistry*, pp. F1.2.1–F1.2.13, John Wiley & Sons, 2001.

[35] A. R. Davis, J. Collins, W. W. Fish, Y. Tadmor, C. L. Webber III, and P. Perkins-Veazie, "Rapid method for total carotenoid detection in canary yellow-fleshed watermelon," *Journal of Food Science*, vol. 72, no. 5, pp. S319–S323, 2007.

[36] R. E. Wrolstad, T. E. Acree, E. A. Decker et al., *Current Protocols in Food Analytical Chemistry*, John Wiley & Sons, New York, NY, USA, 2004.

[37] H. K. Lichtenthaler and C. Buschmann, "UNIT F4.3 chlorophylls and carotenoids: measurement and characterization by UV-VIS spectroscopy," in *Current Protocols in Food Analytical Chemistry*, pp. F4.3.1–F4.3.8, John Wiley & Sons, New York, NY, USA, 2001.

[38] A. I. O. Barba, M. C. Hurtado, M. C. S. Mata, V. F. Ruiz, and M. L. S. de Tejada, "Application of a UV-vis detection-HPLC method for a rapid determination of lycopene and β-carotene in vegetables," *Food Chemistry*, vol. 95, no. 2, pp. 328–336, 2006.

[39] P. Carli, S. Arima, V. Fogliano, L. Tardella, L. Frusciante, and M. R. Ercolano, "Use of network analysis to capture key traits affecting tomato organoleptic quality," *Journal of Experimental Botany*, vol. 60, no. 12, pp. 3379–3386, 2009.

[40] M. Mecozzi, R. Acquistucci, M. Amici, and D. Cardarilli, "Improvement of an ultrasound assisted method for the analysis of total carbohydrate in environmental and food samples," *Ultrasonics Sonochemistry*, vol. 9, no. 4, pp. 219–223, 2002.

[41] S. S. Nelsen, "Phenol-sulfuric acid method for total carbohydrates," in *Food Analysis Laboratory Manual*, chapter 6, Springer, New York, NY, USA, 2010.

[42] M. L. Garbelotti, D. A. P. Marsiglia, and E. A. F. S. Torres, "Determination and validation of dietary fiber in food by the enzymatic gravimetric method," *Food Chemistry*, vol. 83, no. 3, pp. 469–473, 2003.

[43] Association of Official Analytical Chemists, *Official Method of Analysis. Cereal Foods Supplements*, AOAC, Arlington, Va, USA, 1995.

[44] SAS, *SAS/INSIGHT User's Guide*, SAS Online Documents, Version 8, SAS Institute, Cary, NC, USA, 1999.

[45] J. A. Labate, S. Grandillo, T. Fulton et al., "Tomato," in *Genome Mapping and Molecular Breeding in Plants*, C. Kole, Ed., Springer, Berlin, Germany, 2007.

[46] C. F. I. Agency, "Guidelines for Highlighted Ingredients and Flavours," http://www.inspection.gc.ca/food/labelling/food-labelling-for-industry/composition-and-qualityclaims/highlight-edconsultation/highlightedguidelines/eng/1339422936843/1339422996038?chap=0.

[47] J. Gajc-Wolska, H. Skapski, and J. A. Szymczak, "Physical and sensory characteristics of the fruits of eight cultivars of field grown tomato," *Acta Physiologiae Plantarum*, vol. 22, no. 3, pp. 365–369, 2000.

[48] L. F. D'Antuono, S. Elementi, and R. Neri, "Exploring new potential health-promoting vegetables: glucosinolates and sensory attributes of rocket salads and related *Diplotaxis* and *Eruca* species," *Journal of the Science of Food and Agriculture*, vol. 89, no. 4, pp. 713–722, 2009.

[49] B. George, C. Kaur, D. S. Khurdiya, and H. C. Kapoor, "Antioxidants in tomato (*Lycopersium esculentum*) as a function of genotype," *Food Chemistry*, vol. 84, no. 1, pp. 45–51, 2004.

[50] P. Y. Niizu and D. B. Rodriguez-Amaya, "New data on the carotenoid composition of raw salad vegetables," *Journal of Food Composition and Analysis*, vol. 18, no. 8, pp. 739–749, 2005.

[51] D. B. Rodriguez-Amaya, "Assessment of the provitamin A contents of foods—the Brazilian experience," *Journal of Food Composition and Analysis*, vol. 9, no. 3, pp. 196–230, 1996.

[52] M. Petro-Turza, "Flavor of tomato and tomato products," *Food Reviews International*, vol. 2, no. 3, pp. 309–351, 1986.

[53] R. A. Jones and S. J. Scott, "Improvement of tomato flavor by genetically increasing sugar and acid contents," *Euphytica*, vol. 32, no. 3, pp. 845–855, 1983.

[54] T. M. M. Malundo, R. L. Shewfelt, and J. W. Scott, "Flavor quality of fresh tomato (*Lycopersicon esculentum* Mill.) as affected by sugar and acid levels," *Postharvest Biology and Technology*, vol. 6, no. 1-2, pp. 103–110, 1995.

[55] K. K. Petersen, J. Willumsen, and K. Kaack, "Composition and taste of tomatoes as affected by increased salinity and different salinity sources," *Journal of Horticultural Science and Biotechnology*, vol. 73, no. 2, pp. 205–215, 1998.

[56] U. L. Opara and M. R. Al-Ani, "Antioxidant contents of prepacked fresh-cut versus whole fruit and vegetables," *British Food Journal*, vol. 112, no. 8, pp. 797–810, 2010.

[57] L. Jirovetz, D. Smith, and G. Buchbauer, "Aroma compound analysis of *Eruca sativa* (Brassicaceae) SPME headspace leaf samples using GC, GC-MS, and olfactometry," *Journal of Agricultural and Food Chemistry*, vol. 50, no. 16, pp. 4643–4646, 2002.

[58] Z. Kotíková, J. Lachman, A. Hejtmánková, and K. Hejtmánková, "Determination of antioxidant activity and antioxidant content in tomato varieties and evaluation of mutual interactions between antioxidants," *LWT—Food Science and Technology*, vol. 44, no. 8, pp. 1703–1710, 2011.

[59] E. A. F. da Silva Torres, M. L. Garbelotti, and J. M. Moita Neto, "The application of hierarchical clusters analysis to the study of the composition of foods," *Food Chemistry*, vol. 99, no. 3, pp. 622–629, 2006.

Colonization and Spore Richness of Arbuscular Mycorrhizal Fungi in Araucaria Nursery Seedlings in Curitiba, Brazil

Carlos Vilcatoma-Medina [ID],[1] Glaciela Kaschuk [ID],[2] and Flávio Zanette[1]

[1]*Programa de Pós-Graduação em Agronomia-Produção Vegetal, Departamento de Fitotecnia, Federal University of Paraná, Curitiba, PR, Brazil*
[2]*Programa de Pós-Graduação em Ciência do Solo, Departamento de Solos e Engenharia Agrícola, Federal University of Paraná, Curitiba, PR, Brazil*

Correspondence should be addressed to Carlos Vilcatoma-Medina; carlos.vilcatoma@ufpr.br

Academic Editor: Mathias N. Andersen

Araucaria or Paraná pine [*Araucaria angustifolia* (Bertol.) Kuntze, 1898] is an endangered timber tree species of Atlantic Forest that naturally forms symbiosis with arbuscular mycorrhizal fungi (AMF). The objective of this experiment was to evaluate AMF colonization and spore AMF richness in araucaria seedlings produced in nursery at the metropolitan region of Curitiba, Brazil, with the interest of identifying a taxonomical AMF group. For that, soil and fine roots of 6-month-, 1-year-, 2-year-, 3-year-, and 5-year-old araucaria seedlings were sampled and evaluated. Evaluations indicated that araucaria seedlings were well colonized by AMF (with rates varying from almost 50 to over 85%) and produced an abundant number of mycorrhizal spores (from 344 to 676 spores per seedling). Samples contained spores of the species *Acaulospora scrobiculata*, *Dentiscutata heterogama*, and *Glomus spinuliferum* and unidentified species of genera *Gigaspora* and *Glomus*. The *Glomus* genus was the most abundant kind of AMF spores found under nursery conditions. Therefore, the experiment evidenced that *Glomus* is a promising genus candidate for being used as AMF inoculant in production of araucaria seedlings.

1. Introduction

Araucaria or Paraná pine [*Araucaria angustifolia* (Bertol.) Kuntze, 1898] is an endangered tree species of the Mixed Ombrophilous Forest, commonly known as "Mata das Araucárias," located in southern and the southern east of Brazil and in portions of Argentina and Paraguay [1, 2]. The araucaria is a dioecious plant, and its propagation by seeds (pinhões) is conditioned to the existence of intraspecific genetic richness in natural habitat. The problem is that 97% of the Mata das Araucarias ecosystem has disappeared [3] and reminiscent araucaria trees are isolated from each other and may not form viable seeds. In addition, seeds are rich in carbohydrates and proteins may be consumed by wild animals and rapidly lose vigor when left on dry soil [4]. Reforestation by transplanting seed-formed seedlings is suitable strategy, but it takes too long before nursery seedlings reach adequate height for transplantation. Yet, adding fertilizer to the substrate to increase growth rates may result in lower root, shoot ratio, and consequently decrease the rates of survival in the field after transplantation.

Araucaria naturally establishes symbiosis with arbuscular mycorrhizal fungi (AMF) [4], in which the plant supplies photosynthates in exchange for competitive advantages of growth and survival [5–8]. Pioneer studies already reported AMF in fine araucaria roots [9]; however, it was only recently that the AMF has been considered as a strategy to increase araucaria growth under nursery conditions [10, 11]. In fact, the studies performed so far show two important facts: first, that araucaria is colonized by a large and diverse number of AMF species [10, 12]; and second that the araucaria positively respond to inoculation of some AMF species, both in high and low P supply [13]. However, we do not know precisely the most efficient and competitive AMF species under nursery seedling production [14].

TABLE 1: Chemical attributes of soils used in the nursery experiment with araucaria in Curitiba, Brazil.

Soil attribute*	Seedling age				
	6 months	1 year	2 years	3 years	5-years
pH $CaCl_2$	5.5 ± 0.240	4.9 ± 0.081	5.5 ± 0.122	5.2 ± 0.173	4.3 ± 0.394
Al^{3+} ($cmol/dm^3$)	0.0 ± 0.0	0.9 ± 0.176	0.0 ± 0.0	0.0 ± 0.042	2.7 ± 1.494
$H + Al^{3+}$ ($cmol/dm^3$)	5.1 ± 0.678	13.2 ± 1.327	7.4 ± 0.903	7.2 ± 0.537	13.9 ± 2.591
Ca^{2+} ($cmol/dm^3$)	4.3 ± 0.221	6.1 ± 0.522	9.4 ± 0.685	7.1 ± 1.784	3.0 ± 1.338
Mg^{2+} ($cmol/dm^3$)	1.8 ± 0.078	3.7 ± 0.355	4.3 ± 0.269	2.9 ± 0.930	1.6 ± 1.338
K^+ ($cmol/dm^3$)	1.1 ± 0.260	0.3 ± 0.151	1.0 ± 0.286	0.7 ± 0.443	0.2 ± 0.128
P (mg/dm^3)	16.7 ± 5.390	78.2 ± 16.967	144.5 ± 16.408	87.5 ± 30.295	79.1 ± 42.421
C (g/dm^3)	12.7 ± 3.457	85.3 ± 27.451	64.3 ± 11.419	46.7 ± 16.486	61.4 ± 55.447

*pH measured in $CaCl_2$ 0.01 mol L^{-1}; Ca^{2+}, Mg^{2+}, and Al^{3+} extracted with KCl 1 mol L^{-1}; $H^+ + Al^{3+}$ extracted with calcium acetate 0.5 mol L^{-1}; K^+ and P extracted with the Mehlich-1 extractor; C determined with $K_2Cr_2O_7$.

Therefore, the objective of this experiment was to evaluate AMF colonization and spore AMF richness in nursery soil samples at the metropolitan region of Curitiba, Brazil, with the interest of identifying the most abundant taxonomical AMF groups that grow under those conditions.

2. Material and Methods

Araucaria seedlings were produced from seeds in the nursery of the Universidade Federal do Paraná, Curitiba (25°25′47″S and 49°16′19″W, 950 m), Brazil, under entirely random design, five replicates, and five treatments. The substrate used was a mixture of soil and organic residues (crop residues, organic compounds, and others) collected in the metropolitan area of the city. The experimental design was completely randomized, with five replications in each treatment. The seedlings were at the moment of harvesting, according to the respective treatments: 6-month, 1-year, 2-year, and 5-year old. Seedlings grew outdoors, under full light natural conditions, and were well watered during the whole period of experiment. The climate in the region is classified as Köppen Cfb [15] and the eventual climate events such as frost and storms did not damage any stage of plant growth.

Soil samples were submitted to chemical analyses, and the pH was measured in $CaCl_2$ 0.01 mol L^{-1}; Ca^{2+}, Mg^{2+}, and Al^{3+} were extracted with KCl 1 M; $H^+ + Al^{3+}$, with $Ca(C_2H_3O_2)_2$ 0.5 M; K^+ and P, with the Mehlich-1 extractor; carbon (C), with the $K_2Cr_2O_7$ (Table 1).

Mycorrhizal spores were extracted with minor modifications with the procedures of [16, 17] as follows: 100 g of soil sample was suspended in 100 ml and left resting for 60 min; soil suspensions were liquefied for 10 s and sequentially sieved through 500 and 53 μm mesh sieves; spores were recovered in tubes with 70% sucrose solution and centrifuged for four min twice and then sieved for 250 and 50 μm mesh sieves. Spores with similar morphology were assembled; then, they were spread on microscope slide and treated with polyvinyl alcohol-lactic acid-glycerol [18] and the Melzer reagent [19]. Spores were individually identified under microscope (Primo Star/Zeiss) according to the morphological characteristics using the criteria of taxonomical classification proposed by Invam [20].

Root fragments were preserved in lactoglycerol (lactic acid, glycerol, and water in the proportion 1:2:1) before

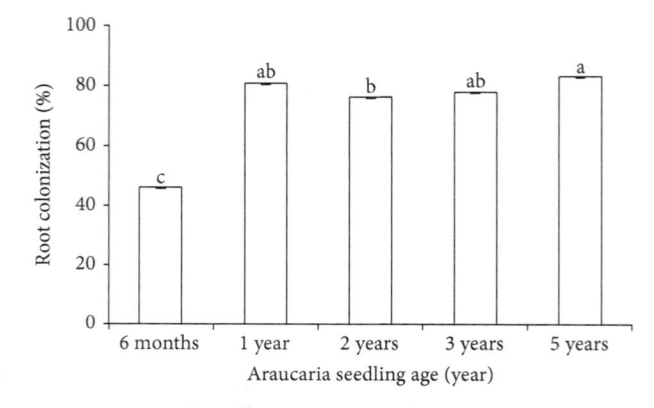

FIGURE 1: Root arbuscular mycorrhizal fungi (AMF) colonization araucaria seedlings with ages between 6 months and 5 years grown in a nursery in field in Curitiba, Brazil. Means followed by the same letter are not significantly different according to Tukey's test ($p \leq 0.05$). The error bars represent the standard error of the mean. $n = 5$.

measurements. The rates of mycorrhizal colonization were estimated by observing cleared and dyed root fragments under microscope (Stemi 305/Zeiss), considering a grid of 1 cm × 1 cm [21], with minor modifications, as follows: fine roots were soaked in KOH 10% at room temperature during 24 hours and then, in water bath at 80°C for an hour; roots were sequentially washed with H_2O_2 100% and water twice and dyed with blue-pen ink in water bath at 80°C for five minutes.

Variance analyses and mean comparisons considered the statistical design of entirely random design, five replicates, and five treatments. Redundancy analyses were performed with Canoco for Windows 4.5 package 43, considering the abundance of spore taxonomic group and chemical attributes as the explanatory variables for the differences between the treatments.

3. Results

Araucaria seedlings with ages between 6 months and 5 years produced under nursery conditions in Curitiba were well colonized by AMF (rates varying from almost 50 to over 85%; Figure 1) and contained an abundant number of mycorrhizal

TABLE 2: Mean number of mycorrhizal spores of araucaria seedlings with ages between 6 months and 5 years grown in a nursery field in Curitiba, Brazil.

Mycorrhizal fungal species	Seedling age (year)				
	6 months	1 year	2 years	3 years	5 years
Acaulospora scrobiculata	13[b]	46[ab]	44[ab]	70[a]	33[b]
Dentiscutata heterogama	11[a]	16[a]	15[a]	17[a]	15[a]
Gigaspora sp.	9[b]	29[a]	8[b]	12[b]	11[b]
Glomus spinuliferum	31[b]	114[a]	53[b]	129[a]	118[a]
Glomus sp. 1	40[ab]	28[b]	40[ab]	62[a]	34[ab]
Glomus sp. 2	91[a]	142[a]	55[a]	161[a]	141[a]
Glomus sp. 3	101[c]	187[a]	82[c]	111[bc]	176[ab]
Glomus sp. 4	48[b]	153[a]	57[b]	73[b]	148[a]
Total of species	344[b]	715[a]	355[b]	635[a]	676[a]

Means followed by the same letter in the lines do not differ by the Tukey test at $p < 0,05$.

spores (varying from 344 to 676; Table 2). Samples contained spores of the species *Acaulospora scrobiculata*, *Dentiscutata heterogama*, and *Glomus spinuliferum* and of four unidentified species of the genera *Gigaspora* (one species) and *Glomus* (four different species).

Mycorrhizal fungi colonization increases from 40 to around 80% in seedlings of 6 months–1 year old. From 1 year onwards, regardless of chemical differences in the soil used, rates of AMF colonization remained at levels of 80% over the years. At first sight, there was a trend in which older seedlings (3- and 5-year old) contained more spores than younger (e.g., 6 months and 2 years old) seedlings, but because 2-year old araucaria seedlings contained the largest number of spores, age was not the main factor affecting the number of spores in this experiment.

The dominant AMF species in the 6-month, 1-year, 2-year, and 5-year old seedlings was *Glomus* sp. 3 and in the 3 years old seedlings, it was *Glomus* sp. 2 (Table 2). The least abundant spores belonged to the family Gigasporaceae (*Dentiscutata heterogama* and *Gigaspora* sp.), and their numbers did not differ between treatments.

Redundancy analyses showed that AMF spore richness in 6-month old seedlings was not affected by soil chemical attributes, but it was so in older seedlings (Figure 2). Attributes related to potential soil acidity (Al^+ and $H+Al^+$) grouped AMF species *Glomus* sp. 3, *Glomus* sp. 4, and *Gigaspora* sp. Attributes related to good soil fertility (Ca, Mg, and P) grouped mycorrhizal fungi species *Glomus* sp. 1, *Dentiscutata heterogama*, *Acaulospora scrobiculata*, and *Glomus* sp. 1 (Figure 2).

4. Discussion

Perennial plants with slow growth like araucaria are not commonly taken into long-term experiments involving AMF because of the obvious difficulty of waiting for a long time for the first results (e.g., compared to annual crops) and also because araucaria is not the first choice of reforestation by companies. Therefore, although AMF colonization has been studied long ago [9], our experiment demonstrated that AMF colonization occurs under nursery conditions to seedlings from very young age up to 5-year old seedlings, without systematic inoculation and regardless of chemical soil attributes (Figure 1; Table 1). This implies that, in addition to field natural conditions [10, 22], araucaria seedlings are also benefited from AMF colonization under nursery conditions.

A great limitation in the application of AMF symbiosis is the fact that we do not have established recommendations of which AMF species are better adapted to nursery and field conditions. Previous studies indicated that araucaria is capable of supporting AMF symbiosis with a greater number of different fungal species [11, 22–25] than that we found in this experiment. To date, araucaria AMF spore richness varied from 8 [23] to 13 [11] in forests of the State of Rio Grande do Sul and from 26 [24] to 58 [25] in forests of Sao Paulo. The variation in spore AMF richness has been frequently attributed to heterogeneity in soil habitat and to the variable capacity of AMF species to adapt to climate and soil in different regions [10, 13]. Therefore, although the AMF richness is great under field conditions, only few species are probably well adapted to nursery conditions.

The genus *Glomus* represented 86% of the spores analyzed in our experiment (Table 2). Previously, it was observed that this genus represented between 40 and 60% of AMF spores in forests in the State of Sao Paulo [10, 22]. In fact, soil acidity associated with exchangeable soil Al (Al^+) may affect spore germination, growth of germinative tube, hyphae, and root [26]. Furthermore, concentrations higher than 100 μm Al^+ affect sporulation of certain species [27]. In this experiment, potential acidity determined by high concentrations of Al^+ and $H + Al^+$ was associated with *Glomus* sp. 3 and *Glomus* sp. 4 (Figure 2). The *Glomus* genus is the most commonly found and probably the most adapted AMF to Brazilian acidy soils [28, 29]. This dominancy indicates that *Glomus* is highly competitive among native spores and it has a high reproductive capacity [10].

The *Acaulospora* genus was the second most abundant AMF spore in the experiment (Table 2), corroborating with the fact that the *Acaulospora* AMF has been found in moderate acidy soils (pH lower than 6.2) [30] and very acidy forest soils [24]. The least abundant genus in this experiment belonged to the genus *Gigaspora* (Table 2). *Gigaspora* has been

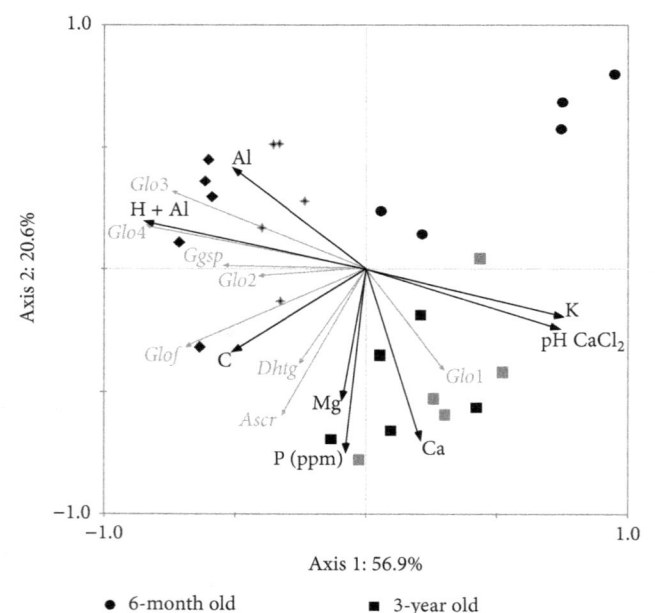

	Redundancies	
	Axis 1	Axis 2
pH CaCl$_2$	0.63	−0.18
Al (cmol/dm^3)	−0.43	0.31
H + Al (cmol/dm^3)	−0.72	0.14
Ca (cmol/dm^3)	0.18	−0.52
Mg (cmol/dm^3)	−0.08	−0.40
K (cmol/dm^3)	0.64	−0.14
P (ppm)	−0.07	−0.55
C (g/dm^3)	−0.44	−0.25

FIGURE 2: Redundancy analyses with several species of arbuscular mycorrhizal fungi in relation to chemical soil attributes in soil used to produce araucaria seedlings in a nursery of Curitiba, Brazil. Note: *Ascr*: *Acaulospora scrobiculata*; *Dhtg*: *Dentiscutata heterogama*; *Ggsp*: *Gigaspora* sp.; *Glof*: *Glomus spinuliferum*; *Glo1*: *Glomus* sp. 1; *Glo2*: *Glomus* sp. 2; *Glo3*: *Glomus* sp. 3; *Glo4*: *Glomus* sp. 4.

associated with moderate acidy soils and high levels of phosphate availability [13, 14]. In our experiment, however, *Gigaspora* was not related to phosphate but it was slightly associated with potential soil acidity (Al$^+$ and H + Al$^+$; Figure 2).

The traditional paradigm describes regulation of AMF symbioses in relation to phosphate availability, in which plants suppress AMF colonization once phosphate is available in enough quantities for being taken from the soil solution. However, this paradigm neglects some aspects of AMF physiology. Reference [31] gave evidence that after the AMF has colonized root wheat (*Triticum aestivum* L.) plants, it does not let it down, even if phosphate is available in enough quantities. It does not imply necessarily that AMF is always efficient regardless of P availability. In fact, P availability determines different AMF compositions and, therefore, different AMF efficiencies [32]. Earlier studies than that of [31] demonstrated that plant growth responses to AMF inoculation under low and high phosphate availability depend very much on the dominant AMF [33]. In that aspect, the genus *Glomus* had a much greater phosphate uptake efficiency than *Gigaspora* [33]. Interestingly, greater efficiency of *Glomus* over *Gigaspora* (increased mycorrhizal dependence of plants) was also measured on araucaria seedlings produced with 150 mg dm^{-3} soil of phosphate fertilizer [14]. Previous studies indicated that mycorrhizal colonization occurs when plants are grown in soils that contain less than 40 mg dm^{-3} [32]. In the present study, the substrate contained 62 mg of P dm^{-3}, and even so Araucaria seedlings were well colonized by mycorrhizae, especially with the genus *Glomus*. This is an observation that should derive more research in our group.

Considering the greater abundance of *Glomus* in relation to other genera, our experiment suggested that other studies should search within *Glomus* diversity to identify AMF inoculants that promote growth of araucaria seedlings. The next step is selecting strains, checking for consistent plant growth responses, and evaluating the viability in inoculant production at commercial scale. As *Glomus* was not related to P soil concentrations in soil samples (Figure 2), it is important that selection occurs under reasonable soil chemical conditions. As a matter of fact, it has been shown that AMF symbioses may be favored by better plant growth [34] and high potential acidity (high Al$^+$) inhibits root growth, damages cellular structure, and hampers water and nutrient [35, 36].

5. Conclusions

Araucaria seedlings are intensively colonized by AMF at least until the age of 5 years old, even when they are not systematically inoculated with specific AMF strains.

Glomus genus is the most abundant kind of AMF spores found under nursery conditions, and therefore other studies should search within *Glomus* diversity to identify AMF inoculants that promote growth of araucaria seedlings.

Acknowledgments

Authors thank Dr. Celso Auer, Dr. Sidney Stürmer, and Dr. Raul M. Cesar, respectively, for the suggestions on the experimental setup, training on spore identification, and statistical analyses. Carlos Vilcatoma-Medina received a CAPES (Coordenação de Aperfeiçoamento de Pessoal de Nível Superior) scholarship to pursue his Ph.D. studies.

References

[1] P. Thomas, "Araucaria angustifolia," *The IUCN Red List of Threatened Species*, Article ID e.T32975A2829141, 2013.

[2] G. F. Paludo, M. B. Lauterjung, M. S. dos Reis, and A. Mantovani, "Inferring population trends of Araucaria angustifolia (Araucariaceae) using a transition matrix model in an old-growth forest," *Southern Forests*, vol. 78, no. 2, pp. 137–143, 2016.

[3] V. M. Stefenon, O. Gailing, and R. Finkeldey, "Genetic structure of *Araucaria angustifolia* (Araucariaceae) populations in Brazil: implications for the in situ conservation of genetic resources," *The Journal of Plant Biology*, vol. 9, no. 4, pp. 516–525, 2007.

[4] R. B. Zandavalli, L. R. Dillenburg, and P. V. D. De Souza, "Growth responses of *Araucaria angustifolia* (Araucariaceae) to inoculation with the mycorrhizal fungus Glomus clarum," *Applied Soil Ecology*, vol. 25, no. 3, pp. 245–255, 2004.

[5] S. Gianinazzi, A. Gollotte, M.-N. Binet, D. van Tuinen, D. Redecker, and D. Wipf, "Agroecology: the key role of arbuscular mycorrhizas in ecosystem services," *Mycorrhiza*, vol. 20, no. 8, pp. 519–530, 2010.

[6] S. E. Smith and F. A. Smith, "Roles of arbuscular mycorrhizas in plant nutrition and growth: New paradigms from cellular to ecosystem scales," *Annual Review of Plant Biology*, vol. 62, pp. 227–250, 2011.

[7] S. E. Smith and F. A. Smith, "Fresh perspectives on the roles of arbuscular mycorrhizal fungi in plant nutrition and growth," *Mycologia*, vol. 104, no. 1, pp. 1–13, 2012.

[8] S. C. Jung, A. Martinez-Medina, J. A. Lopez-Raez, and M. J. Pozo, "Mycorrhiza-induced resistance and priming of plant defenses," *Journal of Chemical Ecology*, vol. 38, no. 6, pp. 651–664, 2012.

[9] F. R. Milanez and H. Monteiro Neto, "Nota prévia sobre micorriza no pinho do Paraná," *Arquivos do Serviço Florestal*, vol. 4, pp. 87–93, 1950.

[10] M. Moreira-Souza, S. F. B. Trufem, S. M. Gomes-Da-Costa, and E. J. B. N. Cardoso, "Arbuscular mycorrhizal fungi associated with *Araucaria angustifolia* (Bert.) O. Ktze," *Mycorrhiza*, vol. 13, no. 4, pp. 211–215, 2003.

[11] R. B. Zandavalli, S. L. Stürmer, and L. R. Dillenburg, "Species richness of arbuscular mycorrhizal fungi in forests with Araucaria in Southern Brazil," *Hoehnea*, vol. 35, no. 1, pp. 63–68, 2008.

[12] M. Moreira, M. I. Zucchi, J. E. Gomes, S. M. Tsai, A. Alves-Pereira, and E. J. B. N. Cardoso, "*Araucaria angustifolia* aboveground roots presented high arbuscular mycorrhizal fungal colonization and diversity in the Brazilian Atlantic Forest," *Pedosphere*, vol. 26, no. 4, pp. 561–566, 2016.

[13] M. Moreira, D. Baretta, and E. J. Cardoso, "Doses de fósforo determinam a prevalência de fungos micorrízicos arbusculares em *Araucaria angustifolia*," *Ciência Florestal*, vol. 22, no. 4, pp. 813–820, 2012.

[14] M. Moreira-Souza and E. J. Cardoso, "Dependência micorrízica de *Araucaria angustifolia* (Bert.) O. Ktze. sob doses de fósforo," *Revista Brasileira de Ciência do Solo*, vol. 26, no. 4, pp. 905–912, 2002.

[15] C. A. Alvares, J. L. Stape, P. C. Sentelhas, J. L. De Moraes Gonçalves, and G. Sparovek, "Köppen's climate classification map for Brazil," *Meteorologische Zeitschrift*, vol. 22, no. 6, pp. 711–728, 2013.

[16] J. Gerdemann and T. H. Nicolson, "Spores of mycorrhizal endogone species extracted from soil by wet sieving and decanting," *Transactions of the British Mycological Society*, vol. 46, pp. 235–244, 1963.

[17] W. R. Jenkins, "A rapid centrifugation technique for separating nematodes from soil," *Plant Disease Report*, vol. 48, article 692, 1964.

[18] J. B. Morton, S. P. Bentivenga, and W. W. Wheeler, "Germ plasm in the International Collection of Arbuscular and Vesicular-Arbuscular Mycorrhizal Fungi (INVAM) and procedures for culture development, documentation and storage," *Mycotaxon*, vol. 48, pp. 491–528, 1993.

[19] R. E. Koske and B. Tessier, "A convenient, permanent slide mounting medium," *Mycological Society of America Newslette*, vol. 34, article 59, 1983.

[20] INVAM, "International culture collection of (vesicular) arbuscular mycorrhizal fungi," https://invam.wvu.edu/.

[21] M. Giovannetti and B. Mosse, "An evaluation of techniques for measuring vesicular arbuscular mycorrhizal infection in roots," *New Phytologist*, vol. 84, no. 3, pp. 489–500, 1980.

[22] M. Moreira, M. A. Nogueira, S. M. Tsai, S. M. Gomes-Da-Costa, and E. J. B. N. Cardoso, "Sporulation and diversity of arbuscular mycorrhizal fungi in Brazil Pine in the field and in the greenhouse," *Mycorrhiza*, vol. 17, no. 6, pp. 519–526, 2007.

[23] M. Breuninger, W. Einig, E. Magel, E. Cardoso, and R. Hampp, "Mycorrhiza of Brazil pine (*Araucaria angustifolia* [Bert. O. Ktze.])," *The Journal of Plant Biology*, vol. 2, no. 1, pp. 4–10, 2000.

[24] M. Moreira, D. Baretta, S. M. Tsai, and E. J. B. N. Cardoso, "Arbuscular mycorrhizal fungal communities in native and in replanted Araucaria forest," *Scientia Agricola*, vol. 66, no. 5, pp. 677–684, 2009.

[25] J. A. Bonfim, R. L. F. Vasconcellos, T. Gumiere, D. de Lourdes Colombo Mescolotti, F. Oehl, and E. J. B. Nogueira Cardoso, "Diversity of arbuscular mycorrhizal fungi in a Brazilian Atlantic Forest Toposequence," *Microbial Ecology*, vol. 71, no. 1, pp. 164–177, 2016.

[26] A. Seguel, J. R. Cumming, K. Klugh-Stewart, P. Cornejo, and F. Borie, "The role of arbuscular mycorrhizas in decreasing aluminium phytotoxicity in acidic soils: a review," *Mycorrhiza*, vol. 23, no. 3, pp. 167–183, 2013.

[27] K. Klugh-Stewart and J. R. Cumming, "Organic acid exudation by mycorrhizal *Andropogon virginicus* L. (broomsedge) roots in response to aluminum," *Soil Biology & Biochemistry*, vol. 41, no. 2, pp. 367–373, 2009.

[28] N. C. Johnson and D. A. Wedin, "Soil carbon, nutrients, and mycorrhizae during conversion of dry tropical forest to grassland," *Ecological Applications*, vol. 7, no. 1, pp. 171–182, 1997.

[29] C. M. R. Pereira, D. K. A. D. Silva, A. C. D. A. Ferreira, B. T. Goto, and L. C. Maia, "Diversity of arbuscular mycorrhizal fungi in Atlantic forest areas under different land uses," *Agriculture, Ecosystems & Environment*, vol. 185, pp. 245–252, 2014.

[30] S. L. Sturmer and M. M. Bellei, "Composition and seasonal variation of spore populations of arbuscular mycorrhizal fungi in dune soils on the island of Santa Catarina, Brazil," *Botany*, vol. 72, no. 3, pp. 359–363, 1994.

[31] H. Li, S. E. Smith, R. E. Holloway, Y. Zhu, and F. A. Smith, "Arbuscular mycorrhizal fungi contribute to phosphorus uptake

by wheat grown in a phosphorus-fixing soil even in the absence of positive growth responses," *New Phytologist*, vol. 172, no. 3, pp. 536–543, 2006.

[32] C. D. Collins and B. L. Foster, "Community-level consequences of mycorrhizae depend on phosphorus availability," *Ecology*, vol. 90, no. 9, pp. 2567–2576, 2009.

[33] S. E. Smith, F. A. Smith, and I. Jakobsen, "Mycorrhizal fungi can dominate phosphate supply to plants irrespective of growth responses," *Plant Physiology*, vol. 133, no. 1, pp. 16–20, 2003.

[34] J. A. Mensah, A. M. Koch, P. M. Antunes, E. T. Kiers, M. Hart, and H. Bücking, "High functional diversity within species of arbuscular mycorrhizal fungi is associated with differences in phosphate and nitrogen uptake and fungal phosphate metabolism," *Mycorrhiza*, vol. 25, no. 7, pp. 533–546, 2015.

[35] L. V. Kochian, M. A. Piñeros, J. Liu, and J. V. Magalhaes, "Plant adaptation to acid soils: the molecular basis for crop aluminum resistance," *Annual Review of Plant Biology*, vol. 66, pp. 571–598, 2015.

[36] Z. Rengel, J. Bose, Q. Chen, and B. N. Tripathi, "Magnesium alleviates plant toxicity of aluminium and heavy metals," *Crop & Pasture Science*, vol. 66, no. 12, pp. 1298–1307, 2015.

Effect of Irrigation and Mulch on the Yield and Water use of Strawberry

Rahena Parvin Rannu [id],[1] Razu Ahmed,[1] Alamgir Siddiky,[1] Abu Saleh Md. Yousuf Ali,[2] Khandakar Faisal Ibn Murad,[3] and Pijush Kanti Sarkar[3]

[1]*Soil & Water Management Section, Horticulture Research Centre, Bangladesh Agricultural Research Institute (BARI), Gazipur 1701, Bangladesh*
[2]*Regional Agricultural Research Station, Bangladesh Agricultural Research Institute (BARI), Gazipur 1701, Bangladesh*
[3]*Irrigation & Water Management Division, Bangladesh Agricultural Research Institute (BARI), Gazipur 1701, Bangladesh*

Correspondence should be addressed to Rahena Parvin Rannu; rannu_bau@yahoo.com

Academic Editor: Othmane Merah

To investigate the effect of irrigation and mulch on the yield of strawberry (line FA-007), this study was conducted at the experimental fields of Pomology Division, BARI, Gazipur and RARS, BARI, Rangpur during Rabi season of 2012-13 and 2013-14. The experiment was conducted followed by the split-plot design with two mulches (black plastic and rice straw), three levels of irrigation (5, 10, and 15 days interval), and three replications. A significant difference was observed for most of the parameters among different treatments for both locations. Irrigation at 5 days interval with rice straw mulch in Gazipur and irrigation at 5 days interval with black polythene mulch in Rangpur were performed better for most of the yield-contributing characters and yield among all other treatments for both the years. But in respect to water productivity, 10 days irrigation interval with rice straw mulch in Gazipur and 10 days interval with black polythene mulch in Rangpur showed highest results among all other treatments. However, it can be concluded that any irrigation practices from the abovementioned options could be adopted by the farmers based on their feasibility and water availability for strawberry cultivation in Bangladesh.

1. Introduction

Strawberry (*Fragaria × annanasa* Dutch.) is one of the most popular fruits in the world. Besides their popularity as fresh fruits, strawberries are used to produce different food products. It is an excellent source of vitamin C, β-carotene, dietary fiber, and some other nutrients for human health and nutrition [1]. It has been recently introduced in Bangladesh, and day by day, it is becoming very popular. The most important thing is that just few years before it was not only impossible for people to taste strawberry but also unpredictable for them to cultivate in their land or in home garden. But it became true, and now a lot of farmers and villagers have taken up the activity to grow strawberry. Strawberry cultivation has already gained much popularity among the farmers in Rangpur, Thakurgaon, Dinajpur, Nilphamari, Gaibandha, Lalmonirhat, Kurigram, Panchagarh, Joypurhat, Naogaon, and other northern districts in Bangladesh [2]. There is no current statistical information regarding strawberry-cultivated area and production in Bangladesh. However, Monda [1] reported that 25000 tonnes of strawberry had been produced in 2009. According to Bangladesh Strawberry Association, around 6,500 bighas of land had been brought under strawberry cultivation throughout the country during the year of 2015-2016 where Rajshahi district is ahead of other districts for strawberry cultivation [3]. The northern region of Bangladesh may be suitable for strawberry cultivation for its relatively cool weather compared to other parts of Bangladesh. The crop is grown in Bangladesh during winter when there is low precipitation and high evapotranspiration. Crop cultivation during this dry period usually requires irrigation. Strawberry is relatively a shallow-rooted plant and susceptible to the water stress condition. Water stress may affect the photosynthetic activity and reduce the potential growth of the plant [4], and Kirnak et al. found that it caused reduction of fruit yield, fruit size, leaf nutrient compositions, and normal plant growth parameters in strawberry except water-soluble dry matter concentrations in fruits [5]. Limited soil moisture affects growth, development, yield, and existence of the strawberry plant in the winter season [6]. Irrigation plays an important role on the total yield, berry weight, runner production, and leaf area of strawberry [7].

So, frequent irrigation is necessary at different stages of strawberry plants. Water unavailability can affect all physiological processes of plant ultimately that has effect on mortality of plant [8]. Yield losses up to 60 to 100% are reported due to long spell of drought stress in different crop species [9]. Some fertilizers and pesticides may be applied through the irrigation system, thus reducing the need and cost to enter the field with equipment. Existing water resources either surface water or groundwater is at risk of near depletion, pollution, and being heavily degraded. There is also strong evidence for the possibility of decreasing rainfall year by year due to climate change [10]. So, it is the high time to develop efficient irrigation management techniques that can minimize water use or maximize the water use efficiency. Increasing water shortage worldwide has led irrigation scheduling to achieve an optimum water supply for productivity [11].

To minimize water use, strawberry planting must be mulched in the fall for good winter survival and maximum yields. Mulch plays an important role to regulate soil temperature, conserve moisture, restrict evaporation losses, and suppress weed growth, reducing the number of dirty and diseased berries [12, 13], enhancing nutrient uptake, and improving water use efficiency and yield [14]. The mulching technique increases the vegetative growth, flowering of plants, yield, and quality of strawberry [15] and creates loose soil surface that increases the total intake of water and reduces surface runoff [16]. Kirnak et al. [5] found that mulching can mitigate negative effects of water stress on the plant growth and fruit yield of strawberry under the field-grown condition especially in semiarid regions. Root zone temperatures play an important role in plant growth and development through the uptake of water and nutrients by roots. Pandey et al. [17] found beneficial effects of mulching on root zone temperature which were reflected in an increased number of fruits and fruit size. He recommended mulch for effective root zone temperature, weed control, and moisture regulation to grow strawberry. In spite of the inorganic plastic mulches (transparent, black or yellow, and others), other types of organic mulches are also used in field conditions in some parts of the world (wheat and rice straw, forest leaf, etc.). Normally, the locally available mulch is rice straw in Bangladesh. In the recent years, plastic mulches are also used for crop production. Organic mulches support diversity of beneficial soil macroinvertebrates in the soil [16] and reduce disease levels in the field plots [18]. On the other hand, black plastic mulch is used to warm the soil early [19] in the soil which ensures faster growth and development of plants. Bakshi et al. found that black polythene mulch also showed significant superiority in reducing weed population over other mulching treatments [20].

However, the information on the judicious use of irrigation water to different mulch materials on strawberry production is scanty in Bangladesh. Therefore, this study was undertaken to investigate the response of strawberry to irrigation and different mulch and to determine an appropriate irrigation schedule for strawberry production, in the agroenvironmental conditions of Bangladesh and similar areas of cultivation.

2. Materials and Methods

2.1. Experimental Site. The experiment was carried out during two successive growth seasons of 2012-13 and 2013-14, at two locations: (1) Horticulture Research Centre (HRC), Joydebpur, Gazipur which is located at 34 m altitude, 23.98 N latitude, and 90.42 E longitude and (2) Regional Agricultural Research Station (RARS), Burirhat, Rangpur which is at 11 m altitude, 25.82 N latitude, and 89.23 E longitude. Both locations are under Bangladesh Agricultural Research Institute (BARI). Between these two locations, RARS, Burirhat, Rangpur location contains more coldness in winter than HRC, Joydebpur, Gazipur location due to geographical features. The average temperature and annual rainfall are 25.8°C and 2036 mm, respectively, for Gazipur, whereas they are 24.9°C and 2192 mm for Rangpur. In Bangladesh, Rangpur is located at the north western part that is faced with prolonged dry season than the rest of the country [21]. For this study, temperature and rainfall data for the entire crop growing season of both locations were recorded (Table 1).

2.2. Treatments and Experimental Design. The treatments were as follows:

Factor A (main plot): M_1 = black polythene and M_2 = rice straw.

Factor B (subplot): I_1 = irrigation at 5 days interval after plant establishment (PE), I_2 = irrigation at 10 days interval after PE, and I_3 = irrigation at 15 days interval after PE.

The experiment was laid out in a split-plot design (SPD) with three replications.

2.3. Planting Materials and Cultural Practices. The Strawberry line FA-007 was used as the test crop because BARI released varieties that were not available enough on that time. FA-007 was a promising line on that time under the supervision of relevant scientists which is released later as a variety (BARI Strawberry-03).

The texture of soil was loamy having a bulk density of 1.40 g/cc. The water content at the wilting point (WP) and field capacity (FC) were 14% and 29%, respectively. Seedlings were transplanted in the first and last weeks of November in the two consecutive years at the age of 1 month of plant on raised bed. The unit plot size and spacing were 2.5 m × 1.0 m and 50 cm × 50 cm, respectively. Each mulch had three rows (each row contained two lines of plants) 1 m apart from each other, and 1.5 m spacing was kept in between black polythene

Table 1: Average monthly weather data during both of experimental seasons for both locations.

Month	Maximum temperature (°C)		Minimum temperature (°C)		Rainfall (mm)	
	Joydebpur, Gazipur	Burirhat, Rangpur	Joydebpur, Gazipur	Burirhat, Rangpur	Joydebpur, Gazipur	Burirhat, Rangpur
December	26.92	23.55	13.87	13.31	4.04	0
January	25.17	22.19	11.42	10.71	0.13	0
February	29.15	25.83	15.05	13.52	19	16.38
March	33.15	30.86	19.95	17.8	5.77	2.67

Table 2: Initial soil nutrient status for both experimental sites.

Nutrient element	Soil test value		Critical level	Soil test interpretation	
	Joydebpur, Gazipur	Burirhat, Rangpur		Joydebpur, Gazipur	Burirhat, Rangpur
pH	6.2	5.6	—	Slightly acidic	Medium acidic
Organic matter (%)	0.95	1.23	—	Low	Low
Ca (meq/100 g soil)	1.12	0.82	2.0	Low	Low
Mg (meq/100 g soil)	0.60	0.25	0.5	Low	Low
Total N (%)	0.08	0.06	0.12	Low	Low
Available P (ppm)	9	11.1	10	Low	Low
K (meq/100 g soil)	0.17	0.11	0.12	Low	Low
S (ppm)	10	2.05	10	Low	Low
Zn (ppm)	1.4	0.41	0.6	Medium	Low
B (ppm)	0.10	0.03	0.2	Low	Low

mulch and rice straw mulch to prevent the water from going to other plots and affecting them. The initial soil characteristics were analyzed (Table 2) for both locations. Fertilizer was applied based on the soil test results and Fertilizer Recommendation Guide, Bangladesh, 2012, at the rate of 120, 40, 110, 25, 4, and 2 kg/ha of N, P, K, S, Zn, and B, respectively. The entire amount of cow dung (5 t/ha), half MoP, and all of other fertilizers except urea were applied as basal and were incorporated into the soil during final land preparation. The urea was applied as top dress in three equal installments at 15, 30, and 45 DAT (days after transplanting), respectively. Rest half of MoP was applied at 45 DAT. Common irrigations (on an avg. 65 mm for Gazipur and 72 mm for Rangpur) were applied till plant establishment. Intercultural operations such as weeding and earthing up were done as and when necessary. For mulching, 13.5 kg straw mulch was used where the thickness of straw was 2 cm. Black polythene mulch with 25-micron thickness were used to cover the entire plot. Both mulches were installed manually on the entire bed. Black plastic mulch with small holes was stretched tightly on soil surface after 40 days from transplanting for strawberry seedlings to grow up. The crop was first harvested on the first week of February in both years.

2.4. Soil Moisture and Irrigation Water Requirement. Initial soil moisture was measured using the gravimetric method. Soil moisture prior to each irrigation and at the time of harvest was determined by the same method. Irrigation water was applied to bring the soil moisture up to field capacity, considering the effective root zone depth. Irrigation treatments began after the plant establishment. Irrigation was applied to each plot in basins by the hose pipe based on the calculated irrigation water.

Irrigation water was calculated using the following equation [22]:

$$d = \frac{\text{FC} - \text{MC}_i}{100} \times A_s \times D, \quad (1)$$

where d is the depth of irrigation (cm), FC is the field capacity of the soil (%), MC_i is the moisture content of the soil before irrigation (%), A_s is the apparent specific gravity of the soil, and D is the depth of the effective root zone (cm).

Seasonal water requirement was calculated using the water balance equation as follows:

Seasonal water requirement (mm) = total irrigation water applied (mm) + seasonal effective rainfall (mm) + soil water contribution (mm).

2.5. Data Collection and Data Analysis. Data on yield, yield-contributing parameters, soil moisture, soil temperature, irrigation water requirement, and so on were recorded time to time. Growth and yield data were statistically analyzed using MSTAT-C program, and the treatment means were separated by DMRT at 5% level of probability.

3. Results and Discussion

3.1. Available Soil Moisture. The moisture content was varied with the irrigation interval for different treatments. The plot with frequent interval irrigation was comparatively in more moisturized condition than others. Black polythene mulch retained higher moisture than rice straw mulch in most of the times (Figure 1). This finding agrees with Kumar and Dey [14]. This might be the fact that black polythene mulch acts as an insulating substance that condenses the

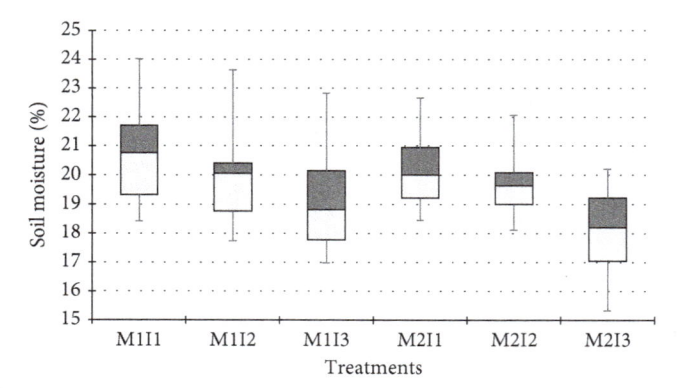

FIGURE 1: Effect of irrigation and mulches on the soil moisture content in different soil layers. M_1 = black polythene mulch; M_2 = rice straw mulch; I_1 = irrigation up to field capacity at 5 days interval after plant establishment (PE); I_2 = irrigation up to field capacity at 10 days interval after PE; I_3 = irrigation up to field capacity at 15 days interval after PE.

TABLE 3: Soil temperature (°C) under both mulches collected once in a week (December–March).

Black polythene mulch										
M_1I_1	15	23	21	22	22	22	24	26	25	28	28
M_1I_2	13	24	21	21	22	22	24	27	26	29	28.5
M_1I_3	14	24	22	22	23	22	24	26	25	29.5	29
Rice straw mulch										
M_2I_1	14.5	22	19	19	20	20	22	23	24	26	26
M_2I_2	13	21	19	19	20	20	21	25	24	26	27
M_2I_3	13	23	20	20	21	21	23	23	23	28	28

M_1 = black polythene mulch; M_2 = rice straw mulch; I_1 = irrigation up to field capacity at 5 days interval after plant establishment (PE); I_2 = irrigation up to field capacity at 10 days interval after plant establishment (PE); I_3 = irrigation up to field capacity at 15 days interval after plant establishment (PE).

evaporating soil moisture inside the mulch and again drops it down to the soil surface [23]. Improvement of water infiltration and higher water retention capacity is also resulted from mulches reported by Swenson et al. [24] and Headu and Kumar [25]. In Figure 1, the black part indicates that most of the value of soil moisture was in this range during the entire cropping season. The possibility of soil moisture was high in the black polythene mulch plot compared to rice straw mulch plot. There was also a possibility of low moisture level in rice straw mulch with a longer irrigation interval.

3.2. Soil Temperature.

Soil temperature was measured below the soil surface (1 cm below) once in a week. Black polythene mulch treatment always retains higher soil temperature compared to rice straw mulch (Table 3). Increase in soil temperature under black plastic mulch and minimum soil temperature under rice straw mulch had been reported by Tariq et al. [23]. This finding agrees with Diaz-Perez et al. [26]. Kumar and Dey found that, on the other hand, hay mulch is more effective than black polythene mulch for raising minimum soil temperature and lowering maximum soil temperature [27].

3.3. Combined Effect of Irrigation and Mulch.

The yield and all yield-contributing characters were performed well for both locations. Different levels of irrigation water and mulch (black polythene and rice straw) had significant effect on all yield-contributing characters and yield for both locations.

3.3.1. Growth Parameters

(1) Plant Growth, Leaf, Flowering, and Harvesting. For Joydebpur, Gazipur, the taller plant, higher number of leaves per plant, and longer leaf length were observed when irrigating field at 5 days irrigation interval (Tables 4 and 5) compared to 10 and 15 days irrigation interval of same mulch. This could be because of the fact that the more moisturized condition due to frequent irrigation helps in enhancing the vegetative growth. The combined effect of irrigation and mulch showed that black polythene mulch with 5 days irrigation interval (M_2I_1) performed better than rice straw mulch (M_1I_1) with 5 days irrigation interval in case of plant height, number of leaves per plant, and leaf length for the two consecutive seasons. Tariq et al. [23] found that black polythene mulch performed better in the case of vegetative growth and yield followed by rice straw mulch for strawberry cultivation. Availability of soil moisture and optimum soil temperature under the mulched condition enhanced the plant height during the vegetative period. Days to 50% flowering and days of first harvest were always earlier in black polythene mulch compare to rice straw mulch. Similar findings were observed from Soliman et al. [28]. Kumar et al. found that plants mulched with black polythene have significantly better growth, early flowering and fruiting, and produced larger fruit and higher yield of strawberry [29]. Similar observation was found by Pandey et al. [30].

For Burirhat, Rangpur, taller plant, higher number of leaves, and longer leaf length were found in black polythene mulch with 5 days irrigation interval (Table 6). Plants mulched with black polythene recorded maximum plant height and maximum number of leaves per plant over other mulching materials [20]. Kaur and Kaur found that vegetative growth parameters of strawberry increased when plants mulched with black polythene [31]. In both the years, early flowering, fruiting, and harvesting were found in black polythene mulch (Table 7) compared to rice straw mulch because black polythene raises soil temperature by absorbing heat in the planting bed that promotes faster development and earlier yields. Besides this, black polythene suppressed weeds as there was absent of light for photosynthesis under plastic film [32].

Since irrigation intervals in treatment I_3 (15 days irrigation interval) were comparatively longer, the plant faced water stress, and as a result of treatment I_3 gave the lowest results for both mulches.

TABLE 4: Effect of irrigation and mulch on plant growth, flowering, and harvesting time of strawberry for Joydebpur, Gazipur.

Treatments	Year 1			Year 2		
	Plant height (cm)	Days to 50% flowering	Days of 1st harvest	Plant height (cm)	Days to 50% flowering	Days of 1st harvest
Black polythene mulch						
M_1I_1	16.53^a	63^c	84.33^d	16.22^a	52.33^c	71^d
M_1I_2	14.20^{bc}	68^{ab}	90^c	15.00^{bc}	56.33^b	75^c
M_1I_3	12.87^c	70^{ab}	95^{ab}	14.44^d	58.67^a	79^b
Rice straw mulch						
M_2I_1	16.27^{ab}	66^{bc}	85.50^d	16.67^a	56.67^b	74.67^c
M_2I_2	14.33^{abc}	70^{ab}	92.5^{bc}	15.45^b	59^a	81b
M_2I_3	12.53^c	72^a	96.50^a	14.67^{cd}	60^a	86.33^a
CV (%)	7.97	3.36	2.15	1.61	1.84	1.85

Means having same or without letter(s) do not differ significantly at 5% level of probability. M_1 = black polythene mulch; M_2 = rice straw mulch; I_1 = irrigation up to field capacity at 5 days interval after plant establishment (PE); I_2 = irrigation up to field capacity at 10 days interval after PE; I_3 = irrigation up to field capacity at 15 days interval after PE.

TABLE 5: Effect of irrigation and mulch on vegetative and yield parameters of strawberry in Joydebpur, Gazipur, for two consecutive seasons.

Treatments	Year 1				Year 2			
	Leaves per plant (number)	Leaves length (cm)	Fruit length (cm)	Fruit diameter (cm)	Leaves per plant (number)	Leaves length (cm)	Fruit length (cm)	Fruit diameter (cm)
M_1I_1	21^a	16.33^a	4.64^a	3.12	16.22^a	15.78^a	4.48^{ab}	3.22^{abc}
M_1I_2	15.07^{cd}	13.67^{cd}	3.93^b	2.93	14.33^{ab}	12.78^{bc}	4.02^b	3.08^{bc}
M_1I_3	13.13^d	12.33^d	3.89^b	2.88	13.56^b	12.14^c	4.17^b	3.01^c
M_2I_1	17.73^b	15.27^{ab}	4.44^a	3.13	15.33^{ab}	14.00^b	4.78^a	3.37^a
M_2I_2	16.13^{bc}	13.93^{bc}	4.42^a	3.15	15.56^{ab}	13.00^{bc}	4.43^{ab}	3.29^{ab}
M_2I_3	15.87^{bc}	12.33^d	4.36^a	3.07	13.89^{ab}	12.67^{bc}	4.23^b	3.01^c
CV (%)	7.95	5.25	5.01	4.79	7.78	5.01	5.84	3.75

Means having same or without letter(s) do not differ significantly at 5% level of probability. M_1 = black polythene mulch; M_2 = rice straw mulch; I_1 = irrigation up to field capacity at 5 days interval after plant establishment (PE); I_2 = irrigation up to field capacity at 10 days interval after plant establishment (PE); I_3 = irrigation up to field capacity at 15 days interval after plant establishment (PE).

TABLE 6: Effect of irrigation and mulch on vegetative and yield parameters of strawberry in Burirhat, Rangpur, for two consecutive seasons.

Treatments	Year 1				Year 2			
	Leaves per plant (number)	Leaves length (cm)	Fruit length (cm)	Fruit diameter (cm)	Leaves per plant (number)	Leaves length (cm)	Fruit length (cm)	Fruit diameter (cm)
M_1I_1	18.73^{ab}	16.20^a	3.80^b	2.54^a	35.01^{ab}	14.00	4.12^a	3.50^a
M_1I_2	19.90^a	15.50^{ab}	3.60^c	2.49^a	36.34^a	13.27	3.87^b	3.19^{bc}
M_1I_3	17.00^{bc}	13.78^c	3.50^d	2.46^{ab}	28.12^{bc}	12.47	3.56^c	2.82^d
M_2I_1	16.40^c	15.10^{ab}	3.80^b	2.51^a	27.56^{bc}	13.20	3.90^b	3.37^{ab}
M_2I_2	19.00^{ab}	14.93^{bc}	3.90^a	2.60^a	24.00^c	13.00	3.63^c	3.09^{cd}
M_2I_3	17.07^{bc}	14.26^{bc}	3.63^c	2.34^b	20.56^c	11.77	3.48^c	2.90^d
CV (%)	7.08	4.56	1.47	3.10	14.85	2.45	2.50	4.75

Means having same or without letter(s) do not differ significantly at 5% level of probability. M_1 = black polythene mulch; M_2 = rice straw mulch; I_1 = irrigation up to field capacity at 5 days interval after plant establishment (PE); I_2 = irrigation up to field capacity at 10 days interval after plant establishment (PE); I_3 = irrigation up to field capacity at 15 days interval after plant establishment (PE).

3.3.2. Yield Parameters. In Joydebpur, Gazipur, the combined effect of irrigation and mulch results showed significant differences among different treatments. From all the treatments, it is found that irrigation at 5 days interval with rice straw mulch (M_2I_1) gave the higher yield closely followed by irrigation at 5 days interval with black polythene mulch (M_1I_1) (Table 8). Among different treatments, the plant grown under plastic mulch produced larger number of fruits, fruit weight, and finally contributed to yield. Saeid and Mohammed [33] also found the same results because of better plant growth due to favorable hydrothermal regime and complete weed free environment. Strawberry plants, when mulched with black polythene film, produced more number of fruits which ultimately resulted in highest strawberry production compared to other types of mulches found by Castaneda et al. [34]. Kher et al. found that higher

TABLE 7: Effect of irrigation and mulch on plant growth, flowering, and harvesting time of strawberry in Burirhat, Rangpur.

Treatments	Year 1			Year 2		
	Plant height (cm)	Days to 50% flowering	Days of 1st harvest	Plant height (cm)	Days to 50% flowering	Days of 1st harvest
Black polythene mulch						
M_1I_1	17.60^a	67.11^b	89^d	18.78	59^d	81.67^c
M_1I_2	16.67^{ab}	70.89^{ab}	94^c	17.50	63^{bd}	84^b
M_1I_3	15.13^b	72^{ab}	97^b	14.62	66^{ab}	87.33^a
Rice straw mulch						
M_2I_1	15.90^b	74.63^a	92^c	16.61	61^{cd}	83^{bc}
M_2I_2	16.60^{ab}	73^{ab}	97^b	15.45	65^{bc}	86.67^a
M_2I_3	15.27^b	73.89^a	101^a	14	69.67^a	88.67^a
CV (%)	5.36	10.6	1.36	6.46	3.75	1.35

Means having same or without letter(s) do not differ significantly at 5% level of probability. M_1 = black polythene mulch; M_2 = rice straw mulch; I_1 = irrigation up to field capacity at 5 days interval after plant establishment (PE); I_2 = irrigation up to field capacity at 10 days interval after PE; I_3 = irrigation up to field capacity at 15 days interval after PE.

TABLE 8: Effect of irrigation and mulch on the yield and yield-contributing characters of strawberry for Joydebpur, Gazipur.

Treatments	Year 1			Year 2		
	Fruits per plant (number)	Individual fruit weight (g)	Yield (t/ha)	Fruits per plant (number)	Individual fruit weight (g)	Yield (t/ha)
Black polythene mulch						
M_1I_1	22.67^a	21.77^a	11.68^a	18.27^a	22^a	12.07^a
M_1I_2	12.90^{cd}	17.39^{bc}	5.94^{bc}	15.60^b	19.20^{bc}	10.35
M_1I_3	11.63^d	16.64^c	4.84^c	11.03^c	18.26^c	6.87
Rice straw mulch						
M_2I_1	19.43^b	21.88^a	11.98^a	19.10^a	25.00^a	12.30^a
M_2I_2	18.60^b	20.43^{ab}	10.46^a	15.30^b	21.11^{bc}	11.35^b
M_2I_3	13.53^c	20.19^{ab}	7.43^b	13.70^c	19.19^{bc}	7.93
CV (%)	4.38	8.99	5.62	6.55	8.71	2.3

Means having same or without letter(s) do not differ significantly at 5% level of probability. M_1 = black polythene mulch; M_2 = rice straw mulch; I_1 = irrigation up to field capacity at 5 days interval after plant establishment (PE); I_2 = irrigation up to field capacity at 10 days interval after PE; I_3 = irrigation up to field capacity at 15 days interval after PE.

vegetative growth and the higher number of fruits per plant were observed under black polythene mulch [35]. Kumar and Dey [14] and Bakshi et al. [20] also reported the same results. In case of yield, though there were no statistically significant differences for 5 days irrigation interval between both mulches, there were significant differences for 10 and 15 days irrigation interval for both mulches. Fruit weight decreased from the maximum of 21.77 g with 5 days irrigation intervals to the minimum of 16.64 g with 15 days irrigation interval in the first year. The decrease in the fruit weight from 5 to 15 days irrigation interval was 23.6% in the first year and 17% in the second year for black polythene mulch. For rice straw mulch, it was about 7% in the first year and 23% in the second year. However, the difference in 5 to 15 days irrigation interval had a significant effect on fruit weight for both the years. Yuan et al. [36] found that increasing irrigation frequency causes cell expansion of fruit and that ultimately increases fruit weight. Akhtar and Rab [37] reported that the maximum fresh fruit weight, moisture content, and ascorbic acid content were found when strawberry plants irrigated at 4 days interval compared to longer irrigation interval. On the other hand, Liu et al. [38] found that fresh weight and water content of strawberry had decreased due to longer irrigation intervals or partial root drying.

However, finally it is found that rice straw mulch at 5 days irrigation interval performs better in Joydebpur, Gazipur (Table 8 and Figure 2). Because initially black polythene mulch performed well for the vegetative growth, yield, and all yield-contributing characters, but the favorable condition by using black polythene mulch did not sustain for the rest of the growing period. The possible reason could be that temperature was rising during the crop growing period day by day. This rising temperature possibly created warmer condition beneath the black polythene mulch and resulted in an unfavorable environment for the strawberry plants to survive. Fruits were getting smaller in size, leaves were burned, and finally plants died. Singh and Yadav [16] observed that fruits from the plant covered by black polythene can be damaged in very hot and sunny weather in the strawberry production under field condition. All yield-contributing characters and yield results revealed that 15 days irrigation interval with black polythene mulch (M_1I_3) will not suit for strawberry production in Joydebpur, Gazipur.

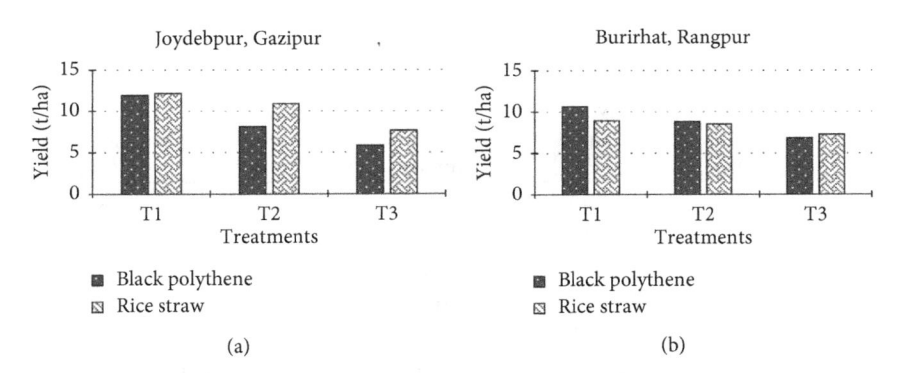

FIGURE 2: Effect of irrigation and mulches on the yield (average of both years) of strawberry for both locations.

TABLE 9: Effect of irrigation mulch on the yield and yield-contributing characters of strawberry in Burirhat, Rangpur.

Treatments	Year 1			Year 2		
	Fruits per plant (number)	Individual fruit weight (g)	Yield (t/ha)	Fruits per plant (number)	Individual fruit weight (g)	Yield (t/ha)
Black polythene mulch						
M_1I_1	14.67^a	16.33^a	9.56^a	24.56^a	17.17	11.74^a
M_1I_2	13.83^{ab}	15.07^b	7.90^b	19.45^{bc}	16.40	9.76^{bc}
M_1I_3	11.83^{cd}	14.13^c	6.66^c	13.00^d	13.50	7.01^d
Rice straw mulch						
M_2I_1	12.17^{cd}	15.10^b	7.35^{bc}	20.17^b	16.30	10.53^{ab}
M_2I_2	12.67^{bc}	15.47^b	7.73^{bc}	17.50^c	15.70	9.27^{bc}
M_2I_3	10.67^d	13.93^c	5.94^d	15.00^d	14.30	8.60^c
CV (%)	6.57	2.23	8.53	6.25	4.14	8.45

Means having same or without letter(s) do not differ significantly at 5% level of probability. M_1 = black polythene mulch; M_2 = rice straw mulch; I_1 = irrigation up to field capacity at 5 days interval after plant establishment (PE); I_2 = irrigation up to field capacity at 10 days interval after PE; I_3 = irrigation up to field capacity at 15 days interval after PE.

But in the case of Burirhat, Rangpur, irrigation at 5 days interval with black polythene mulch (M_1I_1) showed better results considering all yield-contributing characters and yield compared to mulching with rice straw at 5 days irrigation interval for both the years (Table 9 and Figure 2). Kaur and Kaur [31] reported that application of black polythene mulch enhanced yield and yield parameters than other mulches. Sharma and Goel [39] found that the maximum yield was obtained from black polythene mulch compared to wheat and rice straw mulches for the strawberry production. The present findings of this study are in accordance with the previous researchers [23, 28, 30, 40] who reported significantly higher yield under black plastic mulch because of effective soil temperature, weed control, and conservation of soil moisture for the strawberry cultivation. The above results and discussion under Section 3.3 revealed that 5 days interval irrigation in the strawberry field during the growing season brought better vegetative growth, flowering, fruiting, and yield for both locations. Black polythene/rice straw played as a supportive material for the higher yield of strawberry in both locations. In the regions which contain comparatively low temperatures, black polythene mulch may perform well for the entire growing season.

3.4. Water Use and Water Productivity.

Seasonal water use and water productivity of strawberry under different irrigation treatments are presented in Tables 10 and 11 for both locations. The amount of irrigation water applied was increased with the increased irrigation frequency.

From all the treatments (Table 10), total water use was found maximum (498.05 mm in the first year and 490.08 mm in the second year) in treatment M_2I_1, that is, when irrigating field at 5 days interval with rice straw mulch which was followed by irrigating field at 5 days interval with black polythene mulch (474.42 and 467.32 mm for two consecutive years) with the same irrigation frequency. Black polythene absorbs heat and conserves more moisture (Figure 1) than rice straw and for this reason comparatively less water was applied in black polythene mulched plot than rice straw mulched plot. Lowest water use was found in treatment I_3 due to lower frequency of irrigation for both mulches. Although total water use was the highest in treatment M_2I_1, the water productivity in terms of yield per unit water use was found to be the highest from M_2I_2 (3.04 kg/m^3 in the first year, 3.44 kg/m^3 in the second year, and 3.24 kg/m^3 avg. of both years) treatment. However, the lowest water productivity (avg. of both years) was obtained from M_1I_3 treatment in Joydebpur, Gazipur, due to lower irrigation frequency and lower yield.

In Burirhat, Rangpur, the highest irrigation water was used in the rice straw mulched plot followed by the black polythene mulched plot with the same irrigation frequency for both the years (Table 11). The highest water productivity was obtained from treatment M_1I_2 (3.49 kg/m^3) in the

TABLE 10: Irrigation frequency, seasonal water use, and water productivity for Joydebpur, Gazipur.

Treatments	Year 1			Year 2			Average water productivity for both years (kg/m^3)
	Number of irrigation (number)	Seasonal total water use (mm)	Water productivity (kg/m^3)	Number of irrigation (number)	Seasonal total water use (mm)	Water productivity (kg/m^3)	
M_1I_1	25	474.42	2.46	21	467.32	2.58	2.52
M_1I_2	12	321.36	1.85	11	316.15	3.26	2.56
M_1I_3	9	241.05	2.01	7	271.28	2.53	2.27
M_2I_1	25	498.05	2.41	21	490.08	2.51	2.46
M_2I_2	12	343.77	3.04	11	329.78	3.44	3.24
M_2I_3	9	268.36	2.77	7	278.33	2.85	2.81

M_1 = black polythene mulch; M_2 = rice straw mulch; I_1 = irrigation up to field capacity at 5 days interval after plant establishment (PE); I_2 = irrigation up to field capacity at 10 days interval after PE; I_3 = irrigation up to field capacity at 15 days interval after PE.

TABLE 11: Irrigation frequency, seasonal water use, and water productivity for Burirhat, Rangpur.

Treatments	Year 1			Year 2			Average water productivity for both years (kg/m^3)
	Number of irrigation (number)	Seasonal total water use (mm)	Water productivity (kg/m^3)	Number of irrigation (number)	Seasonal total water use (mm)	Water productivity (kg/m^3)	
M_1I_1	25	514.80	1.86	21	450.12	2.61	2.24
M_1I_2	12	365.90	2.16	11	279.27	3.49	2.83
M_1I_3	9	305.08	2.18	7	250.99	2.79	2.49
M_2I_1	25	553.10	1.33	21	489.80	2.15	1.74
M_2I_2	12	394.20	1.96	11	315.75	2.94	2.45
M_2I_3	9	317.30	1.87	7	262.54	3.28	2.58

M_1 = black polythene mulch; M_2 = rice straw mulch; I_1 = irrigation up to field capacity at 5 days interval after plant establishment (PE); I_2 = irrigation up to field capacity at 10 days interval after PE; I_3 = irrigation up to field capacity at 15 days interval after PE.

second year and 2.83 kg/m^3 avg. for both the years, whereas the lowest water productivity (1.74 kg/m^3) was obtained from treatment M_2I_1 because of highest water use.

So, it can be stated that 10 days irrigation interval (I_2) gave the highest water productivity for both locations, but the only difference is in mulching material between these locations.

4. Conclusion

Results revealed that 5 days irrigation interval was performed better with both types of mulches and both locations. In respect of yield, rice straw mulch along with 5 days irrigation interval in Gazipur and black polythene mulch along with 5 days irrigation interval in Rangpur were performed well. But considering water productivity, 10 days irrigation interval with rice straw mulch in Gazipur and 10 days interval with black polythene mulch for Rangpur showed highest results among all other treatments. It can be inferred that if water is available enough, then 5 days irrigation interval with rice straw mulch in Gazipur and 5 days interval with black polythene mulch in Rangpur could be an effective irrigation schedule for strawberry cultivation. But if water is limited and farmers are not capable enough to pay much attention on production, then considering water productivity 10 days irrigation interval with rice straw mulch in Gazipur and 10 days interval with black polythene mulch in Rangpur could be good practice for farmers to cultivate strawberry in Bangladesh and similar strawberry industry conditions.

Disclosure

Rahena Parvin Rannu is currently at Irrigation and Water Management Division, Bangladesh Agricultural Research Institute, Bangladesh. Abu Saleh Md. Yousuf Ali is currently at Regional Horticulture Research Station, Bangladesh Agricultural Research Institute, Bangladesh. Pijush Kanti Sarkar is currently at Department of Agricultural Engineering and Technology, Sylhet Agricultural University, Bangladesh.

Acknowledgments

The authors are grateful to Bangladesh Agricultural Research Institute (BARI), Gazipur, for the financial support and Horticulture Research Centre (HRC), Irrigation and Water Management (IWM) Division, Gazipur, and Regional Agricultural Research Station (RARS), Rangpur, BARI for their facilities to execute this experiment. The authors also thank their colleagues and staff who supported during the study.

References

[1] D. Monda, *Possibility of Strawberry Cultivation in Bangladesh. An Assignment on Strawberry*, Department of Business Administration, Northern University, Dhaka, Bangladesh, October 2017, https://www.scribd.com/doc/.../Possibility-of-Strawberry-Cultivation-in-Bangladesh.

[2] BSS, *Newly Harvested Strawberry Arrives in Northern Markets*, Bangladesh Sangbad Sangstha, Dhaka, Bangladesh, January 2017, http://www.bssnews.net/newsDetails.php?cat=0&id=94716$date=2010-03-14&dateCurrent=2010-03-18.

[3] R. H. Firoz, *Commercial Strawberry Cultivation on the Rise in Rajshahi*, March 2018, http://www.observerbd.com/2016/02/10/135849.php.

[4] A. H. El-Farhan and M. P. Pritts, "Water requirements and water stress in strawberry," *Advances in Strawberry Research*, vol. 15, pp. 5–12, 1997.

[5] H. Kirnak, C. Kaya, D. Higgs, and S. Gercek, "A long-term experiment to study the role of mulches in the physiology and macro-nutrition of strawberry grown under water stress," *Australian Journal of Agricultural Research*, vol. 52, no. 9, pp. 937–943, 2001.

[6] E. Krüger, G. Schmidt, and S. Rasim, "Effect of irrigation on yield, fruit size and firmness of strawberry cv. Elsanta," in *Proceedings of the IV International Strawberry Symposium*, vol. 567, pp. 471–474, Tampere, Finland, 2000.

[7] L. Taparauskienė and O. Miseckaitė, "Effect of mulch on soil moisture depletion and strawberry yield in sub-humid area," *Polish Journal of Environmental Studies*, vol. 23, no. 2, pp. 475–482, 2014.

[8] B. C. Sarker, M. Hara, and M. Uemura, "Proline synthesis, physiological responses and biomass yield of eggplants during and after repetitive soil moisture stress," *Scientia Horticulturae*, vol. 103, no. 4, pp. 387–402, 2005.

[9] M. P. Singh, U. N. Pandey, R. K. Lal, and G. S. Chaturvedi, "Response of *Brassica* species to different irrigation regimes," *Indian Journal of Plant Physiology*, vol. 7, no. 1, pp. 66–69, 2002.

[10] F. Kimura, A. Kitoh, A. Sumi, J. Asanuma, and A. Yatagai, *Downscaling of the Global Warming Projections to Turkey. The Final Report of ICCAP*, Vol. 10, ICCAP Publication, Istanbul, Turkey, 2007.

[11] P. O. Boamah, J. D. Owusu-Sekyere, L. K. Sam-Amoah, and B. Anderson, "Effects of irrigation interval on chlorophyll fluorescence of tomatoes under sprinkler," *Asian Journal of Agricultural Research*, vol. 5, no. 1, pp. 83–89, 2011.

[12] O. A. Khadas, *Effect of Different Irrigation Levels on Growth and Yield of Strawberry under Silver Black Mulch*, Ph.D. dissertation, College of Agricultural Engineering and Technology, Dr. BSKKV, Dapoli, India, 2014.

[13] R. R. Sharma, *Growing Strawberries*, International Book Distributing Co., Lucknow, India, 2002.

[14] S. Kumar and P. Dey, "Effect of different mulches and irrigation methods on root growth, nutrient uptake, water use efficiency and yield of strawberry," *Scientia Horticulturae*, vol. 127, no. 3, pp. 318–324, 2011.

[15] A. Angrej and G. S. Gaur, "Effect of mulching on growth, fruit yield and quality of strawberry (*Fragaria × ananassa* Duch.)," *Asian Journal of Horticulture*, vol. 2, no. 1, pp. 149–151, 2007.

[16] B. K. Singh and K. S. Yadav, "Response of mulching on strawberry under field condition," *Journal of Bio Innovation*, vol. 6, no. 5, pp. 761–767, 2017.

[17] S. Pandey, J. Singh, and I. B. Maurya, "Effect of black polythene mulch on growth and yield of winter dawn strawberry (*Fragaria × ananassa*) by improving root zone temperature," *Indian Journal of Agricultural Sciences*, vol. 85, no. 9, pp. 1219–1222, 2015.

[18] M. V. S. Coelho, F. R. Palma, and A. C. Café-Filho, "Management of strawberry anthracnose by choice of irrigation system, mulching material and host resistance," *International Journal of Pest Management*, vol. 54, no. 4, pp. 347–354, 2008.

[19] A. K. Singh and S. Kamal, "Effect of black plastic mulch on soil temperature and tomato yield in mid hills of Garhwal Himalayas," *Journal of Horticulture and Forestry*, vol. 4, no. 4, pp. 77–79, 2012.

[20] P. Bakshi, D. J. Bhat, V. K. Wali, A. Sharma, and M. Iqbal, "Growth, yield and quality of strawberry (*Fragaria × ananassa* Duch.) cv. Chandler as influenced by various mulching materials," *African Journal of Agricultural Research*, vol. 9, no. 7, pp. 701–706, 2014.

[21] S. Shahid, "Spatial and temporal characteristics of droughts in the western part of Bangladesh," *Hydrological Processes*, vol. 22, no. 13, pp. 2235–2247, 2008.

[22] A. M. Michael, *Irrigation: Theory and Practice*, Vikas Publishing House Pvt. Ltd., New Delhi, India, 1978.

[23] M. S. Tariq, A. Bano, and K. M. Qureshi, "Response of strawberry (*Frageria annanasa*) cv. Chandler to different mulching materials," *Science, Technology and Development*, vol. 35, no. 3, pp. 117–122, 2016.

[24] J. A. Swenson, S. A. Walters, and S. K. Chong, "Influence of tillage and mulching systems on soil water and tomato fruit yield and quality," *Journal of Vegetable Crop Production*, vol. 10, no. 1, pp. 81–95, 2004.

[25] N. K. Headu and M. Kumar, "Effect of different mulches on yield, plant height, nitrogen uptake, weed control, soil moisture and economics of tomato cultivation," *Progressive Horticulture*, vol. 34, no. 2, pp. 208–210, 2002.

[26] J. C. Diaz-Perez, D. Granberry, D. Bertrand, and D. Giddings, "Tomato plant growth during establishment as affected by root zone temperature under colored mulches," in *Proceedings of the XXVI International Horticultural Congress: Issues and Advances in Transplant Production and Stand Establishment Research*, vol. 631, pp. 119–124, Toronto, ON, USA, August 2002.

[27] S. Kumar and P. Dey, "Influence of soil hydrothermal environment, irrigation regime, and different mulches on the growth and fruit quality of strawberry (*Fragaria × ananassa*L.) plants in a sub-temperate climate," *Journal of Horticultural Science and Biotechnology*, vol. 87, no. 4, pp. 374–380, 2012.

[28] M. A. Soliman, H. A. Abd El-Aal, M. A. Ramadan, and N. N. Elhefnawy, "Growth, fruit yield and quality of three strawberry cultivars as affected by mulch type and low tunnel," *Alexandria Science Exchange Journal*, vol. 36, no. 4, pp. 402–414, 2015.

[29] R. Kumar, V. Tandon, and M. M. Mir, "Impact of different mulching material on growth, yield and quality of strawberry (*Fragaria × ananassa* Duch.)," *Progressive Horticulture*, vol. 44, no. 2, pp. 234–236, 2012.

[30] S. Pandey, G. S. Tewari, J. Singh, D. Rajpurohit, and G. Kumar, "Efficacy of mulches on soil modifications, growth, production and quality of strawberry (*Fragaria × ananassa* Duch.)," *International Journal of Science and Nature*, vol. 7, no. 4, pp. 813–820, 2016.

[31] P. Kaur and A. Kaur, "Effect of various mulches on the growth and yield of strawberry cv. Chandler under sub-tropical conditions of Punjab," *International Journal of Recent Trends in Science and Technology*, vol. 25, no. 1, pp. 21–25, 2017.

[32] T. Yaghi, A. Arslan, and F. Naoum, "Cucumber (*Cucumis sativus*, L.) water use efficiency (WUE) under plastic mulch and

drip irrigation," *Agricultural Water Management*, vol. 128, pp. 149–157, 2013.

[33] A. I. Saeid and G. H. Mohammed, "The Effect of color plastic mulches on growth, yield and quality of two hybrids of summer squash (cucurbita pepo l.)," *Science Journal of University of Zakho*, vol. 3, no. 1, pp. 113–118, 2015.

[34] L. M. F. Castaneda, L. E. C. Antunes, N. C. Ristow, and S. Carpenedo, "Utilization of different mulching types in strawberry production," in *Proceedings of the VI International Strawberry Symposium*, vol. 842, pp. 111–113, Huelva, Spain, March 2008.

[35] R. Kher, J. A. Baba, and P. Bakshi, "Influence of planting time and mulching material on growth and fruit yield of strawberry cv. Chandler," *Indian Journal of Horticulture*, vol. 67, pp. 441–444, 2010.

[36] B. Z. Yuan, J. Sun, and S. Nishiyama, "Effect of drip irrigation on strawberry growth and yield inside a plastic greenhouse," *Biosystems Engineering*, vol. 87, no. 2, pp. 237–245, 2004.

[37] I. Akhtar and A. Rab, "Effect of irrigation intervals on the quality and storage performance of strawberry fruit," *Journal of Animal and Plant Sciences*, vol. 25, no. 3, pp. 669–678, 2015.

[38] F. Liu, S. Savić, C. R. Jensen et al., "Water relations and yield of lysimeter-grown strawberries under limited irrigation," *Scientia Horticulturae*, vol. 111, no. 2, pp. 128–132, 2007.

[39] V. K. Sharma and A. K. Goel, "Effect of mulching and nitrogen on growth and yield of strawberry," *International Journal of Science, Environment and Technology*, vol. 6, no. 3, pp. 2074–2079, 2017.

[40] A. A. Shokouhian and A. Asghari, "Study the effect of mulch on yield of some strawberry cultivars in Ardabil condition," in *Proceedings of the International Conference on Agriculture, Ecology and Biological Engineering*, Antalya, Turkey, September 2015.

Phenetic Analysis of Cultivated Black Pepper (*Piper nigrum* L.) in Malaysia

Y. S. Chen,[1] M. Dayod,[2] and C. S. Tawan[3]

[1]*Malaysian Pepper Board, Jalan Utama, Pending Industrial Area, P.O. Box 1653, Kuching, Sarawak 93916, Malaysia*
[2]*Agriculture Research Centre, Semongok, P.O. Box 977, Kuching, Sarawak 93720, Malaysia*
[3]*Faculty of Resource Science and Technology, Universiti Malaysia Sarawak, Jalan Datuk Mohd Musa, Kota Samarahan, Sarawak 94300, Malaysia*

Correspondence should be addressed to Y. S. Chen; yschen@mpb.gov.my

Academic Editor: Nesibe E. Kafkas

Phenetic analysis of all the black pepper cultivars in Malaysia is crucial to determine the morphological difference among them. The objective of this study is to ascertain the morphological distinctness and interrelationships among the cultivars to ensure registration of each variety under the Plant Variety Protection Act, as a prerequisite toward implementation of a monovarietal farm policy in the future. Cluster analysis revealed that cultivars "Semongok Aman" and "Semongok 1" have high distinctness values for identification; thus, varietal diagnosis for the two cultivars is easy. Cultivars "Nyerigai," "India," "Semongok Perak," and "Semongok Emas" were grouped in the most diverse clusters among the ten cultivars studied. The four cultivars have a similarity index as high as 92%; however, investigation of leaf width, leaf width-length ratio, seed weight, and conversion rate (fresh to black pepper) gives the ability to determine the characteristic differences. Cultivars "Lampung Daun Lebar" and "Yong Petai" have a similarity of 96%; however, the two showed distinctive differences in leaf width, leaf length-width ratio, spike thickness, and spike length characteristics. On the contrary, cultivars "Kuching" and "Sarikei" showed the highest similarity index, at 98%, and thus are among the most difficult cultivars to diagnose the morphological difference. However, the principle component analysis showed that the fruit size and seed diameter were the important diagnostic key characteristics. Overall, the leaf width, leaf width-length ratio, fruit spike, and conversion rate characteristics are among the key characteristics to differentiate among cultivars of black pepper in Malaysia. At the same time, the principle component analysis carried out has enlightened some interrelationships on the morphological characteristics between cultivars. This information is crucial for the future of the plant varietal improvement program in Malaysia.

1. Introduction

Black pepper, scientifically called *Piper nigrum* L. from the family of Piperaceae, is the most important spice in the world. In Malaysia, the crop has been highlighted as one of the national commodities based on its substantial contribution to the economy of the country. However, the production of black pepper has been diminishing since the early 1980s mainly due to pest and disease occurrence and labour constraints [1]. Thus, the government strategized a new policy to ensure sustainability of the industry by strengthening the quality of peppercorn. A monovarietal farm concept was believed to be able to strengthen the quality of peppercorn.

Black pepper germplasm assemblage has been established in Sarawak, Malaysia, since the 1980s. Since that time, there have been 47 accessions of black pepper varieties and 46 accessions of unidentified species of *Piper* [2]. In Malaysia's current black pepper farms, most are multivarietal and planted because farmers are unaware of monovarietal importance and lack knowledge on varietal identification. Based on a manual entitled "Pepper Production Technology in Malaysia," released by the Malaysian Pepper Board in 2011 [3], seven cultivated varieties have been described as common cultivars, including cv. "Semongok Aman," cv. "Semongok Emas," cv. "Kuching," cv. "Semongok Perak," cv. "Uthirancotta," cv. "Nyerigai," and cv. "PN129." However, in 2007, Sim reported the existence of other cultivars in a Malaysian farm, namely, cv. "Lampung Daun Lebar" and cv. "Lampung Daun Kecil" [4]. In 2000, Ravindran also reported three cultivars in Malaysia, namely, cv. "Kuching," cv. "Sarikei," and cv. "Miri" [5]. The number of cultivars existing in a Malaysian black pepper farm is thus unidentified.

A morphological study on Malaysian cultivated black pepper has been reported by Sim in 1979 [6]. The preliminary identification of the collection was based on Ridley's The Flora of the Malay Peninsula [7]. In her study, morphological descriptions include dioecious and monoecious classification, branching behavior, and leaf, stem, and flower spikes. The assessment on fruit development of three selected cultivars, that is, cv. "Kuching," cv. "Semongok Emas," and cv. "Hybrid 10," has also been reported by Sim et al. in 1996 [8], while Chen in 2011 reported the floral biology study on $P.\ nigrum$ [9] and also an apomixis study in 2013 which is related to floral morphology [10]. In India, Ravindran [5], George et al. [11], Parthasarathy et al. [12], and Krishnamoorthy and Parthasarathy [13] have reported the description of the morphology of black pepper. Besides, morphometrical analysis of forty-four cultivars [14] and multivariate analysis of fifty cultivars have been carried out in India [15].

Phenetic analysis of black pepper cultivars is novel in Malaysia. To date, none of the existing cultivars have been registered under the Plant Variety Protection (PVP) Act. This study ascertains the morphological distinctness and interrelationships among cultivars to ensure registration of each variety under the PVP Act, as a prerequisite toward implementation of a monovarietal farm policy in the future.

2. Materials and Methods

2.1. Sampling Site and Experimental Design. This experiment was initiated in January 2015 with the aim of collecting morphological data on important cultivars of black pepper in Malaysia. The field-grown vine was established at three locations, namely, Kampung Jagoi, Serikin; Kampung Karu, Padawan; and Kampung Belawan, Sri Aman; and one potted vine experiment was carried out under the controlled environment at the Agriculture Research Center (ARC) Semongok, Department of Agriculture Sarawak.

The field experiment was laid out in the Randomized Complete Block Design (RCBD) having ten treatments with 5 replications, which are T1: "Semongok Aman" vine; T2: "Kuching" vine, T3: "Semongok Emas" vine; T4: "Semongok Perak" vine; T5: "Semongok 1" vine; T6: "Nyerigai" vine; T7: "India" vine; T8: "Lampung Daun Lebar" vine; T9: "Sarikei" vine; and T10: "Yong Petai" vine. Each trial plot at different locations containing ten treatments consists of 50 vines. The planting procedure followed the standard practice as described by Paulus et al. [3] in "Pepper Production Technology in Malaysia." The planting material used is pepper cutting of 5 nodes, planted with a spacing of $1.8\ m \times 2.0\ m$ (between vine × between row). Whilst, the pot experiment was based on the Completely Randomized Design (CRD) that consists of a total of 50 potted vines, with 10 replicates for each treatment, that is, T1: "Semongok Aman" vine; T2: "Kuching" vine, T3: "Semongok Emas" vine; T4: "Semongok Perak" vine; T5: "Semongok 1" vine; T6: "Nyerigai" vine; T7: "India" vine; T8: "Lampung Daun Lebar" vine; T9: "Sarikei" vine; and T10: "Yong Petai" vine. The pot was arranged $1\ m \times 1\ m$ (between vine × between row). The data collection was initiated on a 2-year-old vine.

Vine growing morphology or vigour was assessed on field-grown vines in the three field experimental plots, while leaf, inflorescence, fruit, and seed morphology studies were based on samples collected from potted plants grown under the controlled environment. Data collection was carried out from January to December 2017. Microscopy assessment and data analysis were performed at the Malaysian Pepper Board.

2.2. Data Collection and Measurement. Ten cultivated varieties were selected in this study as the operational taxonomic unit (OTU) for the phenetic analysis. The ten OTUs were the cultivars "Semongok Aman" (SA), "Kuching" (KCH), "Semongok Emas" (SE), "Semongok Perak" (SP), "Semongok 1" (S1), "Nyerigai" (NYE), "India" (IND), "Lampung Daun Lebar" (LDL), "Sarikei" (SAR), and "Yong Petai" (YP). A total of 35 characteristics covering important parts of the plant were targeted for assessment. The characteristics listed in the Black Pepper Test Guideline in 2009 were mostly included in the assessment [2]. Details of characteristics and data collection methods are listed in Table 1.

2.3. Data Analysis. Measurement of morphological characteristics as variables is performed in this analysis. Characteristics used for phenetic analysis all tested to be significantly different between at least two cultivars under the ANOVA test, using SPSS, and are further analyzed by the Duncan test for significant character state differences ($P < 0.05$). Two analyses were performed in this study, cluster analysis and principle component analysis (PCA). The cluster analysis measured the similarity indexes between the OTUs using the Pearson correlation and average linkage with phenogram as the final output [16], while the PCA helped to extract the valuable information from a multivariate data table and express this information as new variables [17].

3. Results and Discussion

3.1. Cluster Analysis. Morphological characteristics used in the phenetic analysis of black pepper cultivars include all parts of the black pepper plant, including the leaf, inflorescence, fruit, seed, and shoot tips. The components of the characteristics used are shown in Table 2, while the data matrix for phenetic character states is shown in Table 3. For all quantitative data, the mean, range, and standard deviation were estimated (Figures 1–19). Characteristics used were all tested and showed significant differences between at least two cultivars under the ANOVA test using SPSS. The Duncan test proved that the characteristics of leaf length-width ratio (Figure 4) and number of flowers per

TABLE 1: Morphological characteristics used in the phenetic analysis of black pepper cultivars.

Morphological characteristics	Measurement methods
Leaf characteristics	
(1) Leaf shape, leaf apex, and leaf base	Description based on the UPOV standard
(2) Leaf area (cm^2), blade width (w, mm), blade length (L, mm), and blade length-width ratio (Lw^{-1})	Measured by using the WinFOLIA image analysis system
(3) Leaf colour (fully expanded leaf)	RHS colour codes used
Inflorescence characteristics	
(1) Inflorescence length at the stigma withering stage (cm) and inflorescence thickness at the stigma withering stage (mm)*	Measured by using a vernier calliper
(2) Inflorescence colour	RHS colour codes used
(3) Number of flowers per inflorescence	Counted via stereomicroscope
(4) Number of inflorescences (spikes) per branch per node	Counted manually
Fruit characteristics	
(1) Fruit spike length (cm) and fruit size in diameter (mm)	Measured by using a vernier calliper
(2) Fruit weight (single fresh berry) (g)	Measured by using an analytical balance
(3) Fruit colour (hard dough stage)	RHS colour codes used
(4) Percent fruit set	Counted manually. Percent = (number of developed fruits)/(number of developed fruits + number of underdeveloped fruits) × 100
(5) Conversion rate (fresh to black pepper) (%)	Measured by using an analytical balance (drying specification: oven-drying at 40°C; moisture content ≤12%)
(6) Conversion rate (fresh to white pepper) (%)	Measured by using an analytical balance (drying specification: oven-drying at 40°C; moisture content ≤12%)
(7) Pericarp thickness (mm)	Measured by using a vernier calliper (horizontal diameter of a fresh berry–horizontal diameter of the seed)
Seed characteristics	
(1) Seed diameter (mm)	Measured by using a vernier calliper (horizontal diameter of the seed)
(2) Seed weight (g)	Measured by using an analytical balance
Vigour	
(1) Branch column	By observation
(2) Internode length (cm)	Measurement by using a ruler (node-to-node distance)
(3) Number of nodes/foot of the stem	Counted manually
Shoot tips	
(1) Anthocyanin: absent or present	By observation of shoot tip colouration. Green colour = absence of anthocyanin; purple colour = presence of anthocyanin

*Withering stage [6].

inflorescence (Figure 7) have seven significant difference groups, among the characteristics with highest distinctness values. Characteristics such as leaf area (cm^2) (Figure 1), leaf width (cm) (Figure 2), inflorescence or spike thickness (mm) (Figure 6), number of inflorescences or spikes per branch per node (Figure 8), seed weight (g) (Figure 17), percent fruit set (Figure 12), internode length (cm) (Figure 18), and number of nodes per foot of the stem (Figure 19) have six significant groups among ten cultivars. Meanwhile, characteristics such as leaf length (cm) (Figure 3), inflorescence length (cm) (Figure 5), number of flowers per spike/inflorescence (Figure 7), conversion rate from fresh to dried white pepper (Figure 14), pericarp thickness (mm) (Figure 15), and seed diameter (mm) (Figure 16) have five significant groups. The Duncan test also showed that characteristics such as single berry size in diameter (mm) (Figure 10), conversion rate from fresh to dried black pepper (Figure 13), and fruit weight (single berry) (g) (Figure 11) have only four significant groups. The assessment of characteristics such as number of branches per node, hilum-micropyle distance, petiole length (mm), blade thickness (mm), colour of an immature fruit, colour of a ripened fruit, seed shape, and venation pattern do not show a significant difference and thus are not included in the analysis.

Cluster analysis is based on measuring the similarity between the operational taxonomic units (OTUs). In this study, ten OTUs, the cultivars "Semongok Aman" (SA), "Kuching" (KCH), "Semongok Emas" (SE), "Semongok Perak" (SP), "Semongok 1" (S1), "Nyerigai" (NYE), "India" (IND), "Lampung Daun Lebar" (LDL), "Sarikei" (SAR), and "Yong Petai" (YP), were compared. Each OTU will be clustered into groups based on a similarity index of morphological characteristics. The greater the value of the similarity index, the closer the OTU unit or grouping, and vice versa [18].

Referring to the phenogram in Figure 20, black pepper cultivars were clustered into five groups, with a similarity index ≥85%. Cluster A is composed of two cultivars, "Kuching" and "Sarikei," with a similarity index as high as 99%. The two cultivars share 15 similar characteristics out of

TABLE 2: Characteristics and character states used for phenetic analysis.

Number	Characteristics and character states
	Leaf characteristics
1	Leaf shape: (1) lanceolate; (2) lanceolate-ovate; (3) ovate
2	Leaf apex: (1) acute; (2) obtuse; (3) rounded
3	Leaf base: (1) acute; (2) oblique; (3) rounded; (4) ovate
4	Leaf area (cm^2): (1) <40; (2) 40–60; (3) 60–80; (4) 80–100; (5) >100
5	Leaf colour (fully expanded leaf)*: (1) green group 137 series (moderate olive green); (2) green group NN137 series (greyish olive green); (3) green group 139 series (dark yellowish green)
6	Blade width (w, mm): (1) <6; (2) 6–8; (3) 8–10; (4) 10–12; (5) >12
7	Blade length (L, mm): (1) <10; (2) 10–11; (3) 11–12; (4) 12–13; (5) >13
8	Blade width-length ratio (Lw^{-1}): (1) <1.4; (2) 1.4–1.7; (3) 1.7–2.0; (4) 2.0–2.3; (5) >2.3
	Inflorescence characteristics
9	Inflorescence length at the withering stage (cm): (1) <7; (2) 7–8; (3) 8–9; (4) 9–10; (5) >10
10	Inflorescence colour*: (1) green group 144 series (strong yellowish green); (2) yellow-green group N144 series (strong yellowish green); (3) yellow-green group 145 series (strong yellowish green)
11	Inflorescence thickness at the withering stage (mm): (1) <2.8; (2) 2.8–3.2; (3) 3.2–3.6; (4) 3.6–4.0; (5) >4.0
12	Number of flowers per inflorescence: (1) <80; (2) 80–90; (3) 90–100; (4) >100
13	Number of inflorescences (spikes) per branch per node: (1) <20; (2) 20–30; (3) 30–40; (4) >40
	Fruit characteristics
14	Fruit spike length (cm): (1) <7; (2) 7–9; (3) 9–11; (4) >11
15	Fruit size (single berry) in diameter (mm): (1) <6; (2) 6–7; (3) >7
16	Fruit weight (single berry) (g): (1) <0.12; (2) 0.13–0.18; (3) >0.18
17	Fruit colour (hard dough stage)*: (1) green group NN137 series (greyish olive green); (2) green group 139 series (dark yellowish green); (3) green group 141 series (deep yellowish green)
18	Percent fruit set: (1) <60; (2) 60–70; (3) >70
19	Conversion rate (fresh to black pepper) (%): (1) <40; (2) 40–50; (3) >50
20	Conversion rate (fresh to white pepper) (%): (1) <20; (2) 20–30; (3) >30
21	Pericarp thickness (mm): (1) <1.6; (2) 1.6–1.8; (3) 1.8–2.0; (4) 2.0–2.2; (5) >2.2
	Seed characteristics
22	Seed diameter (mm): (1) <3.5; (2) 3.5–4.0; (3) >4.0
23	Seed weight (g): (1) <4.8; (2) 4.8–5.0; (3) 5.0–5.2; (4) 5.2–5.4; (5) >5.4
	Vigour
24	Branch column types: (1) erect; (2) horizontal; (3) drooping
25	Internode length (cm): (1) <8; (2) 8–9; (3) 9–10; (4) 10–11; (5) >11
26	Number of nodes/foot of the stem: (1) 1; (2) 2; (3) 3; (4) 4; (5) 5
	Shoot tip
27	Anthocyanin: (1) absent; (2) present

*RHS colour codes used.

27 and 10 minor dissimilarity characteristics (one state difference). However, the distinctions are found in Characteristic 21 (pericarp thickness) and Characteristic 23 (seed weight). The second cluster (B) has the greatest population among all clusters, comprising four cultivars, "Nyerigai," "India," "Semongok Perak," and "Semongok Emas." The group was further clustered into two subgroups, with a similarity index of 97% for "Nyerigai" and "India" (Subgroup 1) and 95% for "Semongok Perak" and "Semongok Emas" (Subgroup 2). The distinctive characteristics that differentiate Subgroup 1 are Characteristic 8 (blade width-length ratio) and Characteristic 21 (pericarp thickness), while for Subgroup 2, Characteristic 8 (blade width-length ratio), Characteristic 17 (fruit colour at the hard dough stage), and Characteristic 23 (seed weight) are the important distinctive characteristics. Cluster C only consists of one cultivar, "Semongok Aman." This cultivar was clustered alone with plenty of distinct characteristics, even when compared to the closest cluster (Cluster B), especially in Characteristic 2 (leaf apex), Characteristic 3 (leaf base), Characteristic 10 (inflorescence colour at the withering stage), and Characteristic 18 (percent fruit set). Cluster D has two cultivars in its group, "Lampung Daun Lebar" and "Yong Petai." The two cultivars share a 94% similarity; however, they show dissimilarity in characteristics such as Characteristic 8 (blade width-length ratio), Characteristic 10 (inflorescence colour at the withering stage), Characteristic 11 (inflorescence thickness at the withering stage), and Characteristic 14 (fruit spike length). Cultivar "Semongok 1" is another cultivar clustered alone. This cultivar showed great morphologically distinct values in Characteristic 3 (leaf base), Characteristic 6 (blade width), Characteristic 23 (seed weight), and Characteristic 25 (number of nodes per foot of the stem), if compared to the nearest cluster (Cluster D).

Ho et al. [19] support the outcome of this analysis in their research on evaluation of genetic relatedness among black pepper (*Piper nigrum* L.) accessions using direct amplification of the minisatellite-region DNA (DAMD). That team included the same ten cultivars in their study, and the DAMD-based clustering is tallied with the phenetic-based clustering in this study.

3.2. *Principle Component Analysis (PCA)*. In this study, principle component analysis was performed using 27 morphological characteristics. Seven components were

TABLE 3: Data matrix of phenetic character states corresponding to Table 1.

Cultivars	Characteristics and character states																											
	1	2	3	4	5	6	7	8	9	10	11	12	13	14	15	16	17	18	19	20	21	22	23	24	25	26	27
SA	2	3	3	2	3	2	2	3	2	1	3	2	1	2	2	3	1	3	1	2	3	3	5	2	5	4	2
KCH	2	1	1	1	3	1	2	4	2	2	3	1	4	2	2	2	1	2	2	3	4	3	3	2	2	5	2
SE	2	2	1	2	1	1	5	5	2	2	3	2	1	2	2	2	3	2	2	3	4	3	5	3	5	4	2
SP	2	2	1	1	2	2	5	3	1	3	4	1	2	1	2	3	1	2	1	2	5	3	2	2	4	3	2
S1	3	2	4	5	3	5	5	2	3	2	4	4	1	4	3	3	1	2	2	2	5	3	5	1	2	4	1
NYE	2	2	1	2	3	2	4	3	2	2	2	1	3	2	2	2	1	2	2	3	5	3	1	1	3	4	2
IND	1	1	1	2	3	1	5	5	2	2	2	2	2	2	2	2	1	2	2	3	2	1	2	3	4	2
LDL	2	2	2	4	2	3	5	2	2	1	4	4	1	2	2	2	2	1	1	2	4	3	1	3	5	4	2
SAR	2	1	1	1	3	1	2	4	1	2	3	1	3	1	1	3	1	2	1	2	2	2	1	2	3	4	2
YP	2	2	1	4	3	2	5	4	3	3	2	3	1	4	3	3	1	1	1	2	5	3	2	2	5	3	2

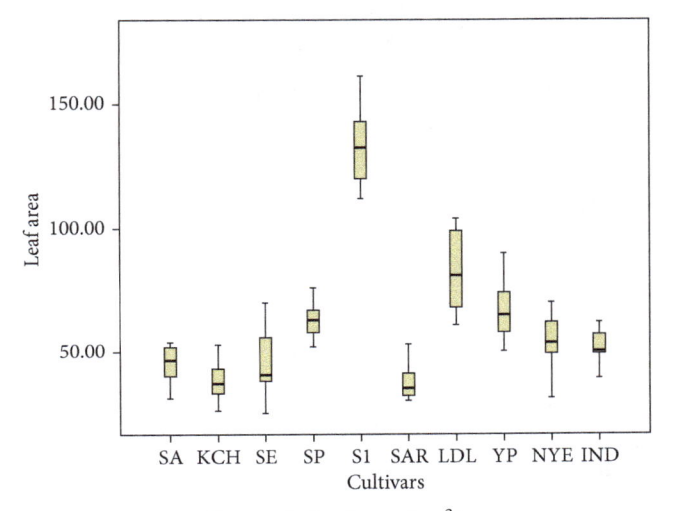

FIGURE 1: Leaf area (cm^2).

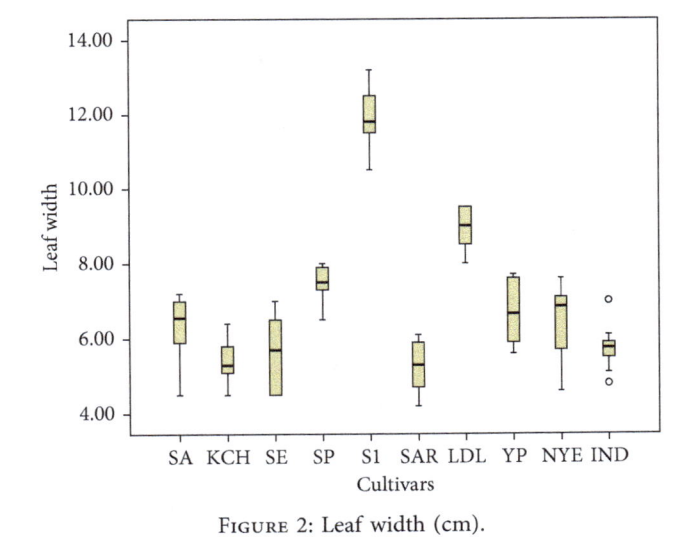

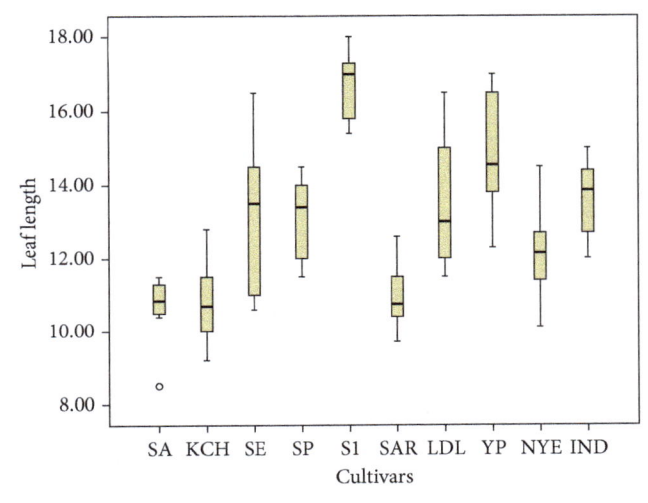

FIGURE 2: Leaf width (cm).

FIGURE 3: Leaf length (cm).

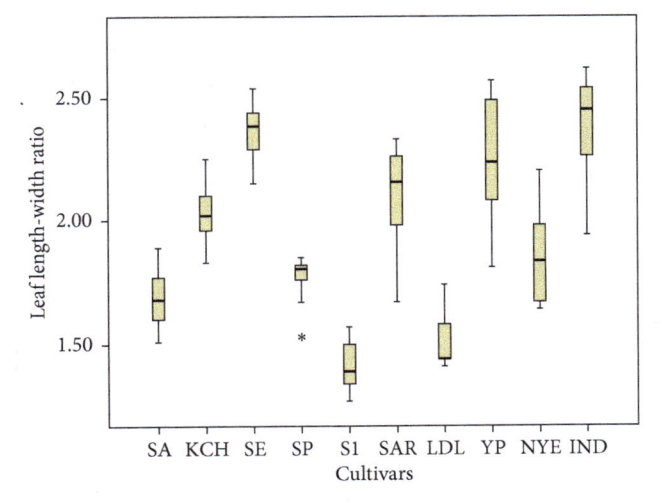

FIGURE 4: Leaf length-width ratio.

extracted as meaningful factors, with eigenvalues >1 (Figure 21). These components explained 95.79% of the total variance, fulfilling the 95% confidence level mentioned by Jackson [20] (Table 4). The first principle component (PC1) explained 33.71% of the total variation, the second component (PC2) explained 17.85% of the variation, and the third component (PC3) explained 13.43% of the variation. The other principle components (PC4–PC7) explained the additional 30.80% of the variation. Pasagi et al. [21] explained that the PCA value can be categorized into three levels: the first level, with the component value $X \geq 0.75$, has a very strong influence on the grouping; the second level,

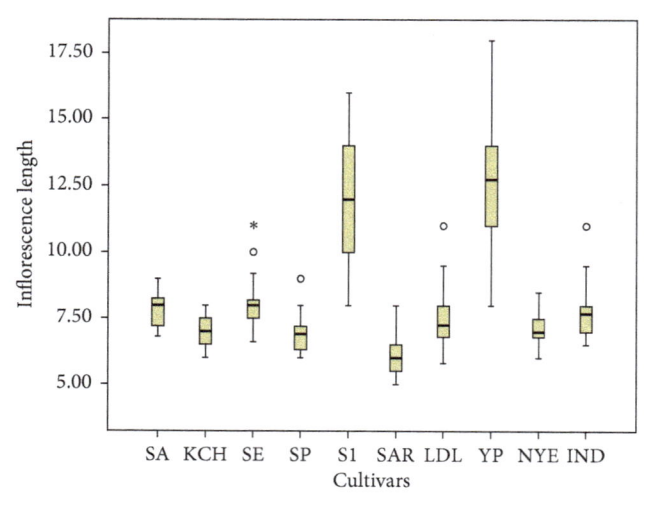

FIGURE 5: Inflorescence length (cm).

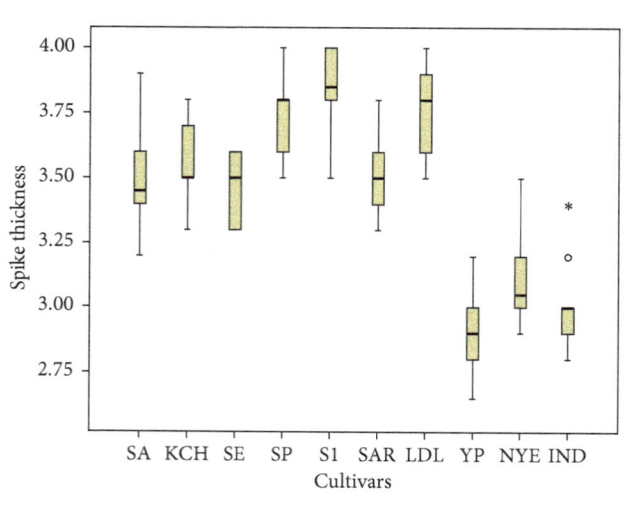

FIGURE 6: Spike or inflorescence thickness (mm).

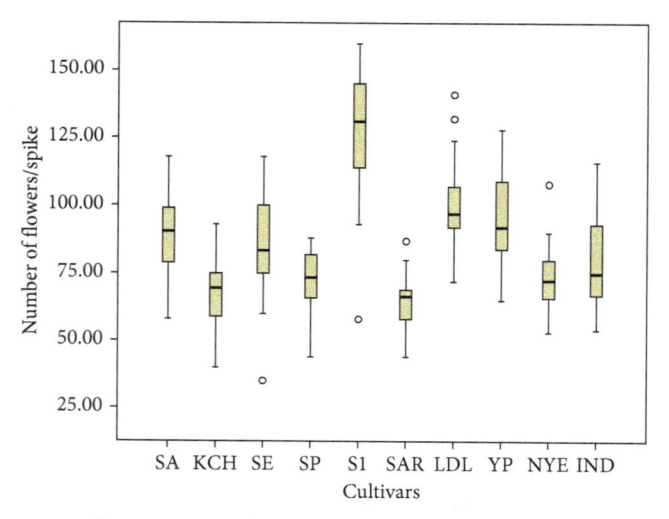

FIGURE 7: Number of flowers per inflorescence.

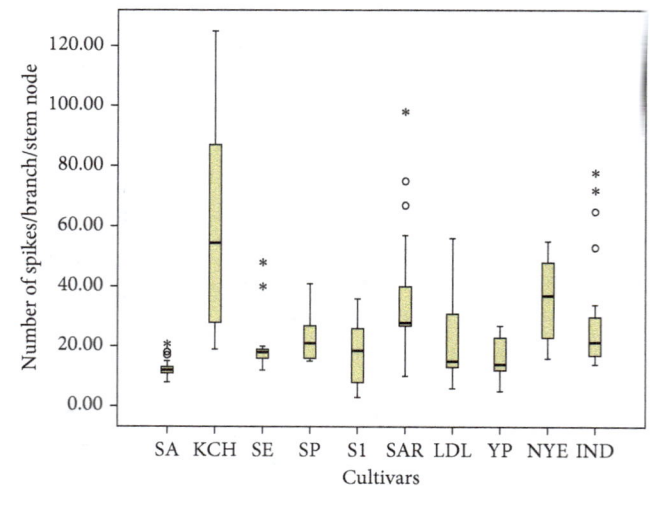

FIGURE 8: Number of inflorescences (spikes) per branch per node.

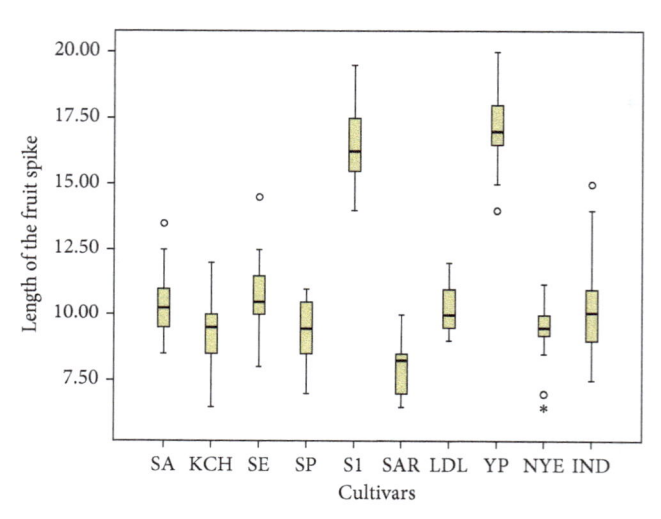

FIGURE 9: Fruit spike length (cm).

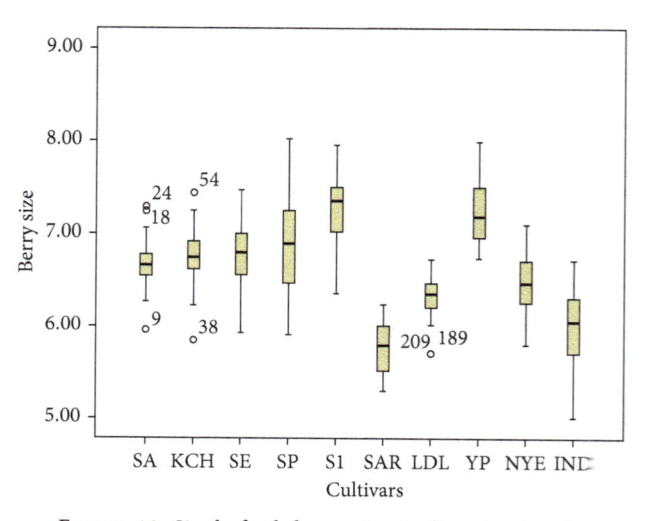

FIGURE 10: Single fresh berry size in diameter (mm).

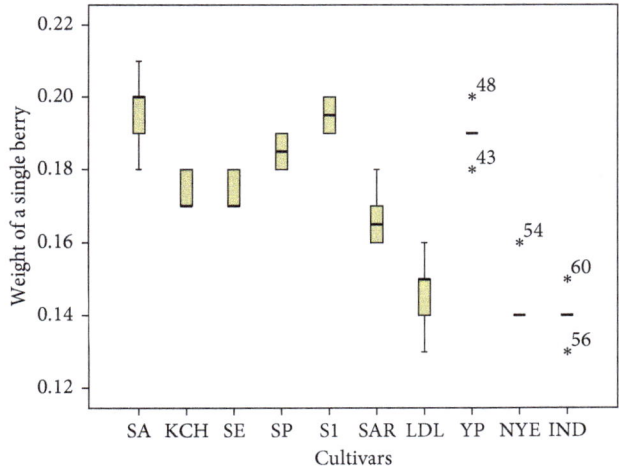

FIGURE 11: Fruit weight (single berry) (g).

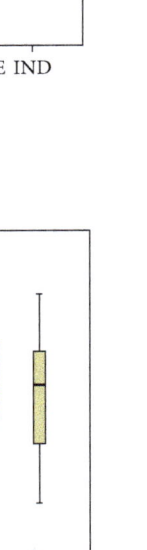

FIGURE 12: Percent fruit set.

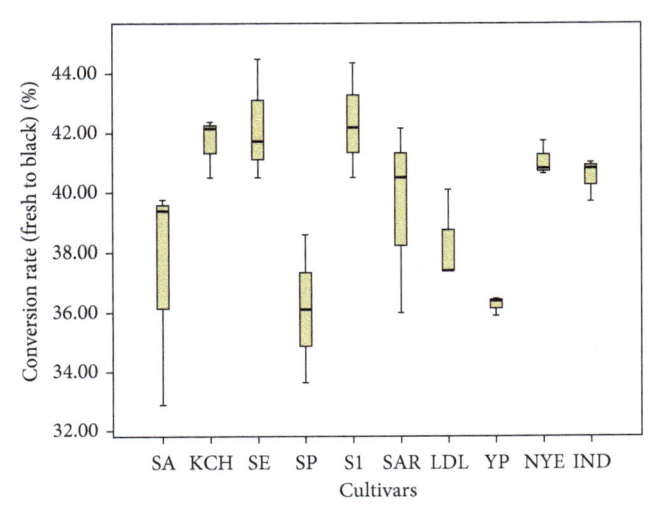

FIGURE 13: Conversion rate (fresh to dried black pepper) (%).

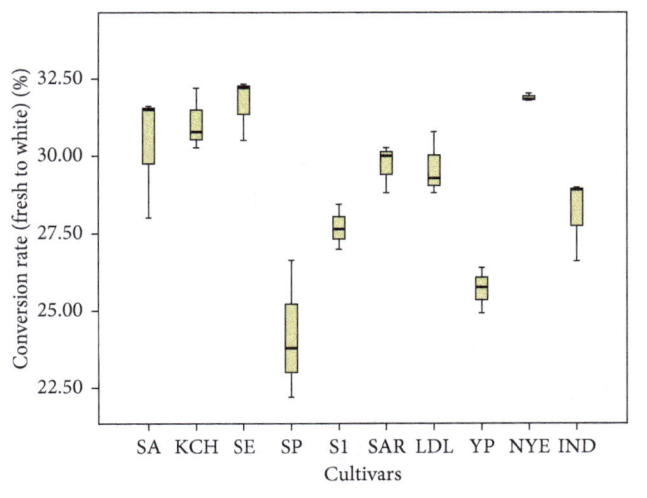

FIGURE 14: Conversion rate (fresh to dried white pepper) (%).

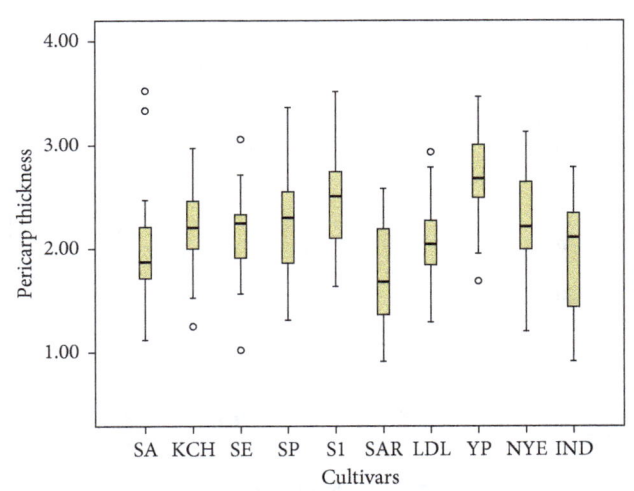

FIGURE 15: Pericarp thickness (mm).

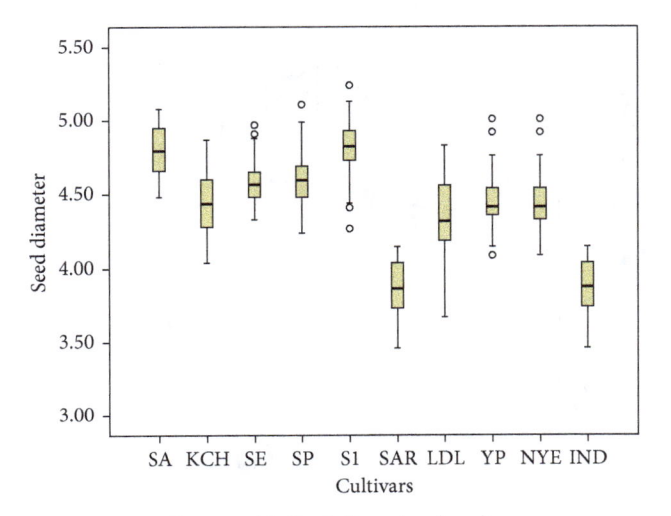

FIGURE 16: Seed diameter (mm).

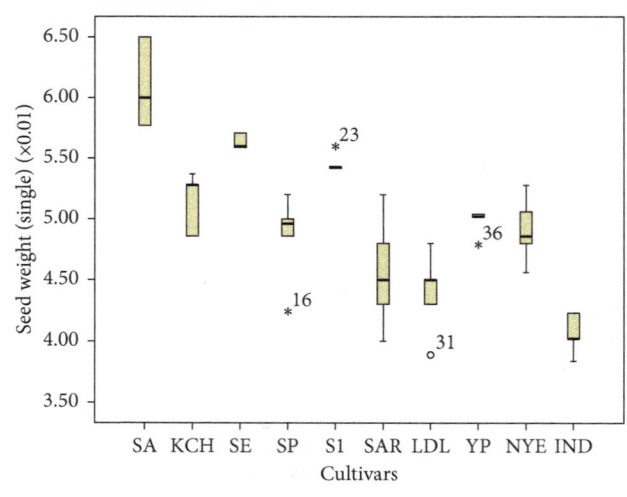

FIGURE 17: Seed weight (10^{-2}) (g).

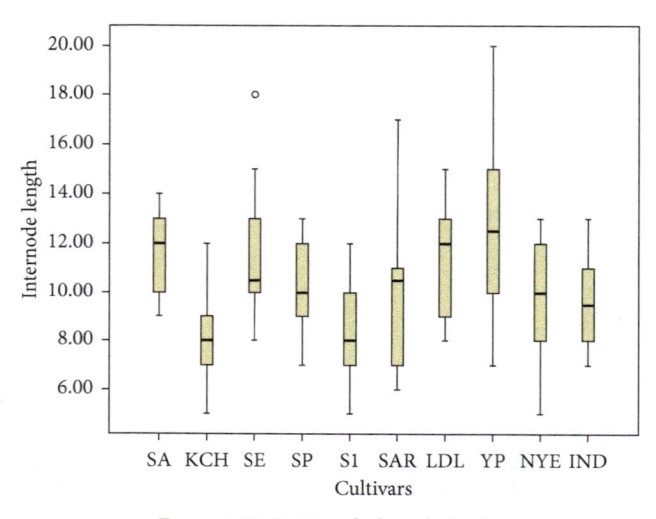

FIGURE 18: Internode length (cm).

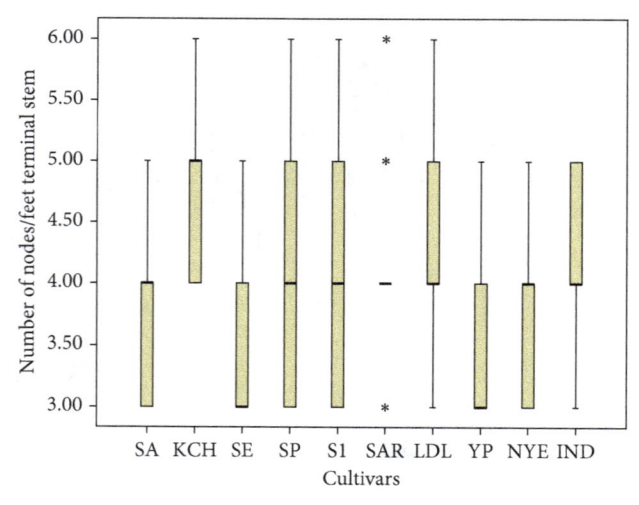

FIGURE 19: Number of nodes per foot of the stem.

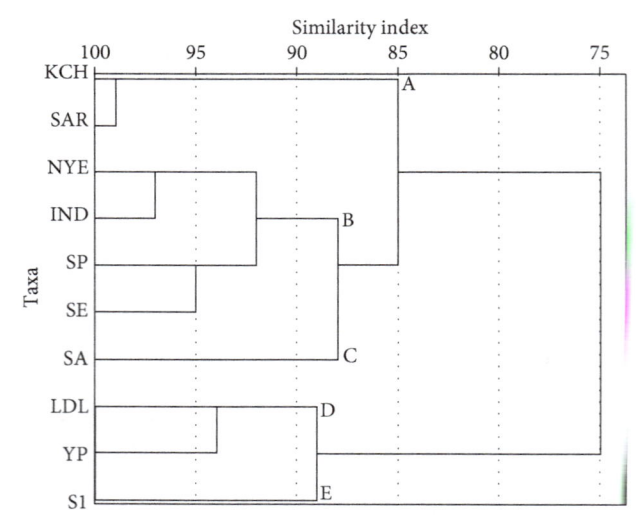

FIGURE 20: Phenogram presented based on the average linkage (between groups) using the squared Euclidean distance method.

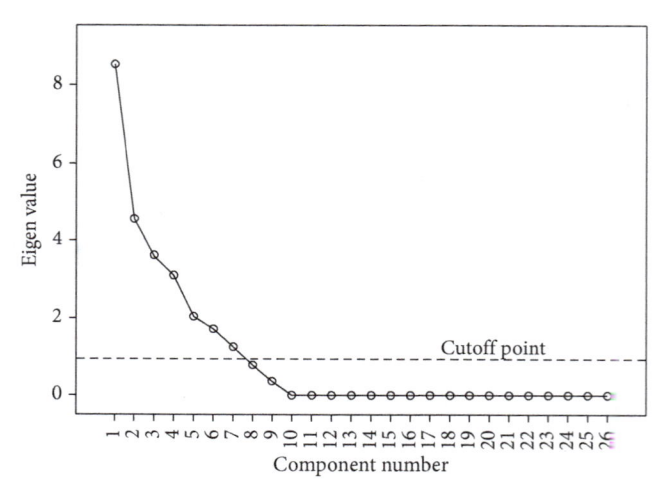

FIGURE 21: Scree plot showing the cutoff point of extracted components for PCA.

with a component value $0.50 \leq X < 0.75$, has a secondary influence on the separation of OUT; the third level, with the component value less than 0.50, has the least influence on separation at a minimal level. Based on PCA (Table 5), among the 27 characteristics used for analysis, 18 characteristics have a strong influence on the grouping of black pepper cultivars, with the PCA value ranging from 0.78 to 0.96. These characteristics include leaf shape, leaf area, blade width, leaf colour, inflorescence length, inflorescence colour, inflorescence thickness, number of flowers/inflorescence, fruit spike length, fruit size in diameter, fruit colour (hard dough stage), percent fruit set, conversion rate percent of fresh to black pepper, conversion rate percent of fresh to white pepper, pericarp thickness, seed weight, number of nodes per foot of the stem, and anthocyanin colouration. This analysis also revealed that leaf, inflorescence, and fruit

TABLE 4: Total variance obtained using principle component analysis (PCA).

PC	Initial eigenvalues			Extraction sums of squared loadings			Rotation sums of squared loadings		
	Total	Variance (%)	Cumulative (%)	Total	Variance (%)	Cumulative (%)	Total	Variance (%)	Cumulative (%)
1	9.102	33.713	33.713	9.102	33.713	33.713	6.213	23.011	23.011
2	4.820	17.850	51.563	4.820	17.850	51.563	5.176	19.169	42.180
3	3.626	13.431	64.994	3.626	13.431	64.994	3.967	14.693	56.873
4	3.133	11.602	76.596	3.133	11.602	76.596	3.789	14.034	70.907
5	2.048	7.587	84.183	2.048	7.587	84.183	2.457	9.099	80.006
6	1.805	6.684	90.867	1.805	6.684	90.867	2.393	8.862	88.868
7	1.329	4.921	95.788	1.329	4.921	95.788	1.869	6.921	95.788
8	0.785	2.907	98.696	—	—	—	—	—	—
9	0.352	1.304	100.000	—	—	—	—	—	—

TABLE 5: Matrix component value for all the distinguishing characteristics.

	Principle component (PC)						
	1	2	3	4	5	6	7
Leaf shape	0.204	**0.807**	−0.101	0.029	0.078	0.402	0.043
Leaf apex	0.279	0.159	0.213	−0.455	−0.151	**0.632**	0.432
Leaf base	0.453	**0.696**	−0.107	−0.167	−0.367	0.047	0.360
Leaf area (cm^2)	**0.868**	0.380	0.072	−0.166	−0.080	0.039	−0.219
Leaf colour	0.167	−0.060	**−0.955**	−0.058	−0.172	−0.154	0.017
Blade width (cm)	**0.538**	**0.793**	−0.090	−0.150	−0.037	0.113	−0.036
Blade length (cm)	**0.531**	0.102	0.497	−0.056	0.485	−0.125	−0.212
Blade width-length ratio	−0.119	**−0.737**	0.167	0.296	0.261	−0.337	0.179
Inflorescence length (cm)	**0.948**	0.048	−0.126	0.154	−0.026	0.195	−0.029
Inflorescence colour	−0.011	−0.125	−0.171	−0.033	**0.957**	0.019	−0.150
Inflorescence thickness (mm)	−0.220	**0.849**	0.334	−0.191	−0.093	0.057	−0.099
Number of flowers/inflorescence	**0.780**	0.393	0.258	−0.233	−0.233	−0.070	−0.234
Number of inflorescences/branch	**−0.621**	−0.155	−0.523	0.515	0.020	−0.045	−0.220
Fruit spike length (mm)	**0.923**	0.150	−0.193	0.036	0.132	0.153	−0.057
Fruit size in diameter (mm)	**0.848**	0.189	−0.080	−0.009	0.298	0.294	−0.027
Fruit weight (single berry) (g)	0.011	0.343	−0.355	**−0.689**	0.300	0.035	0.277
Fruit colour (hard dough stage)	0.050	−0.048	**0.932**	0.219	−0.098	0.105	−0.023
Percent fruit set	−0.340	0.060	−0.196	0.103	−0.203	0.020	**0.881**
Conversion rate (fresh to black pepper) (%)	0.240	0.012	−0.002	**0.922**	0.084	−0.136	0.209
Conversion rate (fresh to white pepper) (%)	−0.189	−0.231	0.132	**0.824**	0.024	0.439	0.029
Pericarp thickness (mm)	0.236	0.276	0.187	0.056	0.043	**0.890**	−0.033
Seed diameter (mm)	0.423	0.032	0.092	0.059	0.477	**0.566**	**−0.500**
Seed weight (g)	−0.169	−0.279	**0.791**	−0.180	−0.246	−0.039	−0.256
Branch column types	0.122	−0.357	**0.601**	**−0.604**	−0.099	0.341	0.023
Internode length (cm)	−0.167	0.065	−0.129	**0.713**	**−0.628**	−0.076	0.019
Number of nodes/foot of the stem	−0.496	**−0.789**	0.177	−0.161	−0.109	0.148	−0.185
Anthocyanin	0.496	**0.789**	−0.177	0.161	0.109	−0.148	0.185

Note. Extraction method: principle component analysis; rotation method: varimax with Kaiser normalisation.

are three important parts in diagnosing the morphological difference among black pepper cultivars. Meanwhile, the most dominant characteristic influencing the grouping is the inflorescence colour, with a PCA value as high as 0.957. On the contrary, the less important characteristics assisting in grouping are blade length and seed diameter, both with a PCA value lower than 0.60.

This analysis also revealed some interesting interactions among morphology characteristics of black pepper cultivars. In PC1, the positive loading of characteristics, such as leaf area, inflorescence length, number of flowers per inflorescence, fruit spike length, and fruit size, proves the proportional relationship among these characteristics, while the characteristic of number of inflorescences per branch has a negative loading in this PC. This explains that if the black pepper vine has a bigger leaf area, the inflorescence length, number of flowers per inflorescence, fruit spike length, and fruit size will have a greater value, while the number of inflorescences per branch will perform the opposite. This reveals that greater leaf areas do not contribute to better cultivar yield, even though the flower intensity per spike, inflorescence length, and fruit spike have positive relationship with fruit sizes. Gazzoni and Moscardi support this finding in their soybean study [22], as do Subedi and Ma in their maize study [23]. On the contrary, Heuvelink et al. reported a reduction of leaf area can contribute to the yield of tomato [24]. PC2 showed a positive loading in characteristics such as leaf shape, leaf base, blade width, inflorescence thickness, and anthocyanin, whilst a negative loading in blade width-length ratio and number of nodes per foot of the stem. This PC reveals that if the cultivars have a very distinct leaf shape compared to other cultivars, the cultivar also tends to have a distinct leaf base, distinct leaf width, greater inflorescence thickness, and anthocyanin present, but smaller width-length ratio and number of nodes per foot of the stem. This PC explains Cluster C and Cluster E (Figure 20), both with single cultivars in the cluster because of their high distinctive values. The identification of these two cultivars is easy. PC3 has a negative loading for two characteristics, leaf colour and number of inflorescences per branch, and a positive loading for fruit colour, seed weight, and branch column type. This explains that when the fruit colour is greener, the seed weight tends to increase and the column type becomes denser, but when the leaf colour seems to be lighter, the number of inflorescences per branch will respond negatively. PC4 to PC7 showed fewer interrelations among the morphology characteristics. In PC4, two types of conversion rate studies, fresh to dried black pepper and fresh to dried white pepper, show proportionate relationships. This result is partially supported by Paulus et al., who reported only on cultivars SA, KCH, and SE [3]. Fruit weight and branch column type also have proportionate relationships, both with negative loadings. This relationship could be interpreted as the less-dense column type of cultivar yielding less-weight fruit, whilst PC5 to PC7 do not show an interesting interrelationship.

4. Conclusions

Phenetic cluster analysis revealed that cultivars "Semongok Aman" and "Semongok 1" have high distinctive values for identification; thus, varietal diagnosis could be very easy. Cultivars "Nyerigai," "India," "Semongok Perak," and "Semongok Emas" were grouped in the most diverse cluster among all clusters. The four cultivars have a similarity index as high as 92%; however, investigation on leaf width, leaf width-length ratio, seed weight, and conversion rate (fresh to black pepper) can determine the characteristic differences. Cultivars "Lampung Daun Lebar" and "Yong Petai" have a similarity of 96%; however, the two showed distinctive differences in leaf width, leaf length-width ratio, spike thickness, and spike length characteristics. In Cluster A (Figure 20), cultivars "Kuching" and "Sarikei" showed the highest similarity index and thus are among the most difficult cultivars to diagnose morphological differences. However, the principle component analysis showed that the fruit size and seed diameter were the key diagnostic characteristics. Overall, the leaf width, leaf width-length ratio, fruit spike, and conversion rate characteristics are among the key characteristics to differentiate among cultivars of black pepper in Malaysia. At the same time, principle component analysis has been carried out, discerning some interrelationships within the morphological characteristics between cultivars. The overall phenetic analysis showed that all the selected cultivars have distinct values enabling registration under the Malaysian Plant Variety Protection Act. This information is also crucial for plant varietal improvement programs in the future.

Acknowledgments

The authors would like to thank Mr. Kevin Muyang Anak Tawie for assistance in morphological assessment via the WinFOLIA image analysis system. Appreciation also goes to Mr. Wan Ambi and Mr. Juvian Jacob, research assistants of the Malaysian Pepper Board, for their excellent technical assistance in the field. This project was fully funded by the Malaysian Pepper Board, sourced from the Economic Planning Unit (EPU) of Malaysia.

References

[1] A. D. Paulus, "Development of superior genotypes and cultural practices for improving productivity of pepper in Sarawak, Malaysia: progress, achievements and research needs," in *Proceedings of 2007 Conference on Plantation Commodities*, pp. 149-155, Kuala Lumpur, Malaysia, July 2007.

[2] S. L. Sim, A. D. Paulus, J. Rosmah, D. Maclin, and E. Lily, *Guidelines for the Conduct of Tests for Distinctness, Uniformity and Stability on Black Pepper (Piper nigrum L.): Plant Variety Protection*, Ministry of Agriculture (MOA), Putrajaya, Malaysia, 2009.

[3] A. D. Paulus, S. L. Sim, L. Eng, G. Megir, and J. Rosmah, *Pepper Production Technology in Malaysia: Botany and Varietal Improvement*, Malaysia Pepper Board, Kuching, Sarawak, 2011.

[4] S. L. Sim, "Varietal improvement of black pepper (*Piper nigrum* L.) in Malaysia," in *Proceedings of 2007 Conference on Plantation Commodities*, pp. 156-169, Kuala Lumpur, Malaysia, July 2007.

[5] P. N. Ravindran, "Fruit set, fruit development and fruit maturity," in *Black Pepper: Botany and Crop Improvement of Black Pepper*, p. 142, Hardwood Academy Publishers, India, 2000.

[6] S. L. Sim, "A preliminary report on floral biology and mode of pollination in pepper," in *Proceedings of 14th Annual Research Officers' Conference*, pp. 35-46, Sarawak, Malaysia, 1979.

[7] H. N. Ridley and J. Hutchinson, *The Flora of the Malay Peninsula. Volume 3, Apetalae*, Asher, Amsterdam, Netherlands, 1967.

[8] S. L. Sim, T. H. Wong, S. P. Chin, and O. Fatimah, "Fruit development in black pepper (*Piper nigrum* L.)," in *Proceedings of 33rd Annual research Officers' Conference*, pp. 9-30, Sarawak, Malaysia, 1996.

[9] Y. S. Chen, "A study on interspecific hybridization between *Piper nigrum* and *Piper colubrinum*," Master thesis, Department of Plant Sciences and Environmental Ecology, University Malaysia Sarawak, Kota Samarahan, Malaysia, 2011.

[10] Y. S. Chen, "An investigation on possible occurrence of apomixis in black pepper (*Piper nigrum* L.)," in *Proceedings of*

2nd International Conference on Environment, Agriculture and Food Sciences, pp. 109–111, Kuala Lumpur, Malaysia, 2013.

[11] C. K. George, A. Abdullah, and K. Chapman, *Black Pepper Production Guide for Asia and the Pacific*, Joint effort of the International Black pepper Community (IPC) and Food and Agricultural Organization of United Nations (FAO), Bangkok, Thailand, 2005.

[12] A. Parthasarathy, B. Sasikumar, N. Ramakrishnan, and K. Johnson, "Black pepper: botany and horticulture," *Horticultural Reviews*, Westport Then, New York, NY, USA, pp. 173–266, 2007.

[13] B. Krishnamoorthy and V. A. Parthasarathy, "Improvement of black pepper," *CAB Reviews Perspectives in Agriculture Veterinary Science Nutrition and Natural Resources*, vol. 4, pp. 1749–8848, 2009.

[14] P. N. Ravindran, R. Balakrishnan, and K.N. Babu, "Morphometrical studies on black pepper (*P. nigrum* L.). I. Cluster analysis of black pepper cultivars," *Journal of Spices and Aromatic Crops*, vol. 6, no. 1, pp. 9–20, 1997.

[15] P. J. Mathew, P. M. Mathew, and V. Kumar, "Multivariate analysis in fifty cultivars/landraces of 'black pepper' (*Piper nigrum* L.) occurring in Kerala, India," in *Proceedings of International Symposium on Breeding Research on Medicinal and Aromatic Plants*, pp. 180–185, Campinas, Brazil, 2006.

[16] R. S. David, "Quantitative taxonomy and biostratigraphy of Middle Cambrian trilobites from Montana and Wyoming," in *Mathematical Geology*, Plenum Publishing Corporation, pp. 149–164, New York, NY, USA, 1975..

[17] I. T. Jolliffe, *Principle Component Analysis: Chapter 7-Principle Component Analysis and Factor Analysis*, University of Aberdeen, Springer publisher, London, UK, 2002.

[18] P. R. Peres-Neto, D. A. Jackson, and K. M. Somers, "Giving meaningful interpretation to ordination axes: assessing loading significance in principal component analysis," *Ecology*, vol. 84, no. 9, pp. 2347–2363, 2003.

[19] W. S. Ho, E. T. Lau, H. R. Jafar, S. L. Sim, and A. D. Paulus, "Evaluation of genetic relatedness among black pepper (*Piper nigrum* L.) accessions using direct amplification of minisatellite-region DNA (DAMD)," in *Proceedings of 6th National Genetic Congress*, pp. 299–302, Kuala Lumpur, Malaysia, 2005.

[20] D. A. Jackson, "Stopping rules in principal components analysis: a comparison of heuristical and statistical approaches," *Ecology*, vol. 74, no. 8, pp. 2204–2214, 1993.

[21] R. Pasagi, H. Hamidah, and Junairiah, "Analisis hubungan kekerabatan varietas pada belimbing (*Averrhoa carambola* L.) melalui dendekatan morfologi," *Journal of Biological Sciences*, vol. 2, pp. 26–33, 2014.

[22] D. L. Gazzoni and F. Moscardi, "Effect of defoliation levels on recovery of leaf area, on yield and agronomic traits of soybeans," *AGRIS*, vol. 4, no. 33, pp. 411–424, 1998.

[23] K. D. Subedi and B. L. Ma, "Ear position, leaf area, and contribution of individual leaves to grain yield in conventional and leafy maize hybrids," *Crop Science*, vol. 45, no. 6, pp. 2246–2257, 2005.

[24] E. Heuvelink, M. J. Bakker, A. Elings, R. C. Kaarsemaker, and L. F. M. Marcelis, "Effect of leaf area on tomato yield," *Acta Horticulturae*, vol. 691, pp. 43–50, 2005.

Agronomic Evaluation of Bunching Onion in the Colombian Cundiboyacense High Plateau

Carlos H. Galeano Mendoza,[1] Edison F. Baquero Cubillos,[2] José A. Molina Varón,[2] and María del Socorro Cerón Lasso[2]

[1]*Colombian Corporation for Agricultural Research (Corpoica), CI Palmira, Palmira, Colombia*
[2]*Colombian Corporation for Agricultural Research (Corpoica), CI Tibaitatá, Mosquera, Colombia*

Correspondence should be addressed to Carlos H. Galeano Mendoza; galeanomendoza@gmail.com

Academic Editor: Allen Barker

Bunching onion (*Allium fistulosum* L.) is a strategic crop for Colombia due to its economic relevance within fresh and processed food markets, and therefore, there is a demand for high yielding genotypes adapted to specific regions. For this reason, after carrying out a clonal selection process including 62 genotypes, ten of these, including a regional control, were evaluated for six different traits in Boyacá (Colombia) during 2012 and 2013. These traits were grouped into agronomic, yield, and processing categories. In general, these showed significant differences ($p \leq 0.01$) for genotypes, location, and genotypes × location interaction. Compared with the regional control and based on the multienvironmental analysis the genotypes Clone 30 and Clone 38 were the most promising new cultivars identified in this study. These two clones showed comparative advantages on earliness and yield, and they moreover showed some level of resistance to downy mildew and root rot, the most limiting diseases for Boyacá's bunch onion farmers. Therefore, Clone 30 and Clone 38 were registered as new bunching onion cultivars for the Cundiboyacense High Plateau region under the names Corpoica Aquitania-1 and Corpoica Tota-1, respectively. Finally, further approaches and initiatives on bunching onion breeding are discussed.

1. Introduction

Within the *Allium* genus, bulb onion (*Allium cepa* L.), garlic (*A. sativum* L.), and bunching onion (*A. fistulosum* L.) are economically the most important species worldwide. Bunching onion, also known as Welsh onion or scallion, is one of the most important crops in Japan, China, and Korea. It is a diploid ($2n = 16$) and allogamous perennial species. The hypothesis on its place of origin suggests that bunching onion comes from northwestern China, but today it is widely adapted and cultivated around the world [1]. Therefore, the highest genetic diversity resources and formal breeding programs are mainly located in China and Japan. Bunching onion is mostly appreciated for its high nutritional value and unique flavor, due to its substantial amounts of vitamins, macro- and micronutrients, volatile oils, and flavonoids that also give it strong antioxidant properties [2, 3]. Moreover, roots, bulbs, and leaves of *A. fistulosum* have traditionally been used in this region for the treatment of febrile diseases, headache, abdominal pain, and diarrhea and are widely used for their antifungal, antioxidative, antiplatelet, antihypertensive, and antiobesity activities [4–6].

Bunching onion is a typical outcrossing crop due to protandry [1]. It is self-compatible but susceptible to severe inbreeding depression [7]. Commercial F1 hybrid varieties are very important in countries that cultivate bunching onion. However, Colombia's production is based on clonal propagation due to its temperate origin and vernalization requirements for bolting [8]. Currently, bunching onion research has mostly been focused on yield [8], freezing tolerance [9], resistance to *Fusarium oxysporum* [10, 11], extraction of medicinal compounds [6, 12], and pungency reduction [13].

Although bunching onion is the second largest vegetable crop produced in Colombia with 17,000 hectares producing 327,000 tons, limited efforts have been made to focus on

TABLE 1: Experimental location used to evaluate bunching onion.

Trial	Town	Location	Latitude	Longitude	m.a.s.l.	Planting	Harvest
T1	Aquitania	Hato Viejo	5°31'34,198"N	72°52'31,227"W	3074	14/09/2013	28/12/2014
T2	Aquitania	Vargas	5°31'18,156"N	72°53'31,24"W	3015	26/04/2013	20/2/2014
T3	Tota	Toquecha	5°31'58,89"N	72°59'11,644"W	2773	10/07/2012	5/2/2014
T4	Tota	Cardón	5°31'30,121"N	72°58'31,332"W	3024	31/05/2013	12/02/2014
T5	Cuitiva	Llano de Alarcón	5°32'13,4"N	72°59'45,4"W	3020	10/07/2013	12/2/2014
T6	Aquitania	Hato Viejo	5°31'34,198"N	72°52'31,227"W	3074	10/07/2012	5/2/2014

crop improvement. In terms of genetic resources, the first initiative led by ICA was carried out around the mid-eighties where some accessions were collected, conserved, and characterized [14]. However, since then and due to inconstant public funding of research schemes and unclear agricultural research vision, no improved bunching onion cultivar has been introduced to any market so far.

Boyacá was the department that stood out during 2010 to 2015 with 70% of the national production, and the productive chain linked from small-scale farmers, to the agroindustry, transportation actors, and the final market [15]. Therefore, bunching onion is an important element for the region's rural development. However, the cultivated area and yield have over the years been reduced due to the incidence of pest and diseases. Consequently, efforts in developing high yielding cultivars with specific and broad adaptability, precocity, resistances to various pest and diseases, and good marketing quality are needed to support the national bunching onion production in the department and in the whole country. Additionally, improving agronomic practices is required; for example, the broadly used practice of using uncomposted chicken manure increases pathogen problems as root rot caused by the *Burkholderia cepacia-Ditylenchus dipsaci* complex, which is the most limiting problem in the Boyacá region [16].

For the reasons mentioned above, this research aims at the identification, evaluation, and selection of promising clones from the Colombian Germplasm Bank. The main goal is to identify elite materials adapted to the current bunching onion growing areas in Colombia, with tolerance to diseases and fitted to farmers and market preferences.

2. Materials and Methods

A total of 62 accessions of the Colombian Germplasm Bank were evaluated in six onion growing areas. Four clonal section cycles were carried out based on the Mamá and Bebé approach [17] (data not shown). Of these, nine clones (Clone 14, Clone 17, Clone 18, Clone 19, Clone 27, Clone 28, Clone 30, Clone 33, and Clone 38) and a regional variety known as "Pastusa" (control) were selected based on agronomic performance and farmers' preferences. The trials were carried out following the Colombian regulation for new cultivars using an Agronomic Performance Test supervised by Instituto Colombiano Agropecuario (ICA) according to resolution number 3168.

The trials were carried out during 2012 and 2014 in six common onion growing locations in towns near Lake Tota in the Cundiboyacense High Plateau (Table 1). The experimental design was a random complete block with four repetitions. The experimental unit included four rows of 8 m with distances of 0.40 m and 0.80 m between plants and rows, respectively. The plants evaluated were measured from the two inner rows. Crop management, that is, fertilization, irrigation, and weeds, pests, and diseases control, was carried out according to the onion's technical manual elaborated by Sanchez et al. [16].

The agronomic traits evaluated per plant were leaf and pseudostem length (LL, SL) and diameter (LD, SD), days to flowering (DF), plant height (PH), number of flowers (FN), days to harvest (DH), and mildew and root rot severity (MS, RRS). Yield (Y) was evaluated as the sole yielding trait. The processing traits evaluated were dry matter content (DM), pyruvic acid content (APC), and flour and paste production; these analyses were carried out in the Fruits and Vegetables Pilot Plant of La Salle University (protocols available upon request).

For the statistical analysis of the data obtained, a combined variance analysis, a Tukey mean comparison, and the additive main effects and multiplicative interaction (AMMI) were carried out using SAS software v. 9 (SAS Institute, Cary, NC, USA).

3. Results and Discussion

3.1. Agronomic Traits. All agronomic traits evaluated showed significant differences between location, genotypes, and location × genotype interaction based on a variance analysis. Therefore, the environmental effects on the onion's agronomic traits were clearly identified. For instance, the trials T1 and T2 presented on average a higher leaf length value (43.6 cm) in comparison to T3 (16 cm) (Figure 1(a)). Clone 33 showed the highest leaf length value with 44 cm followed by the regional control (40 cm), Clone 38 (37 cm), and Clone 30 (37 cm). The genotype with the lowest leaf length values was Clone 19 (23 cm) (Figure 1(b)). Similarly, trial T5 showed the highest average in leaf diameter value (2.5 cm) in comparison with T3 (0.4 cm) (Figure 1(c)). Clone 33 and Clone 30 showed the highest leaf diameter values of 4.8 cm and 2.9 cm, respectively (Figure 1(d)). Leaf traits are important, despite the fact that these are not edible parts in Colombia, mostly because certain markets use them as visual signs of plant vigor and shelf-life. Interestingly, bunching onion leaves are extremely popular in the Asian cuisine and their nutritional properties are very well known [18]. Therefore, bunching onion leaves might have an alternative

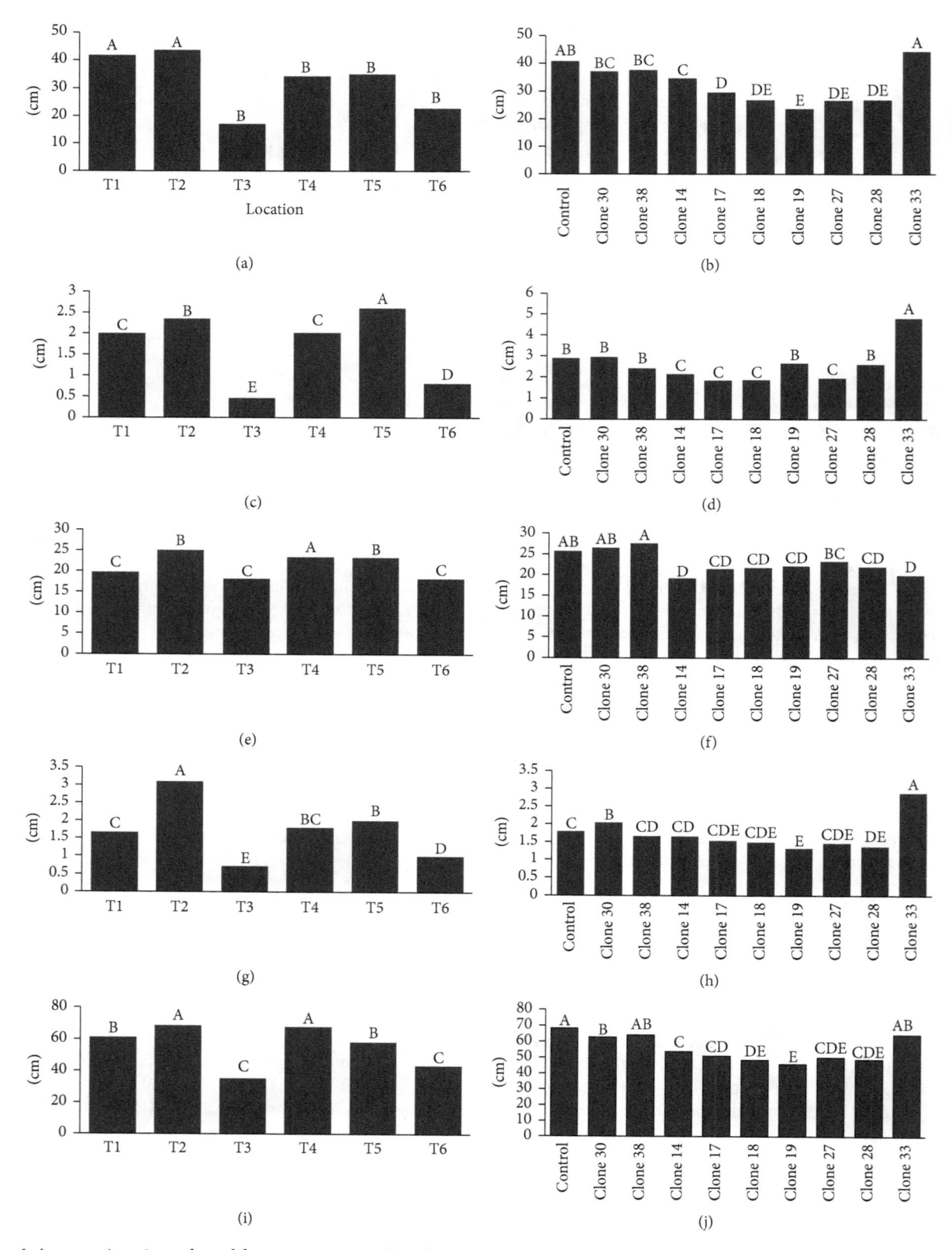

FIGURE 1: Agronomic traits evaluated for ten genotypes of bunching onion on six trials. (a) Leaf length by trials; (b) leaf length by genotypes. (c) leaf diameter by trials; (d) leaf diameter by genotypes; (e) pseudostem length by trials; (f) pseudostem length by genotypes; (g) pseudostem diameter by trials; (h) pseudostem diameter by genotypes; (i) plant height by trials; (j) plant height by genotypes.

and potential use for specific national and international niche markets.

On the other hand, pseudostem length and diameter are extremely important traits in the Colombian bunching onion markets, as the widest genotypes are marketed on local fresh markets, middle size ones are preferred by grocery stores, and the thinnest ones are used by the processing industry.

Trials T2 and T5 showed the highest average in pseudostem length values with 25 cm and 23 cm, respectively (Figure 1(e)). The lowest values were found in T1 and T3 with 18 cm and 17 cm, respectively. The genotype Clone 38 showed the longest pseudostem length value (27.64 cm) across all the locations, followed by Clone 30 (26.5 cm) and the regional control (25.6 cm) (Figure 1(f)). In terms of pseudostem diameter, the T2 trial showed the highest value (3.1 cm) in comparison to T3 (0.71 cm). Clone 33 showed the largest diameter length (2.8 cm) followed by Clone 30 (2.04 cm) and the regional control (1.79 cm).

Plant height values were consequently with leaf and pseudostem length data; therefore the trail T2 presented the highest average value and the tallest genotypes were regional control (68.2 cm), Clone 33 (64.3 cm), Clone 38 (64.1 cm), and Clone 30 (62.7 cm). Normally, farmers prefer higher plants, although the leaf/steam proportion is also very important. For instance, the regional control showed the highest plant length values and had a good leaf length proportion. However, Clone 38 and Clone 30 have, among others, the highest plant length values with an important pseudostem length.

Regarding flowering, in theory onion is a temperate crop requiring a vernalization phase for bolting. Therefore, in tropical conditions with short days, flowering is not expected. However, under specific stress conditions or after several harvest cycles the flowering becomes an important issue under Colombian conditions. Interestingly, Clone 38 and Clone 30 showed more days to flowering as well as a reduced number of flowers compared to the regional control (data not shown).

3.2. Yield and Processing Traits. Evaluation of days to harvest (DTH) showed that on average the trial with less days to harvest was T1 (180 days) in contrast to T3 (203 days) (Figure 2(a)). Similarly, genotype Clone 28 had the longest number of days to harvest (260 DTH) and Clone 14, Clone 30, and Clone 38 were the most precocious ones with 179 DTH, 172 DTH, and 162 DTH, respectively (Figure 2(b)). As bunching onion in Colombia is propagated asexually, the material with less days to harvest or which is precocious is preferred over the less precocious one because it allows harvesting several times during a year. Another advantage of the harvesting earliness is the possibility of escaping abiotic stress conditions such as drought or freezing, very common in onion cultivating areas in Colombia.

The trait yield showed significant differences between trials, genotypes, and genotype × location interaction. Trial T5 showed the highest yield value (34 t·ha^{-1}) and T1 showed the lowest one with 18 t·ha^{-1} (Figure 2(c)). Genotypes Clone 33, Clone 38, Clone 30, and the regional control were considered as high yielding materials with 32.7 t·ha^{-1}, 31.8 t·ha^{-1}, 30.05 t·ha^{-1}, and 28.8 t·ha^{-1}, respectively (Figure 2(d)). Based on the average yield of 13 t·ha^{-1} calculated for bunching onion in Colombia, these materials are a very promising alternative to the ones currently used.

Pungency in these onions is a market-based trait depending on consumers' preferences and it is normally measured by pyruvic acid content [19]. The genotype with the highest pyruvate acid content was Clone 27 (36,152 μmol·g^{-1}) and the one with the lowest value was the regional control (21,717 μmol·g^{-1}) (Figure 2(e)). These values were found to be higher than in previous evaluations carried out in bulb onion by Wall and Corgan [19] and in bunching onion by Levine et al. [20] as a potential resource for the spices industry.

Similarly, the genotype with the highest dry matter content was Clone 19 (10.9%) and the lowest values were found in the regional control and Clone 38 with 7.66 and 7.43, respectively (Figure 2(f)). Additionally, the most efficient genotypes for paste and powder production were Clone 27 (48.1%) and Clone 19 (8.1%), respectively.

These results are interesting because fresh bunching onion market is very much an uncontrolled supply and demand scenario of an inelastic product; that is, there is always an uncertain price paid to farmers. Therefore, the onion agroindustry is an alternative fair-trade market for small-scale farmers, and as a consequence, the characterization and improvement of processing traits will open new market alternatives for small-scale farmers.

3.3. Genotype × Environment Interaction (G × E). The additive main effects and multiplicative interaction (AMMI) analysis was carried out for yield and pyruvic acid, showing significant differences between environments, genotypes, and genotypes × environment. Specifically, the determination and variation coefficients for yield were 0.7% and 44.6%, respectively. The total phenotypic variation is explained by the location (52.9%), genotype (6.2%), and G × E interaction (15.1%). The proportion of the total variation accounted by each of the principal axes of the AMMI model was 71.9% for the first component and 12.1% for the second component.

Based on the graphical representation (Figure 3(a)), the genotypes Clone 17 and Clone 38 were located near to the axes showing a yield stability that was independent from the environment. Interestingly, this plasticity behavior under multiple environment trial evaluations has been identified as related to abiotic stress tolerance, suggesting that genes underlying plasticity could be related to stress response [21]. Additionally, specific adaptations were found to be related to particular trials. For instance, Clone 33, Clone 38, and the regional control showed particular adaptation to T6; and Clone 30 and Clone 14 were adapted to T1 and T3.

The G × E interaction analysis for pyruvic acid reported a determination coefficient of 0.8% and a variation coefficient of 19.5%. The total variation is explained by location (33.5%), genotype (30.5%), and G × E interaction (15.7%). The first and second principal components accounted for 53.2% and 33.7% of the phenotypic variance, respectively. In terms of stability, Clones 28, 33, and 38 showed the highest plasticity on pyruvic acid content. On the other hand, Clone 17 and Clone 28 showed specific adaptability to trials T1 and T6.

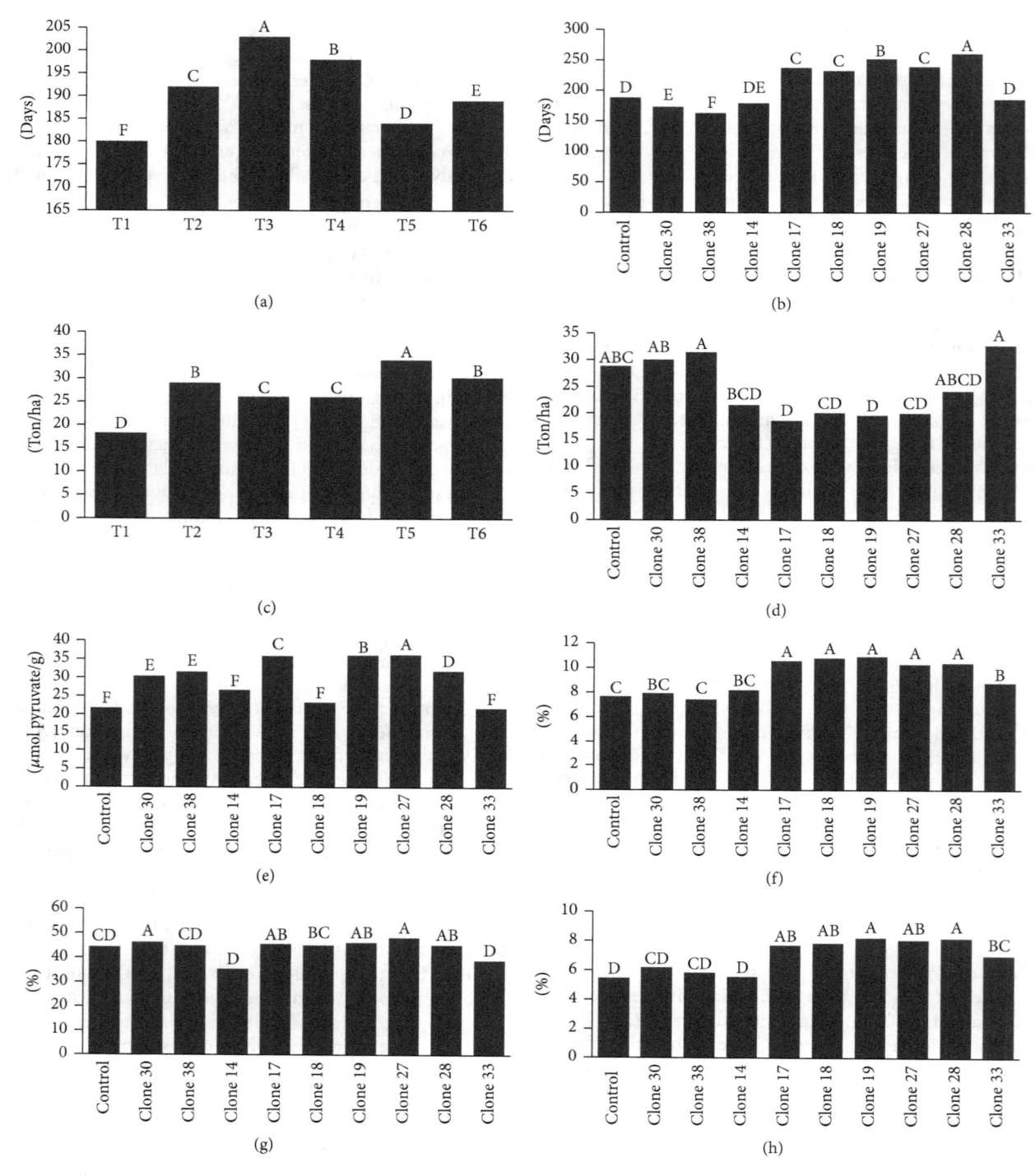

FIGURE 2: Earliness, yield, and processing traits were evaluated for ten genotypes of bunching onion. (a) Days to harvest (DTH) by trials; (b) DTH by genotypes; (c) yield by trials; (d) yield by genotypes; (e) pyruvate concentration; (f) dry matter content; (g) paste production (final paste weight/initial weight); (h) powder production (final powder weight/initial weight).

Interestingly, Clone 14, Clone 18, Clone 19, Clone 27, Clone 30, and the regional control presented a wide variation based on the contribution of each environment evaluated.

3.4. *Diseases Evaluation.* The most limiting bunching onion diseases in the Boyacá region are downy mildew (*Peronospora destructor*) and the root rot (*Burkholderia cepacia-Ditylenchus dipsaci* complex). The results showed that the regional control is susceptible to mildew with a 49.1% of disease severity. Interestingly, the other genotypes evaluated presented moderate to high resistance. However, the most promising genotypes in terms of agronomic behavior

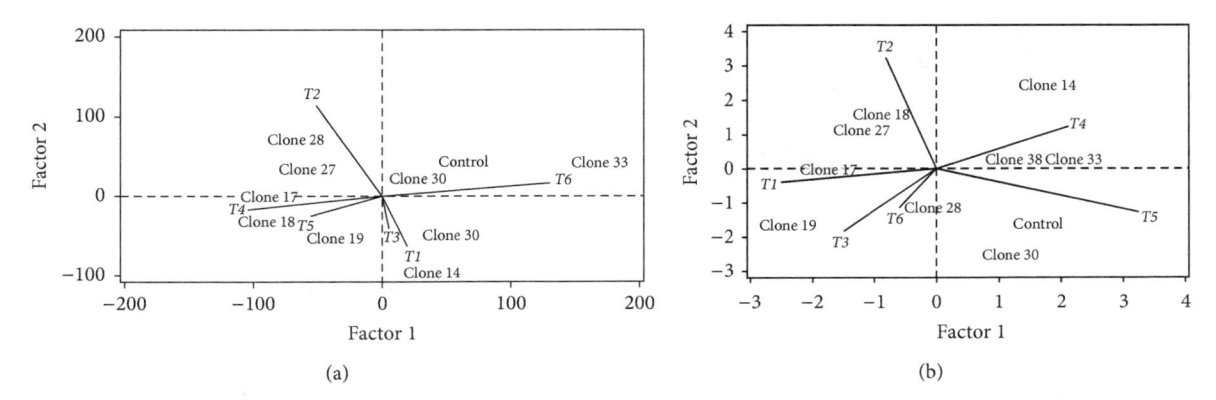

FIGURE 3: AMMI biplot for (a) yield and (b) pyruvic acid content showing genotypes and trials. Any two trials or genotypes are highly correlated if they are located at nearly the same angle from the center of the plot; it will be negatively correlated if opposing and nearly uncorrelated if the angle between them is close to 90 degrees.

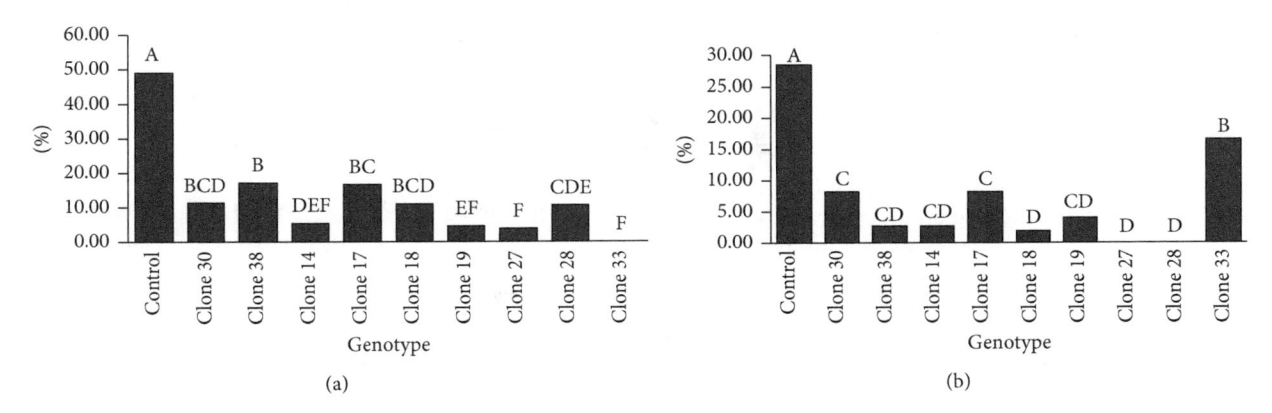

FIGURE 4: Evaluation of ten selected bunching onion genotypes for (a) downy mildew and (b) root rot disease severity.

were Clone 30 and Clone 38 showing 11% and 17% of disease severity, respectively (Figure 4(a)). The use of these promising clones in Colombia's onion growing area would influence directly the farmers' economy as it would reduce the supply costs (less fungicides) and would indirectly benefit the ecosystem by releasing fewer chemicals into the fragile environment of the Lake Tota basin. Similarly, the high severity symptoms for root rot were detected for the regional control with 28.4%; Clone 30 and Clone 38 showed moderate resistance with 8% and 2.7% of disease severity, respectively (Figure 4(b)).

These results revealed a promising variability within Colombian bunching onion germplasm kept in the Colombian Germplasm Bank. However, in order to continue the efforts on plant breeding there are some steps to be followed:

(1) Germplasm collection, introduction, and characterization: there are some specific bunching onion types in the Colombian northern markets called "sonsoneña" that are unrepresented in the germplasm bank, and therefore, germplasm collecting missions have to be carried out to include this and other genotypes still not included. Additionally and based on the fact that Colombia is neither the center of origin nor diversity of A. fistulosum, new germplasm has to be introgressed from other germplasm banks or breeding programs around the world, especially from Asia [22]. Finally, the morphological characterization can be well complemented with genotyping information based on the increased availability of genomic tools such as SSR [23, 24] and transcriptome assembly [25].

(2) Seed propagation initiatives: most of the Colombian bunching onion phytosanitary problems are due to poor asexual seed quality. Therefore, physiological and genetics studies have to be carried out to simulate long day conditions and vernalization based on high environmental effects on bolting [8].

(3) Nutritional value: based on the functional food market tendency, bunching onion plays an important role due to its high vitamin C, carotenoids, macro- and micronutrients content, especially calcium and potassium, and flavonoids that are strong antioxidants [2, 3]. Therefore, these traits as well as their G × E interaction have to be evaluated and offered as a specific and additional value for specific onion markets.

(4) Crop management studies: bunching onion crops in Colombia still have a knowledge gap in terms of nutritional requirements and disease management.

Therefore, multidisciplinary team efforts are necessary to tackle these needs in face of climate change challenges. Additionally, evaluation of other crop systems has to be carried out based on the evidence that greenhouse bunching onions showed yield improvements [26].

(5) Breeding program: a long term initiative, preferably in joint collaboration with different institutions working on this species, has to be developed in order to cross and select recurrently families and genotypes with high frequency of favorable alleles. Additionally, intraspecific crosses and subsequently backcrossing schemes will incorporate well known disease resistance from other species such as *A. cepa* [10, 11]. Finally, the incorporation of cytoplasmic male sterility (CMS) using marker assisted selection [27] will accelerate the hybrid evaluation and trait introgression for new varieties.

4. Conclusions

These results are the first approach carried out in Colombia to offer new and potential bunching onion cultivars to farmers. Therefore and based on their wide adaptation, biotic stress tolerance, earliness, and yield, genotypes Clone 30 and Clone 38 were selected as new cultivars to be released in the Cundiboyacense High Plateau and have been registered as Corpoica Aquitania-1 and Corpoica Tota-1, respectively. However, activities that follow the cultivars adoption in the current cultivation areas are ongoing and altogether will the beginning of a bunching onion breeding program in Colombia.

Acknowledgments

The authors are grateful to the farmers' associations Asoparcela, Asollanos, and Asotoquecha for their kind collaboration and motivation to carry out this study. They thank Juan Clímaco and Jorge Arguelles for their advice and support on the development of the phytopathological and statistical issues, respectively. This research was financed by the Colombian Ministry of Agriculture and Rural Development.

References

[1] B. V. Ford-Lloyd and S. J. Armstrong, "5-Welsh onion: Allium fistulosum L. A2. pp," in *Genetic improvement of vegetable crops*, G. Kalloo and B. O. Bergh, Eds., p. pp, Pergamon Press, Amsterdam, The Netherlands, 1993.

[2] S. Aoyama and Y. Yamamoto, "Antioxidant activity and flavonoid content of welsh onion (Allium fistulosum) and the effect of thermal treatment," *Food Science and Technology Research*, vol. 13, no. 1, pp. 67–72, 2007.

[3] E. Kołota, K. Adamczewska-Sowińska, and C. Uklańska-Pusz, "Yield and nutritional value of Japanese bunching onion (Allium fistulosum L.) depending on the growing season and plant maturation stage," *Journal of Elementology*, vol. 17, no. 4, pp. 587–596, 2012.

[4] S. Sang, A. Lao, Y. Wang, C.-K. Chin, R. T. Rosen, and C.-T. Ho, "Antifungal constituents from the seeds of Allium fistulosum L.," *Journal of Agricultural and Food Chemistry*, vol. 50, no. 22, pp. 6318–6321, 2002.

[5] Y. Yamamoto, S. Aoyama, N. Hamaguchi, and G.-S. Rhi, "Antioxidative and antihypertensive effects of Welsh onion on rats fed with a high-fat high-sucrose diet," *Bioscience, Biotechnology, and Biochemistry*, vol. 69, no. 7, pp. 1311–1317, 2005.

[6] Y.-Y. Sung, T. Yoon, S. J. Kim, W.-K. Yang, and H. K. Kim, "Anti-obesity activity of Allium fistulosum L. extract by down-regulation of the expression of lipogenic genes in high-fat diet-induced obese mice," *Molecular Medicine Reports*, vol. 4, no. 3, pp. 431–435, 2011.

[7] H. Tsukazaki, H. Fukuoka, Y.-S. Song, K.-I. Yamashita, T. Wako, and A. Kojima, "Considerable heterogeneity in commercial F1 varieties of bunching onion (Allium fistulosum) and proposal of breeding scheme for conferring variety traceability using SSR markers," *Breeding Science*, vol. 56, no. 3, pp. 321–326, 2006.

[8] Y. Dong, Z. Cheng, H. Meng, H. Liu, C. Wu, and A. R. Khan, "The effect of cultivar, sowing date and transplant location in field on bolting of Welsh onion (Allium fistulosum L.)," *BMC Plant Biology*, vol. 13, no. 1, article no. 154, 2013.

[9] K. K. Tanino, S. Kobayashi, C. Hyett et al., "Allium fistulosum as a novel system to investigate mechanisms of freezing resistance," *Physiologia Plantarum*, vol. 147, no. 1, pp. 101–111, 2013.

[10] H. Q. Vu, M. A. El-Sayed, S.-I. Ito, N. Yamauchi, and M. Shigyo, "Discovery of a new source of resistance to Fusarium oxysporum, cause of Fusarium wilt in Allium fistulosum, located on chromosome 2 of Allium cepa Aggregatum group," *Genome*, vol. 55, no. 11, pp. 797–807, 2012.

[11] T. Wako, K.-I. Yamashita, H. Tsukazaki et al., "Screening and incorporation of rust resistance from Allium cepa into bunching onion (Allium fistulosum) via alien chromosome addition," *Genome*, vol. 58, no. 4, pp. 135–142, 2015.

[12] H. Ueda, A. Takeuchi, and T. Wako, "Activation of immune responses in mice by an oral administration of bunching onion (Allium Fistulosum) mucus," *Bioscience, Biotechnology, and Biochemistry*, vol. 77, no. 9, pp. 1809–1813, 2013.

[13] H. Tsukazaki, S. Yaguchi, K.-I. Yamashita, M. Shigyo, A. Kojima, and T. Wako, "QTL analysis for pseudostem pungency in bunching onion (Allium fistulosum)," *Molecular Breeding*, vol. 30, no. 4, pp. 1689–1698, 2012.

[14] J. Jaramillo, Y. Palacios, and O. Martinez, "Descripción cuantitativa de la colección colombiana de cebolla de rama *Allium fistulosum*L.," *Revista ICA*, vol. 27, pp. 639-382, 1992.

[15] DANE-ENA, (2015) Encuesta Nacional Agropecuaria ENA 2014. Bogota, Colombia.

[16] G. Sanchez, H. Pinzon, J. Clímaco et al., *Manual de la cebolla de rama*, CORPOICA, Bogotá, Colombia, 2012.

[17] S. De Haan and C. Fonseca, *Guía de evaluación y recolección de datos: metodología Mamá & Bebé para la selección participativa de variedades*, Centro Internacional de la Papa (CIP) - Red LatinPapa, Lima, Perú, 2010.

[18] D. Štajner, N. Milić, J. Čanadanović-Brunet, A. Kapor, M. Štajner, and B. M. Popović, "Exploring Allium species as a source of potential medicinal agents," *Phytotherapy Research*, vol. 20, no. 7, pp. 581–584, 2006.

[19] M. Wall and J. Corgan, "Relationship between pyruvate analysis and flavor perception for onion pungency determination," *Hortscience*, vol. 27, pp. 1029-1030, 1992.

[20] L. Levine, J. Bauer, S. Edney et al., "Scallion (allium fistulosum L.) pungency regulated by genetic makeup and environmental conditions (light and CO2)," *SAE Technical Papers*, 2005.

[21] D. L. Des Marais, K. M. Hernandez, and T. E. Juenger, "Genotype-by-environment interaction and plasticity: Exploring genomic responses of plants to the abiotic environment," *Annual Review of Ecology, Evolution and Systematics*, vol. 44, pp. 5–29, 2013.

[22] J. S. Khosa, J. McCallum, A. S. Dhatt, and R. C. Macknight, "Enhancing onion breeding using molecular tools," *Plant Breeding*, vol. 135, no. 1, pp. 9–20, 2016.

[23] H. Tsukazaki, M. Honjo, K.-I. Yamashita et al., "Classification and identification of bunching onion (Allium Fistulosum) varieties based on SSR markers," *Breeding Science*, vol. 60, no. 2, pp. 139–152, 2010.

[24] L. Yang, C. Wen, H. Zhao et al., "Development of Polymorphic Genic SSR Markers by Transcriptome Sequencing in the Welsh Onion (Allium fistulosum L.)," *Applied Sciences*, vol. 5, no. 4, pp. 1050–1063, 2015.

[25] H. Tsukazaki, S. Yaguchi, S. Sato et al., "Development of transcriptome shotgun assembly-derived markers in bunching onion (Allium fistulosum)," *Molecular Breeding*, vol. 35, no. 1, pp. 1–11, 2015.

[26] M. Tendaj and B. Mysiak, "Usefulness of japanese bunching onion (Allium fistulosum L.) for forcing in greenhouse," *Acta Agrobotanica*, vol. 60, no. 1, pp. 143–146, 2007.

[27] L. M. Gao, Y. Q. Chen, Y. M. Huo et al., "Development of SCAR markers to distinguish male-sterile and normal cytoplasm in bunching onion (*Allium fistulosum* L.)," *The Journal of Horticultural Science & Biotechnology*, vol. 90, no. 1, pp. 57–62, 2015.

Plant Sources, Extraction Methods, and uses of Squalene

M. Azalia Lozano-Grande,[1] Shela Gorinstein,[2] Eduardo Espitia-Rangel,[3] Gloria Dávila-Ortiz[4], and Alma Leticia Martínez-Ayala[1]

[1]*Centro de Desarrollo de Productos Bióticos, Instituto Politécnico Nacional, San Isidro, 62731 Yautepec, MOR, Mexico*
[2]*Institute for Drug Research, School of Pharmacy, Hadassah Medical School, The Hebrew University, Jerusalem 91120, Israel*
[3]*Instituto Nacional de Investigaciones Forestales, Agrícolas y Pecuarias, Campo Experimental Valle de México, Coatlinchan, 56250 Texcoco, MEX, Mexico*
[4]*Instituto Politécnico Nacional, Escuela Nacional de Ciencias Biológicas, Delegación Miguel Hidalgo, 11340 Ciudad de México, Mexico*

Correspondence should be addressed to Alma Leticia Martínez-Ayala; alayala@ipn.mx

Academic Editor: José M. Alvarez-Suarez

Squalene (SQ) is a natural compound, a precursor of various hormones in animals and sterols in plants. It is considered a molecule with pharmacological, cosmetic, and nutritional potential. Scientific research has shown that SQ reduces skin damage by UV radiation, LDL levels, and cholesterol in the blood, prevents the suffering of cardiovascular diseases, and has antitumor and anticancer effects against ovarian, breast, lung, and colon cancer. The inclusion of SQ in the human diet is recommended without causing health risks; however, its intake is low due to the lack of natural sources of SQ and efficient extraction methods which limit its commercialization. Biotechnological advances have developed synthetic techniques to produce SQ; nevertheless, yields achieved are not sufficient for global demand for industrial or food supplement purposes. The effect on the human body is one of the scientific issues still to be addressed; few research studies have been developed with SQ from seed or vegetable sources to use it in the food sector even though squalene was discovered more than half a century ago. The aim of this review is to provide an overview of SQ to establish research focus with special reference to plant sources, extraction methods, and uses.

1. Introduction

Squalene is a linear triterpene synthesized in plants, animals, bacteria, and fungi as a precursor for the synthesis of secondary metabolites such as sterols, hormones, or vitamins. It is a carbon source in the aerobic and anaerobic fermentation of microorganisms [1, 2].

The SQ was discovered by Tsujimoto Mitsumaru in 1916, a Japanese researcher who described the compound as a highly unsaturated molecule, assigning its name to the genus from which was isolated, *Squalus* spp. [3, 4]. The main source of SQ was the liver of marine animals rich in lipids and unsaponifiable matter (50–80%), whose SQ content may comprise up to 79% of the total oil. SQ is considered important in oily extract for the survival of deep-water animals, where oxygen supply is poor and pressures are very high [5].

The use of marine animal oil as a source of SQ has been limited by animal protection regulations and the presence of organic pollutants (POPs) as organochlorine pesticides, polycyclic aromatic hydrocarbons, dioxins, or heavy metals that cause cancer. This has generated interest in finding new natural sources, especially of plant origin [6].

Among the plant sources reporting SQ content are olive oil (564 mg/100 g), soybean oil (9.9 mg/100 g), rice, wheat germ, grape seed oil (14.1 mg/100 g), peanut (27.4 mg/100 g), corn, and amaranth (5942 mg/100 g). Of these species, olive is only used for extracting commercial squalene despite the highest content reported for amaranth.

SQ is also found in the human body, is secreted by the sebaceous glands for skin protection, and forms part of 10–15% of lipids on the skin surface in concentrations of 300–500 μg/g and on internal organs such as the liver and

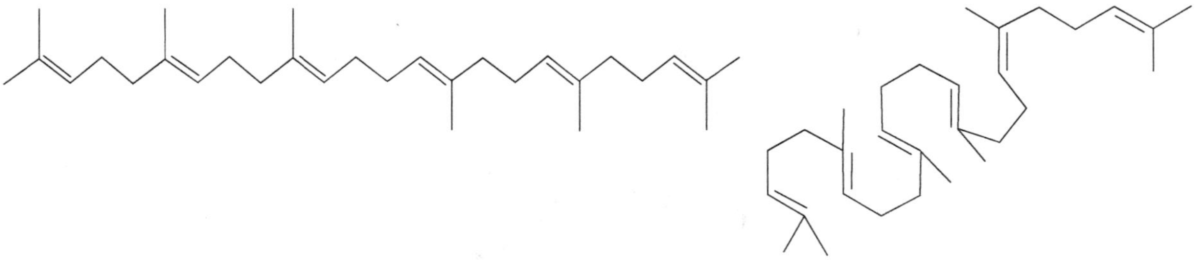

FIGURE 1: Linear (a) and flat trigonal (b) structure of squalene.

small intestine in concentrations of less than 75 μg/g [3, 4, 7].

Several studies have confirmed the health benefits of SQ in nutritional, medicinal, and pharmaceutical aspects. It is considered a potent chemopreventive and chemotherapeutic agent, which inhibits the tumor growth in the colon, skin, lung, and breast, and it stimulates the immune system for the application of drugs in the treatment of diseases such as HIV, H1N1, leukemia, papilloma, and herpes, among others [8–11].

SQ-related research such as pharmaceutical, cosmetic, and food applications are issues that have been addressed independently and do not address commercial expectations. This review provides an overview of the potential natural sources and forms of extraction of this natural lipid.

2. Physicochemical Characteristics of Squalene and Biosynthesis

The SQ (2,6,10,15,19,23-hexamethyl-6,6,10,14,18,20-tetracosahexane) is a hydrocarbon chain formed by six isoprene units (Figure 1); when the units are assembled, they form a triterpene that gives the lipid character. The six carbon double bonds (C=C) allow the molecule to be one of the most unsaturated lipids, and it is sensitive to oxidation [12].

The SQ is physically a transparent oil with the molecular weight of 410.7 g/mol, density of $0.855 \, g/cm^3$, melting temperature of −20°C, and it is soluble in organic solvents and insoluble in water [9, 12].

Due to its unsaturated structure, the SQ is sensitive to oxidation; the double bonds pass to the oxidized form by chain reactions, where the unsaturated carbons join ions producing saturated forms of the molecule. Other oxidation products are generated through self-hydrolytic processes such as peroxides, but the SQ is not susceptible to peroxidation; on the contrary, it has an antioxidant protective effect by trapping oxygen singlets during the reaction processes [4, 13].

Studies on oxidation of SQ were scarce; until the last century, the cyclic diperoxides were reported as oxidation products. After 80 years, with advanced techniques (^{1}H·NMR and ^{13}C·NMR), it was found that oxidation mechanisms and chemical structures of epoxy and alcohol are formed after a prolonged oxidation to 55–150°C [14–17].

Naziri et al. [17] studied squalene oxidation at different temperatures and air conditions, finding the formation of epoxy with prooxidant activity in the oil at 40°C and 62°C. On the other hand, Psomiadou and Tsimidou [18] reported the antioxidant effect of SQ in olive oil, where the antioxidant activity was weak in presence of other fatty acids, possibly due to a competitive oxidation between the same lipids of the extract. As an endogenous compound, the SQ also contributes to oxidative stability, but this has not been completely studied.

3. Function of Squalene in Plants

Plants biosynthesize a wide variety of metabolites (>230 constituents) in response to stress conditions due to drought, predation, or disease. Among these metabolites, those are the fractions unsaponifiable (<1.5%) that contribute to bioactive properties of oils and imparts stability to the cell membrane of plants [13, 19].

Unsaturated lipids are key components in cell membranes, which together with SQ can regulate biophysical properties, diffusion, and dynamic membrane organization. Models of cell membranes have suggested that SQ acts in the lipid bilayer with other fatty acids as inhibitors of proton leakage in alkaline membranes and influence in the synthesis of ATP [12].

By its nonpolar nature, the SQ is located on the hydrophobic center of the lipid bilayer, organized in a spiral or extended conformation, oriented in parallel to the phospholipid chains between the cell membrane (Figure 2) [20].

When the SQ adopts the hexagonal form, it increases the rigidity and the size of the cell membrane. Studies on the lipid-protein interaction in cell membranes have revealed that the SQ increases the polarity and hydrophobic interactions; it is contributing to membrane reconstitution, functional regulation of proteins, and movement of ions. Therefore, the SQ plays an important role in electrochemical cell gradient [13, 21].

Other components such as saponins in interaction with phytosterols and squalene form insoluble complexes, which fix the architecture of the lipid bilayer and generate pores that participate in the permeability of cell membranes [22, 23]. Only 10% of total SQ on the membrane is metabolically active for sterol synthesis, while the rest is stored with triacylglycerides and sterol esters as lipid droplets [21].

The SQ store is used to form important molecules such as β-sitosterol, campesterol, and stigmasterol, which are precursors of hormones for the growth (i.e., brassinosteroids) and plant adaptation to biotic stress [24–26]. The sterols are biosynthesized in plant cells via mevalonate or isoprenoids route, where SQ is also metabolized.

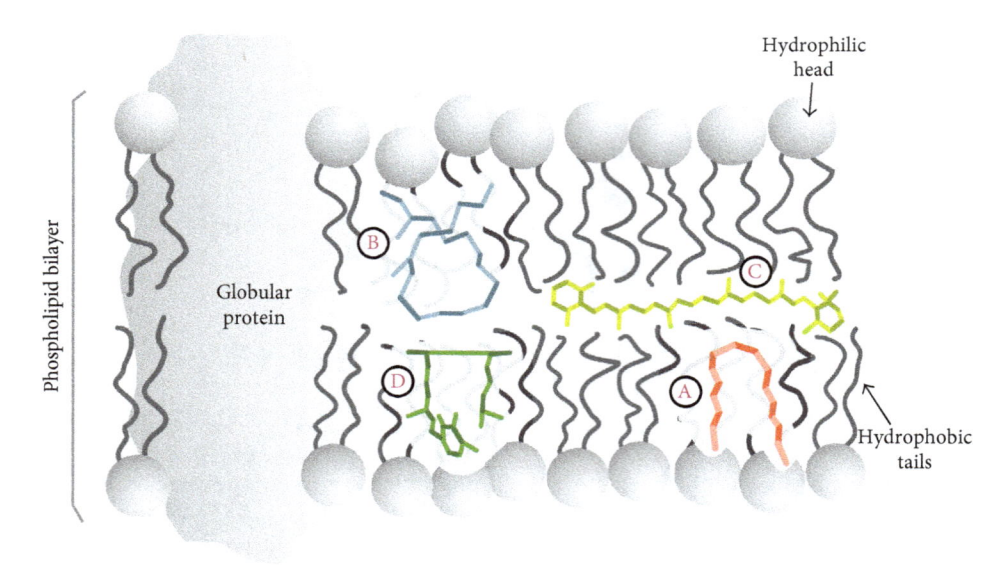

FIGURE 2: Location of minor compounds in the plant cell membrane: oleic acid (A) squalene (B), carotene (C), and tocopherols (D), adapted from López et al. [12].

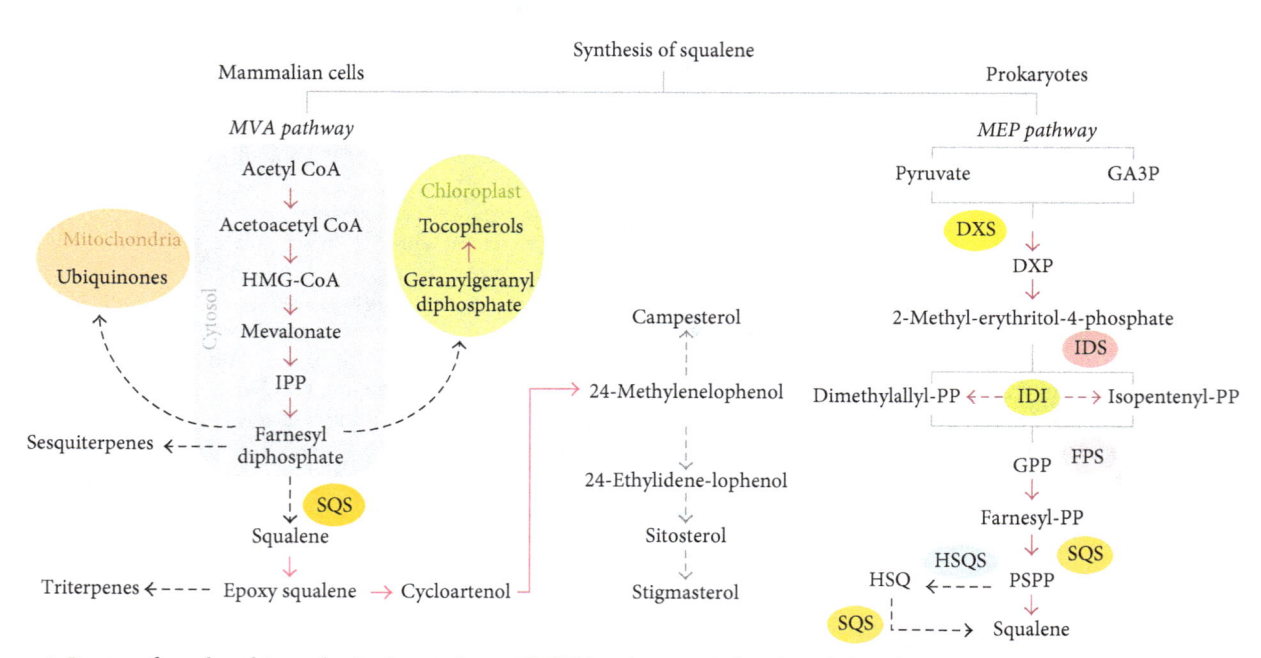

FIGURE 3: Routes of squalene biosynthesis via mevalonate (MVA) and via methylerythritol phosphate (MEP). Acetyl CoA: acetyl coenzyme A; HMG-CoA: hydroxymethyl glutaryl coenzyme A; IPP: isopentenyl diphosphate; SQS: squalene synthase; DXP: desoxicelulose 5-phosphate; DXS: desoxicelulose 5-phosphate synthase; IDI: isopentenyl diphosphate isomerase; IDS: isopentenyl diphosphate synthase; FPS: farnesyl diphosphate synthase; GPP: geranyl diphosphate; HSQS: hydrosqualene.

3.1. Squalene Biosynthesis.

The SQ is an intermediate compound in synthesis of hopanoids, phytosterols, and more than 200 important triterpenes for the cell membrane [27–29]. SQ is synthesized from isopentenyl units, isopentenyl diphosphate (IPP) when the intermediate is mevalonate (MVA), and dimethyl-allyl-diphosphate (DMAPP) when the intermediate is methylerythritol phosphate (MEP). The biosynthesis by IPP begins with conversion of three molecules of acetyl-CoA to MVA through the acetoacetyl-CoA until 3-hydroxy-3-methylglutaryl-CoA (HMG-CoA) is catalyzed by balsamic-acetyl-CoA synthase (AAS) and HMG-CoA synthase (Figure 3).

The reduction of HMG-CoA results in MVA, and this is phosphorylated to MVA-5-diphosphate by the MVA kinase and phospho-MVA-kinase enzymes. The MVA-5-diphosphate in the presence of ATP (adenosine triphosphate) is decarboxylated to form IPP. IPP isomerase (IDI) catalyzes the isomerization of IPP to DMAPP and condensation of two IPP molecules to form GPP (geranyl diphosphate). GPP is condensed with another IPP molecule and creates 15-carbon

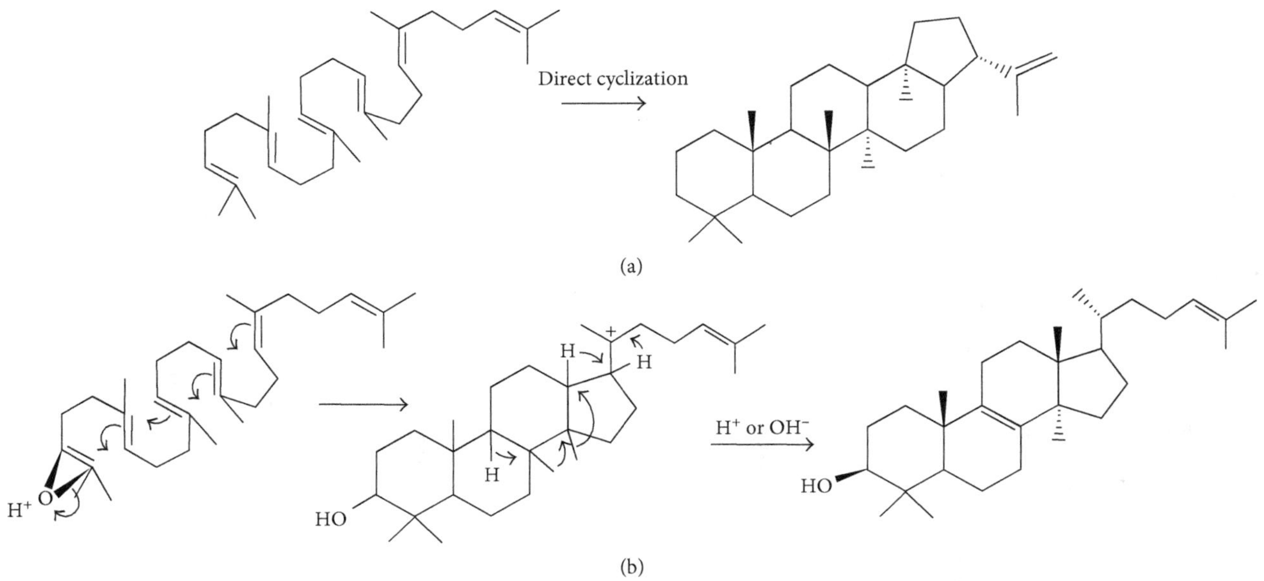

FIGURE 4: Cyclization process of squalene, autohydrolytic (a) or due to involvement ionic species (b).

farnesyl-diphosphate (FPP). Finally, two molecules of farnesyl-diphosphate are reduced via enzymatic reaction to form the SQ which continues its path to the biosynthesis of phytosterols [1, 10, 22, 29].

Phytosterols are the products of cascade cyclization of SQ; these mechanisms are activated from ionic species by deprotonation, hydration (Figure 4), or from oxide-squalene activated by the action of triterpene enzymes such as squalene cyclase (SQs) and oxide-squalene cyclase (OSQ) [17, 30].

SQs and OSQ catalyze reactions that convert the pyrophosphate of geranyl, pyrophosphate of farnesyl, and geranyl-geranyl into 57–66 different terpenes [28]. Despite the differences in sites of action of these enzymes (start or end the carbonated chain), all cycling products obtained from SQ are important for functional components of the cell membrane.

There are other forms of squalene production, those produced by prokaryotes organisms (such as E. coli) or in eukaryotes (fungi and yeasts such as Saccharomyces cereviceae, Torulaspora delbrueckii, and Chlamydomonas reinhardtii). These organisms produce SQ via MEP (2-C-methyl-D-erythritol-4-phosphate) by the merger of two FPP [2, 28, 31].

The biosynthesis in these microorganisms begins with 1-deoxy-D-xylulose-5-phosphate (DXP) formation where DXP-synthase and other enzymes participate to reduce MEP to form building block IPP and DMAPP. This block is isomerized via IDI, and the farnesyl-diphosphate-synthase catalyzes the IPP coupling with DMAPP to result in GPP, eventually forming FPP. Recent studies have proposed the synthesis of SQ by three steps with the participation of the enzymes PSPP-synthase, HSQ-synthase (HSQS), and squalene-synthase (SQS). Not in two consecutive steps some authors mention how the SQS catalyzes the fusion of two FPP molecules to pre-SQ diphosphate (PSPP) and then the arrangement of PSPP with NADPH to form the SQ [32].

MEP biosynthesis is used in metabolic engineering as a new strategy for the commercial production of squalene. This includes fermentation with micro-organisms; however, although these techniques are promising, the reported yields (14.5-160.2 mg/L) have not been sufficient to supply commercial production of squalene, and little mention is made of the costs involved [1, 29].

4. Bioactive Properties of Squalene

Various plant terpenes have excellent bioactive properties with applications such as antimicrobials, antibiotics, supplements, flavorings, or repellents. Some terpenes of interest are limonene, carveol, geraniol, stevioside, β-carotene, and lutein. The case of SQ is considered a triterpene with nutritional and medicinal values with broad expectations for pharmaceutical application. Table 1 shows some properties, and their perspectives will be mentioned.

4.1. Weight and Cholesterol Control. The consumption of SQ from natural sources (olive oil, wheat germ, rice husk, or amaranth) can be part of an integral diet. Oral administration of SQ can produce other benefits when ingested in the body. Between 60% and 80% of the exogenous SQ is absorbed and distributed to various tissues, while the endogenous SQ is first synthesized in the liver and then transported to the skin or organs through blood. It has been established that SQ is absorbed faster in the circulatory system than cholesterol, where it is deposited as excess in adipose tissue or muscle tissue [8, 10, 33].

In rat models, oral SQ was absorbed by lymphatic vessels like cholesterol, with only 20% becoming sterols during transit through the small intestine. While in 102 patients, its consumption was also shown to reduce total cholesterol, LDL

TABLE 1: List of squalene properties and pharmaceutical applications.

	Bioactive property	Application
Squalene	Cardioprotector	Intravenous injection, oral consumption to cholesterol control
	Antioxidant	Topical emulsions, oral administration
	Antibacterial and antifungal	Cream topical, oral medication
	Anticancer	Preventive and chemotherapeutic substances: drugs and vaccines (emulsions, conjugates)
	Detoxifying	Nutritional supplement

cholesterol, and triglyceride levels after 20 weeks with SQ-pravastatin treatments [34].

Other *in vivo* studies have demonstrated the cardioprotective effect of SQ. Farvin et al. [35] tested SQ at a dose of 2% for 45 days and found that it significantly reduced the levels of cholesterol, triglycerides, and fatty acids in plasma and heart tissues. The SQ had an antilipidemic effect on the levels of LDL cholesterol and an increase of HDL cholesterol counteracted lipid peroxidation and maintained the levels in almost normal state of the rats. In another investigation, a dose of 1000 mg/kg reduced leptin levels in blood plasma after 4 weeks, as well as cholesterol, triglycerides, and glucose, and increased testicular weight in rats [36]. The consumption of exogenous SQ has been studied for up to 75 days in high doses without finding differences from the fourth week; however, the effects of excess consumption are still unknown. In this regard, more studies are required for dietary supplement.

4.2. Antioxidant. The six double bonds of SQ in transposition impart reactivity to compound; however, it is very stable to peroxidation and other reactions related to unsaturated compounds. Evidence indicates that the antioxidant activity comes from its capacity to trap oxygen singlets during the autohydrolytic reaction processes and oxidation products (peroxides, SQ-OOH, SQ-OH, and isomers of SQ oxides) [4].

In the skin, it has the capacity to absorb up to a quarter of its weight in oxygen, an important factor that prevents the development of cutaneous flora and peroxidase forms that lead to the development of skin diseases, acne, comedogenics, and wrinkles [15].

In vitro and *in vivo* studies have determined the by-products of oxidation of SQ and the way of interaction on the physiology of the skin and flora, both under normal conditions and under disorders such as acne. Ozone, UV rays, and cigarette smoke are oxidizing agents of SQ; photoluminescence studies with mass spectrometry (NMR) have elucidated that SQ double bonds react with oxygen on the surface of the skin and prevent photooxidation by the rays [37].

On the contrary, the antioxidant activity of the SQ in terms of its capacity to absorb oxygen radicals (ORACs) in food is less investigated. Tikekar et al. [38] studied the antioxidant activity of SQ in amaranth oil, under different temperature conditions (125°C–150°C for 20 min) and bursting (250°C–290°C). In all conditions, the oxidation of the SQ was low (12.7%) even under extreme conditions of toasting and bursting. On the contrary, pure SQ shows weak antioxidant capacity, compared to the complete amaranth extract; the authors suggest that the antioxidant property of SQ is linked to other components of the oil, such as tocopherols and tocotrienols.

Kohno et al. [39] demonstrated the entrapment velocity of O^- radicals by SQ, and this was superior compared to other lipids even compared to additive hydroxytoluene (HBT) and food antioxidant frequently used. Similar studies in olive oil have established their antioxidant potential, both by the content of SQ (196 mg/kg) and by interaction with other minor compounds: polyphenols, secoroids, lignans, and vitamin E [40, 41]. Hrncirik and Fritsche [42] and Psomiadou and Tsimidou [18] demonstrated that the antioxidant capacity of SQ oil is not only a function of concentration but also of temperature conditions (<62°C) and interaction with other acids fatty acids, which reduce the oxidation rate due to competitive oxidation [43].

4.3. Emollient and Moisturizer. SQ is a natural antioxidant molecule that protects cells from oxidative damage by exposure to ultraviolet light and other external sources; this molecule participates as a defense mechanism for the internal and external tissues of the skin in the human body.

Kohno et al. [39] found that SQ is not susceptible to peroxidation like other lipids; it is stable against the attack of peroxide radicals which suggests that the chain reaction of lipid peroxidation is spread with adequate levels of SQ on the surface of the skin. However, this protective effect is modified when the SQ undergoes exacerbated oxidations, becoming a source of lipid peroxides. Investigations have suggested a link between comedogenesis and acne where exposure to UV rays, pollution, and cigarette smoke induce higher peroxide formation which impact on sunspots of the skin [44].

Through high-performance thin layer chromatography (HPTLC) methods, SQ oxides have been described in acne comedons and skin lipids, showing that there is a decrease in pure SQ and an increase in oxidized by-products. This suggests that excessive oxidation of SQ can lead to the production of comedogenic molecules and have no antibacterial properties that prevent the development of the existing flora. Ultimately, they break the comedonal pockets and propagate inflammatory reactions of the dermis that aggravate the effects of acne [4, 45].

In contrast, other studies suggest the SQ relieves skin irritations, through topical applications which prevent photoaging and cancer induced by UV light. Treatments of SQ with oils offer protection to burns, and topical applications of creams added with antioxidants (vitamin E, Co-Q10, and SQ) can increase the bait of the skin and reduce the attack of bacteria and fungi [7, 9, 46].

Thus, it is evident that SQ is not only the main component on the surface of the skin (12%); its importance also lies in the cutaneous physiological function as an antioxidant agent, humectant, and emollient as well as against seborrhea disorders, acne, dermatitis, psoriasis, among other skin diseases.

4.4. Detoxifying. SQ, by its nonpolar nature, has an affinity to nonionized drugs that allows it to function as a purifier of xenobiotic substances in the human body. It has been found that SQ improves the elimination of hexachlorobenzene (HCB, organochlorine xenobiotic) through feces when SQ is supplied in 8% concentrations in the diet. Other xenobiotic substances such as theophylline and strychnine can also be eliminated in feces when the intake of SQ is greater [44].

Other studies in pediatric patients have determined that SQ helps stimulate liver detoxification enzymes, such as the P450 enzyme or for detoxification with lead and other toxic substances. The consumption of SQ in infants also showed greater growth in height and better neuromotor development [47].

Few references [48–50] abound on toxicological investigations of SQ for metals and toxic substances; however, SQ has been classified as a detoxifying agent of the ideal human body. This is an area for further investigations.

4.5. Anticancer. Cancer is a system of genetic changes and cell proliferation that occurs in different stages of development. Each stage involves genetic changes caused by chemical agents, UV light, and reactive species that damage DNA, modify genetic expression, and alter the cellular defense system [51].

The consumption of SQ has been shown to have effects on the incidence of cancer; epidemiological models in animals showed that an oral consumption of 1% SQ in the diet influences mammary cells and colon cancer. A hypothesis on the mechanism of inhibition in anticancer activity indicates that SQ reduces the activity of the HMG-CoA reductase enzyme by limiting the steps towards normal cholesterol synthesis and intermediate stages where geranyl diphosphate (GDP) and farnesyl diphosphate (FDP) produce important substrates for the biosynthesis of ubiquinones and for the prenylation (farnesylation) of proteins. Inhibition of these proteins inactivates and reduces signal transduction in the proliferation and differentiation of active cells such as oncogenes and GTP-binding proteins [51–53].

An increase in SQ consumption may reduce the development of oncogene-dependent tumors that require prenylation for activation, such as colon, breast, pancreatic, and melanoma tumors. It has been shown that when endogenous SQ is increased, FPP production is reduced and the oncogene is inactivated via prenylation [11].

Similarly, Smith [51] proposed three mechanisms of SQ with a protective effect against carcinogenesis. The first is the inhibition of the activity of HMG-CoA reductase towards cholesterol biosynthesis via the MVA, the second is by mechanisms of regulation of the SQ on the enzymes involved in the metabolism of xenobiotic substances that alter the metabolic activation of carcinogenesis, and the third occurs on the elimination of free radicals and reactive oxygen that produce mutagenic lesions in the DNA, lipids, and proteins that lead to carcinogenesis.

Warleta et al. [54] investigated the properties of SQ on cell proliferation, apoptosis, the level of reactive oxygen species (ROS), and oxidative damage to DNA in human breast cells. They found that SQ did not have significant activity in cell proliferation rates; however, it exerted effects on epithelial cells in a dose-dependent manner of SQ. They observed that SQ reduced levels of intracellular ROS and oxidative damage induced by H_2O_2.

Other reports have suggested a relationship between the low incidence of cancer with a feeding style based on SQ-rich products, antioxidants, and fiber. Such is the case of Mediterranean diet, where there is a high consumption of SQ and phenolic compounds from the consumption of fish, vegetables, and fruits, which influence a low incidence of diseases of degenerative diseases [40, 55].

Other hypotheses have been established about the possible anticancer effect of SQ; however, there are still few *in vitro* and *in vivo* studies that specifically describe the role of SQ in antitumor activity among other mechanisms.

5. Potential Applications of Squalene

5.1. Drug Administration Agent. During the experimental studies carried out to verify the potential effect against cancer and antitumor treatments, it was observed that SQ in combination with other compounds improves the effectiveness of drugs and the immune response to the antigen. In this sense, scientific research was conducted in the search for specific treatments to elucidate the way of acting and specific location for the delivery of drugs in the human body.

The nontoxic chemical nature of lipids is considered excellent carriers for their ability to permeate the cell membrane. The SQ due to its lipidic nature has been efficient in the preparation of emulsions and conjugates for the release of drugs, with a prolonged effect on shelf life. The SQ can be prepared in emulsions, alone or as a secondary ingredient. Water-SQ emulsions with polysorbate 80 have been proposed for influenza vaccines and lecithin-squalene emulsions with tween 80 effective for the induction of antibodies [56].

Wang et al. [57] reported SQ emulsions with phosphatidylethanolamine or Pluronic® F68 that prolong the release of morphine and maintain analgesic activity in animal models *in vitro*. Other substances such as aluminum hydroxide, aluminum sulfate, and mineral oils have been used in previous decades for the preparation of vaccines, but have been inefficient in the action against the antigen, causing a variety of pathologies such as the formation of granulomas at the site of injection or tumor development [58, 59].

A promising emulsion, known as MF59, has been developed by the company Novartis®, which is formulated on oil in water (o/w) with SQ (4.3% dispersed phase), surfactant Span85, tween 80, and citrate in the continuous phase. This

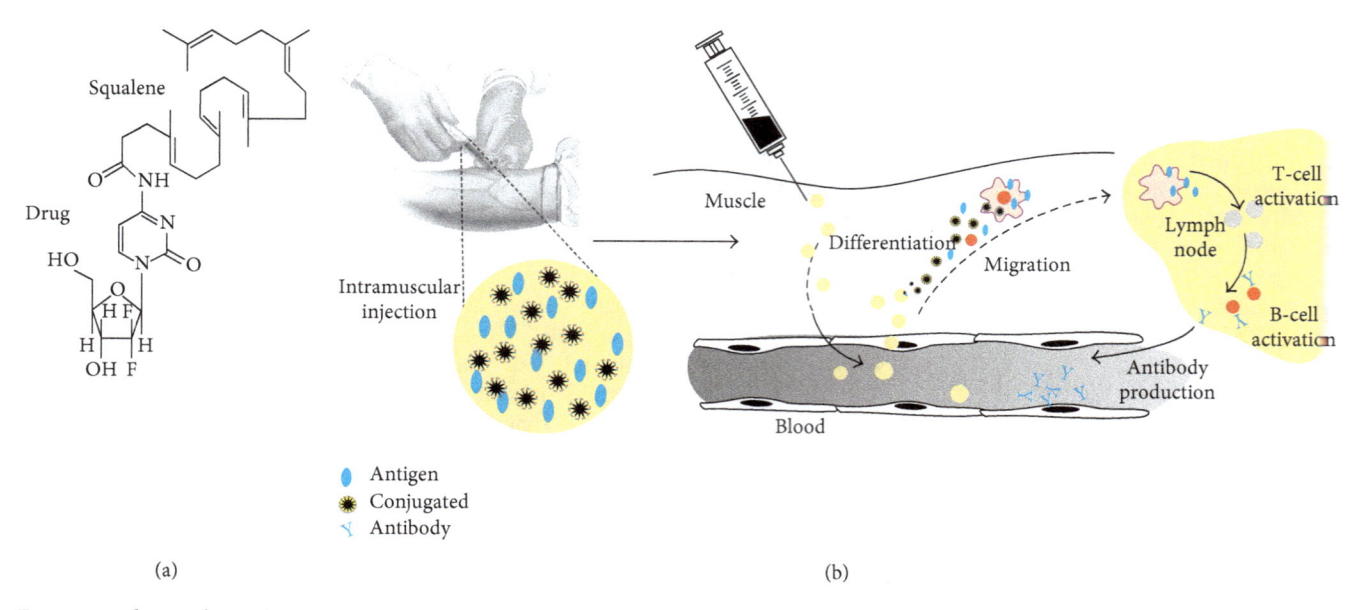

FIGURE 5: Chemical coupling of a conjugate with SQ (squalene-gemcitabine) (a) and immune response mode in human body of squalene vaccine (b), adapted from Seubert et al. [61].

emulsion has been developed as an aid and stimulant of the immune system. Its effectiveness has been proving in several vaccines such as malaria, hepatitis B, hepatitis C, herpes, cytomegalovirus, and even in HIV and the pandemic H1N1 virus [56, 60–62].

The mechanisms which SQ contributes to immune response are not yet clear; recent reports indicate that after injection, MF59 increases the immune response causing an influx of phagocytic cells in the vaccination site, allowing a more efficient transport of the antigen to the lymph nodes, and this improves the pre-activation of immune response, Figure 5 [61, 62].

Other advantages of SQ have been found in therapeutic emulsions to carry and supply poorly soluble drugs since they modify the biodistribution and reduce the toxicity, facilitating the targeting of the drug. These lipid conjugate formed by covalent bonds have gained importance in the market such as the case of docosahexaenoic acid conjugated with paclitaxel (Taxoprexin®) or cardiolipin conjugated with gemcitabine, which has been shown to improve the kinetics of the drug and increase the therapeutic index [63, 64].

The process of carrying the drug has been called *squalenylation*, a technique based on the property of SQ to protect (or coat) the anticancer and antiviral compounds to bring them into the cell and induce their cytotoxic activity [65].

Squalenylation has allowed the formation of nanoassemblies (100–300 nm) when assembled in water without the addition of surfactants. The anticancer effects of *squalenylation* have been demonstrated superior in in vitro human cancer cells and *in vivo* of murine cells with leukemia [65]; however, much remains to be investigated in *in vivo* models.

5.2. Skin Protection. Given the characteristics of SQ as a natural emollient, it is considered an important component in the formulation of cosmetics and moisturizing agents for skin protection. It is a compound of efficient absorption on the surface of the skin, restoring it without leaving oily residues.

It has been reported that SQ is used as a fixing for perfumes and the elaboration of lipsticks because it accelerates the dispersion of the dye and produces greater brightness. When SQ is applied to hair and skin exposed to the sun, it helps restore lost oils and easily forms emulsions with other lipophilic substances which allows not to oxidize quickly [9].

The SQ is a natural constituent of the skin, which has a moisturizing effect that counteracts the appearance of wrinkles and burns through the fixation of water molecules on the surface of the skin. This effect was demonstrated with a synthetic substance such as vernix caseosa (fatty material found in the skin of a newborn) composed of SQ in mixture with other lipids, cholesterol, triglycerides, ceramides, and fatty acids. Its application was successful and recommended as a cream against barrier effect to psoriasis [15].

Other conditions that can be treated with topical SQ applications are seborrheic dermatitis and acne, which controls the amount of unsaturated fatty acids in the skin reducing the condition [9].

The use of SQ in the pharmaceutical and cosmetic industry is broad, although more research is needed related to adverse effects in cutaneous applications.

6. Methods of Extraction of Squalene

The statistics of Global Market Insights for 2016 establish that the world production of SQ exceeded 5,900 tons with a commercial value of USD 111.9 million. By 2022, it is expected its value and production will have a significant increase (9%), attributed to the greater consumption of products with health benefits, cholesterol control medications, food supplements, as well as cosmetics and

TABLE 2: Content of fatty acids and squalene in different sources naturally extracted with various extraction methods.

Source	Fatty acids (% w/w)				MI (% p/p)	Squalene (mg/100 g oil)	Extraction method	Reference
	16:0	18:0	18:1	18:2				
Vegetable								
Olive	44.0	4.0	39.0	11.0	0.7	150–747	DI, DD, P	[67–69]
Amaranth	22.0–42.0	2.7–3.5	29.0	7.5–45.0	5.9	6000–8000	DI, ScCO$_2$	[70–73]
Seed of grape	6.2–8.2	3.6–5.2	12.7–18.4	67.5–73.2	0.2–0.3	2.7–14.1	DI, ScCO$_2$	[74–76]
Pistachio	—	—	—	—	—	1.1–2.2	DI	[77]
Walnuts, macadamia	5.8–8.3	2.7–3.4	65.1–79.3	2.3–10.3	<1.0	0.9–18.6	—	[74, 78]
Peanuts	11.1	2.6	38.4	44.6	—	9.8	DI	[78]
Maize oil	10.4	1.9	27.5	57.2	0.9	10–27	—	[79]
Sunflower oil	—	—	—	—	5.1–7.1	2.2–2.6	DD	[80]
Palm oil	35.0–38.0	10.0–11.0	48.0–50.0	—	1.9	0.1–1300	ScCO$_2$, DD	[81–83]
Soybean	10.5	3.9	23.3	53.0	0.4–7.7	1.2–180	DI, DD	[84]
Animal								
Shark (liver oil)	15.1–15.9	—	23.7–27.7	0.2–1.9	16.4	2300–8,400	DI	[85–87]
Yeast and fungi								
Saccharomyces cereviceae	—	—	—	—	—	40*	—	
Auranthiochytrium sp.	—	—	—	—	—	900–6940**	DI	[1, 29, 88]
Pseudozyma sp.	—	—	—	—	—	340.5**	—	

MI: unsaponifiable matter, ScCO$_2$: CO$_2$ extraction as supercritical fluid, DI: solvent extraction, DD: deodorization distillate, P: pressure extraction, (*) mg/g dry biomass, and (**) mg/L.

pharmaceutical products: moisturizing creams, lotions, lipsticks, bronzers, and conditioners for hair, among others.

An innovative product in recent years is the application of SQ as a coadjuvant carrier in vaccines, a patent of the world-renowned laboratory brand, which allows an area of opportunity for further expansion of the market. Europe and Asia are the most attractive markets, and Germany, France, the United Kingdom, Italy, China, and India are the countries with the highest demand for SQ [66].

The global demand for SQ is mainly covered by three sources of extraction: animal, vegetable, and synthetic methods. The SQ extracted from shark liver oil is the most appreciated for its high yield (up to 40% of the weight of the organ); however, the extraction of plant sources is becoming increasingly important, given the protection of marine species in danger of extinction and the release of prohibitive norms on the extraction of SQ from marine species. Even transnational food and cosmetics companies (L'Oreal® and Unilever®) have declared their products are free of SQ from marine animals; this has increased the demand for SQ from other sources.

Biotechnology has developed techniques for the industrial production of SQ, and however, the yields obtained from *Saccharomyces cerevisiae*, *Botrycoccus braunii*, *Aurantiochytrium* sp., and *E. coli* are lower than those from plant sources (5–15 mg/g dry matter, 4.1–340.5 mg/L) and only reach less than 10% of the world [1].

For this reason, they have dedicated themselves to researching potential sources of plant origin, among which are *Olea europea*, *Amaranthus* sp., *Glycine max*, and *Zea mays*. The concentration of SQ varies depending on the species, harvest season, postharvest conditions, the extraction method, physicochemical treatments to the extract is subjected, and the removal of minor compounds from the oil [3].

In vegetable sources, the yield of squalane varies depending on the extraction method. Table 2 shows that amaranth has the highest content followed by olive, walnut, and other seeds.

Oils are traditionally extracted by mechanical pressure methods or organic solvents, but require refining processes to eliminate undesirable compounds, pigments, free fatty acids, phospholipids, and so on. These processes reduce the yield of SQ to less than 80%; for example, it can decrease from 13 to 7% just by a discoloration process. However, the extraction process by mechanical pressure is still preferred because it produces little modification of the compounds compared with other processes which use physicochemical treatments [3].

The solvent method shows good performance in the extraction of vegetable oils, up to 98%; however, it is difficult to obtain high-purity SQ because of the low concentrations present in the unsaponifiable fraction. Czaplicki et al. [89] reported amaranth SQ concentrations of 5740, 6000, and 6500 mg/100 g of oil, extracted by cold pressure, solvent, and supercritical fluid, respectively. The difference is attributed to the extraction conditions and low sterol content, which are affected when the cell wall is broken during the extraction process.

Other investigations establish supercritical fluid extraction (ScCO$_2$) as a suitable method for oil processing, which facilitates separation of squalene at low temperatures without leaving traces of organic solvents. Generally, ScCO$_2$ is a recommended technique for nonpolar compound extraction with molecular weights less than 500 g/mol soluble in CO$_2$, such as SQ. The extraction time reduces considerably; Krulj et al. [90] found differences between the solvent and ScCO$_2$ method using petroleum ether for the squalene extraction from three amaranth genotypes. The yield obtained was 450 and 350 mg/100 g of oil, respectively. The variability of squalene performance depended on extraction conditions and genotype as well.

He et al. [91] found SQ yields extracted by $ScCO_2$ of 4770 mg/100 g oil based on temperature, pressure, and moisture content of the sample (>10%) as determining factors for extraction efficiency.

Wejnerowska et al. [70] also reported SQ yields in amaranth by $ScCO_2$ extraction, of 1200–9000 mg/100 g oil, the best conditions depending on particle size (0.08–0.5 mm), pressure (20 MPa), temperature (130°C), and time (30 min). They obtained SQ extract with 60% purity while higher purity (>90%) was obtained at longer extraction times (120 min).

The disadvantage of $ScCO_2$ is the high cost of equipment, the complexity of the operating parameters (temperature-pressure), and in some cases, the operating times can be very long. Other techniques such as direct distillation have been tried, although this method is not recommended for SQ because it is a thermolabile compound [3].

Another process to favor SQ extraction is the distillate of deodorization stage during oil refining. Squalene content increases by 15–30% because it produces higher unsaponifiable fraction (sterols) and other components. However, considerable care must be taken with temperature, pressure, and residence time because it influences the quality of the distillate. Soybean and sunflower oil deodorization distillates are by-products most appreciated for the high quality of SQ and tocopherols.

Pramparo et al. [80] reported the best SQ extraction conditions of soybean and sunflower deodorization distillate; they obtained at 140°C the largest amount of distillate and SQ. They observed at higher evaporation temperature greater amount volatile fatty acids were associated to increase in sterols, tocopherols, and SQ.

The composition of unsaponifiable fraction is a factor influencing the squalene yield (0.5–2.0% of the total weight). Olive oil studies have showed that the SQ can represent up to 50% of unsaponifiable fraction, depending on the extraction method and oil refining process. SQ extracted from olive oil yields 5.1–9.6 g of SQ per liter of oil with ~75% purity, but when SQ is refined, the content reduces significantly [92–94].

Table 2 shows the yields of SQ obtained for vegetable species, in which yield recorded for olive and amaranth varied, depending on the extraction methods, the method of quantification, and other factors such as the time of harvest, variety, and even geographical area [67].

Other species, such as rice, wheat, maize, and grapes, contain SQ in a low proportion (<1.0%) therefore not considered potential crops. Amaranth is widely reported with highest SQ content, more than any other plant species, but only two cultivated species are reported where there are more than 75 species worldwide [91].

6.1. Other Methods of SQ Production.
Currently, advances in biotechnology have developed alternatives for commercial production of SQ. In 1920, when the biosynthesis of cholesterol was discovered, it was suspected that SQ could be produced via isoprene, and after the 90s with biotechnology tools, SQ was produced via modified genetic material [95].

Laboratories (Amyris Biotechnologies, Arista Industries Inc., and Nucelis LLC.) have developed patents for obtaining SQ, manufacturing terpene from fermentable sugars using common yeast. This technology mimics natural processes to provide β-farnesene (precursor of squalene), removes the yeast, and produces a simple chemical coupling which avoids the need for isolation of lipophilic and oxidatively unstable squalene from the fermentation biomass. Hydrogenation and purification technologies are used in the final stage to get high-purity squalene [96].

Even though the method is reproducible, the manufacturing process must be strictly controlled to ensure quality. Nevertheless, surveys in marketing ensure preference towards natural products, mainly the European market.

6.2. Methods of Quantification.
The SQ as raw material for the cosmetics and medicinal industry thus requires quality and purity, which depends on the extraction and analytical methods. The most common methods of analysis include gas chromatography (GC) and high-resolution liquid chromatography (HPLC) [5].

These analyses propose preview processes of separation that include saponification of sample, separation of the unsaponifiable fraction, and fractionation by chromatography. These methods usually are laborious, causing modification of SQ during saponification and give false results in the chromatography analysis [97].

Other sample preparation techniques include methylation with KOH in MeOH (2 N) and washed with ethanol : water, in minutes, getting the sample ready to be analyzed by chromatography avoiding interference from minority components (methyl esters of triglycerides). This method results in a real concentration of SQ it has been tested in olive oils, sunflower, soybean, maize, rapeseed, and peanut [97].

Among the tools for separating, identifying, and quantifying components include spectroscopy of masses coupled to gas chromatography (GC-MS), flame ionization (GC-FID), and spectroscopy of masses coupled to liquid chromatography (LC-MS); however, these are not specific for separating SQ [76, 97].

For this purpose, is required a combination of high resolution and sensitivity given by different detectors such as photoionization (UV), infrared (IR), or photodiodes (DAD). However, direct quantification is difficult, so it requires pretreat samples to remove interference [98].

An efficient method was reported by Salvo et al. [77], where squalene concentration was analyzed in olive oil by UPLC/PDA chromatography. The sample was pretreated with hexane, purified on column SPE, and analyzed by UPLC; the method was efficient in the yield.

On the process of extraction and analysis, isomerization potential and squalene oxidation should be considered. Few works have been developed identifying the oxidation products derived from the extraction process. Mountfort et al. [99] reported oxidation products from squalene by LC/APCI-MS locating the epoxy formation and hydroperoxides of SQ.

Among the industrial techniques for calculating SQ is the oil refraction index, which is a quick and economical method, but least precise since it is temperature-sensitive and does not provide information on oil composition. Another recent technology for SQ determination is Raman spectrometry, which is based on density calculation of unusual spectra in band intensity at 1670 cm^{-1} due to an intensity accumulated by the symmetry of six double bonds of squalene structure. This analytical technique results useful in olive oil, facilitating the study of squalene characteristics even at concentrations less than 1% in lipid material [87].

7. Conclusions

The SQ is an important compound for human health; from its discovery to date, the biosynthesis routes, biological activity, and different extraction methods have been described.

The most important biological effects of squalene stand out as a cancer inhibitor antitumor and antioxidant agent in the skin; that is why its applications in pharmaceutical and cosmetic industries will demand it more in coming decades. However, there is a lack of aspects that require further attention, such as the adverse effect on human consumption, the mechanisms of release, assimilation, and the mode of action against carcinogenesis on the skin.

Nevertheless, the next decade squalene will be a compound of the wide application as an adjuvant in vaccines, which has opened a new unprecedented landscape in the pharmaceutical industry. It is, therefore, necessary to study other potential sources of plants, innovative techniques that guarantee quality and yield but above all the development of commercial-scale crops that guarantee the production of SQ to meet the global demand.

Acknowledgments

The authors acknowledge support of CONACYT with 258081 scholarship granted to M. Azalia Lozano-Grande and Instituto Politécnico Nacional for financial support.

References

[1] G. P. Ghimire, H. T. Nguyen, N. Koirala, and J. K. Sohng, "Advances in biochemistry and microbial production of squalene and its derivatives," *Journal of Microbiology and Biotechnology*, vol. 26, no. 3, pp. 441–451, 2016.

[2] M. Rohmer, M. Seemann, S. Horbach, S. Bringer-Meyer, and H. Sahm, "Glyceraldehyde 3-phosphate and pyruvate as precursors of isoprenic units in an alternative non-mevalonate pathway for terpenoid biosynthesis," *Journal of the American Chemical Society*, vol. 118, no. 11, pp. 2564–2566, 1996.

[3] O. Popa, N. E. Băbeanu, I. Popa, S. Niţă, and C. E. Dinu-Pârvu, "Methods for obtaining and determination of squalene from natural sources," *BioMed Research International*, vol. 2015, Article ID 367202, 16 pages, 2015.

[4] D. M. Pham, B. Boussouira, D. Moyal, and Q. L. Nguyen, "Oxidization of squalene, a human skin lipid: a new and reliable marker of environmental pollution studies," *International Journal of Cosmetic Science*, vol. 37, no. 4, pp. 357–365, 2015.

[5] A. Andrieş, I. Popa, N. Elena Băbeanu, S. Niţă, and O. Popa, "Squalene- natural resources and applications," *Farmacia*, vol. 62, no. 5, pp. 840–862, 2014.

[6] G. M. Turchini, W. K. Ng, and D. R. Tocher, *Fish Oil Replacement and Alternative Lipid Sources in Aquaculture Feeds*, CRC Press, Boca Raton, FL, USA, 2011.

[7] V. Kostyuk, A. Potapovich, A. Stancato et al., "Photo-oxidation products of skin surface squalene mediate metabolic and inflammatory responses to solar UV in human keratinocytes," *PLoS One*, vol. 7, no. 8, Article ID e44472, 2012.

[8] F. Esra Güneş, M. Üniversitesi Sağlık Bilimleri Fakültesi, B. Diyetetik Bölümü, and İ. Türkiye, "Medical use of squalene as a natural antioxidant derleme/review," *Journal of Marmara University Institute of Health Sciences*, vol. 33, no. 44, pp. 220–228, 2013.

[9] K. Wołosik, M. Knaś, A. Zalewska, and M. Niczyporuk, "The importance and perspective of plant-based squalene in cosmetology," *Journal of Cosmetic Science*, vol. 64, no. 1, pp. 19–65, 2013.

[10] L. H. Reddy and P. Couvreur, "Squalene: a natural triterpene for use in disease management and therapy," *Advanced Drug Delivery Reviews*, vol. 61, no. 15, pp. 1412–1426, 2009.

[11] H. L. Newmark, "Squalene, olive oil, and and cancer risk: a review and hypoyhesis," *Carcinogenesis*, vol. 6, no. 12, pp. 1101–1103, 1997.

[12] S. Lopez, B. Bermudez, S. Montserrat-de la Paz et al., "Membrane composition and dynamics: a target of bioactive virgin olive oil constituents," *Biochimica et Biophysica Acta-Biomembranes*, vol. 1838, no. 6, pp. 1638–1656, 2014.

[13] J. Prades, S. S. Funari, P. V. Escribá, and F. Barceló, "Effects of unsaturated fatty acids and triacylglycerols on phosphatidylethanolamine membrane structure," *Journal of Lipid Research*, vol. 44, no. 9, pp. 1720–1727, 2003.

[14] L. Xu and N. A. Porter, "Free radical oxidation of cholesterol and its precursors: implications in cholesterol biosynthesis disorders," *Free Radical Research*, vol. 49, no. 7, pp. 835–849, 2015.

[15] Z. R. Huang, Y. K. Lin, and J. Y. Fang, "Biological and pharmacological activities of squalene and related compounds: Potential uses in cosmetic dermatology," *Molecules*, vol. 14, no. 1, pp. 540–554, 2009.

[16] H. Yin, E. Niki, and K. Uchida, "Special issue on "recent progress in lipid peroxidation based on novel approaches"," *Free Radical Research*, vol. 49, no. 7, pp. 813–815, 2015.

[17] E. Naziri, R. Consonni, and M. Z. Tsimidou, "Squalene oxidation products: monitoring the formation, characterisation and pro-oxidant activity," *European Journal of Lipid Science and Technology*, vol. 116, no. 10, pp. 1400–1411, 2014.

[18] E. Psomiadou and M. Tsimidou, "On the role of squalene in olive oil stability," *Journal of Agricultural and Food Chemistry*, vol. 47, no. 10, pp. 4025–4032, 1999.

[19] P. V. Escribá, J. M. González-Ros, F. M. Goñi et al., "Membranes: a meeting point for lipids, proteins and therapies: translational medicine," *Journal of Cellular and Molecular Medicine*, vol. 12, no. 3, pp. 829–875, 2008.

[20] T. Hauß, S. Dante, N. A. Dencher, and T. H. Haines, "Squalane is in the midplane of the lipid bilayer: implications

for its function as a proton permeability barrier," *Biochimica et Biophysica Acta-Bioenergetics*, vol. 1556, no. 2-3, pp. 149-154, 2002.

[21] M. Spanova, D. Zweytick, K. Lohner et al., "Influence of squalene on lipid particle/droplet and membrane organization in the yeast *Saccharomyces cerevisiae*," *Biochimica et Biophysica Acta*, vol. 1821, no. 4, pp. 647-653, 2012.

[22] T. Moses, K. K. Papadopoulou, and A. Osbourn, "Metabolic and functional diversity of saponins, biosynthetic intermediates and semi-synthetic derivatives," *Critical Reviews in Biochemistry and Molecular Biology*, vol. 49, no. 6, pp. 439-462, 2014.

[23] J. M. Augustin, V. Kuzina, S. B. Andersen, and S. Bak, "Molecular activities, biosynthesis and evolution of triterpenoid saponins," *Phytochemistry*, vol. 72, no. 6, pp. 435-457, 2011.

[24] H. T. M. Nguyen, A. K. Neelakadan, T. N. Quach et al., "Molecular characterization of Glycine max squalene synthase genes in seed phytosterol biosynthesis," *Plant Physiology and Biochemistry*, vol. 73, pp. 23-32, 2013.

[25] T. A. Woyengo, V. R. Ramprasath, and P. J. H. Jones, "Anticancer effects of phytosterols," *European Journal of Clinical Nutrition*, vol. 63, no. 7, pp. 813-820, 2009.

[26] H. Schaller, "The role of sterols in plant growth and development," *Progress in Lipid Research*, vol. 42, no. 3, pp. 163-175, 2003.

[27] M. Spanova and G. Daum, "Squalene-biochemistry, molecular biology, process biotechnology, and applications," *European Journal of Lipid Science and Technology*, vol. 113, no. 11, pp. 1299-1320, 2011.

[28] R. Xu, G. C. Fazio, and S. P. T. Matsuda, "On the origins of triterpenoid skeletal diversity," *Phytochemistry*, vol. 65, no. 3, pp. 261-291, 2004.

[29] W. Xu, X. Ma, and Y. Wang, "Production of squalene by microbes: an update," *World Journal of Microbiology and Biotechnology*, vol. 32, no. 12, 2016.

[30] Y.-S. Chi and S. Kay Obendorf, "Aging of oily soils on textiles. Chemical changes upon oxidation and interaction with textile fibers," *Journal of Surfactants and Detergents*, vol. 1, no. 3, pp. 371-380, 1998.

[31] M. Rohmer, M. Knani, P. Simonin, B. Sutter, and H. Sahm, "Isoprenoid biosynthesis in bacteria: a novel pathway for the early steps leading to isopentenyl diphosphate.," *Biochemical Journal*, vol. 295, pp. 517-524, 1993.

[32] J.-J. Pan, J.-O. Solbiati, G. Ramamoorthy et al., "Biosynthesis of squalene from farnesyl diphosphate in bacteria: three steps catalyzed by three enzymes," *ACS Central Science*, vol. 1, no. 2, pp. 77-82, 2015.

[33] H. Relas, H. Gylling, and T. Miettinen, "Fate of intravenously administered squalene and plant sterols in human subjects," *Journal of Lipid Research*, vol. 42, no. 6, pp. 988-994, 2001.

[34] P. Chan, B. Tomlinson, C. B. Lee, and Y. S. Lee, "Effectiveness and safety of low-dose pravastatin and squalene, alone and in combination, in elderly patients with hypercholesterolemia," *Journal of Clinical Pharmacology*, vol. 36, no. 5, pp. 422-427, 1996.

[35] K. H. S. Farvin, R. Anandan, S. Hari Senthil Kumar et al., "Cardioprotective effect of squalene on lipid profile in isoprenaline-induced myocardial infarction in rats," *Journal of Medicinal Food*, vol. 9, no. 4, pp. 531-536, 2006.

[36] Y. Liu, X. Xu, D. Bi et al., "Influence of squalene feeding on plasma leptin, testosterone & blood pressure in rats," *Indian Journal of Medical Research*, vol. 129, no. 2, pp. 150-153, 2009.

[37] S. Ekanayake Mudiyanselage, M. Hamburger, P. Elsner, and J. J. Thiele, "Ultraviolet A induces generation of squalene monohydroperoxide isomers in human sebum and skin surface lipids in vitro and in vivo," *Journal of Investigative Dermatology*, vol. 120, no. 6, pp. 915-922, 2003.

[38] R. V. Tikekar, R. D. Ludescher, and M. V. Karwe, "Processing stability of squalene in amaranth and antioxidant potential of amaranth extract," *Journal of Agricultural and Food Chemistry*, vol. 56, pp. 10675-10678, 2008.

[39] Y. Kohno, Y. Egawa, S. Itoh, S.-i Nagaoka, M. Takahashi, and K. Mukai, "Kinetic study of quenching reaction of singlet oxygen and scavenging reaction of free radical by squalene in n-butanol," *Biochimica et Biophysica Acta (BBA)/Lipids and Lipid Metabolism*, vol. 1256, no. 1, pp. 52-56, 1995.

[40] R. W. Owen, A. Giacosa, W. E. Hullc et al., "Olive-oil consumption and health: the possible role of antioxidants," *The Lancet Oncology*, vol. 1, no. 2, pp. 107-112, 2000.

[41] R. W. Owen, W. Mier, A. Giacosa, W. E. Hull, B. Spiegelhalder, and H. Bartsch, "Phenolic compounds and squalene in olive oils: the concentration and antioxidant potential of total phenols, simple phenols, secoiridoids, lignansand squalene," *Food and Chemical Toxicology*, vol. 38, no. 8, pp. 647-659, 2000.

[42] K. Hrncirik and S. Fritsche, "Relation between the endogenous antioxidant system and the quality of extra virgin olive oil under accelerated storage conditions," *Journal of Agricultural and Food Chemistry*, vol. 53, no. 6, pp. 2103-2110, 2005.

[43] Ryzard Amarowicz, "Squalene: a natural antioxidant?," *European Journal of Lipid Science and Technology*, vol. 111, no. 5, pp. 411-412, 2009.

[44] G. S. Kelly, "Squalene and its potential clinical uses," *Alternative Medicine Review*, vol. 44, no. 11, pp. 29-36, 1999

[45] T. Tochio, H. Tanaka, S. Nakata, and H. Ikeno, "Accumulation of lipid peroxide in the content of comedones may be involved in the progression of comedogenesis and inflammatory changes in comedones," *Journal of Cosmetic Dermatology*, vol. 8, no. 2, pp. 152-158, 2009.

[46] D. Saint-Leger, A. Bague, E. Cohen, and M. Lchivot "A possible role for squalene in the pathogenesis of acne. I. In vitro study of squalene oxidation," *British Journal of Dermatology*, vol. 114, no. 5, pp. 535-542, 1986.

[47] A. L. Ronco, E. De Stéfani, and A. Ronco, "Squalene: a multitask link in the crossroads of cancer and aging," *Functional Foods in Health and Disease*, vol. 3, no. 12, pp. 462-476, 2013.

[48] H. Kamimura, N. Koga, K. Oguri, and H. Yoshimura, "Enhanced elimination of theophylline, phenobarbital and strychnine from the bodies of rats and mice by squalane treatment," *Journal of pharmacobio-dynamics*, vol. 15, no. 5, pp. 215-221, 1992.

[49] E. Richter, B. Fichtl, and S. G. Schäfer, "Effects of dietary paraffin, squalane and sucrose polyester on residue disposition and elimination of hexachlorobenzene in rats," *Chemico-Biological Interactions*, vol. 40, no. 3, pp. 335-344, 1982.

[50] I. A. E. Queirolo and M. G. Morales, "Plombemia y squalene," *Tendencias, Actualidad en Pediatría*, vol. 10, pp. 131-137, 2007.

[51] T. J. Smith, "Squalene: potential chemopreventive agent," *Expert Opinion on Investigational Drugs*, vol. 9, no. 8, pp. 1841-1848, 2000.

[52] C. V. Rao, H. L. Newmark, and B. S. Reddy, "Chemopreventive effect of squalene on colon cancer," *Carcinogenesis*, vol. 19, no. 2, pp. 287-290, 1998.

[53] H. L. Newmark, "Squalene, olive oil, and cancer risk: a review and hypothesis," *Cancer Epidemiology, Biomarkers & Prevention*, vol. 6, no. 12, pp. 1101–1103, 1997.

[54] F. Warleta, M. Campos, Y. Allouche et al., "Squalene protects against oxidative DNA damage in MCF10A human mammary epithelial cells but not in MCF7 and MDA-MB-231 human breast cancer cells," *Food and Chemical Toxicology*, vol. 48, no. 4, pp. 1092–1100, 2010.

[55] R. W. Owen, R. Haubner, G. Würtele, W. E. Hull, B. Spiegelhalder, and H. Bartsch, "Olives and olive oil in cancer prevention," *European Journal of Cancer Prevention*, vol. 13, no. 4, pp. 319–326, 2004.

[56] S. K. Kim and F. Karadeniz, "Biological importance and applications of squalene and squalane," *Advances in Food and Nutrition Research*, vol. 65, pp. 223–233, 2012.

[57] J. J. Wang, K. C. Sung, C. H. Yeh, and J. Y. Fang, "The delivery and antinociceptive effects of morphine and its ester prodrugs from lipid emulsions," *International Journal of Pharmaceutics*, vol. 353, no. 1-2, pp. 95–104, 2008.

[58] L. Reddy and P. Couvreur, "Novel approaches to deliver gemcitabine to cancers," *Current Pharmaceutical Design*, vol. 14, no. 11, pp. 1124–1137, 2008.

[59] L. H. Reddy, C. Dubernet, S. L. Mouelhi, P. E. Marque, D. Desmaele, and P. Couvreur, "A new nanomedicine of gemcitabine displays enhanced anticancer activity in sensitive and resistant leukemia types," *Journal of Controlled Release*, vol. 124, no. 1-2, pp. 20–27, 2007.

[60] G. Lippi, G. Targher, and M. Franchini, "Vaccination, squalene and anti-squalene antibodies: facts or fiction?," *European Journal of Internal Medicine*, vol. 21, no. 2, pp. 70–73, 2010.

[61] A. Seubert, E. Monaci, M. Pizza, D. T. O'Hagan, and A. Wack, "The adjuvants aluminum hydroxide and MF59 induce monocyte and granulocyte chemoattractants and enhance monocyte differentiation toward dendritic cells," *Journal of Immunology*, vol. 180, no. 8, pp. 5402–5412, 2008.

[62] D. T. O'Hagan, G. S. Ott, and G. Van Nest, "Recent advances in vaccine adjuvants: the development of MF59 emulsion and polymeric microparticles," *Molecular Medicine Today*, vol. 3, no. 2, pp. 69–75, 1997.

[63] P. Chen, P.-Y Chien, A. R. Khan et al., "In-vitro and in-vivo anti-cancer activity of a novel gemcitabine-cardiolipin conjugate.," *Anti-Cancer Drugs*, vol. 17, no. 1, pp. 53–61, 2006.

[64] M. Payne, P. Ellis, D. Dunlop et al., "DHA-paclitaxel (Taxoprexin) as first-line treatment in patients with stage IIIB or IV non-small cell lung cancer: report of a phase II open-label multicenter trial.," *Journal of Thoracic Oncology*, vol. 1, no. 9, pp. 984–990, 2006.

[65] P. Couvreur, B. Stella, L. Harivardhan Reddy et al., "Squalenoyl nanomedicines as potential therapeutics," *Nano Letters*, vol. 6, no. 11, pp. 2544–2548, 2006.

[66] F. Dent and S. Clarke, "State of the global market for shark products," *FAO Fishereis and Aquaculture Technical Paper*, vol. 590, p. 187, 2015.

[67] A. Fernandez-Cuesta, L. Leon, L. Velasco, and R. De la Rosa, "Changes in squalene and sterols associated with olive maturation," *Food Research International*, vol. 54, no. 2, pp. 1885–1889, 2013.

[68] A. Salvo, G. L. La Torre, A. Rotondo et al., "Determination of squalene in organic extra virgin olive oils (EVOOs) by UPLC/PDA using a single-step SPE sample preparation," *Food Analytical Methods*, vol. 10, no. 5, pp. 1377–1385, 2017.

[69] E. Naziri, S. B. Glisic, F. T. Mantzouridou, M. Z. Tsimidou, V. Nedovic, and B. Bugarski, "Advantages of supercritical fluid extraction for recovery of squalene from wine lees," *Journal of Supercritical Fluids*, vol. 107, pp. 560–565, 2016.

[70] G. Wejnerowska, P. Heinrich, and J. Gaca, "Separation of squalene and oil from *Amaranthus* seeds by supercritical carbon dioxide," *Separation and Purification Technology*, vol. 110, pp. 39–43, 2013.

[71] E. Naziri, F. Mantzouridou, and M. Z. Tsimidou, "Squalene resources and uses point to the potential of biotechnology," *Lipid Technology*, vol. 23, no. 12, pp. 270–273, 2011.

[72] H. P. He and H. Corke, "Oil and squalene in *Amaranthus* grain and leaf," *Journal of Agricultural and Food Chemistry*, vol. 51, no. 27, pp. 7913–7920, 2003.

[73] J. Ortega, M. Martínez Zavala, M. Hernández, and J. Díaz Reyes, "Analysis of trans fatty acids production and squalene variation during amaranth oil extraction," *Open Chemistry*, vol. 10, no. 6, pp. 1773–1778, 2012.

[74] M. M. Wall, "Functional lipid characteristics, oxidative stability, and antioxidant activity of macadamia nut (*Macadamia integrifolia*) cultivars," *Food Chemistry*, vol. 121, no. 4, pp. 1103–1108, 2010.

[75] X. Wen, M. Zhu, R. Hu et al., "Characterisation of seed oils from different grape cultivars grown in China," *Journal of Food Science and Technology*, vol. 53, no. 7, pp. 3129–3136, 2016.

[76] T. H. J. Beveridge, B. Girard, T. Kopp, and J. C. G. Drover, "Yield and composition of grape seed oils extracted by supercritical carbon dioxide and petroleum ether: varietal effects," *Journal of Agricultural and Food Chemistry*, vol. 53, no. 5, pp. 1799–1804, 2005.

[77] A. Salvo, G. L. La Torre, V. Di Stefano et al., "Fast UPLC/PDA determination of squalene in Sicilian P.D.O. pistachio from Bronte: optimization of oil extraction method and analytical characterization," *Food Chemistry*, vol. 221, pp. 1631–1636, 2017.

[78] L. S. Maguire, S. M. O'sullivan, K. Galvin, T. P. O'connor, and N. M. O'brien, "Fatty acid profile, tocopherol, squalene and phytosterol content of walnuts, almonds, peanuts, hazelnuts and the macadamia nut," *International Journal of Food Sciences and Nutrition*, vol. 55, no. 3, pp. 171–178, 2004.

[79] Y. Yamamoto and S. Hara, "Novel fractionation method for squalene and phytosterols contained in the deodorization distillate of rice bran oil," *Scientific, Health and Social Aspects of the Food Industry Benjamin Valdez: IntechOpen*.

[80] M. Pramparo, S. Prizzon, M. A. Martinello, and M. A. Martinello, "Study of purification of fatty acids, tocopherols and sterols from deodorization distillate.," *Grasas y Aceites*, vol. 56, no. 3, pp. 228–234, 2005.

[81] S. Norhidayah, B. M. Hamed, and I. S. M. Zaidul, "Squalene recovery from palm fatty acid distillate using supercritical fluid extraction," *International Food Research Journal*, vol. 19, no. 4, pp. 1661–1667, 2012.

[82] H. L. N. Lau, C. W. Puah, Y. M. Choo, A. N. Ma, and C. H. Chuah, "Simultaneous quantification of free fatty acids, free sterols, squalene, and acylglycerol molecular species in palm oil by high-temperature gas chromatography-flame ionization detection," *Lipids*, vol. 40, no. 5, pp. 523–528, 2005.

[83] S. H. Goh and K. Lumpur, "Minor constituents of palm oil," *Journal of the American Oil Chemists' Society*, vol. 62, no. 2, pp. 237–240, 1985.

[84] S. Gunawan, N. S. Kasim, and Y. H. Ju, "Separation and purification of squalene from soybean oil deodorizer distillate," *Separation and Purification Technology*, vol. 60, no. 2, pp. 128–135, 2008.

[85] Sunarya, M. Hole, and K. D. A. Taylor, "Methods of extraction composition and stability of vitamin A and other components in dogfish (*Squalus acanthias*) liver oil," *Food Chemistry*, vol. 55, no. 3, pp. 215-220, 1996.

[86] M. Tsujimoto, "Squalene: a highly unsaturated hydrocarbon in shark liver oil," *Industrial and Engineering Chemistry*, vol. 12, no. 1, pp. 63-72, 1920.

[87] D. W. Hall, S. N. Marshall, K. C. Gordon, and D. P. Killeen, "Rapid quantitative determination of squalene in shark liver oils by Raman and IR spectroscopy," *Lipids*, vol. 51, no. 1, pp. 139-147, 2016.

[88] A. Nakazawa, Y. Kokubun, H. Matsuura et al., "TLC screening of thraustochytrid strains for squalene production," *Journal of Applied Phycology*, vol. 26, no. 1, pp. 29-41, 2014.

[89] S. Czaplicki, D. Ogrodowska, R. Zadernowski, and D. Derewiaka, "Characteristics of biologically-active substances of amaranth oil obtained by various techniques," *Polish Journal of Food and Nutrition Sciences*, vol. 62, no. 4, pp. 235-239, 2012.

[90] J. Krulj, T. Brlek, L. Pezo et al., "Extraction methods of *Amaranthus* sp. grain oil isolation," *Journal of the science of food and agriculture*, vol. 96, no. 10, pp. 3552-3558, 2016.

[91] H. P. He, Y. Cai, M. Sun, and H. Corke, "Extraction and purification of squalene from *Amaranthus* grain," *Journal of Agricultural and Food Chemistry*, vol. 50, no. 2, pp. 368-372, 2002.

[92] D. Boskou, *Olive Oil: Minor Constitutents and Health*, CRC Press, Boca Raton, FL, USA, 2008.

[93] C. Nergiz and D. Çelikkale, "The effect of consecutive steps of refining on squalene content of vegetable oils," *Journal of Food Science and Technology*, vol. 48, no. 3, pp. 382-385, 2011.

[94] M. Murkovic, S. Lechner, A. Pietzka, M. Bratacos, and E. Katzogiannos, "Analysis of minor components in olive oil," *Journal of Biochemical and Biophysical Methods*, vol. 61, no. 1, pp. 155-160, 2004.

[95] R. Steiner, M. Drescher, M. Bonakdar, and W. Johannisbauer, "Method for producing squalene," US Patent US20040015033A1, 2004.

[96] D. Mcphee, A. Pin, L. Kizer, and L. Perelman, "Squalane from sugarcane," *Cosmetics & Toiletries magazine*, vol. 129, no. 6, pp. 1-6, 2014.

[97] A. Lanzón, A. Guinda, T. Albi, and C. De la Osa, "Método rápido para la determinación de escualeno en aceites vegetales," *Grasas y Aceites*, vol. 46, no. 4-5, pp. 276-278, 1995.

[98] J. A. Pérez-Serradilla, M. C. Ortiz, L. Sarabia, and M. D. L. De Castro, "Focused microwave-assisted Soxhlet extraction of acorn oil for determination of the fatty acid profile by GC-MS. Comparison with conventional and standard methods," *Analytical and Bioanalytical Chemistry*, vol. 388, no. 2, pp. 451-462, 2007.

[99] K. A. Mountfort, H. Bronstein, N. Archer, and S. M. Jickells, "Identification of oxidation products of squalene in solution and in latent fingerprints by ESI-MS and LC/APCI-MS," *Analytical Chemistry*, vol. 79, no. 7, pp. 2650-2657, 2007.

Microorganisms in Soils of Bovine Production Systems in Tropical Lowlands and Tropical Highlands in the Department of Antioquia, Colombia

Licet Paola Molina-Guzmán,[1,2] Paula Andrea Henao-Jaramillo,[1] Lina Andrea Gutiérrez-Builes,[1] and Leonardo Alberto Ríos-Osorio[2]

[1]Grupo Biología de Sistemas, Facultad de Medicina, Universidad Pontificia Bolivariana, Calle 78B No. 72A-109, Medellín, Colombia
[2]Grupo de Investigación Salud y Sostenibilidad, Escuela de Microbiología, Universidad de Antioquia, Calle 67 No. 53-108, Medellín, Colombia

Correspondence should be addressed to Leonardo Alberto Ríos-Osorio; leonardo.rios@udea.edu.co

Academic Editor: Glaciela Kaschuk

Studies on the physical and chemical effects of extensive grazing on soils have been performed in Colombia, but the effects of dairy cattle rearing on the biological properties of soils are not well known. The objective of this study was to evaluate microorganisms in 48 soils from livestock farms in the highland and lowland tropics in the Northern and Magdalena Medio subregions of the Department of Antioquia (Colombia). Principal component analysis demonstrated differences in the edaphic compositions of the soils, with increased percentages of root colonization by arbuscular mycorrhizal fungi and the density of microorganisms in farms that have soils with moderate phosphorus and nitrogen contents, low potassium content, and a moderately acidic pH. Agglomerative cluster analysis showed two groups for the highland tropic soils and six groups for the lowland tropic soils based on their population densities and interactions with the studied parameters. These results represent a first attempt to describe the density of microorganisms and the effect of soil physicochemical parameters on colonization by arbuscular mycorrhizal fungi in areas with determinant agroecological conditions, microbial functional diversity, and the presence of mycorrhizal fungi in livestock farm soils in Colombia.

1. Introduction

Livestock production is the most widespread economic activity worldwide, with one-third of terrestrial land destined for pasture and animal feed cultivation, and uses approximately eight percent of the global freshwater. One of the largest environmental impacts of this activity is the emission of greenhouse gas. An estimated 7.1 billion tonnes of carbon dioxide-equivalent were generated in 2005 by livestock production activities, representing approximately 14 percent of the total human emissions that year contributing to erosion caused by wind and water. This process is associated with compaction and increased apparent density of soils, changes in humidity and pH depending on the grazing intensity, loss and mineralization of nutrients, and increased decomposition of organic matter due to the breakdown of soil aggregates by trampling [1].

The effects of grazing on soils are reflected in their physical, chemical, and biological properties since this activity contributes to changes in the enzymatic activity and the microflora and macroflora [1]. Microorganisms that are part of the soil microflora are essential to its ecosystemic function since they are involved in processes that regulate biogeochemical cycles, make nutrients available for plant growth, control pathogens, improve the soil structure, and decompose organic matter [2].

Previous studies reported that grazing might also have positive effects on soil species richness, because the decay of animal faeces and urine facilitates the diversification of these organisms and their substrates. Grass cutting has been reported to increase activity in the rhizosphere, and trampling can help disperse microbial communities [3]. However, this same phenomenon may negatively affect certain microbial functional groups, such as phosphate-solubilizing microorganisms [4]. Despite the existing studies on the subject, the effects of grazing on the density and diversity of microorganisms are not generalizable since the environmental conditions, type of soil and its properties, type of plant cover, grazing intensity, type of animal, and management practices also have significant influences over the microbial communities [4].

In Colombia, there are two main systems of livestock production: tropical highland and tropical lowland. The former includes systems specialized in dairy production and is characterized by proximity to urban zones and colder climates with an undulating topography, whereas the lowland tropics include double-purpose cattle and are characterized by high temperatures, remoteness of markets, and flat topography. Both production systems present between 80 and 90% pasture, with a predominance of *Pennisetum clandestinum* in the highland tropics and *Brachiaria* spp. in the lowland tropics. These systems present only approximately 10% forested areas. These systems are also radically different in their soil fertilization processes; in highland tropics with dairy production cattle, pastures are irrigated and fertilized with chemicals and receive large amounts of nitrogen, whereas these processes are rarely performed in the lowland tropics [5].

Although livestock production is becoming more specialized in Colombia, these techniques have been practiced extensively, and activities such as overgrazing, burning, and mechanization have contributed to soil erosion, loss of biodiversity, and decreases in productivity [6]. The physical and chemical properties of the soil in areas destined for beef were studied by Jiménez et al. [7], but the biological components were not evaluated. Cañón-Cortázar et al. [8] reported the density and diversity of microorganisms associated with the nitrogen cycle in soils in paramo plateaus used for double-purpose livestock (mainly dairy production) combined with potato cultivation. The authors found that livestock might favour the density of these microorganisms due to the use of fertilizers in this production system, the mechanical action on the soil, and the deposition of faecal matter and urine. These studies have focused their analyses on microbial communities in the soils of production systems with agroforestry and intensive silvopastoral systems and have demonstrated that microbial populations of fungi and bacteria are favoured [9] and that the microbial communities generally tend to be similar to the communities found in native forests [10].

Since studies that reflect the impact of livestock on the microbiological properties of soils are scarce, the objective of the present study was to evaluate microorganisms involved in the carbon, nitrogen, and phosphorus cycles in soils used for bovine production systems in tropical lowlands and tropical highlands in the Department of Antioquia, Colombia, offering support to future investigations directed at evaluating their quality and encouraging good management practices to take advantage of this resource. These results represent a first attempt to describe the density of microorganisms and the effect of soil physicochemical parameters on colonization by arbuscular mycorrhizal fungi in areas with determinant agroecological conditions, microbial functional diversity, and the presence of mycorrhizal fungi in livestock farm soils in Colombia.

2. Materials and Methods

2.1. Description of the Area and Study Type. An absolute experimental study was performed by comparative observation in two subregions of Antioquia, Colombia. The first group in the northern subregion included the municipalities of Entrerríos, Belmira, and San Pedro de los Milagros, situated between 2,300 and 2,475 metres above sea level with average temperatures between 14 and 16°C. The second group in the Magdalena Medio subregion included the municipalities of Puerto Berrio, Puerto Triunfo, and Puerto Nare, located between 140 and 125 metres above sea level with an average temperature of 28°C. A survey was designed as an instrument for the collection of farm information, including general information on the production system and aspects related to the use of fertilizers in the soil.

2.2. Soil Sample Collection. A zigzag track was drawn on the delimited terrain; and subsamples were taken at each vertex where the direction of the course changed. One soil sample was collected for each sampling unit; this sample consisted of a "composite sample" composed of ten subsamples taken at random in the field. At each sampling site, plants and fresh leaf litter were removed (1-2 cm) from a 40 cm × 40 cm area, and the soil was extracted to a 20 cm depth. Approximately 200 g of soil was transferred to a recipient to remove gravel, thick roots, worms, and insects. Then, the soil samples were crumbled and mixed manually according to recommendations by the Instituto Colombiano Agropecuario [11]. Finally, one kilogram of sample was transferred to a clean, closed plastic bag marked with the name or number of the farm, the date, and the municipality.

2.3. Microbiological Analysis of the Soils. Microorganisms responsible for the biogeochemical processes that occurred in the soil were evaluated. These groups of microorganisms account for the soil functional diversity and are directly associated with its health and fertility (i.e., cellulolytic, proteolytic, amylolytic, phosphate solubilizers, nitrogen-fixing bacteria with glucose as a carbon source, nitrogen-fixing bacteria with malate as a carbon source, Actinobacteria, total fungi, and heterotrophic bacteria). Isolation of the microorganisms was performed by seeding serial dilutions onto the surface of culture media specific for each group, culture medium for cellulolytic [12], culture medium for amylolytic [13], culture medium for proteolytic [14], PK medium for phosphate-solubilizing bacteria [15], Burk culture medium for fixing bacteria of N2-glucose source of C [16], Nfb culture medium for N2-fixing bacteria-malate C source [17], and starch-casein agar culture medium for actinomycetes [18]. Ten grams of soil was taken and diluted in 90 mL of 0.85% sterile saline solution, stirred vigorously for 20 minutes, and allowed to settle; then, 1 mL of sample was transferred to 9 mL of saline. This procedure was repeated up to the 10-6 dilution. For seeding, we used the 10-4, 10-5, and 10-6 dilutions, which were stirred, dispersed, and homogenized with 0.1 mL of sample over the culture medium using a Drigalski bacterial loop. Each dilution was seeded in duplicate and incubated at 28°C for the time necessary for the growth of each group of microorganisms as recommended by the same authors [14]. The growth of each group was reported as the colony forming units per gram of dry soil (CFU/gs.s.).

2.4. Determination of the Percentage of Root Colonization by Arbuscular Mycorrhizal Fungi (AMF). To evaluate the functioning of the mycorrhizal fungi, the presence of colonization in the roots must be determined and quantified. The percent colonization was determined by staining fungal structures inside the root. The roots were initially discoloured with a strong base (10% KOH), neutralized with an acid (10% HCl), and exposed to 0.05% Trypan blue for three days. Then, the roots were observed under a microscope (NIKON E200, NY, USA) to determine the presence of arbuscules, vesicles, and endospores. The percentage of root colonization was calculated as described by Phillips and Hayman [19].

2.5. AMF Spore Count. Spores of AMF are considered the most important reproductive structures and are generally the most resistant to adverse conditions compared to other propagules. Spores are also considered a relative indicator of the abundance of the AMF populations. For this determination, 10 g of soil was added to 20 mL of 5% hydrogen peroxide and stirred every three minutes for 15 minutes; then, a preliminary screening with 0.250 and 0.045 mm sieves was performed, and the spores were extracted by gradient centrifugation for three minutes at 2640 rpm using 80% sucrose. Three-quarters of the tube contents were deposited onto the 0.045 mm sieve, washed for three minutes to remove the sucrose, and transferred to filter paper in a Petri dish. The spore count was performed in a stereoscope (NIKON SMZ445, Melville, NY 11747-3064USA), and the results are expressed as the number of spores per gram of soil following the description of García et al. (2012).

2.6. Physical-Chemical Analysis of the Soils. Analysis of the physicochemical properties was performed using professional testing equipment for soil analysis (model STH-14 code 5010-01, LaMotte Company, MD, USA). The nitrate, nitrite, ammoniacal nitrogen, phosphorus, magnesium, and aluminium concentrations were evaluated through colorimetric tests for determination of available minerals and the calcium; this test measures the amount of calcium present in the base exchange complex and available potassium concentrations were evaluated using turbidity tests. The values for all parameters are expressed as mg/kg according to the manufacturer's recommendations. Additionally, the moisture content was determined as described in ASTM D4959-07 [20] with some modifications; the amount of soil used was 10 g per sample, and the drying time was 24 hours.

2.7. Statistical Analysis. Summary measures were calculated for the quantitative variables. The percent variability of the microorganism density, root colonization, and number of AMF spores and the physicochemical parameters of the sampled farms were determined using a multivariate principal component analysis. The classification of the sampled farms, which took into account information collected in the principal component analysis, was performed with an agglomerative cluster analysis of squared Euclidean distances with Ward's method. The normality of the dataset was verified based on the Shapiro-Wilk test to determine the statistical relationships between variables ($p \leq 0.05$). Subsequently, the Spearman correlation coefficient (rho) and the coefficient of determination ($r2$) were estimated in both regions and by region. All statistical analyses were performed in the Statistical Package for the Social Sciences (SPSS) version 22 for Windows XP.

3. Results and Discussion

3.1. Management of the Studied Cattle Farms. The predominant pasture in all analysed soils in the northern subregion (highland tropics) was kikuyu *(Pennisetum clandestinum)*. Fertilizer use was a management practice in all evaluated farms; chemical fertilization combined with organic fertilization (52.2%) predominated, with a frequency of 45 days or less in 12 of the 24 farms (RIC 40–49). The production of organic manure was not a predominant activity in this highland tropic region (16.7%). Pest control was common in most cattle farms (91.3%), with a frequency of 45 days or less in 12 of the 24 farms (RIC 32–45), and the predominant production system was dairy. Regarding soil management in the farms of the Magdalena Medio subregion (lowland tropics), the predominant pasture species were in the genus *Brachiaria*, especially *B. decumbens*, *B. humidicola*, and *B. brizantha*, and the main production systems were reproduction and double-purpose. The farms of the Magdalena Medio subregion, fertilizer use, manure production, and pest control were not reported as management practices.

Several studies have revealed that the application of fertilizers and irrigation play important roles in increasing the production of green fodder. However, the excessive use of fertilizers, such as ammonium sulphate, might change soil conditions, such as the pH and the populations of some microorganisms [10, 21]. The application of organic materials stimulates the growth of the microbial populations in charge of the nitrification processes by increasing the levels of soluble nutrients for plants, water retention that facilitates nutrient exchange from organic matter, and the solubility of minerals such as potassium [14].

3.2. Population Density of Soil Microorganisms. Regarding the population density of the nine groups of microorganisms in the soils of the farms in this study, the average density of the heterotrophic bacteria (3.0×10^5 CFU/g dry soil) in Northern Antioquia was higher than the average density of the farms in the Magdalena Medio subregion (average 1.3×10^4 CFU/g dry soil). Similar results were obtained when evaluating the densities of the amylolytic bacteria, proteolytic bacteria,

Actinobacteria, nitrogen-fixing bacteria that use malate as a carbon source, and nitrogen-fixing bacteria that use glucose as a carbon source, with average values of 2.2×10^5, 3.0×10^4, 1.5×10^4, 2.0×10^5, and 9.2×10^5 CFU/g dry soil in the northern subregion compared with 2.3×10^3, 2.2×10^3, 2.2×10^3, 8.3×10^3, and 3.0×10^3 CFU/g dry soil obtained in the Magdalena Medio subregion, respectively.

However, the average values obtained for the fungal counts (1.5×10^6 CFU/g dry soil) and cellulolytic bacteria (2.0×10^5 CFU/g dry soil) were higher in Magdalena Medio than in the northern subregion (average values of 7.1×10^3 and 6.6×10^3 CFU/g dry soil, resp.). The average densities of phosphate-solubilizing bacteria in both subregions were similar with 2.5×10^4 and 2.2×10^4 CFU/g dry soil for the Northern and Magdalena Medio subregions, respectively.

The number of CFUs of these groups in each municipality varied as follows: for the farms located in the northern subregion, specifically in Entrerríos and Belmira, the values obtained for heterotrophic bacteria, proteolytic bacteria, nitrogen-fixing bacteria that use glucose as a carbon source, nitrogen-fixing bacteria that use malate as a carbon source, and Actinobacteria were, on average, 2.6×10^5, 5.8×10^2, 6.3×10^5, 5.4×10^4, and 7.5×10^3 CFU/g of dry soil, respectively, for both municipalities. These values were lower than the values found in the municipality of San Pedro de los Milagros, with average numbers of 4.0×10^5, 6.7×10^4, 1.5×10^6, 2.3×10^5, and 3.0×10^5 CFU/g of dry soil, respectively. The population behaviour of the cellulolytic bacteria (5.0×10^3 CFU/g dry soil) in the municipalities of San Pedro de los Milagros and Entrerríos was lower compared to Belmira (1.2×10^5 CFU/g dry soil). The opposite trend was observed for the population density of amylolytic bacteria, with higher numbers found in San Pedro de los Milagros and Entrerríos (3.0×10^5 CFU/g dry soil) and lower numbers found in Belmira (9.4×10^4 CFU/g dry soil). Belmira and San Pedro de los Milagros presented lower average counts of fungi and phosphate-solubilizing bacteria (5.0×10^4 and 4.2×10^2 CFU/g dry soil, resp.) than Entrerríos (1.1×10^4 and 2.2×10^4 CFU/g dry soil, resp.).

For the municipalities in the Magdalena Medio subregion, the numbers of heterotrophic bacteria, cellulolytic bacteria, and Actinobacteria were higher on average in the municipalities of Puerto Berrio and Puerto Nare, with 1.7×10^5, 3.0×10^5, and 3.3×10^3 CFU/g of dry soil, respectively, compared to the population densities obtained in the municipality of Puerto Triunfo, with 5.0×10^4, 2.0×10^4, and 1.0×10^1 CFU/g of dry soil, respectively. The population numbers of the phosphate-solubilizing bacteria, amylolytic bacteria, and proteolytic bacteria were higher in the municipality of Puerto Berrio (3.2×10^4, 7.4×10^4, and 6.7×10^2 CFU/g of dry soil, resp.) compared to the values obtained for Puerto Triunfo and Puerto Nare (5.0×10^3, 7.0×10^2, and 1.0×10^1 of dry soil, resp., for both municipalities). The values obtained for nitrogen-fixing bacteria that use malate as a carbon source were similar for all three municipalities, with an average count of 3.0×10^4 CFU/g dry soil. The opposite trend was observed for nitrogen-fixing bacteria that use glycose as a carbon source, with different values obtained for each municipality [the highest in Puerto Berrio (7.3×10^4 CFU/g dry soil), followed by Puerto Triunfo (1.4×10^4 CFU/g dry soil), and finally, Puerto Nare (3.3×10^2 CFU/g dry soil)].

Comparing the results obtained in the present study and the values reported in the literature and taking into account the edaphic changes that occurred within each subregion, different strains of bacteria, such as *Azospirillum brasilense*, showed significant increases in growth and total incorporation of nutrients. Likewise, different strains exposed to different concentrations of fertilizers improved the assimilation of nutrients either by changing the root structure or with the aid of the bacterial enzyme nitrate reductase. However, this effect may be due to the involvement of the plants in symbiotic associations with the bacteria, resulting in the nutrients being obtained by biological fixation [15].

3.3. Percent Root Colonization by AMF and Spore Density.

When evaluating the percent root colonization by AMF in the soils of the farms under study, similar average values were obtained for colonization in both the Northern and Magdalena Medio subregions. In the northern subregion municipalities, the highest value for colonization was found in the municipality of Entrerríos (63.4%), followed by Belmira (48.3%) and San Pedro de los Milagros (34.2%). In the Magdalena Medio subregion, the municipality of Puerto Berrio presented the highest colonization rate (65.5%), followed by Puerto Nare (48%) and Puerto Triunfo (43%).

However, variability was observed within each municipality for each subregion. For farms in the northern subregion in the municipality of Entrerríos, AMF colonization varied between 11 and 91%, with farms F3 and F5 presenting the highest percent colonization (91 and 90.5%, resp., which represented a significant difference ($p = 0.04$) compared to farm F6 with 11%). In the municipality of Belmira, the percent colonization varied between 16.5 and 86%, with farm F9 presenting the highest value of 86%, which represented a significant difference ($p = 0.05$) compared to farms F14 and F16, with colonization rates of 16.5%. In the municipality of San Pedro de los Milagros, the observed percent colonization varied between 16 and 51%, with the highest values in farm F20 (51%) and the lowest in farm F17 (16%). The results for the farms in this municipality are statistically similar among themselves.

In the Magdalena Medio subregion, colonization by AMF in the municipality of Puerto Berrio was between 38 and 86%, with farms F25 and F31 presenting the highest percent colonization (86 and 83%, resp.); this percent colonization was significantly different ($p = 0.03$) from farm F30, with 38% colonization. In the municipality of Puerto Triunfo, the percent colonization varied between 4.0 and 80%; the highest values were obtained in farm F38 (80%) and were significantly different ($p = 0.01$) relative to farm F34, with 4.0% colonization. The lowest percent root colonization values among all municipalities in the study were observed in the municipality of Puerto Nare, where colonization varied between 67 and 4.0%; farm F44 presented 67% AMF colonization and farm F48 presented 4% AMF colonization, which represented a significant difference ($p = 0.001$).

The highest number of AMF spores was found in the soils of farms in the municipalities in the northern subregion of Antioquia. The values obtained for the municipality of Entrerríos gave an average of 363 spores/g soil, followed by the soils from Belmira and San Pedro de los Milagros, with 216 and 215 spores/g soil, respectively. For the municipalities in the Magdalena Medio subregion, the highest values were obtained from Puerto Triunfo, with 80 spores/g soil, followed by the municipality of Puerto Triunfo, with 56 spores/g soil, and the municipality of Puerto Berrio, with 51 spores/g soil.

The numbers of spores varied among the municipalities in the northern subregion. The values varied between 47 and 683 spores/g of soil for the farms in Entrerríos, with farm F7 presenting the highest density (683 spores/g soil) and a significant difference relative to the other farms ($p = 0.045$). For the municipality of Belmira, these values ranged between 22 and 706 spores/g soil, with the highest density found in farm F11 (706 spores/g soil); this value presented a significant difference ($p = 0.03$) relative to the other farms. For the municipality of San Pedro de los Milagros, the number of spores varied between 67 and 293 spores/g soil, with the highest spore density found in farm F21; this value was significantly different ($p = 0.035$) compared to farm F23 (67 spores/g soil).

In the Magdalena Medio subregion, the soils from the municipality of Puerto Berrio presented a density between 29 and 110 spores/g soil, with farm F27 presenting the highest values (110 spores/g soil) and representing a significant difference ($p = 0.01$) compared to the other farms. For the municipality of Puerto Nare, the counts ranged between 26 and 195 spores/g soil; farm F42 presented the highest density, which was significantly different ($p = 0.03$) compared to farm F46 (34 spores/g soil). The soils from the municipality of Puerto Triunfo presented a density between 18 and 97 spores/g soil, with farm F40 presenting the highest values (97 spores/g soil), which was significantly different ($p = 0.04$) compared to farms F36 and F37 (18 and 20 spores/g soil, resp.).

3.4. Physical and Chemical Properties of the Soils. The humus, nitrite, potassium, aluminium, and ammoniacal nitrogen concentrations were constant in all of the evaluated municipalities (low, 1 mg/kg, 50 mg/kg, 86 mg/kg, and 5 mg/kg, resp.). The pH varied from moderately acidic to slightly acidic for the farms located in the northern subregion, whereas the pH values for the farms in the Magdalena Medio subregion were strongly acidic.

The percent moisture in the soils was higher in the municipalities located in the northern subregion than in the soils located in the Magdalena Medio subregion. The average humidity was 85% in the soils from San Pedro de los Milagros, 73% in the soils from Entrerríos, 68% in the soils from Belmira, 25% in the soils from Puerto Triunfo, 20% in the soils from Puerto Nare, and 18% in the soils from Puerto Berrio.

The nitrate and calcium concentrations were similar among farms located in the same subregion but differed between subregions. The nitrate and calcium concentrations were approximately 13 mg/kg and 3458 mg/kg for the northern subregion and 5 mg/kg and 2791 mg/kg for Magdalena Medio, respectively. The opposite trend was observed for the average phosphorus and magnesium concentrations, with the farms in the Magdalena Medio subregion presenting higher values (67 mg/kg and 30 mg/kg, resp.) compared to the concentrations from the soils from the northern subregion (29 mg/kg and 12 mg/kg, resp.).

3.5. Relationships between the Population Density of Microorganisms, Percentage of Root Colonization, Numbers of Spores, and Physicochemical Properties. The principal component analysis of the population densities of microorganisms in livestock farms in the subregion of the north (Figure 1) showed that the first component ("x-"axis) contrasted the microorganism population density (cellulolytic, amylolytic, proteolytic, and nitrogen-fixing bacteria that use glucose as a carbon source, fungi, heterotrophic bacteria, and phosphate solubilizers) with the percentage of colonization, with a component value of $\ddot{e}_1 = 6.2$ that explained 62% of the total variability of the data. Farm F3 presented high numbers of microorganisms, the highest percent colonization values (90%) and spore counts (509 spores/g soil), and a moderately acidic pH. Farms F34 and F36 presented high phosphorus contents (375 mg/kg), low colonization values (4.1 and 6.3%, resp.), and lower population densities (6.7×10^2 CFU/g dry soil).

The second component ("y-"axis) was associated with the population density of fungi, nitrogen-fixing bacteria that used malate as a carbon source, and Actinobacteria, with a component value of $\ddot{e}_2 = 3.0$ that explained 30% of the total variability of the data. For example, farms F18 and F23 had higher population densities of Actinobacteria (4×10^5 CFU/g dry soil) and nitrogen-fixing bacteria that used malate as a carbon source (1.4×10^6 CFU/g dry soil), high AMF spore counts (683 spores/g soil), and low percentages of root colonization when high phosphorus contents were present (100 mg/kg). Both components explained 92% of the variation of the data.

The analysis of microorganism population densities compared to the physicochemical parameters by agglomerative cluster analysis (Figure 2) based on the principal component analysis for population densities (Figure 1) for the northern subregion localities showed two groups of farms based on their microorganism counts, percentage of root colonization, and physic chemical parameters of the soil (Figure 2(a)). Group 1 contained farms associated with a moderately acidic pH; low phosphorus, ammoniacal nitrogen, and magnesium concentrations; low counts of heterotrophic bacteria, proteolytic bacteria, and fungi; allowing a percentage of mycorrhizal colonization; and moderate numbers of spores per gram of soil. Group 2 consisted of a farm with a slightly acidic pH; low phosphorus, potassium, and aluminium concentrations; a low ammoniacal nitrogen concentration; high microorganism numbers; and higher percentages of colonization by AMF.

Five groups were formed in the Magdalena Medio subregion (Figure 2(b)). Group 1 presented farms associated with moderate microorganism population densities and similar phosphate, nitrate, aluminium, and calcium concentrations. Group 2 contained farms associated with low phosphate and magnesium concentrations, a moderate calcium concentration, a low percentage of colonization, and low humidity.

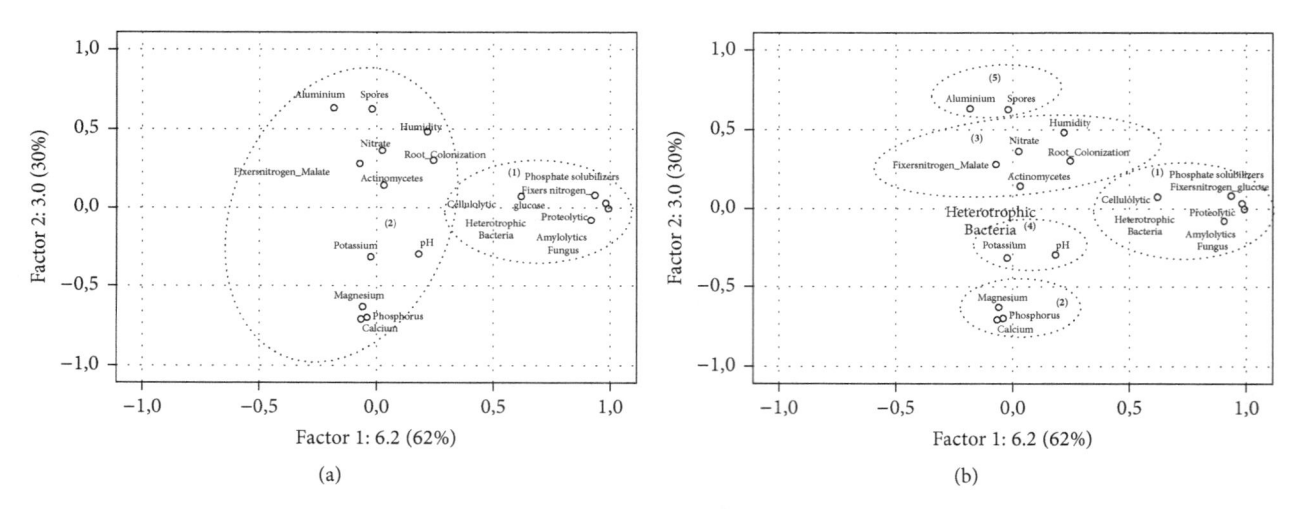

FIGURE 1: Principal component analysis of the population density of microorganisms, percentage of colonization, and soil physicochemical parameters on the cattle farms under study. (a) (1)-(2): farm groups related to parameters measured in Northern Antioquia (Colombia). (b) (1)-(5): farm groups related to parameters measured in Magdalena Medio Antioquia (Colombia).

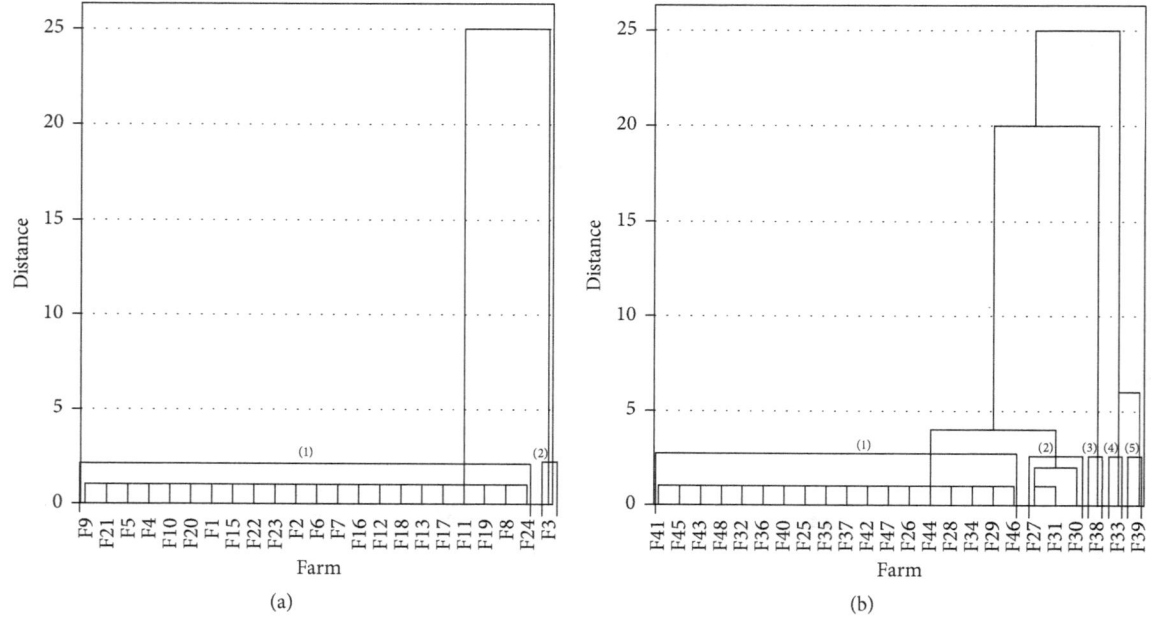

FIGURE 2: Agglomerative cluster analysis between the microorganism population density, percentage of colonization, and soil physicochemical parameters for cattle farms. (a) (1)-(2): farm groups related to parameters measured in Northern Antioquia (Colombia). (b) (1)-(5): farm groups related to parameters measured in Magdalena Medio, Antioquia (Colombia).

Group 3 consisted of one farm that presented a null density of Actinobacteria, low nitrogen-fixing bacteria that use malate as a carbon source count, a low nitrate concentration, low humidity, a moderate percentage of colonization, and a strongly acidic pH. Group 4 contained one farm that presented moderately acidic pH, low potassium and aluminium concentrations, a low percentage of colonization, and low humidity. Group 5 contained one farm with a high aluminium concentration, high spore count, high fungal population density relative to the other microorganisms, and a strongly acidic pH.

The correlation analysis of the soil chemical parameters indicated that significant differences existed between the macroelements (phosphorus, aluminium, calcium, and nitrate) and humidity evaluated in this study. A positive correlation was observed in both subregions between the humidity and the population densities of proteolytic bacteria, nitrogen-fixing bacteria that use glycose as a carbon source, nitrogen-fixing bacteria that use malate as a carbon source, and Actinobacteria and the density of spores per gram of soil. A positive correlation was also observed with the percentage of root colonization by AMF, the aluminium concentration,

TABLE 1: Spearman's correlation (Rho) coefficients between physicochemical properties, microorganism, and biological activity of endomycorrhizal fungi in the soil from 48 farms in the North and Magdalena Medio subregions in Antioquia, Colombia.

	Humidity percentage	Phosphorus (mg/kg)	Aluminium (mg/kg)	Calcium (mg/kg)	Magnesium (mg/kg)	Nitrate (mg/kg)
Spores/g soil	0.530**	−0.272	0.312*	−0.48**	−0.496**	0.330*
Percentage of root colonization by AMF	−0.208	−0.550**	0.428**	−0.266	−0.133	−0.266
Proteolysis	0.401*	0.055	−0.187	−0.048	−0.349*	0.444**
Nitrogen-fixing bacteria glycose as a carbon source	0.363*	−0.115	−0.081	−0.105	−0.427**	0.280
Nitrogen-fixing bacteria malate as a carbon source	0.389**	0.135	−0.085	0.007	−0.281	0.406**
Actinobacteria	0.404**	0.111	−0.074	0.186	−0.222	0.437**

*Correlation is significant at $p = 0.05$. **Correlation is significant at $p = 0.01$.

and the densities of bacteria, Actinobacteria, and nitrogen-fixing bacteria that use malate as a carbon source with the nitrate concentration. Furthermore, the population densities of proteolytic microorganisms and nitrogen-fixing bacteria that use glycose as a carbon source were negatively correlated with the magnesium concentration. The percentage of root colonization by AMF presented a significant negative correlation with the soil phosphorous concentration (Table 1).

The correlation analysis of the soil chemical parameters indicated that significant differences existed between the macroelements by region, for the concentration values of elements such as Ca, Mg, K, and P and pH obtained in this study and a strong significant negative correlation was found between these parameters and the spore count for the Magdalena subregion; however, in the subregion of the north, the correlation was significant negative for the calcium concentration and for the other parameters evaluated no correlation was found. Additionally, when evaluating the correlation of the parameters with the percentage of colonization, a slight positive correlation was observed with the concentration of phosphorus and potassium in the subregion of the north; for the other parameters evaluated and the subregion of Magdalena Medio, no correlation was found. Regarding the percentage of colonization and the concentration of K and P a slight positive correlation was observed in the subregion of north (Table 2).

The influence of K on the number of spores is probably due to the fact that the concentrations of this element in the soil were high (50 mg/kg,) for 97.5% of the sampled farms, which explains that having the soil higher levels of this element decreases the number of spores in the rhizosphere. In the case of P, the greater content in the soil, the lower number of spores in the soil. These data confirm what was obtained by Mofidi et al. [1], where it was determined that increasing the concentration of P in the soil increases the electrical conductivity and modifies the osmotic potentials that can affect both the diversity and the number of spores (Table 2).

Likewise, when considering the subregion of Magdalena Medio with a ground of low elevation mountains that undergo processes of laminar erosion, subjected to the use in extensive cattle ranching, microbial biomass and respiration index show very low levels. Additionally, very high counts of mycorrhizal spores are possible, since the strongly acidic pH tends to favour the development of fungi, in this case, the formation of arbuscular mycorrhizae and the production of their reproductive structures [11].

AMF (Glomeromycetes) represent between 5 and 50% of the total biomass of soil microorganisms [16]. Symbiotic associations with AMF increase nutrient assimilation, especially phosphorus, because the diameter and length of the hyphae allow the plant to explore a larger volume of the edaphic environment. These relationships also allow the formation of micro- and macroaggregates that improve the physical characteristics of the soil [3, 17]. Additionally, AMF play an important role in the acquisition of nitrogen for the plant; according to Zubek et al. (2012), mycorrhizal fungi have a direct effect on the absorption of nutrients in the symbiotic system formed by the fungus and the plant roots [2].

These associations in ecosystems are influenced by the relationships among organic and inorganic nutrients, hydric relationships, and the carbon cycle in plants, as well as the edaphic conditions, such as the chemical composition, humidity, temperature, pH, cation exchange capacity, and biotic and abiotic factors [22], and can affect plant endosymbiosis and AMF. For instance, an excess temperature can affect the germination of spores and induce moisture deficiency, which can inhibit the formation of endosymbiosis.

The phosphorus concentration favours the structure of the microbial community by significantly increasing the relative abundance of AMF, which obtains carbon from its host plants in exchange for mineral nutrients [23]. Conversely, if the phosphorus availability increases, a decrease in the abundance of AMF is expected [24]. Our results are consistent with the expectation that the abundance of P suppresses the inversion of the plant in mycorrhizal symbiosis (see Table 1), which was the case in the lowland tropics, whereas adding phosphorus may increase mycorrhizal growth. Because they are more efficient as soil nutrient cleaners than plant roots, the threshold for nutrient limitation may be lower for mycorrhizal fungi than for plants. Another possible explanation is that an increase in the soil pH is often associated with an increase in the AMF biomass in the soil [25].

TABLE 2: Spearman's correlation *(Rho)* coefficients between the spore counts, the percentage of colonization, and the physical-chemical properties of the soils studied by region.

Subregion	pH		phosphorus (mg/kg)		potassium (mg/kg)		Calcium (mg/kg)		Magnesium (mg/kg)	
	N	MM	N	MM	N	MM	N	MM	N	MM
Spores/g soil	0.530**	−0,735*	−0.272	−0,703*	0.312*	−0,807**	−0.48**	−0,659*	0.496**	−0,715*
Percentage of root colonization by AMF	−0.208	0.370	−0.550**	−0.770	0.428**	0.650	−0.266	0.214	−0.133	0.230

*Correlation is significant at $p = 0.05$. **Correlation is significant at $p = 0.01$. N: north subregion (tropical highlands) and MM: Magdalena Medio subregion (tropical lowlands).

4. Conclusions

The AMF-plant-microorganism relationship is not considered specific, since any AMF or bacterial species can colonize or form a symbiotic relationship. However, some of these microorganisms may benefit a certain host to a higher or lower degree under certain soil and climate conditions.

Mycorrhizal fungi are a biological resource whose management and conservation effects on plant productivity generate environmental benefits by improving the physical-chemical and biological conditions of the soil. The benefits from a biological perspective derive from their interactions with the various macrogroups and microorganisms of the rhizosphere, such as those involved in the cycling of nutrients (nitrogen-fixing bacteria and phosphate-solubilizing microorganisms). Furthermore, fungi that interact with the microorganisms are involved in the biological control of pathogens present in the soil, demonstrating that different types of interactions exist with arbuscular mycorrhizal fungi. The pH, soil moisture, and nutrient availability influence not only the colonization but also the number of spores produced by AMF. AMF are found in all types of soils and can colonize any plant that establishes symbiosis with them; however, the physical-chemical conditions of the soil could generate some degree of specificity with respect to the host plants based on the responses shown by the plants to certain AMF species.

Few studies have demonstrated the effect of environmental conditions on the establishment of AMF and microbial communities in different ecosystems. This study is one of the first to focus on the highland and lowland tropics of Antioquia, Colombia, and shows certain physic chemical parameters (i.e., pH, phosphorus, nitrogen, and sodium) of soils from livestock farms in the Department of Antioquia that have a direct effect on the establishment of AMF on roots and some soil microorganism groups.

Acknowledgments

This research was supported by Departamento Administrativo de Ciencia, Tecnología e Innovación (COLCIENCIAS), Grant no. 121056934576; CIDI-UPB no. 211B-02/14-65. The authors acknowledge COLCIENCIAS for the support to Paula Henao in the National Call for Young Researchers and Innovators 645-2014 (Convention no. 0208-2014). They thank the owners of the herds chosen for this research and the administrative and technical staff in the farms for their invaluable cooperation in the field and also acknowledge the regionals cooperatives societies of livestock farmers (COLANTA, COLETRIUNFO, APAGRONAR (Asociación de Productores Agropecuarios de Puerto Nare), and COREGAN (Comite Regional de Ganaderos de Puerto Berrio) because their important logistical support make it possible to execute the field work in this research. They want to also specially thank Carlos Adrian Lopera, project manager of Ecosphaira Colombia, for his valuable advice in the analyses of percent root colonization and spore density for this paper.

References

[1] M. Mofidi, M. Rashtbari, H. Abbaspour, A. Ebadi, E. Sheidai, and J. Motamedi, "Impact of grazing on chemical, physical and biological properties of soils in the mountain rangelands of Sahand, Iran," *The Rangeland Journal*, vol. 34, no. 3, pp. 297–303, 2012.

[2] S. Zubek, A. M. Stefanowicz, J. Błaszkowski, M. Niklińska, and K. Seidler-Łozykowska, "Arbuscular mycorrhizal fungi and soil microbial communities under contrasting fertilization of three medicinal plants," *Applied Soil Ecology*, vol. 59, pp. 106–115, 2012.

[3] R. S. Shange, R. O. Ankumah, A. M. Ibekwe, R. Zabawa, and S. E. Dowd, "Distinct soil bacterial communities revealed under a diversely managed agroecosystem," *PLoS ONE*, vol. 7, no 7, Article ID e40338, 2012.

[4] N. L. Olivera, L. Prieto, M. B. Bertiller, and M. A. Ferrero, "Sheep grazing and soil bacterial diversity in shrublands of the Patagonian Monte, Argentina," *Journal of Arid Environments*, vol. 125, pp. 16–20, 2016.

[5] F. A. Holmann, L. Rivas, J. Carulla et al., "Evolución de los sistemas de producción de leche en el trópico latinoamericano y su interrelación con los mercados: un análisis del caso colombiano," *CIAT*, p. 53, 2003.

[6] V. E. Vallejo Quintero, "Importancia y utilidad de la evaluación de la calidad de suelos a través del componente microbiano: Experiencias en sistemas silvopastoriles," *Colombia Forestal*, vol. 16, no. 1, p. 83, 2013.

[7] Y. Jiménez, C. Martínez, and N. Mancera, "Caracteristicas fisicas y quimicas del suelo en diferentes sistemas de uso y manejo en el centro agropecuario Cotové, Santa Fe de Antioquia, Colombia," *Suelos ecuatoriales*, vol. 40, no. 2, pp. 176–188, 2010.

[8] R. Cañón-Cortazar, M. Avellaneda-Torres, and E. Torres-Rojas, "Associated microorganisms to the nitrogen cycle in soils under three systems of use: potato crop, livestock and páramo, in Los Nevados National Natural Park," *Acta Agronómica*, vol. 61, no. 4, pp. 339–347, 2012.

[9] V. E. Vallejo, Z. Arbeli, W. Terán, N. Lorenz, R. P. Dick, and F. Roldan, "Effect of land management and Prosopis juliflora (Sw.) DC trees on soil microbial community and enzymatic activities in intensive silvopastoral systems of Colombia," *Agriculture, Ecosystems & Environment*, vol. 150, pp. 139–148, 2012.

[10] A. M. Cubillos, V. E. Vallejo, Z. Arbeli et al., "Effect of the conversion of conventional pasture to intensive silvopastoral systems on edaphic bacterial and ammonia oxidizer communities in Colombia," *European Journal of Soil Biology*, vol. 72, pp. 42–50, 2016.

[11] Instituto Colombiano Agropecuario, *Fertilización en diversos cultivos: quinta aproximación*, Santa Fé de Bogotá, 5th edition, 1992.

[12] P. J. Wood, "Specificity in the interaction of direct dyes with polysaccharides," *Carbohydrate Research*, vol. 85, no. 2, pp. 271–287, 1980.

[13] G. Pontecorvo, J. A. Roper, L. M. Chemmons, K. D. Macdonald, and A. W. J. Bufton, "The Genetics of Aspergillus nidulans," *Advances in Genetics*, vol. 5, no. C, pp. 141–238, 1953.

[14] G. Andrade, "Role of functional groups of microorganisms on the rhizosphere microcosm dynamics," *Plant Surface Microbiology*, pp. 51–69, 2008.

[15] L. Beltrán-Pineda and M. Lizarazo-Forero, "Grupos funcionales de microorganismos en suelos de páramo perturbados por incendios forestales," *Revista de Ciências*, vol. 17, no. 2, pp. 121–136, 2013.

[16] S. Wilson and P. Knight, *Experiments in bacterial physiology*, Burguess, Publishing Co, Minneapolis. USA, 1952.

[17] J. Dobereiner and J. Day, "Associative symbiosis and free-living systems," in *Proceedings of the 1st International symposium on Nitrogen fixation*, W. E. Newton and C. J. Nyman, Eds., pp. 518–538, Washington state University. Press. Pullman., 1976.

[18] E. Küster and S. T. Williams, "Selection of media for isolation of streptomycetes," *Nature*, vol. 202, pp. 928-929, 1964.

[19] J. M. Phillips and D. S. Hayman, "Improved procedure for clearing roots and staining parasitic and vesicular-arbuscular mycorrhizal fungi for rapid assessment of colonization," *Fungal Biology*, vol. 55, no. 1, pp. 158–161, 1970.

[20] Astm, "D4959 - Standard Test Method for Determination of Water (Moisture) Content of Soil By Direct Heating," *Astm D4959*, pp. 5–8, 2000.

[21] M. R. L. Garcia, A. A. M. Sampaio, and E. Nahas, "Impact of different grazing systems for bovine cattle on the soil microbiological and chemical characteristics," *Revista Brasileira de Zootecnia*, vol. 40, no. 7, pp. 1568–1575, 2011.

[22] H. Wang, J. Mo, X. Lu, J. Xue, J. Li, and Y. Fang, "Effects of elevated nitrogen deposition on soil microbial biomass carbon in major subtropical forests of southern China," *Frontiers of Forestry in China*, vol. 4, no. 1, pp. 21–27, 2009.

[23] S. A. Billings and S. E. Ziegler, "Altered patterns of soil carbon substrate usage and heterotrophic respiration in a pine forest with elevated CO_2 and N fertilization," *GCB Bioenergy*, vol. 14, no. 5, pp. 1025–1036, 2008.

[24] K. K. Treseder and M. F. Allen, "Direct nitrogen and phosphorus limitation of arbuscular mycorrhizal fungi: a model and field test," *New Phytologist*, vol. 155, no. 3, pp. 507–515, 2002.

[25] B. F. T. Brockett, C. E. Prescott, and S. J. Grayston, "Soil moisture is the major factor influencing microbial community structure and enzyme activities across seven biogeoclimatic zones in Western Canada," *Soil Biology & Biochemistry*, vol. 44, no. 1, pp. 9–20, 2012.

Diversity among Modern Tomato Genotypes at Different Levels in Fresh-Market Breeding

Krishna Bhattarai,[1] Sadikshya Sharma,[2] and Dilip R. Panthee[1]

[1]Department of Horticultural Science, North Carolina State University, Mountain Horticultural Crops Research and Extension Center, Mills River, NC 28759, USA
[2]Department of Environmental Horticulture, University of Florida, 2550 Hull Rd, Gainesville, FL 32611, USA

Correspondence should be addressed to Dilip R. Panthee; dilip_panthee@ncsu.edu

Academic Editor: Allen Barker

Cultivated tomato has been in existence for about 400 years and breeding activities have been conducted for only eight decades. However, more than 10,000 tomato cultivars have already been developed. Ninety-one tomato genotypes were characterized for twenty-one morphological traits using developmental, vegetative, and fruit traits. Correlation, principal component, and cluster analysis between the traits were carried out. Higher correlations between fruit traits including fruit shape, fruit size, and fruit types were observed. These correlations indicate that specific fruit types require specific traits like branched inflorescence and a greater number of fruits per inflorescence are beneficial only for smaller fruit sizes like cherry and grape tomatoes. Contrastingly, traits like determinate growth habit and fruit maturity are preferred in all fruit types of tomato for better cultivation practices and longer production duration and hence showed lower correlations. Principal component analysis clustered tomato genotypes into three main clusters with multiple subgroups. Similar tomato genotypes were placed into one or more clusters confirming the results from correlation analysis. Involvement of private breeding programs in cultivar development has increased the competition on introgression of novel and desired traits across new cultivars. Understanding the diversity present in modern cultivars and potential traits identification in related wild species can enhance tomato diversity and improve quality and production.

1. Introduction

Tomato is one of the most important vegetables produced all over the world. It belongs to the diverse Solanaceae family along with potato, pepper from the South American origin, and eggplant from Asia. Tomato breeding in the modern era has focused on increased production and adaptation to different consumption need. With Latin American origin, the domesticated tomato was cultivated in South America with selection for edible fruits, attractive red color, and increased fruit size as compared to the wild tomato. There have been several bottlenecks over the ages leading to severe reduction in its genetic diversity during domestication, introduction to Europe, and introduction in the US [1]. Due to the broad adaptation of tomato from Alaska in summer to tropical conditions, there have been adaptive and morphological variations. However, introgression of highly desirable characteristics and higher selection pressure to select those traits in tomato population has decreased the genetic diversity [2]. Phenotypic characteristics have been used to study genetic differences and for genetic diversity analysis and cultivar development [3–6]. Assessment of diversity in commercial cultivars provides the status of a crop in relation to its wild and domesticated relatives and assists in creating a breeding scheme to bring in the lost alleles due to extreme selection and bottlenecks during domestication or introduction [7]. In the absence of genetic variation, breeding efforts in crops remain without any gain [8]. This may lead to severe crop failure with the onset of new biotic or abiotic stress. Most of the disease resistant genes in modern varieties are not only the result of selection from these varieties but also originated from the wild species [9].

In this study, we characterized ninety-one tomato genotypes for twenty-one phenotypic traits based on tomato descriptors developed by International Plant Genetic Resources Institute [10] and additional methods with some modifications. Tomato genotypes that are actively used in developing modern varieties by different breeding programs are studied to understand their morphological characteristics and develop a scenario on the level of diversity available among these lines that develop into or give rise to new cultivars in the near future.

2. Materials and Methods

2.1. Plant Materials. Ninety-one tomato genotypes from ten public and private breeding programs including breeding lines, hybrids, and advanced lines were selected and studied in the breeding programs (Table 1). These genotypes were sown in flatbed metal trays in a soilless seeding mixture (2:2:1 v/v/v peat moss: pine bark: vermiculite) with macro- and micronutrients (Van Wingerden International Inc., Mills River, NC). The germinated seedlings were transplanted to 72-cell flat trays after ten days of sowing and were grown for four weeks before they were transplanted in the field. These plants were planted in the field at Mountain Horticultural Crops Research and Extension Center, Mills River, North Carolina, with complete randomized block design with two blocks. Six plants per genotype were planted in each block with 45 cm space between individuals and 150 cm between rows. Management practices for fertilization, diseases, and insects control and irrigation were applied according to standard recommendations for tomato production in North Carolina [11].

2.2. Data Collection. Vegetative and reproductive traits that are related to tomato production and quality were measured using quantitative or a rated scale. Data for each plot were collected by averaging the performance of six individuals present in each plot. All the traits measured and the scale are illustrated in Table 2. Quantitative traits like seeding and plant heights were measured using a measuring tape, chlorophyll content was measured by SPAD 502 plus chlorophyll meter (Spectrum Technologies, Aurora, IL, USA), and fruits per inflorescence were measured by counting number of completely developed fruits in a cluster manually. Rating for bacterial spot disease was done by using Horsfall and Barrett scale and late blight severity was rated by using percentage rating converted to a scale of one to five where one indicated the least late blight severity. Both bacterial spot and late blight diseases were not inoculated artificially but developed naturally despite regular application of recommended control measures. Therefore, leaf spot and late blight diseases severities were rated at the fruit harvest to assess the genotypic performance of each tomato line. Morphological foliar and fruit traits were rated using the scale recommended on tomato descriptor developed by International Plant Genetic Resources Institute [10]. There were 11 fruits related traits and four vegetative traits that were rated using tomato descriptor scales.

TABLE 1: List of fresh market tomato genotypes including breeding lines, advanced lines, and selections and hybrids that are being used or tested in tomato breeding programs for the development of new varieties.

Genotype	Company[α]	Type[β]
Carolina Gold	NCSU	HY
Cherokee	NCSU	BL
Fletcher	NCSU	HY
Monte Verde	NCSU	BL
Mountain Belle	NCSU	HY
Mountain Fresh	NCSU	HY
Mountain Glory	NCSU	HY
Mountain Gold	NCSU	BL
Mountain Magic	NCSU	HY
Mountain Majesty	NCSU	HY
Mountain Merit	NCSU	HY
Mountain Pride	NCSU	HY
Mountain Spring	NCSU	HY
Mountain Supreme	NCSU	HY
NC1 Grape	NCSU	BL
NC109	NCSU	BL
NC1C	NCSU	BL
NC 714	NCSU	BL
NC1CELBR	NCSU	BL
NC1CS	NCSU	BL
NC1rinEC	NCSU	BL
NC1Y	NCSU	BL
NC2 Grape	NCSU	BL
NC25P	NCSU	BL
NC2C	NCSU	BL
NC2CELBR	NCSU	BL
NC2rinEC	NCSU	BL
NC3 Grape	NCSU	BL
NC30P	NCSU	BL
NC50-7	NCSU	BL
NC8276	NCSU	BL
NC8288	NCSU	BL
NC84173	NCSU	BL
NC946	NCSU	BL
NCEBR1	NCSU	BL
NCEBR2	NCSU	BL
NCEBR3	NCSU	BL
NCEBR5	NCSU	BL
NCEBR6	NCSU	BL
NCEBR7	NCSU	BL
NCEBR8	NCSU	BL
NCHS1	NCSU	BL
Piedmont	NCSU	BL
Plum Crimson	NCSU	HY
Plum Dandy	NCSU	HY
Plum Regal	NCSU	HY
Smarty	NCSU	HY
Summit	NCSU	BL

TABLE 1: Continued.

Genotype	Company[α]	Type[β]
Sun Leaper	NCSU	HY
Mountain Vineyard	NCSU	HY
Mountain Honey	NCSU	HY
Mountain Rouge	NCSU	HY
NC 161L	NCSU	BL
NC 4Grape	NCSU	BL
NC 5Grape	NCSU	BL
UG 1104113	UG	HY
UG 1155013	UG	HY
XTM0017	SSA	HY
XTM7262	SSA	HY
PSGH-13-16	PSU	HY
PSGH-13-33	PSU	HY
PSFH-13-19	PSU	HY
PSFH-13-47	PSU	HY
Iron Lady	HMOS	HY
Ridge Runner	SYN	HY
Monticello	SYN	HY
Brickyard	SYN	HY
Summerpick	SYN	HY
Richmond	SYN	HY
NC33EB1 x CUTR 1st Gen CUTR	CU	HY
NC33EB1 x CUTR 2nd Gen 124116-4	CU	HY
VFT Vendor x CUTR 2nd Gen 124116-4	CU	HY
NC132S x PAU351-Y	NCL	HY
NC132S x PAU332	NCL	HY
PAU332 x NC2rinEC	NCL	HY
YCBT-1 x NC2rinEC	NCL	HY
335 x NC132S	NCL	HY
NC132S x C23-3	NCL	HY
335 x NC2CELBR	NCL	HY
NC2CELBR x C23-3	NCL	HY
YCBT-1 xN C132S	NCL	HY
NC2CELBR x PAU332	NCL	HY
NC2CELBR x PAU351-Y	NCL	HY
141G	NCSU	BL
143G	NCSU	BL
145G	NCSU	BL
147G	NCSU	BL
148G	NCSU	BL
Fla 8624H	UFL	HY
Fla 8638B	UFL	HY
Fla 8902LBR	UFL	HY

[α]NCSU: North Carolina State University; UG: United Genetics, SSA: Sakata Seed America; PA: Pennsylvania State University; HMOS: High Mowing Organic Seeds; SYN: Syngenta; CU: Cornell University; NCL: Nutrifert Chile Ltda; UFL: University of Florida; [β]BL: breeding line; HY: hybrid.

2.3. Statistical Analysis. Data collected were analyzed using SAS statistical software version 9.4 (SAS Institute Inc., Cary, NC) using GLM procedure. Analysis of variance was done using randomized complete block design. Correlation analysis was done using Pearson correlation with three levels of significance at $p < 0.05, 0.01$, and 0.001. Principle component analysis and cluster analysis were performed using chord distance coefficient and average linkage method as described by Mazzucato et al. [12].

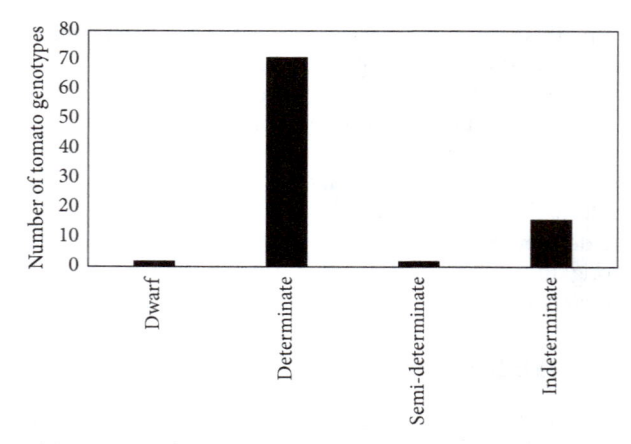

FIGURE 1: Distribution of tomato genotypes based on their growth habit.

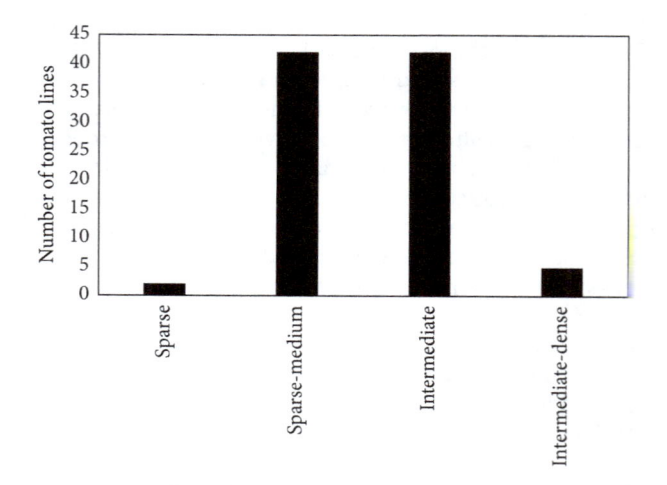

FIGURE 2: Foliage density distribution of tomato genotypes.

3. Results

3.1. Diversity in Morphological Traits. A wide range of diversity among the genotypes was observed for the traits measured in the study. There were 71 (78%) of determinate lines, 16 (17.6%) indeterminate lines, and 2 (2.2%) dwarf and semi-determinate lines (Figure 1). Different growth type tomato lines started to show variation in their development from seedling stage ranging from 2.15 to 7.4 inches when they were still in 72-cell trays (Table 3). All the tomato lines exhibit a linear growth pattern, however, with different rates, after being transplanted in the field until flowers. Foliage densities of these lines also showed variation with a range of sparse (2.1%), medium-sparse (46.2%), intermediate (46.2%), and intermediate dense (5.5%) as a result of selections for the ease of harvesting and protection of fruits from direct sunlight (Figure 2). The extremes of the foliar density as described in tomato descriptor are difficult to find since they either

TABLE 2: Measuring/rating scale used for traits phenotyped for tomato genotypes.

Traits	Measuring/rating scale						
Fruit shape*	Flattened (1)	Slightly flattened (2)	Rounded (3)	High rounded (4)	Heart shaped (5)	Cylindrical (6)	Pyriform (7)	Ellipsoid (8)
Leaf shape*	Dwarf (1)	Potato leaf type (2)	Standard (3)	*Peruvianum* (4)	*Pimpinellifolium* (5)	*Hirsutum* (6)		
Green fruit color*	Greenish-white (3)	Light green (4)	Green (5)	Dark green (6)	Very dark green (7)	Dark (8)		
Fruit size*	Very small (1)	Small (2)	Intermediate (3)	Large (4)	Very large (5)			
Late blight	20% (1)	40% (2)	60% (3)	80% (4)	100% (5)			
Fruit shoulder shape*	Flat (1)	Slightly depressed (3)	Moderately depressed (5)	Strongly depressed (7)				
Growth type*	Dwarf (1)	Determinate (2)	Semi-determinate (3)	Indeterminate (4)				
Fruit type*	Large (1)	Plum (2)	Grape (3)	Cherry (4)				
Ribbing*	Very weak (1)	Weak (3)	Intermediate (5)	Strong (7)				
Ripening*	Very early (1)	Early (1)	Standard (3)	Late (4)				
Green shoulder*	Uniform green	Slight	Intermediate	Dark				
Leaf color*	Light green (1)	Standard (2)	Dark green (3)					
Foliage density*	Sparse (3)	Intermediate (5)	Dense (7)					
Ripe fruit color*	Red (1)	Pink (2)	Yellow (3)					
Inflorescence	Generally uniparous (1)	Both uniparous and multiparous (2)						
Pedicel*	Jointless (0)	Jointed (1)						
Bacterial spot	Horsfall-Barette scale							
Height	Quantitative							
Number of fruits/inflorescence	Quantitative							
Chlorophyll	Quantitative							
Seedling height	Quantitative							

*Traits measured using the phenotypic scale developed by IPGRI (1996) with some modifications. Chlorophyll content was measured using SPAD 502 plus chlorophyll meter and height measurements were done by measuring scale.

TABLE 3: Analysis of variance (ANOVA), standard deviation (Std. Dev), and minimum and maximum scores of ninety-one fresh-market tomato genotypes for respective morphological traits measured.

Variable	ANOVA	Std. Dev	Minimum	Maximum
Growth type	2.32***	0.77	1.50	4.00
Seedling height	4.73***	1.08	2.15	7.40
Average height 1	10.23*	2.04	4.13	18.61
Average height 2	17.88*	2.91	8.26	24.03
Chlorophyll	52.18	3.65	42.15	62.05
Leaf color	2.39*	0.41	1.50	3.00
Leaf shape	3.08***	0.32	2.00	4.00
Foliage density	4.34**	0.55	3.00	6.00
Inflorescence	1.33***	0.47	1.00	2.00
Jointless	0.94***	0.24	0	1.00
Fruit size	3.49***	0.75	1.00	4.00
Fruit type	1.67*	1.00	1.00	5.00
Fruit shape	2.41***	2.08	1.00	6.00
Fruit per inflorescence	6.44	3.53	3.00	19.00
Shoulder shape	3.47***	1.51	1.00	6.00
Ribbing	3.63***	1.31	1.00	6.00
Green shoulder	0.26***	0.54	0.00	2.00
Green fruit color	4.59***	1.15	3.00	8.00
Ripening	2.06***	0.54	1.00	4.00
Ripe fruit color	1.25	0.47	1.00	3.00
Bacterial spot	3.23***	1.12	1.00	9.00
Late blight	2.90**	0.65	2.00	4.50

*, **, and * * * indicate level of significance at <0.05, <0.01, and <0.001, respectively.

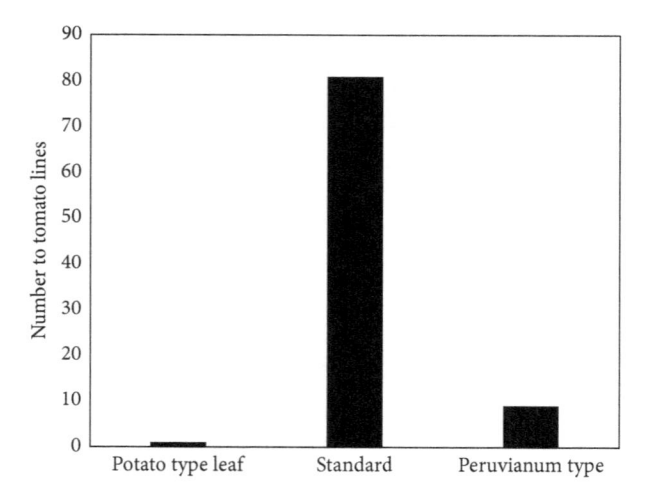

FIGURE 3: Leaf shape distribution of ninety-one fresh-market tomato genotypes.

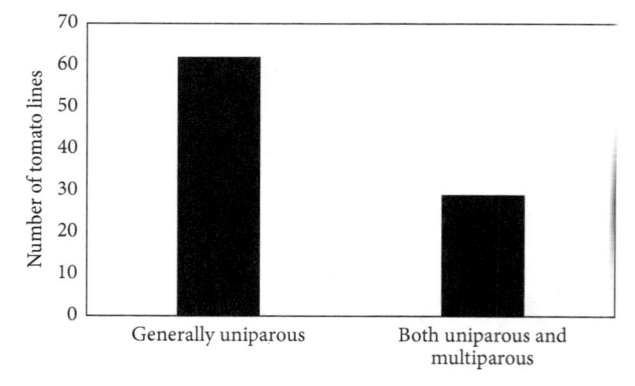

FIGURE 4: Distribution of inflorescence type in ninety-one fresh-market tomato genotypes.

produce insufficient energy due to less foliage or consume excess energy in maintaining the foliar growth. Most leaf types in those lines exhibited standard leaf type 81 (89%), some had *peruvianum* leaf type 9 (9.9%), and only 1 (1.1%) had potato type leaf (Figure 3).

Tomato is grown for fruits and most of the traits measured in this study were related to fruits or associated with fruits. Inflorescence of tomato lines is mainly found as uniparous (68.1%) whereas there were 29 (31.9%) lines that had both uniparous and multiparous inflorescences in them (Figure 4). Tomato lines in the study had a wide range of distribution regarding number of fruits per inflorescence from 3 to 19 (Table 3). Classification of these lines was done in two categories: <10 fruits per inflorescence (86.7%) and lines with 10 to 20 fruits per inflorescence (13.3%) (Figure 5). Pedicels arise from the inflorescence to hold tomato fruits to the plant. With the selection process to facilitate mechanical harvesting, a jointless pedicel varieties are being selected in the breeding programs. In this study, majority of the tomato

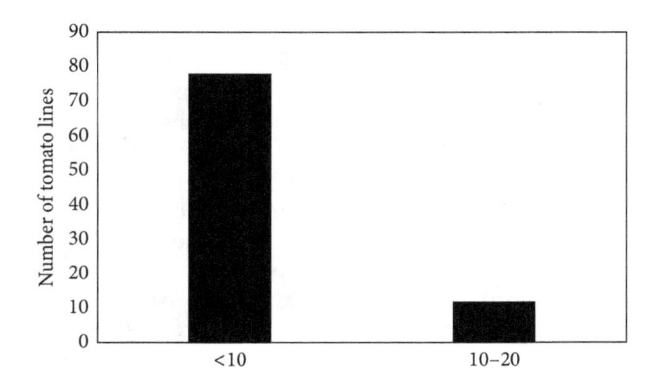

FIGURE 5: Distribution of tomato genotypes in relation to number of fruits per inflorescence.

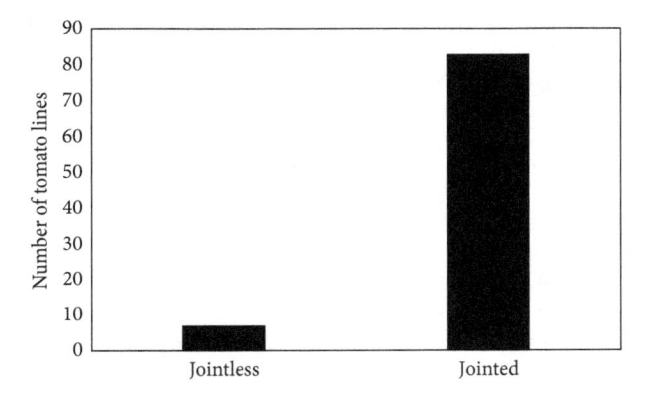

FIGURE 6: Distribution of ninety-one fresh-market tomato genotypes based on pedicel type.

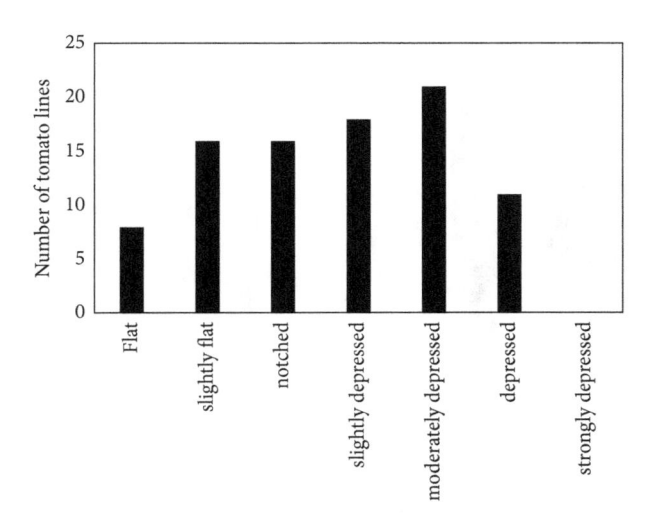

FIGURE 7: Distribution of ninety-one fresh-market tomato genotypes based on shoulder shape of fruits.

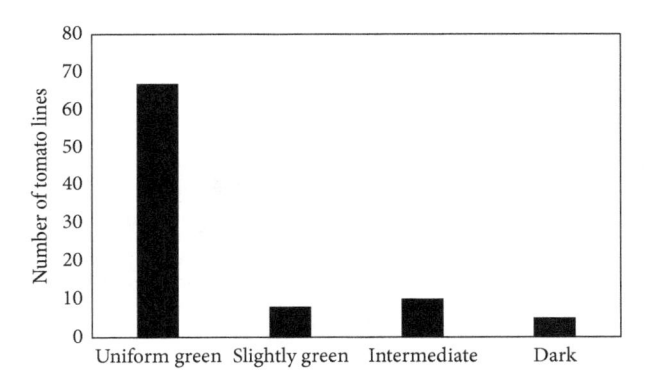

FIGURE 8: Distribution of ninety-one fresh-market tomato genotypes based on green shoulder.

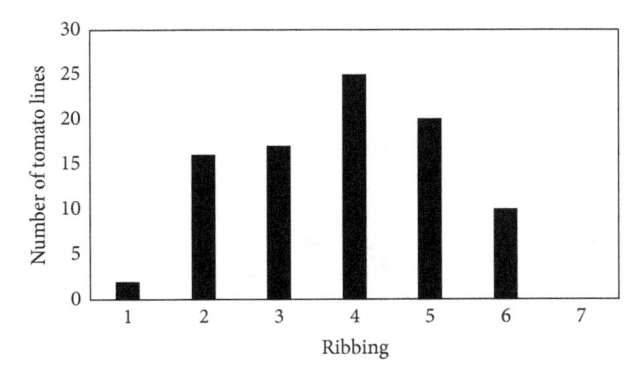

FIGURE 9: Distribution of ninety-one fresh-market tomato genotypes based on ribbing on fruits. The scale ranging from 1 to 7 with 1 being no or very weak ribbing and 7 being the strong.

lines had jointed pedicel (92.2%) in comparison to that of jointless (7.8%) (Figure 6). The connection point of pedicel and the fruit in tomatoes is known as shoulder. Tomato lines were also grouped into various classes based on their fruit shoulder shape ranging from 1 to 7 with flat shoulder shape being 1 to strongly depressed being 7 and the distribution is shown in Figure 7. Most of the tomato lines had at least some depression of 81 (91.1%) lines in their shoulder and just 8 (8.9%) of the lines had smooth, flat shoulder shape. Some of the tomato lines exhibited green shoulder in the basal portion of the fruits. This was classified into five classes: 67 tomato lines did not demonstrate green shoulder (74.4%), 8 lines showed slightly green shoulder (8.9%), 10 lines showed intermediate green shoulder (11.1%), and 5 lines had green shoulder (5.6%) (Figure 8). Tomato lines in the study also exhibited different forms of edges before they got softened after ripening. This trait was classified into seven groups ranging from very weak ribbing as 1 to strong ribbing as 7 (Figure 9). Fresh-market tomatoes have been recently bred for different colors. Tomato lines in this study displayed three predominant colors in tomatoes: red (89.5%), pink (5.8%), and yellow (4.7%) (Figure 10). Along with variation in ripe fruit colors, there is also significant variation in duration of ripening of tomatoes. With selections for early maturity, there were 4 lines that had very early maturity (4.4%), 5 lines with early maturity (5.5%), 77 lines with standard maturity (84.6%), and 5 lines with late maturity (5.5%) (Figure 11).

Fruits were classified into five different categories based on their sizes. There was 1 very small sized fruit (1%), followed by 3 small sized (4%), 20 intermediate (22%), 11

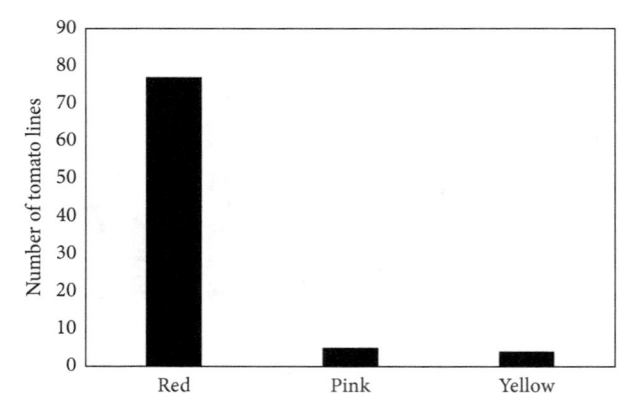

FIGURE 10: Distribution of ninety-one fresh-market tomato genotypes based on ripe fruit color.

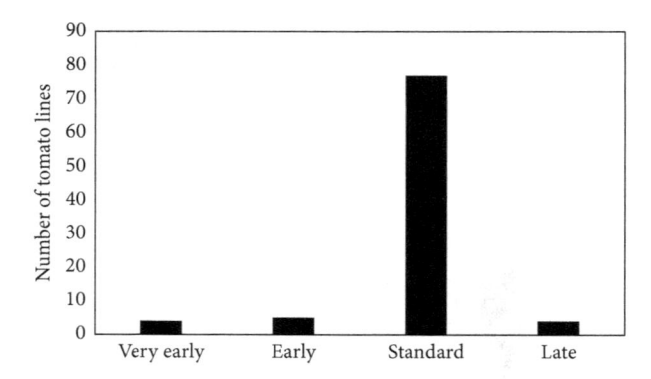

FIGURE 11: Distribution of ninety-one fresh-market tomato genotypes based on fruit ripening.

large (12%), and 55 very large (61%) fruits (Figure 12). Fruits were classified based on their types into four categories: cherry (1%), grape (13%), plum (16%), and large (70%) (Figure 13). According to the shape of fruits, tomato lines were categorized into six classes: flattened (60%), slightly flattened (6%), highly rounded (1%), rounded (8%), heart-shaped (4%), and cylindrical (21%) (Figure 14).

3.2. Correlation Analysis. All the data measured in the study were also analyzed for correlations, and the results are presented in Table 4. There were only a few correlations that were very high, and most of the traits either were not significantly correlated or, if significant, were not very high. This was not surprising because of the long tomato breeding history and moving of genes from one type of tomato into another had significant impact on this. Correlations between the seedling heights and average heights were 0.66 and 0.50, respectively, which were significant at $P < 0.001$. The correlation decreased in the second height in comparison to the first because of the lines that had *br* gene and determinate growth habit which led them to grow normally during the seedling stage but after seedling stage, the internodes of those with *br* gene developed shorter than that of determinate lines and growth of determinate lines were checked after flowering whereas semi-determinate and indeterminate lines continued to grow as flowering occurred. As the trend of plant growth was similar in the postseedling stage, the correlation between the two average heights also increased as expected (0.78, $P < 0.001$). The correlation between the inflorescence and growth type was 0.54 ($P < 0.001$) which indicates that semi-determinate and indeterminate growth types are likely to have branched inflorescence in comparison to that of determinate types. However, the correlation was not very high because of some grape and plum tomatoes that have been developed with determinate growth habit but they exhibit the branched inflorescences. Fruit size was negatively correlated (−0.37, $P < 0.001$), with growth type indicating that semi-determinate and indeterminate growth type plants are likely to have relatively smaller fruit sizes as compared to dwarf and determinate growth type. However, the correlation was comparatively low because of the development of grape and plum type tomatoes that are of determinate types and have smaller fruit sizes. Inflorescence type was significantly negatively correlated with the fruit size (−0.71, $P < 0.001$). This is important to produce different fruit size tomato. If the inflorescence type is branched giving rise to an increased number of flowers, it will be likely to hold more fruits. However, a plant will not be able to handle many larger size tomato fruits in one inflorescence or many in comparison to smaller size tomato fruits. Due to smaller size, a large number of fruits can develop well in a cluster whereas larger fruits have to compete for space, nutrients and protection, or need of sunlight being in the same cluster. Smaller fruits tend to be more of cherry, grape, and plum type whereas larger tomatoes tend to be flat. Hence, fruit size is highly correlated with fruit type (0.75, $P < 0.001$). Another important trait of fruit is fruit shape on which a lot of work has been done in the breeding programs. Despite the development of different types of tomato fruits with various shapes, the correlation between these traits is still significant (0.43, $P < 0.001$). Fruit load per inflorescence is highly dependent on fruit characteristics and growth type. As expected, with an increase in fruit size, the number of fruits per inflorescence decreases significantly (−0.83, $P < 0.001$). Fruit types like cherry and grapes had a higher number of fruits per inflorescence followed by plum and larger fruit types (0.65, $P < 0.001$). Correlation among tomato plant growth type and number of fruits per inflorescence was significant (0.55, $P < 0.001$) although not very high. Shoulder shapes of larger fruits were found to be depressed to a greater intensity in comparison to smaller fruits (0.75, $P < 0.001$) whereas negative correlation was observed between shoulder shape and fruit type and fruit shape (−0.69, $P < 0.001$, and 0.68, $P < 0.001$, resp.). Flat shaped large fruits had a higher intensity of shoulder depression whereas cherry, grape, and plum types showed lower depression in shoulder shape. Shoulder shape was also, although not very high, significantly negatively correlated with number of fruits per inflorescence (0.59, $P < 0.001$). It was also observed that fruits demonstrated higher ribbing when shoulder shape depression was higher (0.76, $P < 0.001$). Fruits that developed into flattened shape after ripening were observed to have high ribbing and those that were heart-shaped or round had low ribbing intensity (0.53, $P < 0.001$). Similarly, smaller fruits showed less ribbing

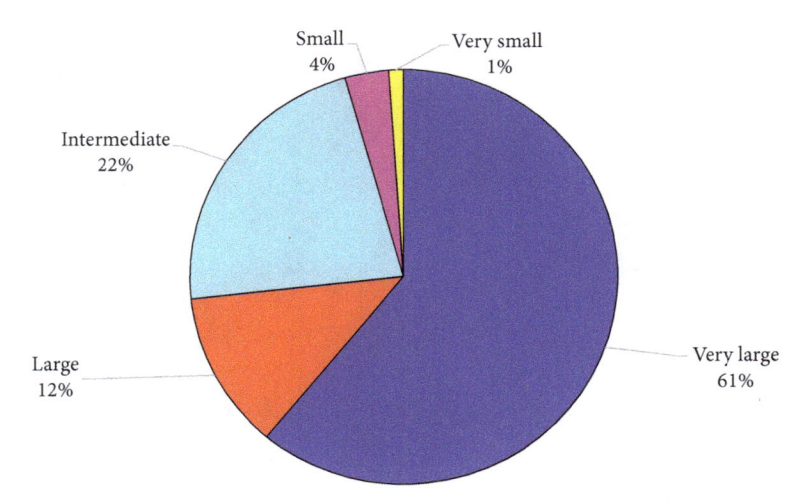

FIGURE 12: Distribution of ninety-one fresh-market tomato genotypes based on fruit size.

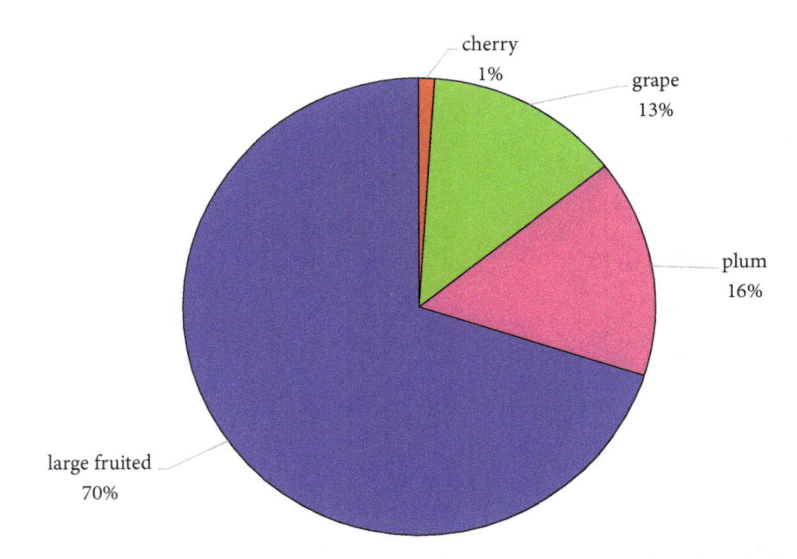

FIGURE 13: Distribution of ninety-one fresh-market tomato genotypes based on fruit type.

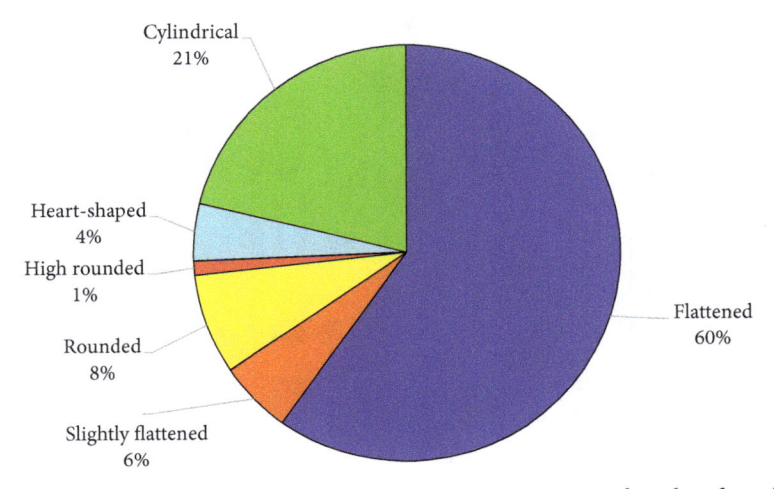

FIGURE 14: Distribution of ninety-one fresh-market tomato genotypes based on fruit shape.

TABLE 4: Correlation analysis of morphological traits of fresh-market tomato genotypes.

Parameters	Growth type	Seedling height	Average height 1	Average height 2	Chlorophyll	Leaf color	Leaf shape	Foliage density	Inflorescence	Fruit size	Fruit type	Fruit shape	Fruit per inflorescence	Shoulder shape	Ribbing	Green shoulder	Green fruit color	Ripening	Ripe fruit color	Bacterial spot
Seedling height	-0.093																			
Average height 1	-0.049	0.663***																		
Average height 2	0.230*	0.503***	0.778***																	
Chlorophyll	0.033	-0.079	0.029	-0.162																
Leaf color	0.112	-0.112	-0.129	-0.201	0.180															
Leaf shape	0.505***	-0.368**	-0.406***	-0.28**	0.135	0.372***														
Foliage density	-0.140	-0.041	-0.188	-0.32**	0.027	0.252*	0.145													
Inflorescence	0.535***	-0.094	-0.067	0.251*	-0.009	0.040	0.402***	0.033												
Fruit size	-0.373**	0.239*	0.281**	0.028	-0.023	-0.289**	-0.551***	-0.274*	-0.711***											
Fruit type	0.289**	-0.166	-0.231*	-0.018	-0.068	0.187	0.429***	0.133	0.559***	-0.748***										
Fruit shape	0.248*	-0.235*	-0.187	0.082	-0.095	0.140	0.354**	0.159	0.634***	-0.765***	0.425***									
Fruit per inflorescence	0.549***	-0.195	-0.211*	0.045	0.034	0.343**	0.607***	0.130	0.588***	-0.830***	0.649***	0.550***								
Shoulder shape	-0.086	0.262*	0.347**	0.150	0.108	-0.186	-0.380**	-0.179	-0.547***	0.748***	-0.686***	-0.684***	-0.585***							
Ribbing	0.031	0.128	0.180	0.027	0.209	-0.139	-0.273**	-0.285**	-0.4***	0.557***	-0.518***	-0.629***	-0.424***	0.758***						
Green shoulder	0.598***	0.195	0.258*	0.408***	-0.009	-0.153	-0.073	-0.125	0.204	0.071	-0.148	-0.080	0.000	0.339**	0.296**					
Green fruit color	0.481***	0.144	0.221*	0.309**	0.078	-0.076	-0.216*	-0.057	0.085	0.109	-0.197	-0.180	-0.099	0.334**	0.371**	0.836***				
Ripening	0.023	-0.272*	-0.254*	-0.175	0.058	0.297**	0.476***	0.231*	0.331*	-0.646***	0.51*	0.407***	0.664***	-0.561***	-0.434***	-0.33**	-0.348**			
Ripe fruit color	0.291**	-0.044	0.056	0.186	0.052	-0.131	0.060	-0.251*	0.121	0.055	-0.125	-0.002	-0.029	0.068	0.131	0.245*	0.268*	-0.24*		
Bacterial spot	-0.277**	-0.080	0.020	0.144	-0.076	-0.276**	-0.160	-0.079	0.101	-0.025	0.060	0.223*	0.036	-0.180	-0.251*	-0.129	-0.130	0.215*	-0.105	
Late blight	-0.306**	-0.048	-0.144	-0.147	-0.099	-0.062	-0.249*	-0.130	-0.240*	0.212*	-0.241*	0.031	-0.297*	0.038	0.018	-0.130	-0.085	-0.130	-0.091	0.386**

*, **, and *** indicate level of significance at <0.05, <0.01, and <0.001, respectively.

TABLE 5: Eigenvalues of the correlation matrix of principal component analysis of ninety-one fresh-market tomato genotypes.

PCA	Eigenvalue	Difference	Proportion	Cumulative
1	9.0367821	3.4435029	0.3012	0.3012
2	5.5932792	2.7534108	0.1864	0.4877
3	2.8398684	1.0918018	0.0947	0.5823
4	1.7480666	0.4301305	0.0583	0.6406
5	1.3179361	0.1266968	0.0439	0.6845
6	1.1912393	0.1131292	0.0397	0.7242
7	1.0781102	0.1113939	0.0359	0.7602
8	0.9667163	0.2271584	0.0322	0.7924
9	0.739558	0.0453556	0.0247	0.8171
10	0.6942024	0.0093849	0.0231	0.8402

while larger fruits had higher degree of ribbing (0.56, $P < 0.001$) and fruit type was significantly negatively correlated to ribbing (-0.52, $P < 0.001$) indicating that ribbing intensity gradually increases from cherry, grape, and plum to large tomato fruits. Green shoulder on the basal end of tomato fruits was found to be correlated with growth type (0.6, $P < 0.001$) indicating dwarf and determinate type of tomato lines are less likely to have a green shoulder in their fruits in comparison to semi-determinate and indeterminate type tomato lines. Fruits with green shoulder resulted in darker green looking fruits with highly significant correlation (0.84, $P < 0.001$) with green fruit color. Fruit ripening time was found to be correlated to different traits at a different level. Increase in fruits per inflorescence was the most highly correlated trait with fruit ripening (0.66, $P < 0.001$) followed by fruit size (0.65, $P < 0.001$) and moderately correlated with shoulder shape (0.56, $P < 0.001$) and fruit type (0.51, $P < 0.01$). Bacterial spot caused by *Xanthomonas perforans* and late blight caused by *Phytophthora infestans* are one of the major problems in fresh-market tomato. Although there were some statistically significant correlations found between other traits and these diseases but those correlations were very low (Table 3). However, correlations between these two diseases per se were the highest between them (0.39, $P < 0.01$). The incidence was measured later in the season when the diseases had enough time to establish themselves in the plants. This could be due to decreased immunity of plants once they are infected by one disease that compromises the immunity of the plants increasing the susceptibility of the plant for other pathogens prevalent in the environment. There were traits that did not show any significant correlations with any other traits like chlorophyll content, leaf color which indicates that selection procedure in modern tomato lines has already led to the desired level of these traits for the current production.

3.3. Principal Component Analysis and Cluster Analysis. Principal component analysis (PCA) of the tomato genotypes indicated that six principal components (PC) described more than 70% of the variation (Table 5). The first two PCs include traits that are of most importance to the tomato production. PC1 includes the seedling and plant heights and fruit traits like green shoulder and fruit color during developmental stage and PC2 contains other fruit traits like inflorescence type, number of fruits per inflorescence, fruit type, and fruit shape and leaf shape. These two PCs explain nearly half of the variation (48.77%) present in the study (Table 5).

Cluster analysis of the tomato genotypes generated two major divisions that separate majority of tomato lines developed at North Carolina tomato breeding program with rest of others including grape type genotypes like Smarty, Mountain vineyard, and Mountain Honey, plum type genotypes like NCEBR3, NCEBR5, NCEBR6, and NC30P, and large type genotypes like NCHS1 from the program (Figure 15). Both clusters contain a combination of different types of genotypes which is a clear indication that different traits have been significantly transferred into the new lines being developed to increase the value of new cultivars, which was expected. These divisions were primarily divided into three clusters which were further divided into groups either constituting tomato genotypes with similar or different characteristics. When the groups within each cluster were observed, a clear distinction between them was observed.

The first cluster consists of tomato lines that are advanced breeding lines, hybrids, and advanced selections from various breeding programs. In this cluster, 141G, 143G, 145G, 147G, and 148G are of indeterminate growth type with branched inflorescence containing up to five large size fruits. Other tomato lines in this group Fla 8638B, Fla 8902LBR, NC2CELBRxPAU351-Y, YCBT-1 x-NC132S, 335 x NC2CELBR, NC2CELBRxPAU332, NC2CELBR x C23-3, NC132S x C23-3, 335 x NC132S, YCBT-1 x NC2rinEC, PAU332xNC2rinEC, NC132SxPAU332, NC132SxPAU351-Y, VFT Vendor x CUTR 2nd Gen 124116-4, NC33EB1 x CUTR 1st Gen CUTR, Summerpick, NC33EB1 x CUTR 2nd Gen 124116-4, Richmond, Brickyard are determinate type with unbranched inflorescence containing up to six fruits/inflorescence. Fruits of these lines are large except VFT Vendor x CUTR 2nd Gen 124116-4 and NC33EB1 x CUTR 2nd Gen 124116-4 which are of intermediate size.

The second cluster consisted of two main subgroups. First subgroup consisted of a combination of different types of tomato lines from different breeding programs. This cluster consists of PSGH-13-33, Monticello, Ridge Runner, PSFH-13-19, PSFH-13-47, PSGH-13-16, XTM0017, XTM262, UF 1104113, UG1155013, NC 161L, and Mountain Rouge. Among these lines PSGH-13-33, PSGH-13-16, and Mountain Rouge are indeterminate types. PSGH-13-33 and PSFH-13-47 had a plum type, and PSGH-13-16 had elongated, grape type fruits whereas rest of all had large fruits.

Second subgroup of tomato lines from second major division consisted of lines NC 5Grape, Mountain Honey, Mountain Vineyard, and Smarty. These lines had indeterminate growth type, branched inflorescence containing more than 13 fruits/inflorescence which were small and slightly round to elongated in shape.

Third subgroup consisted of a combination of a different type of tomato lines regarding fruit characteristics. This included lines Summit, Sun Leaper, Piedmont, NCHS1, Plum Dandy, NCEBR6, Plum Crimson, NCEBR3, NCEBR5, NCEBR7, NCEBR8, and NC30P. All of these lines were

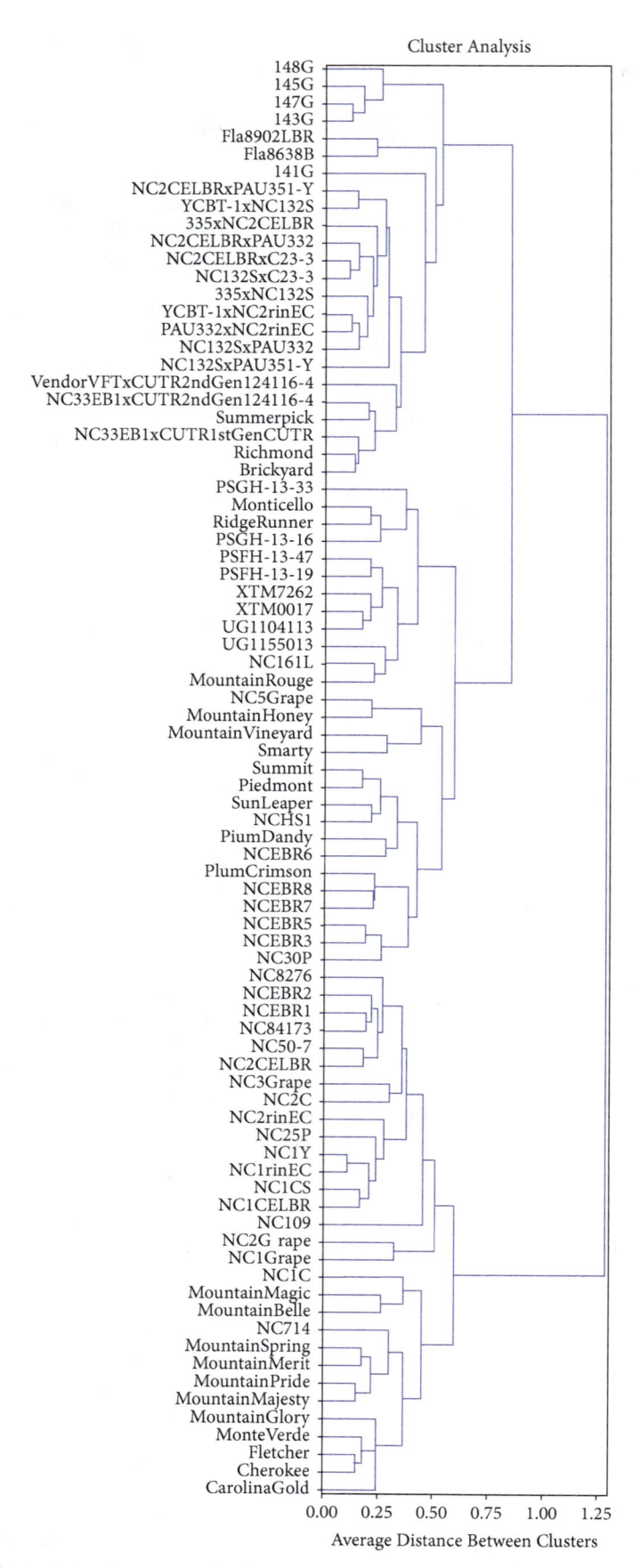

FIGURE 15: Cluster analysis of ninety-one fresh-market tomato genotypes based on average distance method.

determinate. However, among them, Summit, Sun Leaper, Piedmont, and NCHS1 had large, flattened shaped fruits when ripe. Inflorescence of these lines was unbranched with up to 5 fruits per inflorescence. Other lines Plum Crimson, NCEBR3, NCEBR5, NCEBR7, NCEBR8, and NC30P had small to medium sized plum-shaped fruits ranging from five to nine fruits per inflorescence depending on lines.

The third cluster consisted of few small subgroups. First subgroup in this cluster consisted of NC2CELBR, NC50-7, NC8276, NC84173, NCEBR1, and NCEBR2. These lines are of determinate growth type with uniparous inflorescence containing up to five fruits per inflorescence. These fruits were medium to large and of flattened shape when ripe. These lines are also important regarding fungal diseases because of their resistance to fusarium wilt and verticillium wilt. Second subgroup consisted of NC1CELBR, NC1CS, NC1rinEC, NC1Y, NC2rinEC, and NC25P which were determinate lines with up to 5 fruits per inflorescence. All the lines except NC25P are large fruited with uniparous inflorescence; however, the inflorescence of NC-25P was sometimes multiparous with elongated, plum type fruits. Third subgroup of NC 1Grape and NC 2Grape was observed which are of indeterminate growth type containing slightly elongated grape type fruits that are small. Inflorescence of these lines is multiparous having more than 15 fruits in a cluster. NC2C and NC 3Grape formed fourth subgroup which had branched inflorescence with more than ten small sized fruits. However, NC-109 was not clustered in any subgroup which was a determinate line with plum type fruits. Fifth subgroup consisted of tomato lines NC 714, Mountain Spring, Mountain Merit, Mountain Pride, Mountain Majesty, Mountain Glory, Mountain Verde, Fletcher, Cherokee, and Carolina Gold which are of determinate growth types with large fruit types with the mostly flattened shape when ripe. A second cluster consisting of Mountain Magic, NC 1C, and Mountain Belle exhibited different growth habit but have multiparous inflorescence with more than ten fruits per inflorescence, cherry type, very small fruits.

4. Discussion

Domestication and years of selection processes imposed on tomato have led to severely decreased genetic variability [13–15]. Domestication of tomato was based on nontoxic, palatable, and nutritious features. Qualities like flavor, juiciness, and other consumer-desirable characteristics became important after domestication phase. These traits were incorporated by farmers based on selection methods without or with the very minimum breeding approach. Landraces display greater diversity in comparison to modern cultivars because of desirable uniformity to best fit in the production system [16]. It has been estimated that modern tomato cultivars contain <5% of the genetic variation that is present in their wild relatives [17]. It is difficult to increase diversity just with classical breeding that depends on naturally occurring recombination, by crossing two modern cultivars. With the involvement of private breeding companies in the new cultivars production in the tomato industry, it has become very competitive and hence developing new traits using wild germplasm is not feasible with significant linkage drag. Therefore, it is not surprising to see the increasing relativity among the modern cultivars with decreasing diversity as agriculture becomes monoculture rather than diversified. Although modern tomato shows higher phenotypic diversity, there has been a loss of genetic diversity in modern cultivars as compared to the wild or landraces. It has been studied previously that the genetic diversity of tomato lines was low [12, 16]. Morphological traits have been used to assess the diversity in different germplasms, sometimes with molecular markers [12, 17–19]. Ninety-seven tomato accessions from Iran and Turkey were classified into five clusters [17], sixty-seven lines from Argentina were classified into three clusters [18] and fifty lines, including both cultivated and wild, were classified into six clusters [19].

Based on consumption, the tomato has been bred separately for fresh-market and processing industries. As a result, cultivars yielding fruits with various sizes, shapes, and colors and different production techniques were developed [20, 21]. With tomato breeding being widely performed, after 1930, the goal of tomato breeding has been primarily yield and productivity followed by disease resistance to maintain production in challenging environment. Identification of important phenotypic trait in wild or related germplasm, their genetic basis, and introgression of those traits into advanced modern lines have been key to the development of new cultivars in tomato. With fresh-market tomato, some of the important traits that have been of main focus for producer are yield or size, disease and insect resistance, heat and other abiotic stress tolerance, uniform/synchronous ripening, and harvesting with focus in mechanization; for marketing are longer shelf life, minimum handling, and shipping damage; and for consumer are flavor and taste, appearance/color, and forms of use like salad and cooking [20–22]. The jointless (*j2*) allele [23] was introgressed from *S. cheesmanii*. Compact fruit set by determinate growth, *sp* for self-pruning, was discovered from a natural mutation in the early 20th century. Four genes are known to control fruit shape: *sun* [24] and *ovate* [25] control elongated shape like in Roma type whereas *fasciated* (FAS) [26] and *locule number* (LC) [27] control fruit locule and flat shape.

There are different types of fresh-market tomato developed according to the need: large fruited, plum-shaped also known as Roma tomato, grape type, and cherry type. Large fruited and plum type are used in cooking and meals whereas grape and cherry types are primarily used as fresh fruits and in salads. Cherry type tomato developed due to the genetic admixture of *S. pimpinellifolium* and cultivated accessions [28, 29]. Wild tomato is of indeterminate growth type; however advanced tomato cultivars that are commercial grown in fresh-market production system are determinate making it easier to contain them over plastic beds and manage them while pruning, staking and tying with the string, spraying, and harvesting as they demand less management practice in comparison to the wild lines. With the identification of determinate and dwarf growth habit, these traits are highly being incorporated into other cultivars including all types of tomatoes. As a result, there are determinate large fruited, plum, and grape type tomato lines already released. Higher negative correlation between inflorescence and fruit size indicates that despite the introgression of determinate trait in among all the new cultivars, smaller fruit sized tomato lines tend to have semi-determinate or determinate growth habit. One of the reasons for this could be the growth trait linked with fruit size or taste. Small size fruits have high total soluble sugars in comparison to larger fruits. Indeterminate growth may be preferred in smaller sized fruit lines for increased production for a longer season. However, there are varieties like NC-3 Grape which are determinate [30]. Introgression of disease and insect resistant genes in all newly being developed cultivars has also been very common in all types of tomatoes [9, 31]. Cherry or grape type tomato has been developed for consumption like other fruits and needs to contain the similar amount and balanced combination of sugar and acid. With increase in fruit size, the amounts of soluble sugars and acids get diluted and decrease [32]. Plum and large type fruits have been developed for cooking or to be a component of meals so yield remains the focus rather than its constituents. This was also observed in this study and was the reason for higher correlation between fruit size and fruit types. When large fruits ripe, they soften, and due to the size, the fruits become flattened or globular in size whereas ripe plum fruits become smoother but remain cylindrically oblong keeping the elongated shape. Due to smaller shape of cherry and grape types, they remain firm and maintain their elongated and round shape. Because of this nature, correlation between the fruit type and size remained higher although with the breeding techniques large fruits are being bred for firmness that keep the fruit round shaped rather than turn into flattened when ripe like fruits of NC84173 are round to elongated round even at harvesting and when ripe [33]. Number of fruits grown in an inflorescence also highly depends on size of the fruits as was seen in this study. If the fruits are smaller, more number of fruits can be developed in one inflorescence whereas when the fruit size increases, increased energy drain to the inflorescence remains insufficient to develop equal number of fruits compared to that of smaller sizes like grape and cherry type. Larger fruits also compete for space, sunlight, and protection by foliage and therefore limited number of fruits develop into mature fruits even when there are more flowers in an inflorescence during the time of floral development. Shoulder shape of the fruits increases when the fruits development occurs and outgrows the region where the pedicel connects the fruits. This is more prominent in large fruits and relatively less in plum, grape, and cherry type fruits because of the relatively less increase in the size of these types of fruits in comparison to the larger fruits. Fruits that are smaller in size are expected to ripe early because of their faster development and reaching maturity. However, early maturity is also a very desirable trait in all types of tomato and hence is being introgressed into different types of tomato including plum and large for greater value. Similarly, varieties with smaller sized fruits are bred with larger sized fruit varieties to develop fruits with standard ripening time and multiple colors.

There have been many types of researches on finding resistance to diseases like bacterial spot; however, no single cultivar has been developed that can resist all races of *Xanthomonas* causing bacterial spot making this disease a very difficult problem to overcome [34]. This explains the low correlation of bacterial spot with any other traits observed in this study as seen in a previous study [35], whereas resistance to fungal diseases like early blight, late blight, fusarium, and verticillium wilts and nematode resistance have been identified and are already being introgressed in new varieties [36]. Pyramiding the genes for resistance to diseases and pests has been one of the major goals of tomato breeding programs. Modern varieties developed have been reported to be introgressed with up to 12 different diseases and pest resistant genes. This was also observed in the cluster analysis where different tomato types were clustered into one or different groups indicating the presence of common traits across all the tomato lines regardless of their origin or fruit characteristics [37, 38].

5. Conclusion

Morphological characteristics of tomato lines, from different origins being actively used in developing new cultivars, advanced selections for inbred development, or lines being tested as putative cultivars were used in this study to assess diversity present in fresh-market tomato. Higher correlations between fruit traits indicate the necessity of those traits to remain related to higher quality and yield. The greater number of large sized tomato fruits per inflorescence will lead to low-quality fruits whereas a lower number of cherry or grape type fruits per cluster result in lower yield. Lower correlations between some traits and fruit types indicated the introgression of those traits in all fruit types like determinate growth habit is desirable for all types of fruits for better production management, not only early fruit maturity but standard and late maturity are also required for longer seasonal production as these trends could be observed in the study. Grouping of different types of tomato lines in one cluster indicates the most desirable traits, including disease resistance, being transferred into all types of tomato lines being developed by breeding programs, whereas grouping of similar tomato lines with specific traits like grape type tomato indicates the unique features of the cluster that is preferred only to have in those types of tomato lines. As limited number traits are being commonly introgressed into the modern tomatoes across the world, the lost alleles during the domestication and rigorous selections can be brought back by crossing wild lines with the cultivated species. Genomics can play an important role in opening new perspectives and opportunities to layout the divergence over time and help in bringing characteristics that were lost or missed during domestication and selection. With the availability of the genome sequence of tomato and developed molecular marker information, use of wild relatives can be strongly reinforced.

Acknowledgments

This research was supported by the funding grant of North Carolina Tomato Growers' Association and USDA Hatch Project. The authors would like to thank different public and private breeding programs that contributed tomato genotypes for the research.

References

[1] C. M. Rick, "Tomato (family Solanaceae)," in *Evolution of crop plants*, N. W. Simmonds, Ed., pp. 268–273, Longman Publications, 1976.

[2] C. E. Williams and D. A. St. Clair, "Phenetic relationships and levels of variability detected by restriction fragment length polymorphism and random amplified polymorphic DNA analysis of cultivated and wild accessions of Lycopersicon esculentum," *Genome*, vol. 36, no. 3, pp. 619–630, 1993.

[3] A. Khadivi-Khub, Z. Zamani, and N. Bouzari, "Evaluation of genetic diversity in some Iranian and foreign sweet cherry cultivars by using RAPD molecular markers and morphological traits," *Horticulture, Environment and Biotechnology*, vol. 49, pp. 188–196, 2008.

[4] P. J. Terzopoulos and P. J. Bebeli, "Phenotypic diversity in Greek tomato (Solanum lycopersicum L.) landraces," *Scientia Horticulturae*, vol. 126, no. 2, pp. 138–144, 2010.

[5] K. Nikoumanesh, A. Ebadi, M. Zeinalabedini, and Y. Gogorcena, "Morphological and molecular variability in some Iranian almond genotypes and related Prunus species and their potentials for rootstock breeding," *Scientia Horticulturae*, vol. 129, no. 1, pp. 108–118, 2011.

[6] H. Fufa, P. S. Baenziger, B. S. Beecher, I. Dweikat, R. A. Graybosch, and K. M. Eskridge, "Comparison of phenotypic and molecular marker-based classifications of hard red winter wheat cultivars," *Euphytica*, vol. 145, no. 1-2, pp. 133–146, 2005.

[7] N. Ranc, S. Mũos, S. Santoni, and M. Causse, "A clarified position for solanum lycopersicum var. cerasiforme in the evolutionary history of tomatoes (solanaceae)," *BMC Plant Biology*, vol. 8, article no. 130, 2008.

[8] S. D. Tanksley and S. R. McCouch, "Seed banks and molecular maps: Unlocking genetic potential from the wild," *Science*, vol. 277, no. 5329, pp. 1063–1066, 1997.

[9] C. M. Rick and R. T. Chetelat, "Utilization of related wild species for tomato improvement," *Acta Horticulturae*, no. 412, pp. 21–38, 1995.

[10] IPGRI, "Tomato Descriptors," in *Proceedings of the International. Plant Genetic Resources Institute*, Rome, Italy, 1996.

[11] K. L. Ivors and F. J. Louws, *North Carolina Agricultural Chemicals Manual*, College of Agriculture and Life Sciences, North Carolina State University, 2013.

[12] A. Mazzucato, R. Papa, E. Bitocchi et al., "Genetic diversity, structure and marker-trait associations in a collection of Italian tomato (Solanum lycopersicum L.) landraces," *Theoretical and Applied Genetics*, vol. 116, no. 5, pp. 657–669, 2008.

[13] K. Hammer, "Das Domestikationssyndrom," *Die Kulturpflanze*, vol. 32, no. 1, pp. 11–34, 1984.

[14] S. Grandillo and S. D. Tanksley, "QTL analysis of horticultural traits differentiating the cultivated tomato from the closely related species Lycopersicon pimpinellifolium," *Theoretical and Applied Genetics*, vol. 92, no. 8, pp. 935–951, 1996.

[15] Y. Bai and P. Lindhout, "Domestication and breeding of tomatoes: what have we gained and what can we gain in the future?" *Annals of Botany*, vol. 100, no. 5, pp. 1085–1094, 2007.

[16] J. Cebolla-Cornejo, S. Roselló, and F. Nuez, "Phenotypic and genetic diversity of Spanish tomato landraces," *Scientia Horticulturae*, vol. 162, pp. 150–164, 2013.

[17] M. Henareh, A. Dursun, and B. A. Mandoulakani, "Genetic diversity in tomato landraces collected from Turkey and Iran revealed by morphological characters," *Acta Scientiarum Polonorum, Hortorum Cultus*, vol. 14, no. 2, pp. 87–96, 2015.

[18] X. Hu, H. Wang, J. Chen, and A. W. Yang, "Genetic diversity of Argentina tomato varieties revealed by morphological traits, simple sequence repeat, and single nucleotide polymorphism markers," *Pakistan Journal of Botany*, vol. 44, no. 2, pp. 485–492, 2012.

[19] R. Zhou, Z. Wu, X. Cao, and F. L. Jiang, "Genetic diversity of cultivated and wild tomatoes revealed by morphological traits and SSR markers," *Genetics and Molecular Research*, vol. 14, no. 4, pp. 13868–13879, 2015.

[20] G. R. Rodríguez, S. Muños, C. Anderson et al., "Distribution of SUN, OVATE, LC, and FAS in the tomato germplasm and the relationship to fruit shape diversity," *Plant Physiology*, vol. 156, no. 1, pp. 275–285, 2011.

[21] G. Bauchet and M. Causse, "Genetic diversity in tomato (Solanum lycopersicum) and its wild relatives," in *Genetic diversity in plants*, M. Çaliskan, Ed., pp. 133–162, InTechOpen, Rijeka, Croatia, 2012.

[22] Q. Gascuel, G. Diretto, A. J. Monforte, A. M. Fortes, and A. Granell, "Use of natural diversity and biotechnology to increase the quality and nutritional content of tomato and grape," *Frontiers in Plant Science*, vol. 8, article no. 652, 2017.

[23] E. J. Szymkowiak and E. E. Irish, "JOINTLESS suppresses sympodial identity in inflorescence meristems of tomato," *Planta*, vol. 223, no. 4, pp. 646–658, 2006.

[24] E. van der Knaap, A. Sanyal, S. A. Jackson, and S. D. Tanksley, "High-resolution fine mapping and fluorescence in Situ hybridization analysis of sun, a locus controlling tomato fruit shape, reveals a region of the tomato genome prone to DNA rearrangements," *Genetics*, vol. 168, no. 4, pp. 2127–2140, 2004.

[25] H.-M. Ku, S. Doganlar, K.-Y. Chen, and S. D. Tanksley, "The genetic basis of pear-shaped tomato fruit," *Theoretical and Applied Genetics*, vol. 99, no. 5, pp. 844–850, 1999.

[26] B. Cong, L. S. Barrero, and S. D. Tanksley, "Regulatory change in YABBY-like transcription factor led to evolution of extreme fruit size during tomato domestication," *Nature Genetics*, vol. 40, no. 6, pp. 800–804, 2008.

[27] S. Muños, N. Ranc, E. Botton et al., "Increase in tomato locule number is controlled by two single-nucleotide polymorphisms located near WUSCHEL," *Plant Physiology*, vol. 156, no. 4, pp. 2244–2254, 2011.

[28] C. M. Rick and M. Holle, "Andean lycopersicon esculentum var. cerasiforme: genetic variation and its evolutionary significance," *Economic Botany*, vol. 44, no. 3, pp. 69–78, 1990.

[29] I. E. Peralta and D. Spooner, "History, origin and early cultivation of tomato (Solanaceae)," in *Genetic improvement of Solanaceous crops*, M. K. Razdan and A. K. Mattoo, Eds., vol. 2, pp. 1–24, Science Publisher, Enfield (NH), 2007.

[30] R. G. Gardner and D. R. Panthee, "Grape tomato breeding lines: NC 1 grape, NC 2 grape, and NC 3 grape," *HortScience*, vol. 45, no. 12, pp. 1887-1888, 2010.

[31] J. H. Venema, P. Linger, A. W. Van Heusden, P. R. Van Hasselt,

and W. Brüggemann, "The inheritance of chilling tolerance in tomato (*Lycopersiconspp.*)," *The Journal of Plant Biology*, vol. 7, no. 2, pp. 118–130, 2005.

[32] J. McGillivray and L. Clemente, "Effect of tomato size on solids content," in *Proceedings of American Society of Horticultural Science*, pp. 68–466, 1956.

[33] "North Carolina Agricultural Research Service. 1990. Notice of release of NC 84173 tomato line," https://mountainhort.ces.ncsu.edu/wp-content/uploads/2017/03/release-NC84173.pdf?fwd=no.

[34] K. Bhattarai, F. J. Louws, J. D. Williamson, and D. R. Panthee, "Resistance to Xanthomonas perforans race T4 causing bacterial spot in tomato breeding lines," *Plant Pathology*, vol. 66, no. 7, pp. 1103–1109, 2017.

[35] K. Bhattarai, F. J. Louws, J. D. Williamson, and D. R. Panthee, "Differential response of tomato genotypes to Xanthomonas-specific pathogen-Associated molecular patterns and correlation with bacterial spot (Xanthomonas perforans) resistance," *Horticulture Research*, vol. 3, Article ID 16035, 2016.

[36] J. W. Scott and R. G. Gardner, "Breeding for resistance to fungal pathogens," *Genetic improvement of Solanaceous crops*, vol. 2, pp. 421–456, 2006.

[37] K. Bhattarai, F. J. Louws, J. D. Williamson, and D. R. Panthee, "Diversity analysis of tomato genotypes based on morphological traits with commercial breeding significance for fresh market production in eastern USA," *Australian Journal of Crop Science*, vol. 10, no. 8, pp. 1098–1103, 2016.

[38] K. Bhattarai, *Screening for Bacterial Spot (Xanthomonas spp.) Resistance in Tomato (Solanum lycopersicum L.) and Microbe Associated Molecular Patterns. Masters Thesis [Master thesis]*, North Carolina State University, 2016, https://repository.lib.ncsu.edu/bitstream/handle/1840.16/9920/etd.pdf;sequence=2.

Bridging a Gap between Cr(VI)-Induced Oxidative Stress and Genotoxicity in Lettuce Organs after a Long-Term Exposure

Cristina Monteiro,[1] Sara Sario,[2] Rafael Mendes,[2] Nuno Mariz-Ponte,[2] Sónia Silva,[3] Helena Oliveira,[1] Verónica Bastos,[2] Conceição Santos,[2] and Maria Celeste Dias[2,4]

[1]Laboratory of Biotechnology and Cytomics, University of Aveiro, Campo Santiago, 3810-193 Aveiro, Portugal
[2]Department of Biology, LAQV/REQUIMTE, Faculty of Sciences, University of Porto, Rua Campo Alegre, 4169-007 Porto, Portugal
[3]QOPNA, Department of Chemistry, University of Aveiro, 3810-193 Aveiro, Portugal
[4]Center for Functional Ecology (CEF), Department of Life Science, University of Coimbra, Calçada Martim de Freitas, 3000-456 Coimbra, Portugal

Correspondence should be addressed to Conceição Santos; csantos@fc.up.pt

Academic Editor: Glaciela Kaschuk

Chromium (Cr) contamination in arable soils and irrigating water remains a priority, particularly due to the challenges posed to crop production and food safety. Long-term Cr(VI) effects remain less addressed than short-term ones, particularly regarding organ-specific genotoxic profiles. Here we used the crop *Lactuca sativa* growing in a protected horticultural system and irrigated for 21 days with Cr(VI) (up to 200 mg/L). Besides the oxidative stress, the genotoxicity was evaluated. Shoots and roots showed distinctive oxidative stress status and genotoxic effects, in a dose-dependent manner. While 50 mg/L stimulated antioxidant activities and no major genotoxic effects were found, plants exposed to ≥150 showed an increase of oxidative disorders, together with cytostatic and DNA damage effects, and some mitotic impairment. Leaves showed less oxidative signs at 50 mg/L, while at 150/200 mg/L the antioxidant battery was stimulated. In Cr treated plants, the highest dose increased the DNA damage, reinforcing the idea that DNA breaks were related to mitotic disorders in higher doses. In conclusion, long-term exposure data show a highly responsive root, with a quadratic response meaning higher defenses at lower Cr doses, and higher oxidative and DNA damage and cytostatic effect at a higher dose.

1. Introduction

Chromium contamination of soils and irrigation water remains a matter of concern [1], with increased health risks in humans due to the food-chain contamination. A fertilizer plant site usually has a soil contamination of ~700 mg/Kg and in industrial/tannery effluents/soils may reach 5000–45,000 mg/Kg (reviewed by Shahid et al. [1]). Chromium's most stable oxidation forms are Cr(III) and Cr(VI), both having different mobility, bioavailability, and toxicity [2]. The physiological mechanisms of Cr transport in the plant through root absorption point to several ways of Cr input. Cr transport promotes gaps in other types of nutrient absorption and water [3]. Most studies on Cr-plant interactions have addressed Cr transfer from soil to plant and Cr hyperaccumulation for phytoremediation purposes (e.g., [4]) and morphophysiological and biochemical responses (e.g., root surface area, growth, morphology, photosynthesis, water/nutrient interactions, and transpiration) (e.g., [1, 5]). Cr accumulation promotes an impairment of photosynthesis in the chloroplasts [3]; thus, these effects promoted a reduction of plant biomass [6].

Besides the direct interaction of Cr with nucleic acids (e.g., promoting DNA-DNA cross-links), it has also been demonstrated that Cr(VI) may increase the generation of reactive oxygen species (ROS), leading to indirect cytotoxic effects [7–10]. The correlation of the antioxidant status with genotoxic damage remains a matter of debate. Besides, sequestration and/or chelation, a major strategy to limit these oxidation impacts, involve an antioxidant strategy that includes the antioxidant enzymes. Enzymes such as superoxide dismutase (SOD) or catalase (CAT) are essential in ROS molecules scavenging, as SOD is responsible for catalysing the conversion of superoxide radicals to H_2O_2 and CAT [as well as ascorbate peroxidase (APX) and peroxidases that use guaiacol as substrate (GPX)] has an important role in controlling H_2O_2 levels [11, 12]. Some studies on Cr-induced toxicity already reported a decrease in these antioxidant enzymes as a result of exposure to high Cr concentrations, while in lower doses their activity was higher, suggesting that some plants can be Cr-tolerant [e.g., Rice (*Oryza sativa*), Sunflower (*Helianthus annuus* L.), Rapeseed (*Brassica napus* L.), and Indian mustard (*Brassica juncea* L.)] [13–16]. When the activity of these antioxidant enzymes is not sufficient to combat excessive ROS generation, oxidative processes of lipids, proteins, and nucleic acids occur.

Cr(VI) genotoxicity in plants is far less known than on animals or microorganisms. Studies are restricted to short-term exposures (1 h–96 h), and few address the plant's profile after long-term exposure. Short-term exposures to $Na_2Cr_2O_7$ reduced mitosis while inducing chromosomal aberrations in *Pisum* (≤80 ppm) [17]. Also, chromosome fragmentation/aberration and mitotic abnormalities including lagging were reported in *Hordeum vulgare* (<0.5 mM) [18]. Using only 1 h–24 h exposure, Huang et al. [19] also found the disturbance on the up/down levels of transcripts in *Oryza sativa*. In a longer-term exposure, we have demonstrated that *Pisum sativum* exposed to Cr(VI) doses had cell cycle impairment and DNA fragmentation. Micronuclei formation is intimately associated with clastogenicity [20]. In the last years, comet assay and flow cytometry (FCM) have provided several sensitive and robust parameters that may complement the mitotic and chromosomal classical data and the quantitative genetic analyses.

This study used the crop *Lactuca sativa* under a simulated protected culture system to evaluate the effects of a long-term exposure to Cr(VI) on the antioxidant status and genotoxic disorders in both organs (roots and shoots). For that, the antioxidant enzymatic battery was analyzed, together with flow cytometric, micronuclei, mitotic aberrations, and comet assays.

2. Materials and Methods

Growth Conditions and Cr Content. Seedlings with 15 days old of *Lactuca sativa* L. (cv. "Povoa," Viveiros do Litoral, Aveiro, Portugal) were grown in pots with a peat : vermiculite (1 : 3 v/v) mixture, in a greenhouse with 430 μmol m^{-2} s^{-1} light intensity, a 16 h photoperiod, and 23 ± 2°C. During 21 days, plants were irrigated with 15 mL of 1/2 strength Hoagland's solution (pH 5.8) with 0, 50, 150, and 200 mg/L of Cr(VI) supplied by K_2CrO_4 (Sigma-Aldrich, USA). After that, plants' length was measured and plants were analyzed for chlorosis and necrotic spots. Roots were washed in a 5% $Ca(NO_3)_2$ followed by water to remove adsorbed Cr. For total Cr quantification, weighted fresh roots and leaves were dried at 60°C and then treated according to Azevedo et al. [21]. Elemental Cr content was analyzed by ICP-AES (Jobin Ivon JY70 Plus, France). *L. sativa* plants were used to assess the phytotoxicity of Cr, as this species is recommended as an ISO models for toxicological assays.

Oxidative Stress and Cell Membrane Permeability. Frozen leaf and root samples were ground in 0.1 M potassium phosphate buffer (pH 7.8), 1% polyvinylpyrrolidone, 5 mM Na_2EDTA, and 0.2% Triton X-100. After centrifugation (8000 ×g, 15 min), the supernatant was used for enzymatic assessment and soluble protein quantification (MicroTotal Protein Kit, Sigma-Aldrich, USA). APX, GPX, CAT, and SOD activity and glutathione reductase (GR) activities were determined as described by Silva et al. [22] and Mariz-Ponte et al. [12]. Lipid peroxidation was assessed by the malondialdehyde (MDA) method and the membrane stability was assessed by the membrane permeability (CMP) method [12, 23].

Flow Cytometry. Clastogenicity and changes in ploidy level and cell cycle progression were evaluated in roots and leaves, by flow cytometry [24]. Nucleus suspensions were obtained in woody plant buffer. To the nucleus suspensions, 50 μg mL^{-1} propidium iodide and 50 μg mL^{-1} RNAse were added. At least 5000 nuclei were analyzed per sample in a flow cytometer EPICS-XL Coulter Electronics (USA) with an argon laser (15 mW, 488 nm), and data were analyzed with a SYSTEM II software (v. 3.0, Beckman Coulter®). Nucleus populations in phases G_0/G_1, S, and G_2 and changes in cell cycle progression were analyzed. The Index of Cell Proliferation (%IP) was determined according to the following equation: %IP = (%S + %G_2)/(%G_0/G_1 + %S + %G_2), and the S phase fraction (SPF) was also calculated as %SPF = %S/(%G_0/G_1 + %S + %G_2).

Comet Assay. Root and leaf DNA fragmentation was quantified by the comet assay [25]. Tissues were sliced in 0.4 M Tris buffer, pH 7.5. 50 μL of nucleoids and 50 μL of 1% LMPA were spread on slides with a 1% NMPA layer and treated with alkaline buffer (0.30 M NaOH, 1 mM EDTA, pH > 13). Electrophoresis (30 min) took place at 0.74 V cm^{-1} at 4°C. Slides were neutralized with 0.4 M Tris pH 7.5 and stained with ethidium bromide. Comets were analyzed with a fluorescence microscope (Nikon Eclipse 80i; excitation filter: 510–560 nm; barrier filter: 590 nm). The %DNA in the tail (%TDNA) and the tail length (TL) were calculated with the CASP v1.2.2 software.

Micronuclei and Mitotic Aberrations. As micronuclei and mitotic aberrations need mitotic cells, only root meristems were used. Root tips were collected and stained with propidium iodide [26]. Slide preparations were observed under 1000x magnification using a fluorescence microscope (Nikon Eclipse 80i; excitation filter of 510–560 nm; barrier filter of 590 nm). To calculate the micronuclei (%MN), 200 cells were scored [26]. For mitotic aberrations, meristematic tissues were collected and stained with 2% acetic orcein. Samples were observed with a Nikon Eclipse 80i microscope, with a NIS-Elements F 3.00, SP7 software.

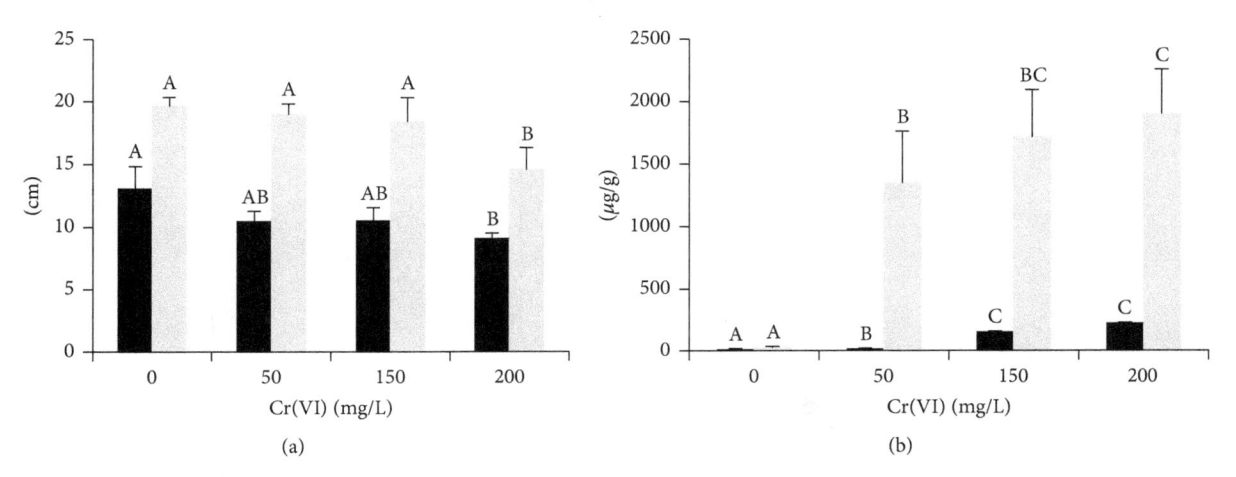

FIGURE 1: Cr(VI) effect on lettuce shoot (black bars) and root (grey bars) (a) growth (cm) and (b) Cr accumulation (μg/gDW). Different letter means significant differences relative to the control ($p < 0.05$).

Statistical Analysis. Sampling of ~six plants was used as individual or as a pool. Data were analyzed by one-way ANOVA ($p < 0.05$), followed by a Holm-Sidak test ($p < 0.05$) to evaluate the significance of differences in the parameters, and transformed to achieve normality. If required the nonparametric test, Kruskal–Wallis one-way ANOVA by ranks, was used. Statistical analyses were performed with Sigma Plot 11.0 (Systat Software Inc., Germany).

3. Results

Plant Growth and Total Cr Content. At the end of exposure some plants treated with 150 and 200 mg/L developed reddish leaves. While growth decreased and senescence in some leaves was evident, after this long-term exposure, survival rates were not affected (~100%). Figure 1 shows how Cr(VI) affected plant growth, with an evident decrease of both root and shoot length with Cr doses. As expected, the lowest shoot length (69% of the control) and root length (74% of the control) were achieved in plants exposed at 200 mg/L (Figure 1(a)). The elemental content of Cr increased in a linear way in both organs ($R^2 = 0.90$ for shoots and $R^2 = 0.83$ for roots) but the increase was much higher ($p < 0.05$) in roots. The absolute accumulation per gDW in leaves was <10x lower than in roots (Figure 1(b)). For the highest exposure, the accumulation of Cr was ~18x in shoots compared with the 75x increment of Cr in roots.

Oxidative Status. Leaves and roots showed different enzymatic profiles, with the roots being more susceptible to peroxidation ($p < 0.05$ for 150 mg/L and 200 mg/L) and to damage in cell membranes damage ($p < 0.05$ for ≥50 mg/L) (Table 1). In leaves, significant lipid peroxidation and increased cell membrane permeability were only significant for 200 mg/L. GPX decreased in both organs for higher doses, while it increased in a quadratic manner ($p < 0.05$) in roots exposed to 50 mg/L. APX increased in shoots exposed to Cr ($p < 0.05$) and in a quadratic manner in roots (increased at 50 mg/L ($p < 0.05$), decreasing thereafter). CAT and GR showed a general trend to increase in higher doses of Cr in leaves and roots ($p < 0.05$ for 200 mg/L). SOD activity increased in shoots reaching maximum values at 150 mg/L, while in roots it followed a quadratic response of maximum values at 50 mg/L.

Flow Cytometry. No significant changes were found in the complexity and size of nuclei. The control histograms presented FPCV values lower than 4%, an indicator of the reliability of the technique and protocol used. To score for clastogenic damage, the %FPCV of the G_0/G_1 peaks were analyzed, with no detectable variations (Table 2).

No ploidy abnormalities (e.g., aneuploidy and polyploidy) were induced by the doses of Cr(VI) tested, and while no changes in the cell cycle dynamics were detected in leaves, roots showed some cytostatic effects. Cr(VI) induced a blockage of the S phase ($p < 0.05$ for the highest concentration), with consequent trends of decreasing the nuclei populations at G_0/G_1 and G_2 (Table 2). Consequently, the root SPF and PI showed a trend to increase with Cr doses.

Comet Assay. Cr(VI) exposure increased DNA damage in leaves at higher concentrations (both TD and TM; 150 mg/L and 200 mg/L) ($p < 0.05$). Roots exposed to Cr(VI) showed a heterogeneity in the results, with higher damage induced in root cells exposed to 50 mg/L of Cr(VI) ($p > 0.05$) (Figure 2).

Mitotic Aberrations and Micronuclei. Micronuclei were absent in control roots and appeared only occasionally (app. 1/1000) in 50 mg/L exposed roots, but their occurrence increased significantly in roots exposed to higher Cr(VI) concentrations (Table 3). Contrarily, the mitotic index (%MI) showed a general trend for a decrease with the increase of Cr(VI), with significant differences for the higher Cr(VI) concentrations. Other occurrences related to mitotic abnormalities increased in the roots exposed to the highest concentration (Table 3).

4. Discussion

Principle Component Analysis. The PCA analyses showed that the analyzed parameters had different correlations in

TABLE 1: Cr(VI) effect on lettuce shoot and root oxidative activity. Different letters indicate for the same parameter significantly different means ($p < 0.05$). Chromium (Cr), guaiacol peroxidase (GPX), ascorbate peroxidase (APX), catalase (CAT), glutathione reductase (GR), superoxide dismutase (SOD), malondialdehyde (MDA), and membrane permeability method (CMP).

Cr(VI) (mg/L)	GPX (nKat.mg TSP^{-1})	APX (nKat.mg TSP^{-1})	CAT (nKat.mg TSP^{-1})	GR (nKat.mg TSP^{-1})	SOD (nKat.mg TSP^{-1})	MDA equivalents (nmol mL^{-1} mgFM^{-1})	CMP (%MD)
Shoot							
0	0.58 ± 0.12^c	37.73 ± 4.61^a	1.91 ± 0.32^a	2.42 ± 0.10^a	982.47 ± 477.15^a	0.67 ± 0.34^a	14.37 ± 3.46^a
50	0.23 ± 0.10^b	46.84 ± 0.94^{ab}	2.40 ± 0.13^a	1.85 ± 0.83^a	2062.51 ± 347.36^b	0.93 ± 0.17^a	19.37 ± 2.89^a
150	0.36 ± 0.03^b	47.89 ± 1.29^b	2.36 ± 0.35^{ab}	2.82 ± 0.62^a	4252.10 ± 266.38^d	2.36 ± 0.25^a	29.00 ± 7.20^a
200	0.13 ± 0.02^a	48.00 ± 2.63^b	5.67 ± 1.21^b	3.36 ± 1.86^b	3029.70 ± 97.89^c	3.10 ± 0.37^b	74.13 ± 11.60^b
Root							
0	1.72 ± 0.49^b	33.46 ± 0.89^a	12.5 ± 2.24^a	14.10 ± 3.81^a	1057.84 ± 672.24^a	0.71 ± 0.24^a	12.60 ± 0.37^a
50	2.15 ± 0.63^c	55.05 ± 2.06^b	18.94 ± 4.92^a	24.29 ± 5.06^{ab}	2349.10 ± 432.61^b	1.15 ± 0.48^a	19.75 ± 3.86^b
150	0.85 ± 0.03^a	37.39 ± 1.30^a	12.30 ± 2.16^a	17.06 ± 4.21^{ab}	453.03 ± 39.10^a	2.72 ± 0.69^{cd}	22.38 ± 5.32^b
200	0.91 ± 0.03^a	33.04 ± 3.35^a	62.28 ± 2.64^b	24.50 ± 1.09^b	725.25 ± 38.61^a	3.46 ± 0.12^d	89.81 ± 2.85^c

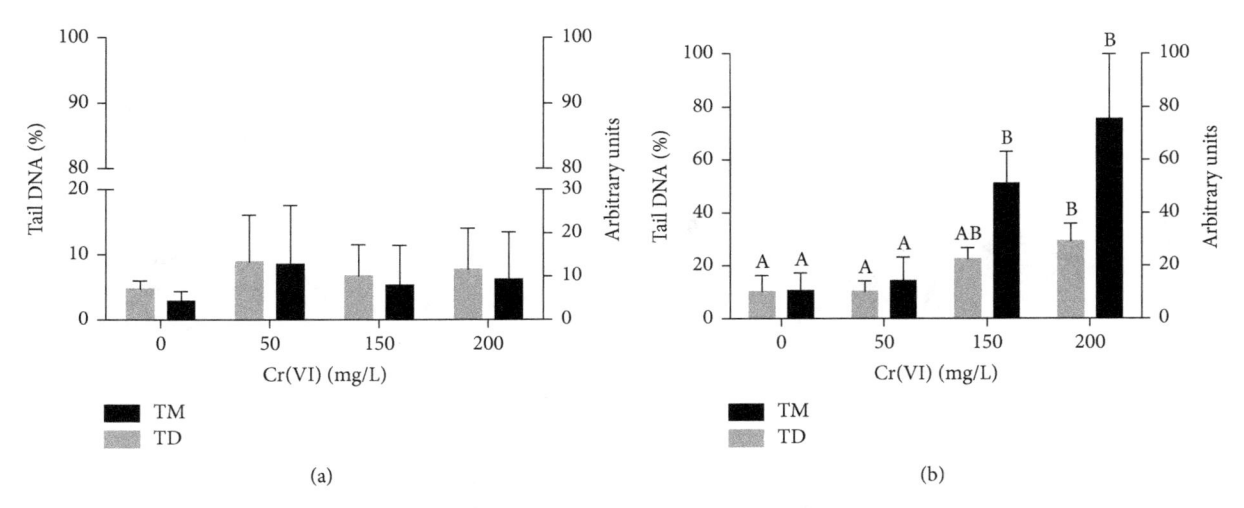

FIGURE 2: Tail DNA (TD) (%) and tail moment (TM) in (a) roots and (b) leaves exposed to Cr(VI). Different letters indicate for the same parameter significantly different means between control and stressed individuals ($p < 0.05$).

TABLE 2: Flow cytometry data for roots of *L. sativa* L. plants exposed to Cr(VI). Values are presented as the mean ± coefficient of variation. Different letters indicate for the same parameter significantly different means ($p < 0.05$).

Cr(VI) (mg/L)	Ploidy/abn	%FPCV	%G0/G1	%S	%G2	%IP
			Shoots			
0	2n/No	2.68 ± 0.58^a	89.34 ± 3.65^a	5.23 ± 1.87^a	5.43 ± 0.65^a	0.11
50	2n/No	2.95 ± 0.62^a	89.25 ± 3.96^a	5.72 ± 1.03^a	5.03 ± 0.45^a	0.11
150	2n/No	2.89 ± 4.02^a	88.56 ± 4.02^a	6.01 ± 1.11^a	5.43 ± 0.72^a	0.12
200	2n/No	2.74 ± 0.43^a	88.24 ± 2.84^a	6.15 ± 1.57^a	5.61 ± 0.59^a	0.13
			Roots			
0	2n/No	3.37 ± 0.74^a	87.62 ± 4.25^a	6.51 ± 2.22^a	4.07 ± 0.94^a	0.12
50	2n/No	3.31 ± 0.87^a	85.96 ± 1.43^a	7.60 ± 1.38^{ab}	4.33 ± 1.58^a	0.13
150	2n/No	2.97 ± 0.37^a	82.72 ± 4.52^a	9.81 ± 4.24^{ab}	5.31 ± 0.94^a	0.17
200	2n/No	3.11 ± 0.49^a	84.84 ± 1.46^a	9.67 ± 1.29^b	3.90 ± 0.76^a	0.15

TABLE 3: Micronuclei and mitotic index and abnormalities in roots of *L. sativa* L. exposed to Cr(VI). Values are presented as the mean ± standard deviation. Different letters indicate for the same parameter significantly different means ($p < 0.05$). Image: microphotograph of an abnormal metaphase in roots (bar = 20 μm).

Cr(VI) (mg/L)	Micronuclei (n/1000)	MI%	Mitotic abnorm. (n/1000)
0	Nd (0)	15 ± 5^a	Nd (0)
50	1 ± 1^a	19 ± 6^a	Nd (0)
150	5 ± 4^{ab}	9 ± 5^{ab}	5 ± 3
200	9 ± 5^b	7 ± 3^b	8 ± 3

leaves versus roots (Figures 3(a) and 3(b)). For roots, PC1 explained 58.8% of the variance, and PC2 explained 26.9% of the variance. In the root PCA, a clear separation between the control (top left quadrant), negatively correlating with most oxidative and genotoxic data, and the other conditions was evident. Also, the population 50 mg/L scores at the down left quadrant positively correlated with GPX, APX, and SOD and with %FPCV and cell cycle parameters (Figure 3(a)). The concentration 150 mg/L scored at the top right positively correlating with a G_2 accumulation/delay and negatively with all the parameters positively correlating with 50 mg/L. Ranking in the right down quadrant, the 200 mg/L samples positively correlated with multiple parameters for peroxidation, membrane degradation, mitotic disorders, and micronuclei.

For leaves, the PC1 explained 65.1% of the variance, and PC2 explained 21.6% of the variance (Figure 3(b)). Similar to roots, control is isolated (left down quadrant), negatively correlating with most oxidative and genotoxic parameters. Samples exposed to 150 and 200 mg/L show more similar

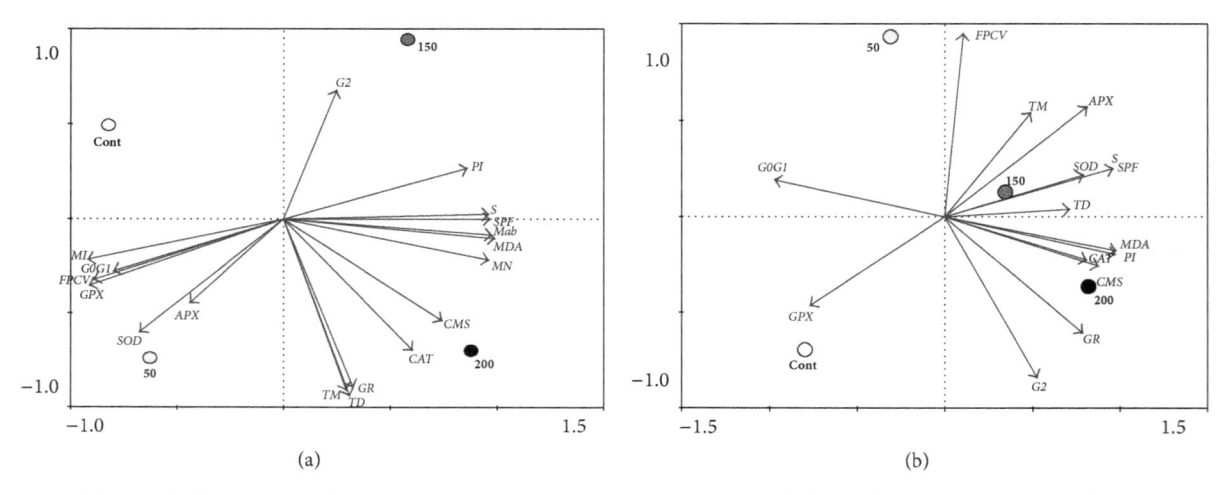

FIGURE 3: PCA analyses of the oxidative stress and genotoxic effects of Cr(VI) in plants: (a) roots and (b) leaves.

profiles, yet distinctive, as 150 mg/L correlated more with antioxidant and comet parameters, and 200 mg/L correlated with peroxidation and cell membrane degradation.

5. Discussion

Compared with animals and bacteria, a gap of information on Cr effects in plants still persists, and even less is known regarding the plants' response after chronical exposures. The observed reduction of the growth of shoots and roots was negatively correlated with the Cr content in those organs. This is in line with the literature, where Cr-associated growth reduction has been addressed and was also associated with a declined photosynthetic activity and nutritional disturbances for this and other species [3, 27].

As a redox metal, Cr can directly generate ROS such as $O_2^{-\bullet}$, H_2O_2 (a second messenger) $OH^{\bullet}$ and $^1O_2$, and oxidative injury via the Haber-Weiss and Fenton reactions, resulting in cell homeostasis disruption [3]. In this work, we elucidate how a long-term exposure to Cr(VI) leads to discriminative responses according to the organ and dose. We also evidence that the antioxidant enzymes are not stimulated similarly in the organs. Moreover, in roots, their response is variable, with all enzymes being particularly stimulated at 50 mg/L, and while some decreased thereafter, others (e.g., CAT) continued to be stimulated with Cr concentration. It should be stressed that 50 mg/L of Cr/VI) stimulated the antioxidant enzymatic battery and no major damage was visualized (e.g., genotoxic and or membrane permeability). Other doses seemed to already compromise some of these enzymes activities. Literature has shown that the antioxidant response to Cr exposure is complex depending on the dose, period, species, and parameters analyzed. For example, in maize plants exposed to ≤300 mM, H_2O_2, lipid peroxidation, and SOD and GPX activities increased regarding the controls [28], but *Camellia sinensis* L. showed lower enzymatic activity (decreased SOD, peroxidase, and CAT) [29]. Contrarily to 50 mg/L, the 150 mg/L exposed roots showed an evident increase of peroxidation and of the loss of membrane integrity. Besides being an indicator highly correlated with lipid peroxidation (as observed here) and eventually with cell death, the increase of CMP is also particularly relevant because, being the absorption of Cr facilitated by a carrier membrane, the increased accumulation of ROS will also compromise this membrane transport [28]. Besides, at the high Cr concentration, there is a correlation with a decrease of the mitotic index, supporting that the cell cycle progression is reduced or delayed. This is also supported by the blockages/delays found in the high Cr doses, as demonstrated by the increase of numbers in S phase and SPF. The decreased root growth due to Cr toxicity could be due to inhibition of root cell division/root elongation or extension of the cell cycle in roots [30, 31], as plant growth depends on both cell division and cell elongation. It is also evidenced here that Cr differently affected the plant phenology, again depending on the organ and dose.

Contrarily to our chronical exposure, several studies have been conducted with germinating/seedlings, while, like in another study, a tendency to occur a cytostatic effect is evident for high doses, which was supported by the decrease in MI in a dose-dependent way. For example, Fozia et al. [32] suggested that the decrease in root length of plants exposed to Cr was due to a *"cell cycle extension triggered by Cr toxicity."* Also, Cr reduced root growth of *Amaranthus* species, which was associated with the inhibition of cell division and oxidative stress [33, 34]. For *L. sativa*, Cr exposure showed a decrease of roots and aerial part length in high concentrations [6]. The Cr dose influenced the cell cycle dynamics, which was related to the DNA clastogenicity/fragmentation and with the delay in the cell cycle progression. In roots, a blockage at S (at expenses of mostly G_0/G_1) was seen for ≥150 mg/L, suggesting that cells suffer an arrest during/after S synthesis, preventing the cell from entering cell division thus limiting mitosis. This hypothesis was accompanied by the increase of comet assay parameters, supporting that higher Cr doses induced breaks in the genetic material as seen for other species [20]. Huang et al. [19] showed in tomato plants that during Cr-24 h exposure around 2,097 genes were more responsive than those responsive to 1 h exposure and that some of those were involved in DNA repair. In roots, in the PCA right half where 150 and 200 mg/L are ranking, there is an evident correlation with S delay, seen in the SPF and PI, together with a correlation with degradative events such as the mitotic abnormalities, micronuclei, and CMP. Samples exposed to 200 mg/L also correlate more with both %TDNA and TL, suggesting higher damage in this concentration. Cr may interact directly with nucleic acids, forming DNA-DNA cross-links and DNA-protein [35]. On the other hand, ROS-induction by Cr may contribute to damage in DNA and fragmentation, specially targeting bases such as guanine, promoting fragmentations and mutations [36]. Also, Cr-induced chromosomal abnormalities will lead to mitotic disorders, thus increasing the micronuclei formation. Abnormalities, such as chromosomal bridge, may be due to sticky characteristics of chromosomes, which compromises their migration towards the cell poles during anaphase [37]. Also, Cr may bind the tubulin carbonyl groups, compromising its structure and function and thus affecting also cell division (e.g., loss of microtubule of spindle fibers).

As recently stressed by Shahid et al. [1] the Cr-induced toxic effects (cytotoxic, genotoxic, and mutagenic) vary according to the plant and organ, so Cr-induced toxicity remains an area to be explored.

6. Conclusions

In conclusion, DNA damage induced by metal stress may be originated by indirect processes, such as the increase of ROS or by direct interaction of the metal with DNA [18, 36] that may lead to breaks and mutations. This is the most comprehensive evidence in long-term Cr-exposed plant genotoxic effects using multiparametric data (flow cytometry, comet, and mitotic parameters), where it is also demonstrated that the response is dependent on the organ and dose. Comparing shoots and roots, these parameters show often more dramatic responses in roots, where defense batteries are particularly active at low Cr doses in roots, and degradation endpoints are evident for higher doses.

Acknowledgments

This work was supported by FCT/MEC through national funds and cofunded by FEDER, PT2020 Partnership Agreement, and COMPETE 2020 (UID/BIA/04004/2013, UID/QUI/00062/2013, POCI/01/0145/FEDER/007265, and UID/QUI/50006/2013). FCT supported M. C. Dias (SFRH/BPD/100865/2014), S. Silva (SFRH/BPD/74299/2010), and R. J. Mendes (SFRH/BD/100865/2017).

References

[1] M. Shahid, S. Shamshad, M. Rafiq et al., "Chromium speciation, bioavailability, uptake, toxicity and detoxification in soil-plant system: A review," *Chemosphere*, vol. 178, pp. 513–533, 2017.

[2] S. K. Panda and S. Choudhury, "Chromium stress in plants," *Brazilian Journal of Plant Physiology*, vol. 17, no. 1, pp. 95–102, 2005.

[3] M. A. D. C. Gomes, R. A. Hauser-Davis, M. S. Suzuki, and A. P. Vitória, "Plant chromium uptake and transport, physiological effects and recent advances in molecular investigations," *Ecotoxicology and Environmental Safety*, vol. 140, pp. 55–64, 2017.

[4] D. K. Patra, C. Pradhan, and H. K. Patra, "An in situ study of growth of Lemongrass Cymbopogon flexuosus (Nees ex Steud.) W. Watson on varying concentration of Chromium (Cr +6) on soil and its bioaccumulation: Perspectives on phytoremediation potential and phytostabilisation of chromium toxicity," *Chemosphere*, vol. 193, pp. 793–799, 2018.

[5] R. A. Gill, B. Ali, P. Cui et al., "Comparative transcriptome profiling of two Brassica napus cultivars under chromium toxicity and its alleviation by reduced glutathione," *BMC Genomics*, vol. 17, no. 1, article no. 885, 2016.

[6] M. C. Dias, J. Moutinho-Pereira, J. Correia et al., "Physiological mechanisms to cope with Cr(VI) toxicity in lettuce: can lettuce be used in Cr phytoremediation?" *Environmental Science and Pollution Research*, vol. 23, no. 15, pp. 15627–15637, 2016.

[7] A. R. Patnaik, V. M. M. Achary, and B. B. Panda, "Chromium (VI)-induced hormesis and genotoxicity are mediated through oxidative stress in root cells of Allium cepa L," *Plant Growth Regulation*, vol. 71, no. 2, pp. 157–170, 2013.

[8] F. Islam, T. Yasmeen, M. S. Arif et al., "Combined ability of chromium (Cr) tolerant plant growth promoting bacteria (PGPB) and salicylic acid (SA) in attenuation of chromium stress in maize plants," *Plant Physiology and Biochemistry*, vol. 108, pp. 456–467, 2016.

[9] E. P. Eleftheriou, I.-D. S. Adamakis, E. Panteris, and M. Fatsiou, "Chromium-induced ultrastructural changes and oxidative stress in roots of Arabidopsis thaliana," *International Journal of Molecular Sciences*, vol. 16, no. 7, pp. 15852–15871, 2015.

[10] R. A. Gill, L. Zang, B. Ali et al., "Chromium-induced physio-chemical and ultrastructural changes in four cultivars of Brassica napus L.," *Chemosphere*, vol. 120, pp. 154–164, 2015.

[11] M. Feng, H. Yin, H. Peng, Z. Liu, G. Lu, and Z. Dang, "Hexavalent chromium induced oxidative stress and apoptosis in Pycnoporus sanguineus," *Environmental Pollution*, vol. 228, pp. 128–139, 2017.

[12] N. Mariz-Ponte, R. Mendes, S. Sario, J. Ferreira de Oliveira, P. Melo, and C. Santos, "Tomato plants use non-enzymatic antioxidant pathways to cope with moderate UV-A/B irradiation: A contribution to the use of UV-A/B in horticulture," *Journal of Plant Physiology*, vol. 221, pp. 32–42, 2017.

[13] M. Adrees, S. Ali, M. Rizwan et al., "Mechanisms of silicon-mediated alleviation of heavy metal toxicity in plants: A review," *Ecotoxicology and Environmental Safety*, vol. 119, pp. 186–197, 2015.

[14] S. Afshan, S. Ali, S. A. Bharwana et al., "Citric acid enhances the phytoextraction of chromium, plant growth, and photosynthesis by alleviating the oxidative damages in Brassica napus L.," *Environmental Science and Pollution Research*, vol. 22, no. 15, pp. 11679–11689, 2015.

[15] M. Farid, S. Ali, M. Rizwan et al., "Citric acid assisted phytoextraction of chromium by sunflower; morpho-physiological and biochemical alterations in plants," *Ecotoxicology and Environmental Safety*, vol. 145, pp. 90–102, 2017.

[16] J. A. Mahmud, M. Hasanuzzaman, K. Nahar, A. Rahman, M. S. Hossain, and M. Fujita, "Maleic acid assisted improvement of metal chelation and antioxidant metabolism confers chromium tolerance in Brassica juncea L.," *Ecotoxicology and Environmental Safety*, vol. 144, pp. 216–226, 2017.

[17] P. Rai and S. Dayal, "Evaluating genotoxic potential of chromium on Pisum sativum," *Chromosome Botany*, vol. 11, pp. 44–47, 2016.

[18] E. Truta, C. Mihai, D. Gherghel, and G. Vochita, "Assessment of the cytogenetic damage induced by chromium short-term exposure in root tip meristems of barley seedlings," *Water, Air, & Soil Pollution*, vol. 225, pp. 1–12, 2014.

[19] T.-L. Huang, L.-Y. Huang, S.-F. Fu, N.-N. Trinh, and H.-J. Huang, "Genomic profiling of rice roots with short- and long-term chromium stress," *Plant Molecular Biology*, vol. 86, no. 1-2, pp. 157–170, 2014.

[20] E. Rodriguez, R. Azevedo, P. Fernandes, and C. Santos, "Cr(VI) induces DNA damage, cell cycle arrest and polyploidization: A flow cytometric and comet assay study in Pisum sativum," *Chemical Research in Toxicology*, vol. 24, no. 7, pp. 1040–1047, 2011.

[21] H. Azevedo, C. G. G. Pinto, and C. Santos, "Cadmium effects in sunflower: Nutritional imbalances in plants and calluses," *Journal of Plant Nutrition*, vol. 28, no. 12, pp. 2221–2231, 2005.

[22] S. Silva, G. Pinto, B. Correia, O. Pinto-Carnide, and C. Santos, "Rye oxidative stress under long term Al exposure," *Journal of Plant Physiology*, vol. 170, no. 10, pp. 879–889, 2013.

[23] C. Santos, S. Fragoeiro, and A. Phillips, "Physiological response of grapevine cultivars and a rootstock to infection with Phaeoacremonium and Phaeomoniella isolates: An in vitro approach using plants and calluses," *Scientia Horticulturae*, vol. 103, no. 2, pp. 187–198, 2005.

[24] G. Brito, J. Loureiro, T. Lopes, E. Rodriguez, and C. Santos, "Genetic characterisation of olive trees from Madeira Archipelago using flow cytometry and microsatellite markers," *Genetic Resources and Crop Evolution*, vol. 55, no. 5, pp. 657–664, 2008.

[25] T. Gichner, I. Žnidar, and J. Száková, "Evaluation of DNA damage and mutagenicity induced by lead in tobacco plants," *Mutation Research - Genetic Toxicology and Environmental Mutagenesis*, vol. 652, no. 2, pp. 186–190, 2008.

[26] C. Monteiro, C. Santos, S. Pinho, H. Oliveira, T. Pedrosa, and M. C. Dias, "Cadmium-induced cyto- and genotoxicity are organ-dependent in lettuce," *Chemical Research in Toxicology*, vol. 25, no. 7, pp. 1423–1434, 2012.

[27] E. Rodriguez, C. Santos, R. Azevedo, J. Moutinho-Pereira, C. Correia, and M. C. Dias, "Chromium (VI) induces toxicity at different photosynthetic levels in pea," *Plant Physiology and Biochemistry*, vol. 53, pp. 94–100, 2012.

[28] S. Maiti, N. Ghosh, C. Mandal, K. Das, N. Dey, and M. K. Adak, "Responses of the maize plant to chromium stress with reference to antioxidation activity," *Brazilian Journal of Plant Physiology*, vol. 24, no. 3, pp. 203–212, 2012.

[29] J. Tang, J. Xu, Y. Wu, Y. Li, and Q. Tang, "Effects of high concentration of chromium stress on physiological and biochemical characters and accumulation of chromium in tea plant (Camellia sinensis L.)," *African Journal of Biotechnology*, vol. 11, no. 9, pp. 2248–2255, 2012.

[30] A. K. Shanker, C. Cervantes, H. Loza-Tavera, and S. Avudainayagam, "Chromium toxicity in plants," *Environment International*, vol. 31, no. 5, pp. 739–753, 2005.

[31] J. Zou, M. Wang, W. Jiang, and D. Liu, "Chromium accumulation and its effects on other mineral elements in Amaranthus viridis L.," *Acta Biologica Cracoviensia Series Botanica*, vol. 48, no. 1, pp. 7–12, 2006.

[32] A. Fozia, A. Z. Muhammad, A. Muhammad, and M. K. Zafar, "Effect of chromium on growth attributes in sunflower (*Helianthus annuus* L.)," *Journal of Environmental Sciences*, vol. 20, no. 12, pp. 1475–1480, 2008.

[33] A. Emamverdian, Y. Ding, F. Mokhberdoran, and Y. Xie, "Heavy metal stress and some mechanisms of plant defense response," *The Scientific World Journal*, vol. 2015, Article ID 756120, 18 pages, 2015.

[34] G. Bashri, P. Parihar, R. Singh, S. Singh, V. P. Singh, and S. M. Prasad, "Physiological and biochemical characterization of two Amaranthus species under Cr(VI) stress differing in Cr(VI) tolerance," *Plant Physiology and Biochemistry*, vol. 108, pp. 12–23, 2016.

[35] K. P. Nickens, S. R. Patierno, and S. Ceryak, "Chromium genotoxicity: A double-edged sword," *Chemico-Biological Interactions*, vol. 188, no. 2, pp. 276–288, 2010.

[36] N. R. Jena, "DNA damage by reactive species: mechanisms, mutation and repair," *Journal of Biosciences*, vol. 37, no. 3, pp. 503–507, 2012.

[37] D. Beyersmann and A. Hartwig, "Carcinogenic metal compounds: recent insight into molecular and cellular mechanisms," *Archives of Toxicology*, vol. 82, no. 8, pp. 493–512, 2008.

Effect of Integrated Inorganic and Organic Fertilizers on Yield and Yield Components of Barley in Liben Jawi District

Tolera Abera [ID], Tolcha Tufa, Tesfaye Midega, Haji Kumbi, and Buzuayehu Tola

Natural Resources Management Research Process, Ambo Agricultural Research Center,
Ethiopia Institute of Agricultural Research Institute, P.O. Box 382, Ambo, West Showa, Oromia, Ethiopia

Correspondence should be addressed to Tolera Abera; thawwii2014@gmail.com

Academic Editor: Yong In Kuk

Barley is an important food and beverage crop in the highlands of Ethiopia, although intensive cultivation and suboptimal fertilizer application have caused nutrient depletion and yield decline. With this in view, integrated inorganic and organic fertilizer sources on yield and yield components of barley were studied. Ten treatments involving the sole NP, vermicompost, conventional compost, and farmyard manure based on N equivalency were laid out in a randomized complete block design with three replications in 2015 and 2016 cropping seasons. Significantly higher grain yield and biomass yield of barley were obtained with the application of sole recommended NP and the integrated use of 50 : 50% vermicompost and conventional compost with recommended NP. Mean grain yield of 2567 and 2549 kg·ha^{-1} barley was obtained from application of 50 : 50% conventional compost and vermicompost based on N equivalence with recommended NP fertilizer rate, which markedly reduce the cost of chemical NP fertilizer required for the production of barley. The economic analysis confirmed the profitability of the integrated use of 50 : 50% conventional compost and vermicompost with recommended NP fertilizer for barley production. Therefore, the integrated use of 50 : 50% conventional compost and vermicompost based on N equivalency with recommended NP fertilizer was recommended for sustainable barley production in Chelia district and similar agroecology.

1. Introduction

Barley (*Hordeum vulgare* L.) ranks fourth among cereals in the world and is grown annually on 48 million hectares in a wide range of environments [1]. Ethiopia, the second largest barley producer in Africa, accounts for about 25 percent of the total production in the country [2]. CSA [3] reported that barley is the fifth most important cereal crop after teff, wheat, maize, and sorghum in total production in the country. The majority of barley production is food barley with its share estimated to be 90 percent compared with malt barley [4]. However, the national and regional yield of barley has remained below 1.3 t·ha^{-1} [3], whereas the potential yield goes up to 6 t·ha^{-1} at experimental plots [5]. This lower yield of food barley is attributed to lack of improved varieties and poor soil fertility management [6]. Soil fertility is the most limiting factor for barley production in the highlands of Ethiopia [7, 8]. Furthermore, in Ethiopia, low productivity of barley is due to low soil fertility and poor agronomic practices [9]. Tolera et al. [10] reported that low productivity of barley is mainly due to traditional methods of production and poor soil fertility. The recommended fertilizer rate for barley production in the central highlands of Ethiopia, in the form of di-ammonium phosphate, is mainly at rates of 9/10 to 18/20 kg·N/P ha^{-1}, which are suboptimal rates for barley production [11]. Woldeyesus et al. [8] reported that the application of P fertilizer increases N use efficiency and thus mines soil N when N is in short supply. However, chemical fertilizer usage by the resource-poor subsistence farmers in Ethiopia is insignificant and inadequate [12]. However, Getachew et al. [13] reported that continuous applications of inorganic fertilizers alone resulted in deterioration of soil health in terms of physical, chemical, and biological properties of the soil. Organic fertilizer application has been reported to improve crop growth by supplying plant nutrients including micronutrients as well as improving physical, chemical, and biological properties of the soil, thereby providing a better environment for root development by improving the soil structure [14].

Thus, the application of inorganic or organic fertilizers alone did not bring a sustainable increase in yields [15]. MoA [16] reported that research efforts on how to use farmyard manure (FYM) together with low rates of mineral fertilizers could be one alternative solution for sustainable fertility management for barley production. Likewise, Shata et al. [17] suggested that by the use of mixed chemical and organic fertilizers not only production can be kept at optimum level, but also the amount of chemical fertilizers to be used can be reduced, which had negative impacts on cost production and environments. Abay and Tesfaye [18] found higher barley biomass yield of 8259 and 8065 $kg \cdot ha^{-1}$, and other agronomic parameters at Adiyo and Ghimbo were obtained with the application of $5 t \cdot ha^{-1}$ FYM in combination with 75% inorganic NP.

Limited agronomic studies in line to integrated nutrient use call to conduct research on integrated nutrient management practices and recommend best practices in order to maximize yield potential of crops such as barley [11]. Integrated use of organic and mineral fertilizers for tackling soil fertility depletion and sustainably increasing crop yields had a paramount importance [9, 13, 19]. Many research findings have shown that neither inorganic fertilizers nor organic sources alone can result in sustainable productivity [20]. Integrated soil fertility management involving the judicious use of combinations of organic and inorganic resources is a feasible approach to overcome soil fertility constraints and contribute high crop productivity in agriculture [21]. Hence, no information is available on the yield potential of improved variety with integrated nutrient management practice in ultisols of Ambo. The integrated use of chemical and organic fertilizer rate is needed to investigate in order to utilize the potential yield of improved barley variety in the area. Therefore, the objective was to determine the effect of different integrated organic and inorganic fertilizers on grain yield and yield components of barley in ultisols of Chelia districts.

2. Materials and Methods

The experiment was conducted in humid highland agroecosystems of western Oromia National Regional State, western Ethiopia. It was executed on farmers' field around Toke Kutaye in 2015 and 2016 cropping seasons. It lies between $8°9'80''N$ N latitude and $37°71'E$ longitude and at an altitude ranged from 2273 meter above sea level, receiving mean annual rainfall of 1040 mm with unimodal distribution [22]. It has a medium cool subhumid climate with the mean minimum, mean maximum, and average air temperatures of 8.9, 27.4, and 18.1°C, respectively [22]. The soil type is brown clay loam ultisols [23].

The experiment was laid out in a randomized complete block design with three replications, and the plot size was 4 m × 3 m. Ten rates of sole and integrated organic and inorganic fertilizer (T1: recommended NP ($64/46 kg \cdot NP\ ha^{-1}$), T2: conventional compost (based on N equivalency of recommended. N), T3: farmyard manure based on N equivalency of recommended N, T4: vermicompost based on N equivalency of recommended N, T5: vermicompost (50% of treatment 4) + conventional compost (50% of treatment 2), T6: vermicompost (50% of treatment 4) + FYM (50% of treatment 3), T7: vermicompost (1/3 t/ha) + conventional compost (1/3 t/ha) + FYM (1/3 recommended), T8: vermicompost (50% of treatment 4) + 50% recommended NP, T9: conventional compost (50% of treatment 2) + 50% recommended NP, and T10: FYM (50% of treatment 3) + 50% recommended NP)) were used for HB-43 barley variety. Sole and integrated inorganic and organic fertilizer rates were applied at planting. All other agronomic management practices (ploughing four times with oxen, time of sowing (early July), seed rate ($150 kg \cdot ha^{-1}$), and two times hand weeding) were applied as per recommendation for the barley production. Different agronomic parameters such as plant height was measured in cm from five plants sampled randomly from the central rows using graduated meter one week before harvesting and summed and divided by the number of plants to get the height of each plant, spike length was measured in (cm) from five plants' spikes sampled randomly from the central rows at harvesting using graduated meter, summed and divided by the number of plants' spikes to get the length of each spike, total number of tillers $plant^{-1}$ was counted from ten plants sampled randomly from the central rows at harvesting, summed and averaged to obtain the number of tillers $plant^{-1}$, dry biomass yield was weighed from above ground harvested plants per plot and converted to dry biomass per hectare in kilogram, grain yield was taken by threshing the harvested plants per plots, and grain yield per plot was weighted and converted to yield in $kg \cdot ha^{-1}$ adjusted to 10% moisture, thousand seed weight was taken by counting 1000 seed number from the bulk of trashed grain weight at 10% moisture level and weighed using a sensitive balance expressed in gram, and harvest index was computed as a ratio of grain yield to dry biomass yield * 100 and expressed in % of barley were collected. The collected data were computerized and analyzed using SAS computer software [24]. Mean separation was done using least significance difference (LSD) at 5% probability level [25].

Economic analysis such as partial budget, value to cost ratio, and marginal rate of return for barley grain yield was valued at an average open market price of EB 1200 per 100 kg for the last 5 years. Labour cost for field operation was EB 28 per man-day. The yield was adjusted down by 10% to reflect actual production conditions [26]. The cost of fertilizers (urea and DAP) was EB 1090 and 1390 per 100 kg with current market price. The cost of farmyard manure, conventional compost, and vermicompost was estimated by considering preparation, application, and its price costs 20, 40, and 30 $EB \cdot kh^{-1}$.

3. Results and Discussion

3.1. Plant Height, Spike Length, and Total Number of Tillers $Plant^{-1}$ of Barley. The mean plant height, spike length, and total number of tillers $plant^{-1}$ of barley are indicated in Table 1. Mean plant height of barley was significantly ($P < 0.05$) affected by the sole and integrated use of inorganic and inorganic fertilizer application. Significantly taller (104 cm) mean plant height of barley was obtained from the

TABLE 1: Effects of the integrated use of organic and inorganic fertilizers on plant height, spike length, and total number of tillers plant^{-1} of barley in 2015-2016 cropping seasons on smallholder farmers' fields in Chelia district.

Treatments	Plant height (cm)			Spike length (cm)			Total tillers plant^{-1}		
	2015	2016	Mean	2015	2016	Mean	2015	2016	Mean
Recommended NP	108	100	104	7	7	7	5	5	5
Conventional compost (N equivalency)	101	91	96	7	7	7	5	5	5
Farmyard manure (N equivalency)	100	71	86	7	6	6	4	4	4
Vermicompost (N equivalency)	105	82	93	7	6	6	5	5	5
50:50% vermicompost:conventional compost	105	86	95	7	6	7	5	5	5
50:50% vermicompost:farmyard manure	104	71	87	7	6	6	5	5	5
33:33:33% vermicompost:conventional compost:farmyard manure	102	81	92	7	6	7	5	5	5
50:50% vermicompost:recommended NP	104	93	99	6	6	6	5	5	5
50:50% conventional compost:recommended NP	107	92	100	6	7	7	5	5	5
50:50% farmyard manure:recommended NP	106	78	92	7	6	7	5	5	5
LSD (%)	6.714	10.16	7.53	NS	1.03	0.64	0.971	0.923	0.57
CV (%)	3.75	7.01	6.87	7.91	9.35	8.39	11.33	11.35	9.92

NS = nonsignificant difference at 5% probability level; numbers followed by same letter in the same column are not significantly different at 5% probability level.

application of recommended NP fertilizer followed by 50:50% vermicompost and conventional compost with NP fertilizer (Table 1). Similarly, Woubshet et al. [27] found significantly higher plant height of barley with the integrated application of lime, balanced fertilizer, and compost in Wolmera district. Correspondingly, Mitiku et al. [28] found a significant effect of combined application of organic and inorganic fertilizers on plant height barley at Adiyo and Ghimbo with the application of 5 t·ha^{-1} farmyard manure + 75% of recommended NP and 5 t·ha^{-1} vermicompost and 75% of recommended NP. Likewise, Getachew [29] reported that the use of organic manures in combination with mineral fertilizers maximized the plant height of barley. Similar result was reported by Ofosu and Leitch [30] and Amanuliah and Maimoona [31]. Mean plant height performance of barley varied across years with higher in 2015 cropping season (Table 1), indicating variations in environmental factors (rainfall and temperature) and soil fertility status across smallholder farmers' fields. Similarly, Kassu et al. [32] found cropping season significantly affected most of the measured variables of malt barley. The mean spike length and total number of tillers plant^{-1} of barley were significantly ($P < 0.05$) affected by the sole and integrated use of NP and organic fertilizer sources (Table 1). Higher spike length of barley was obtained from the application of sole NP and 50:50% NP fertilizer with organic fertilizer sources. Similar result was reported by Kumar [33] and Shantveerayya et al. [34].

Significantly higher total number of tillers plant^{-1} of barley was obtained from the application of sole NP and 50:50% NP fertilizer with organic fertilizer sources. This indicates the easy availability of nutrients from inorganic fertilizers as compared to the gradually release of nutrients from organic fertilizer sources. Getachew et al. [35] found that the integrated application of organic with N fertilizer rate significantly improved productive tillers of barley at Holetta and Robgebeya. Likewise, the application of 5 t·ha^{-1} FYM combined with 75% inorganic NP gave the highest number of productive tiller m^{-2} (227 and 215) [28]. Kumar [33] reported that the number of total tillers plant^{-1} was significantly increased with the application of nitrogen fertilizer. The result was in agreement with the findings of Prystupa et al. [36], Iqtidar et al. [37], Sepat et al. [38], Tariku [39], and Kassu et al. [32] in malt barley.

3.2. Grain Yield and Thousand Seed Weight of Barley. The mean grain yield and thousand seed weight of barley are indicated in Table 2. Mean grain yield of barley was significantly ($P < 0.05$) affected by the integrated use of NP fertilizer and organic fertilizer sources in 2016 and combined mean (Table 2). Significantly higher mean grain yield of barley was obtained with application 50:50% vermicompost and conventional compost with recommended NP fertilizer followed by conventional compost and NP fertilizer indicating better contribution of integrating NP fertilizer with organic fertilizer sources. Similarly, Mitiku et al. [28] reported that the application of inorganic fertilizers (NP or NPK) with FYM gave a better yield of barley than the application of 100% inorganic fertilizers alone. Correspondingly, grain yield of malting barley was significantly increased by the combined application of organic and mineral nutrients [32]. Likewise, Assefa [40] and Getachew et al. [35] also reported similar results on barley. The application of organic fertilizer with inorganic fertilizer could directly increase yield, improve soil fertility status, and reduce the cost production. The improved yields of barley due to combined application of organic and mineral amendments resulted from positive changes to the soil, including increased soil pH, available P and total N, and possibly other macronutrients and micronutrients [32, 35]. Likewise, the combined application of inorganic and organic fertilizers was widely recognized as a way of increasing yield [41]. Higher yields were obtained with the application of compost or compost + biochar, and the highest yields were achieved in combination with moderate rates of applied N [35]. Integrated use of organic and inorganic fertilizer is economizing the input cost for barley production. The chemical fertilizer required to achieve optimum yield levels can be decreased with the application of organic fertilizers [42]. Besides, yield increase, environmental benefits, soil quality, and quality food test with the application of organic fertilizer were higher in different crops. Significant improvement in

TABLE 2: Effects of the integrated use of organic and inorganic fertilizers on grain yield and thousand seed weight of barley in 2015-2016 cropping seasons on smallholder farmers' fields in Chelia district.

Treatments	Grain yield (kg·ha^{-1})			Thousand seed weight (g)		
	2015	2016	Mean	2015	2016	Mean
Recommended NP	3396	1566	2481	45.24	45.23	45.24
Conventional compost (N equivalency)	3502	1563	2533	45.62	45.60	45.61
Farmyard manure (N equivalency)	3276	571	1924	42.89	42.87	42.88
Vermicompost (N equivalency)	3405	1007	2206	45.85	45.83	45.84
50 : 50% vermicompost : conventional compost	3394	1086	2240	45.38	45.37	45.38
50 : 50% vermicompost : farmyard manure	3339	666	2003	43.32	43.3	43.31
33 : 33 : 33% vermicompost : conventional compost : farmyard manure	3377	859	2118	42.62	42.6	42.61
50 : 50% vermicompost : recommended NP	3547	1551	2549	43.99	43.97	43.98
50 : 50% conventional compost : recommended NP	3630	1504	2567	43.86	43.87	43.87
50 : 50% farmyard manure : recommended NP	3372	1178	2275	42.46	42.47	42.47
LSD (%)	NS	524	382.68	NS	NS	2.409
CV (%)	9.77	26.46	14.39	5.28	5.37	4.7

NS = nonsignificant difference at 5% probability level.

the quality characteristics of the soil amended with biochar, compost and compost + biochar while the growth and yield components of barley supplemented with organic amendments and lower N fertilizer rates showing the enhanced synergistic effects of mixed treatments [35]. The second higher grain yield of barley with reduced rate of inorganic fertilizer might be due to improved nutrient use efficiency and stimulated barley growth by organic fertilizer sources, which is in agreement with the results of several findings [35, 43–45]. Therefore, the integrated use of inorganic and organic fertilizer sources has improved yield and other soil and environmental benefits, which enhance sustainable production barley.

Combined mean over years of thousand seed weight of barley was significantly ($P < 0.05$) affected with the integrated use of NP fertilizer and organic fertilizer sources (Table 2). Similarly, Abay and Tesfaye [18] reported that higher thousand grain weights (45 g and 44 g) at Ghimbo and Adiyo were obtained with the application of 5 t·ha^{-1} FYM in combination with 25% recommended rate of inorganic NP and 5 t·ha^{-1} FYM in combination with 75% recommended rate of inorganic NP, and the lowest thousand grain weights were recorded from the control plots for both locations. Likewise, Saidu et al. [46] reported higher 1000 grain weight, from the application of 5 t·ha^{-1} FYM in combination with 50% inorganic NP while the lowest from no fertilizer application. Getachew et al. [35] also found that amendment of organic with N fertilizer rate interaction significantly improved thousand grain weight barley at Robgebeya.

3.3. Biomass Yield and Harvest Index of Barley.

The mean biomass yield and harvest index of barley are indicated in Table 3. Mean biomass yield of barley was significantly ($P < 0.05$) affected by sole and integrated use of NP fertilizer and organic fertilizer sources. Likewise, Kassu et al. [32] reported that the total above-ground biomass of malting barley was significantly ($P < 0.001$) increased by the combined application of organic and mineral nutrients. Significantly higher mean biomass yield of barley was obtained from the sole application of NP fertilizer followed by 50 : 50% vermicompost and conventional compost with NP fertilizer rates. Likewise, higher biomass (11,514 kg·ha^{-1}) yields of malting barley were obtained from the application of recommended rate of NP (36–20 kg·NP ha^{-1}) from mineral sources (DAP and urea) [32]. Similarly, the application of 5 t·ha^{-1} FYM in combination with 75% inorganic NP gave the higher biomass yield of 8259 and 8065 kg·ha^{-1} at Adiyo and Ghimbo [18]. Getachew et al. [35] also found that the interaction of organic amendment by N fertilizer rate significantly improved biomass yield barley at Holetta and Robgebeya. This revealed the easy release of nutrients from NP fertilizer as compared to organic fertilizer sources. Integrated soil fertility management involving the judicious use of combinations of organic and inorganic resources is a feasible approach to overcome soil fertility constraints and contribute high crop productivity in agriculture [21]. Yearly performance variation of biomass yield was obtained between the two cropping seasons, which might be due to the environmental variation and soil fertility status of the two farms since different farm fields were used in two cropping seasons. The result was in agreement with Kassu et al. [32].

The mean harvest index of barley was significantly ($P < 0.05$) affected with the integrated use of NP fertilizer and organic fertilizer sources. The mean harvest index of barley was ranged from 39 to 48% (Table 3). Significantly higher mean harvest index of barley was obtained from 50 : 50% vermicompost with conventional compost followed by conventional compost application. Similar result was found by Chakrawarty [28, 39, 47]. Likewise, Woubshet et al. [27] reported that harvest indices of barley ranged from 37 to 47% and showed significantly an increasing trend with increased combination of applied lime, organic and blended inorganic fertilizers in acid affected soil of Wolmera district. This might be due to increase of nutrient efficiency and supplying different available nutrients which more partitioned to grain when organic fertilizer sources are applied. Similarly, Woubshet et al. [27] stated that the application of lime with increased rate of compost and inorganic fertilizers might have increased the efficiency of barley to partition the dry matter to the seed.

TABLE 3: Effects of the integrated use of organic and inorganic fertilizers on dry biomass yield and harvest index of barley in 2015-2016 cropping seasons on smallholder farmers' fields in Chelia district.

Treatments	Biomass yield (kg·ha^{-1})			Harvest index (%)		
	2015	2016	Mean	2015	2016	Mean
Recommended NP	7967	4330	6148	42.68	35.56	39.12
Conventional compost (N equivalency)	7500	3833	5667	46.70	41.15	43.93
Farmyard manure (N equivalency)	7158	1526	4342	45.66	37.12	41.39
Vermicompost (N equivalency)	7458	2504	4981	45.62	40.42	43.02
50 : 50% vermicompost : conventional compost	7167	2230	4698	47.43	49.59	48.51
50 : 50% vermicompost : farmyard manure	7392	1670	4531	45.21	39.82	42.52
33 : 33 : 33% vermicompost : conventional compost : farmyard manure	7292	2126	4709	46.25	39.84	43.05
50 : 50% vermicompost : recommended NP	7817	3807	5812	45.41	40.79	43.10
50 : 50% conventional compost : recommended NP	7967	3804	5885	45.50	39.56	42.53
50 : 50% farmyard manure : recommended NP	7300	2985	5143	46.14	38.70	42.42
LSD (%)	NS	1191	845	2.831	6.2856	3.41
CV (%)	8.49	24.09	14.01	3.61	9.1	6.84

NS = nonsignificant difference at 5% probability level.

TABLE 4: Effects of the integrated use of organic and inorganic NP fertilizers on economic profitability of barley production on smallholder farmers' fields in Chelia district.

Treatments	Grain yield (kg·ha^{-1})	Adjusted grain yield (kg·ha^{-1})	Gross field benefit (EB·ha^{-1})	TCV (EB·ha^{-1})	Net benefit (EB·ha^{-1})	Value to cost ratio	MRR (%)
Farmyard manure (N equivalency)	1924	1732	20779	1000	19779	19.78	
50 : 50% VC : FYM	2003	1803	21632	1350	20282	15.02	144
33 : 33 : 33 t·VC : CC : FYM ha^{-1}	2118	1906	22874	1583	21291	13.45	433
50 : 50% FYM : RNP	2275	2048	24570	1740	22830	13.12	980
Vermicompost (N equivalency)	2206	1985	23825	1750	22075^D	12.61	
50 : 50% VC : CC	2240	2016	24192	1875	22317^D	11.90	
Conventional compost (N equivalency)	2533	2280	27356	2000	25356	12.68	972
50 : 50% VC : RNP	2549	2294	27529	2115	25414	12.02	50
50 : 50% CC : RNP	2567	2310	27724	2240	25484	11.38	56
Recommended NP	2481	2233	26795	2480	24315^D	9.80	

D = dominated; adjusted grain yield = the grain yield barley was adjusted downwards to 10% to reflect the difference between the experimental plot yield and the yield farmers; TVC = total variable cost; MRR = marginal rate of return; barley grain price = 12 EB·kg^{-1}; NP = 14 EB·kg^{-1}; DAP = 13.9 EB·kg^{-1}; urea price = 10.90 EB·kg^{-1}; vermicompost price = 250 EB·t^{-1}; farmyard manure price = 300 EB·t^{-1}; conventional compost = 350 EB·ha^{-1}; labour cost = 28 EB·day^{-1}; VC = vermicompost; CC = conventional compost; RNP = recommended NP (100/100 kg·urea/DAP·ha^{-1}); FYM = farmyard manure; 1US$ = 27.57 EB.

3.4. Integrated Use of NP Fertilizer and Organic Fertilizer Sources on Economic Profitability of Barley Production.

The value to cost ratio for NP fertilizer and organic fertilizer sources was 9.80 to 19.78 EB per unit of investment (Table 4). Correspondingly, the application of 1 : 1 compost : mineral fertilizer could enable farmers to earn a return of US$7.57 for every US$1.0 investment, which implied high profitability of malt barley due to integration of the locally available organic materials with mineral fertilizers [32]. Barley production with application of 50 : 50% conventional compost with NP fertilizer gave net profit advantage of 25,484 EB with marginal rate return of 56% followed by 50 : 50% vermicompost with NP fertilizer, and conventional compost gave net benefit of 25,414 and 25,356 EB·ha^{-1} with marginal rate return of 50 and 972% (Table 4). Tariku [39] reported that the highest net benefit of EB 58553 ha^{-1} and marginal rate of return 36.45% of barley was obtained from the application of 66.6 : 33.4% NPS : FYM followed by EB 57781 ha^{-1} and marginal rate of return 2153% of barley gained from the application of 100% NPS (Table 4).

Similarly, Mitiku et al. [28] also reported that the application of 5 t·ha^{-1} FYM + 75% inorganic NP gave the highest net return with 15,859 EB·ha^{-1} at Adiyo and 13,108 EB·ha^{-1} at Ghimbo. Likewise, Woubshet et al. [27] reported that the application of lime integrated with compost and blended fertilizer (NPSB) indicated the highest net return of EB 30633 with highest marginal rate return of 667% with values to cost ratio of EB 5.49 profit per unit investment for barley production in Wolmera district. The highest marginal rate of return of 980% was obtained with the application of 50 : 50% farmyard manure : with recommended NP for barley production (Table 4). The economic analyses confirmed that barley production with the application of 50 : 50% conventional compost with NP fertilizer or 50 : 50% vermicompost with NP fertilizer was profitable for Chelia highlands.

4. Conclusion

The integrated use of inorganic NP with organic fertilizer sources significantly improved yield and yield components of barley. Yield and yield components of barley were significantly influenced by cropping seasons indicating variation of climatic factors across different seasons, which may have been the results of temperature and rainfall distribution occurred during growing season. Significantly higher grain yield of barley was obtained from integration of 50:50% vermicompost and conventional compost based on N equivalence with recommended NP fertilizer application. The application of 50:50% conventional compost with NP fertilizer gave net profit advantage of 25,484 EB with marginal rate return of 56% followed by 50:50% vermicompost with NP fertilizer and conventional compost gave net benefit of 25,414 and 25,356 EB·ha^{-1} with marginal rate return of 50 and 972%. Therefore, the application of 50:50% conventional compost and vermicompost based on N equivalence with recommended NP fertilizer rate was recommended for optimum grain yield and economical profitable barley production in Chelia district, West Showa, Ethiopia.

Acknowledgments

The authors thank Natural Resources Management Research Process for funding the experiment. The authors are very grateful to Ambo Agricultural Research Center Management for providing all necessary equipments and logistics during the research work. All the technical and field assistants of Natural Resources Management Research Process are also acknowledged for unreserved effort during executing the experiment. The authors also thank Chelia district farmers for providing them land for field research work and also for their assistance in field management.

References

[1] ICARDA, *Dry Land Cereals: A Global Alliance for Improving Food Sufficiency, Nutrition and Economic Growth for the World's Most Vulnerable Poor*, A CGIAR Research Program submitted by ICRISAT and ICARDA to the CGIAR Consortium Board, 2011.

[2] FAO, *Food Balance Sheets*, FAO STAT, Rome, Italy, 2014.

[3] CSA, *Agricultural Sample Survey 2014/15. I. Report on Area and Production for Major Crops. Statistical Bulletin 578*, CSA, Addis Ababa, Ethiopia, 2015.

[4] D. Alemu, K. Kelemu, and B. Lakew, "Trends and prospects of malt barley value chains in Ethiopia," *Technical Report, International Food Policy Research Institute (IFPRI)*, Washington, DC, USA, 2014.

[5] B. Lakew, H. Gebre, and F. Alemayehu, "Barley production and research in Ethiopia in: barley research in Ethiopia: past work and future prospects," in *Proceedings of 1st Barley Research Review Workshop*, pp. 1–8, IAR/ICARDA, Addis Ababa, Ethiopia, October 1996.

[6] A. Woldeab, T. Mamo, M. Bekelle, and T. Ajema, *Soil Fertility Management Studies on Wheat in Ethiopia in Wheat Research in Ethiopia, A Historical Perspective*, CIMMYT/IAR, Addis Ababa, Ethiopia, 1991.

[7] Y. Chilot, F. Alemayehu, and W. Sinebo, *Barley Livestock Production System in Ethiopia: An Overview in Barley-Based Farming System in the Highlands of Ethiopia*, Ethiopian Agriculture Research Organization, Addis Ababa, Ethiopia, 1998.

[8] S. Woldeyesus, C. Yirga, and R. Fisehaye, "On-farm fertilizer trial on food barley in Wolemera area," in *Proceedings of a Client-Oriented Research Evaluation Workshop on Towards Farmer Participatory Research: Attempts and Achievements in the Central Highlands of Ethiopia*, pp. 266–279, Holetta Agricultural Research Centre, Holetta, Ethiopia, 2002.

[9] Z. Gete, G. Agegnehu, D. Abera, and R. Shahidur, *Fertilizer and Soil Fertility Potential in Ethiopia: Constraints and Opportunities for Enhancing the System*, IFPRI, Addis Ababa, Ethiopia, 2010.

[10] A. Tolera, A. Molla, A. Feyissa et al., "Research achievements in barley cultural practices in Ethiopia," in *Proceedings of the 2nd National Barley Research and Development Review Workshop on Barley Research and Development in Ethiopia*, pp. 113–126, HARC, Holetta, Ethiopia, ICARDA, Aleppo, Syria, 2011.

[11] Y. Chilot, B. Lakew, and F. Alemayehu, "On-farm evaluation of food barley production packages in the highlands of Wolemera and Degem, Ethiopia," in *Proceedings of a Client-Oriented Research Evaluation Workshop on Towards Farmer Participatory Research: Attempts and Achievements in the Central Highlands of Ethiopia*, pp. 176–187, Holetta Agricultural Research Centre, Holetta, Ethiopia, 2002.

[12] A. Getachew, M. Liben, A. Molla, A. Feyissa, A. Bekele, and F. Getaneh, "Research achievements in soil fertility management in relation to barley in Ethiopia," in *Proceedings of the 2nd National Barley Research and Development Review Workshop*, pp. 137–153, HARC, Holetta, Ethiopia, ICARDA, Aleppo, Syria, 2011.

[13] A. Getachew, C. van Beek, and I. B. Michael, "Influence of integrated soil fertility management in wheat and teff productivity and soil chemical properties in the highland tropical environment," *Journal Soil Sciences Plant Nutrition*, vol. 14, pp. 532–545, 2014.

[14] M. Dejene and M. Lemlem, *Integrated Agronomic Crop Managements to Improve Teff Productivity under Terminal Drought, Water Stress*, In Tech Open Science, London, UK, 2012.

[15] A. Getachew, B. Lakew, and N. N. Paul, "Cropping sequence and nitrogen fertilizer effects on the productivity and quality of malting barley and soil fertility in the Ethiopian highlands," *Arch Agronomy Soil Sciences*, vol. 60, no. 9, pp. 1261–1275, 2014.

[16] MoA-Ministry of Agriculture, *Animal and Plant Health Regulatory Directorate. Crop variety register, Issue 15*, Ministry of Agriculture, Addis Ababa, Ethiopia, 2012.

[17] S. M. Shata, A. Mahmoud, and S. Siam, "Improving calcareous soil productivity by integrated effect of intercropping and fertilizer," *Research Journal of Agriculture and Biological Sciences*, vol. 3, no. 6, pp. 733–739, 2007.

[18] A. Abay and D. Tesfaye, "Combined application of organic and inorganic fertilizers to increase yield of barley and improve soil properties at Fereze in Southern Ethiopia," *Innovative Systems Design and Engineering*, vol. 22, no. 1, pp. 25–34, 2012.

[19] A. Getachew and A. Tilahun, "Integrated soil fertility and plant nutrient management in tropical agro-ecosystems: a review," *Pedosphere*, vol. 27, no. 4, pp. 662–680, 2017.

[20] A. S. Godara, U. S. Gupta, and R. Singh, "Effect of integrated nutrient management on herbage, dry fodder yield and quality of oat (*Avena sativa* L.)," *Forage Research*, vol. 38, no. 1, pp. 59–61, 2012.

[21] T. Abedi, A. Alemzadeh, and S. A. Kazemeini, "Effect of organic and inorganic fertilizers on grain yield and protein

[22] NMSA, *Meteorological Data of Chelia Area for 2009-2015*, NMSA, Addis Ababa, Ethiopia, 2015.

[23] A. Tolera, W. Dagne, E. Semu, B. M. Msanya, D. Tolessa, and K. Haekoo, "Pedological characterization, fertility status and classification of the soils under maize production of Bako Tibe and Toke Kutaye districts of Western Showa, Ethiopia," *Ethiopian Journal of Applied Science Technology*, vol. 7, no. 1, pp. 1–17, 2016.

[24] SAS, *SAS Software Syntax, Version 9.0*, SAS Institute, Cary, NC, USA, 2004.

[25] R. G. D. Steel and J. H. Torrie, *Principles and Procedures of Statistics: A Biometrical Approach*, McGraw-Hill, New York, NY, USA, 2nd edition, 1980.

[26] CIMMYT, *From Agronomic Data to Farmer Recommendations: An Economics Training Manual*, CIMMYT, Mexico, 1988.

[27] D. Woubshet, K. Selamyihun, A. Tolera, and R. V. Cherukuri, "Effects of lime, blended fertilizer (NPSB) and compost on yield and yield attributes of Barley (*Hordium vulgare L.*) on acid soils of Wolmera District, West Showa, Ethiopia," *Ethiopian Journal of Applied Science Technology*, vol. 8, no. 2, pp. 84–100, 2017.

[28] W. Mitiku, T. Temedo, T. N. Singh, and M. Teferi, "Effect of integrated nutrient management on yield and yield components of barley (Hordium vulgare L.) in Keffa Zone, South western Ethiopia," *Science Technology Arts Research Journal*, vol. 3, no. 2, pp. 34–42, 2014.

[29] A. Getachew, "Ameliorating effects of organic and inorganic fertilizers on crop productivity and soil properties on reddish-brown soils," in *Proceedings of 10th Conference of the Ethiopian Society of Soil Science*, pp. 127–150, EIAR, Addis Ababa, Ethiopia, March 2009.

[30] A. J. Ofosu and M. Leitch, "Relative efficacy of organic manures in spring barley (*Hordeum vulgare L.*) production," *Australian Journal of Crop Science*, vol. 3, no. 1, pp. 13–19, 2009.

[31] J. Amanuliah and N. Maimoona, "The response of wheat (*Triticum aestivum L.*) to farm yard manure and nitrogen under rain fed condition," *African Crop Science Proceeding*, vol. 8, pp. 37–40, 2007.

[32] T. Kassu, M. Asrat, A. Almaz et al., "Malting barley response to integrated organic and mineral nutrient sources in Nitisol," *International Journal of Recycling of Organic Waste in Agriculture*, vol. 7, no. 2, pp. 125–134, 2018.

[33] A. Kumar, "Response of wheat cultivars of nitrogen fertilization under late sown condition," *Indonesian Journal Agronomy*, vol. 30, no. 4, pp. 464–467, 2009.

[34] M. C. P. Shantveerayya, S. C. Alagundagi, and S. R. Salakinkop, "Productivity of Barley (*Hordeum vulgare L.*) genotypes to integrated nutrient management and broad bed and furrow method of cultivation in watershed area," *International Journal of Agriculture Sciences*, vol. 7, no. 4, pp. 497–501, 2015.

[35] A. Getachew, P. N. Nelson, and M. I. Bird, "Crop yield, plant nutrient uptake and soil physicochemical properties under organic soil amendments and nitrogen fertilization on Nitisols," *Soil Tillage Research*, vol. 160, pp. 1–13, 2016.

[36] P. Prystupa, G. Slafer, and A. Savin, *Leaf Appearance, Tillering and Their Coordination in Response to NxP Fertilization in Barley*, Springer, Netherlands, 2004.

[37] H. Iqtidar, K. M. Ayyaz, and K. E. Ahmad, "Bread wheat varieties as influenced by different nitrogen levels," *Zhejiang Journal of Universal Sciences*, vol. 7, pp. 70–78, 2006.

[38] R. N. Sepat, R. K. Rai, and S. Dhar, "Planting systems and integrated nutrient management for enhanced wheat (*Triticum aestivum*) productivity," *Indian Journal of Agronomy*, vol. 55, no. 2, pp. 114–118, 2010.

[39] D. Tariku, "Effect of integrated nutrient management on growth and yield of food barley (*Hordeum vulgare*) variety in Toke Kutaye district, West Showa Zone, Ethiopia," M.Sc. thesis, Agronomy at Ambo University, Ambo, Ethiopia, 2016.

[40] W. Assefa, "Response of barley (*Hordium vulgare L.*) to integrated cattle manure and mineral fertilizer application in the vertisol areas of south Tigray, Ethiopia," *Journal Plant Sciences*, vol. 3, no. 2, pp. 71–76, 2015.

[41] A. Mahajan, R. M. Bhagat, and R. D. Gupta, "Integrated nutrient management in sustainable rice wheat cropping system for food security in India," *SAARC Journal of Agriculture*, vol. 6, no. 2, pp. 29–32, 2008.

[42] K. Berecz, T. Kismanyoky, and K. Debreczeni, "Effect of organic matter recycling in long-term fertilization trials and model pot experiments," *Communication Soil Science Plant Analysis*, vol. 36, pp. 191–202, 2005.

[43] H. Schulz and B. Glaser, "Effects of biochar compared to organic and inorganic fertilizers on soil quality and plant growth in a greenhouse experiment," *Journal Plant Nutrition Soil Sciences*, vol. 175, pp. 410–422, 2012.

[44] J. A. Alburquerque, P. Salazar, V. Barrón et al., "Enhanced wheat yield by biochar addition under different mineral fertilization levels," *Agronomy Sustainable Development*, vol. 33, pp. 475–484, 2013.

[45] T. T. Doan, T. Henry-des-Tureaux, C. Rumpel, J. L. Janeau, and P. Jouquet, "Impact of compost, vermicompost and biochar on soil fertility, maize yield and soil erosion in Northern Vietnam: a three year mesocosm experiment," *Sciences Total Environment*, vol. 514, pp. 147–154, 2015.

[46] A. Saidu, K. Ole, and B. O. Leye, "Performance of Wheat (*Triticum aestivum L.*) as influenced by complementary use of organic and inorganic fertilizers," *International Journal of Science and Nature*, vol. 5, no. 4, pp. 532–537, 2012.

[47] V. K. Chakrawarty and K. P. Kushwaha, "Performance of barley (*Hordeum valgare*) varieties under sowing date and nutrient levels in Bundelkhand," *Progrssive Research*, vol. 2, no. 2, pp. 163-164, 2009.

Distribution, Density, and Abundance of Parthenium Weed (*Parthenium hysterophorus* L.) at Kuala Muda, Malaysia

C. M. Maszura, S. M. R. Karim, M. Z. Norhafizah, F. Kayat, and M. Arifullah

Faculty of Agro-Based Industry, Universiti Malaysia Kelantan, Jeli Campus, Kota Bharu, Malaysia

Correspondence should be addressed to S. M. R. Karim; rezaul@umk.edu.my

Academic Editor: Maria Serrano

Knowledge of distribution, density, and abundance of weed in a place is a prerequisite for its proper management. Parthenium hazard is a national agenda in Malaysia, and Kedah is the worst infested state in the country. Despite it, the distribution and abundance of the weed is not systematically documented. Periodical weed surveys were conducted at Kuala Muda, Kedah, during March and September 2015 to identify infested locations, to determine density, abundance, and severity of infestation, and to do mapping of weed distribution of the area. Geographic locations were recorded using a GPS. Weed density was measured following the list count quadrat method. The mapping of weed infestation was done by the ArcGIS software using data of GPS and weed density. Different letters were used to indicate the severity of infestation. Results indicated that in Kuala Muda, sixteen sites are infested having average weed density of 10.6 weeds/m^2. The highest density was noted at Kg. Kongsi 6 (24.3 plants/m^2). The relative density was highest at Semeling (27.25%) followed by Kg. Kongsi 6 (23.14%). The average severity of infestation was viewed as the medium. Parthenium abundance and relative density increased by 18.0% and 27%, respectively, in the second survey conducted. The intervention of concerned authority to tackle the weed problem using integrated weed management approach is emphasized.

1. Introduction

The most obnoxious, allergenic, and environmental pollutant weed, *Parthenium hysterophorus* L. (locally called as Rumpai Miang Mexico) is an invasive alien species in Malaysia. The weed scientists of Universiti Malaysia Kelantan detected it for the first time at Batang Kali, Selangor, in 2013 [1–3]. At present, ten states of Malaysia are invaded by the weed, and the state Kedah is the worst infested area [4, 5]. The weed has harmful impacts on crop production, livestock production, human and animal health, and biodiversity [6, 7]. The weed spreads very fast through transport, agricultural implements, crop seeds, compost, and organic manures. It also spreads through flood water, wind pressure, and tyre-carried mud of vehicles [8]. It is essential to know the critical analytical characters such as density, frequency, and abundance of the species if we want to know its dominance in a community. Weed density measures the number of the species in a unit area, sometimes expressed as a percentage. Frequency is the number of times the species occur in the sampling unit, or it is the degree of dispersal of the species. The abundance of a species is the total number of the species present in the weed community and is a relative measure [9]. The severity of weed infestation usually regards the percentage of area covered by the species. The study of weed distribution, abundance, and severity over time are helpful in determining how a population changes over time in response to agronomic practices and agroclimatic conditions [5]. Therefore, it is essential to identify the parthenium infested locations, that is, the geographical range of the weed, its abundance, and severity of infestation in the study site. The primary objective of the study was to accurately identify the areas with parthenium populations and quantify their abundance and severity of infestation. It is intended that the scientists, land managers, and the inhabitants can foresee those areas, potentially subject to parthenium invasion, understand the biology of invasion process, and thereby develop the appropriate weed management plans.

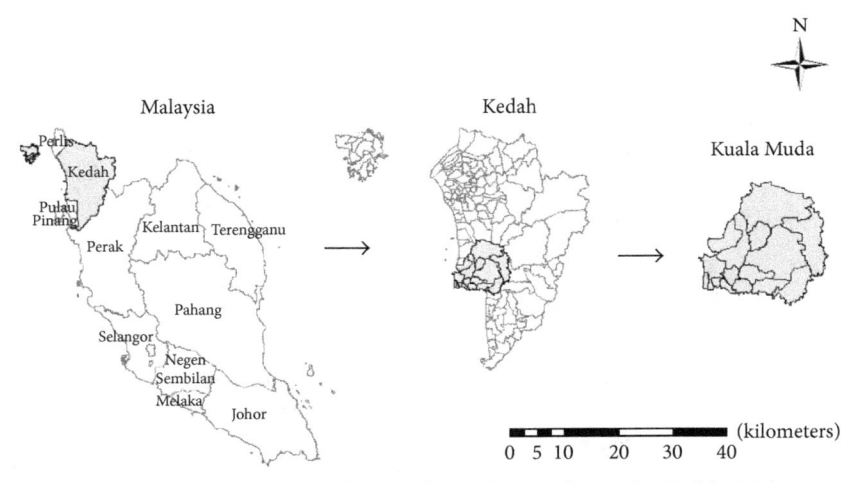

Figure 1: Map showing the study sites located at Kuala Muda, Kedah, Malaysia.

A regular weed monitoring is needed to control the weed sustainably. Weed monitoring involves repetitive surveys to track weed populations over time. The comparisons between different periodic surveys can help to elucidate the effect of new weed control technologies on weed species shift. Our preliminary investigation indicated that the weed is scattered in various areas of Batang Kali (Selangor) and Sungai Petani (Kedah), but detailed information was not documented from elsewhere [2, 5].

2. Materials and Methods

2.1. Location of Weed Survey. The weed survey was conducted at Kuala Muda (05°38.770′N, 100°28.771′E) in Kedah, located in Peninsular Malaysia (Figure 1).

2.2. Procedure of Weed Survey. The field surveys were conducted throughout Kuala Muda subdistrict, Kedah, during March and September 2015. The observation was made alongside the highways and major roads of the study sites with an interval of 500 meters. The farmland, wasteland, and river bank around the survey spots were also considered in the survey following the list quadrat method [9, 10]. The geographical coordinates were recorded using a GPS. When parthenium was observed at a density of at least one plant per $10\,m^2$ area, it was considered as presence of the weed.

2.3. Determination of Relative Density and Abundance of Parthenium Weed. The weed density was measured by placing a quadrat ($1\,m^2$ size) randomly on the survey spots following the list quadrat method [10]. Ten quadrats were placed at each site of the survey spots. Parthenium weed and other weed species within each quadrat were counted [11]. Parthenium weed density and frequency were calculated based on the formulae as applied by Tauseef et al. [12] and Nkoa et al. [5]. Whereas the relative density was determined using the equations of Yakubu et al. [13] and Knox et al. [14]. The abundance of parthenium weed invasion was determined based on the formula of Kilewa and Rashid [15]:

$$\text{density} = \frac{\text{total number of parthenium weed in a quadrat}}{\text{total area of a quadrat}(1\,m^2)} \times 100,$$

$$\text{frequency} = \frac{\text{number of quadrats with parthenium weed}}{\text{total number of quadrats used in each spot}} \times 100,$$

$$\text{relative density} = \frac{\text{total number of parthenium weed}}{\text{total number of all weed species}} \times 100,$$

$$\text{abundance} = \frac{\text{total number of parthenium weed in all quadrats}}{\text{total number of quadrats in which parthenium weed occurred}}.$$

(1)

2.4. Mapping the Distribution of Parthenium Infestation and the Percentage of Area Coverage (Severity Class). The information on parthenium distribution and severity of infestation marked on the base maps of Kedah following the technique of Cooksey and Sheley [16]. The mapping of parthenium weed distribution was done by measuring the percentage of area coverage completed by visual estimate within the quadrat according to Philippoff and Cox [17]. The

TABLE 1: Classification of severity of parthenium infestation.

Symbol used	Severity class	Percent coverage
T	Trace/rare	Less than 1%
L	Low/occasional plants	Between 1 and 5%
M	Moderate/scattered plants	Between 5 and 25%
H	High/fairly dense	Between 25 and 100%

Adopted from the U.S. Department of the Interior [18] and Cooksey and Sheley [16].

severity of infestation is categorized and presented in Table 1.

All the data of infested areas were transferred into a digital map using Geographic Information System (GIS) software, ArcGIS [16]. The distribution, density, and percentage of coverage by the parthenium weed were considered in mapping the weed infestation.

3. Results and Discussion

3.1. Location and Distribution of Parthenium Infestation. The survey results indicated that in Kuala Muda, Kedah, parthenium grew vigorously in 16 villages namely, Ladang Sungai Bongkok, Havard Golf Course area, Kg. Sungai Tok Rawang, Kg. Zainal Abidin, Kg. Semeling, Kg. Kongsi 6, Kg. Telok, Penghulu Him, Taman Kg. Raja, Batu Dua, Taman Songket Indah, Pokok Terap, Taman Nilam Sari, Taman Cahaya Baiduri, Teluk Wang Besar, and Kg. Banggol. Most of the villages have medium (M) level of infestation except the places like Batu Dua, Pokok Terap, Taman Nilam Sari, and Kg. Banggol, where low (L) infestation was recorded, and at Teluk Wang Besar it was regarded as rare, especially during the first survey (Figure 2(a)). The difference in intensity of infestation is due to natural selection as no other extra influence has worked here to make difference.

However, during the second survey, all the villages had "medium" level of infestation signposted by "M," and Kg. Semeling had a high level of infestation designated by "H." It is mentioned here that the personnel from Department of Agriculture (DOA), Kedah, sprayed with glyphosate twice (trade name: Knockdown) between the surveys. Although the above-ground parts of the weed were destroyed, a vast number of seeds remained within the soil and emerged later with higher energy. That is why the infestation level was higher in the second survey (Figure 2(b)). Probably the weed seedbank of parthenium was encouraged after herbicide spraying above ground. Worku [19] stated that parthenium became the major weed in Ethiopia, India, and Australia within a short period, even after herbicide sprays. Therefore, continuous monitoring and control measures are needed for an extended period.

3.1.1. Parthenium Preference according to Habitat. In Kuala Muda, most of the infestations are noticed along roadsides, residential areas, wasteland, and crop farms (Figure 3). When parthenium is prevalent on roadsides, it is subject to quick spread through wind pressure of moving transports [20]. According to Ayele et al. [21], parthenium weed was introduced to Ethiopia in the year 1968 through cereals consignment. However, another form of introduction occurred to eastern Ethiopia in the year 1976 through weed seeds attached to the army transport at the time of Ethio-Somali war.

When parthenium grows in the wasteland, it usually remains undisturbed and leads the plant to produce a higher number of seeds. Parthenium growth in the residential area is dangerous to the inhabitants. The children who play with the weed are prone to be infected by parthenium showy white flowers which cause allergy. People may inhale the pollen from the parthenium flowers during breathing which might adversely affect their respiratory system. The weed surveyors noticed many parthenium plants in front of mosques and Hindu temples, which are also risky for the people who come for worships. The dominance of parthenium weed in crop farms may lead to its contamination with crop seeds which leads to its further spread. Upadhyay et al. [22] commented that parthenium weed commonly grows near the roadsides, in farms, and paddocks. Adnan et al. [1] also noted that most of the habitats infested by the parthenium weed are alongside road, the fields for crop cultivation, and wasteland rather than alongside water canal.

Presence of parthenium near cattle farm is risky for the cattle. If the cattle graze on the parthenium infested land, the animals might be infected with skin dermatitis due to close contact with the weed. When parthenium is abundant along the river bank or watercourses, the spread of the weed may occur due to the mixing of parthenium weed seeds with river water. All these information indicate the risks of the parthenium weed at Kuala Muda.

3.2. Density and Abundance of Parthenium Weed at Different Sites. The density, relative density, frequency, abundance, and other parameters of parthenium weed infestation at Kuala Muda, Kedah, are shown in Table 2.

3.2.1. Weed Density. The parthenium weed density at Kuala Muda, Kedah, significantly varied ($P < 0.05$) at different villages in both the surveys on March and September 2015. During the first survey, the highest density of parthenium weed was found at Kg. Kongsi 6 (18.8 weed/m^2) followed by Taman Cahaya Baiduri (16.0 weed/m^2), Kg. Penghulu Him (15.8 weed/1 m^2), and Kg. Zainal Abidin (13.4 weeds/m^2). The lowest density was noticed at Kg. Teluk Wang Besar (0.4 weeds/m^2) and Kg. Banggol (1.2 weeds/m^2) (Figure 4).

The highest relative density was in Taman Cahaya Baiduri (23.74%) followed by Kg. Penghulu Him (22.19%), Kg. Kongsi 6 (19.85%), and Kg. Zainal Abidin (19.79%) (Table 2).

It is not clear to us that why such a variability in parthenium density exists among different villages. Since the people of all these villages were unaware about this invasive species [4], the discrepancy in weed densities might be due to natural selection. More or less similar ranking of weed frequency was noticed in different survey sites. For example, the highest frequency was noted in Kg. Kongsi 6, Kg. Penghulu Him, Kg. Telok, Taman, and Cahaya Baiduri and

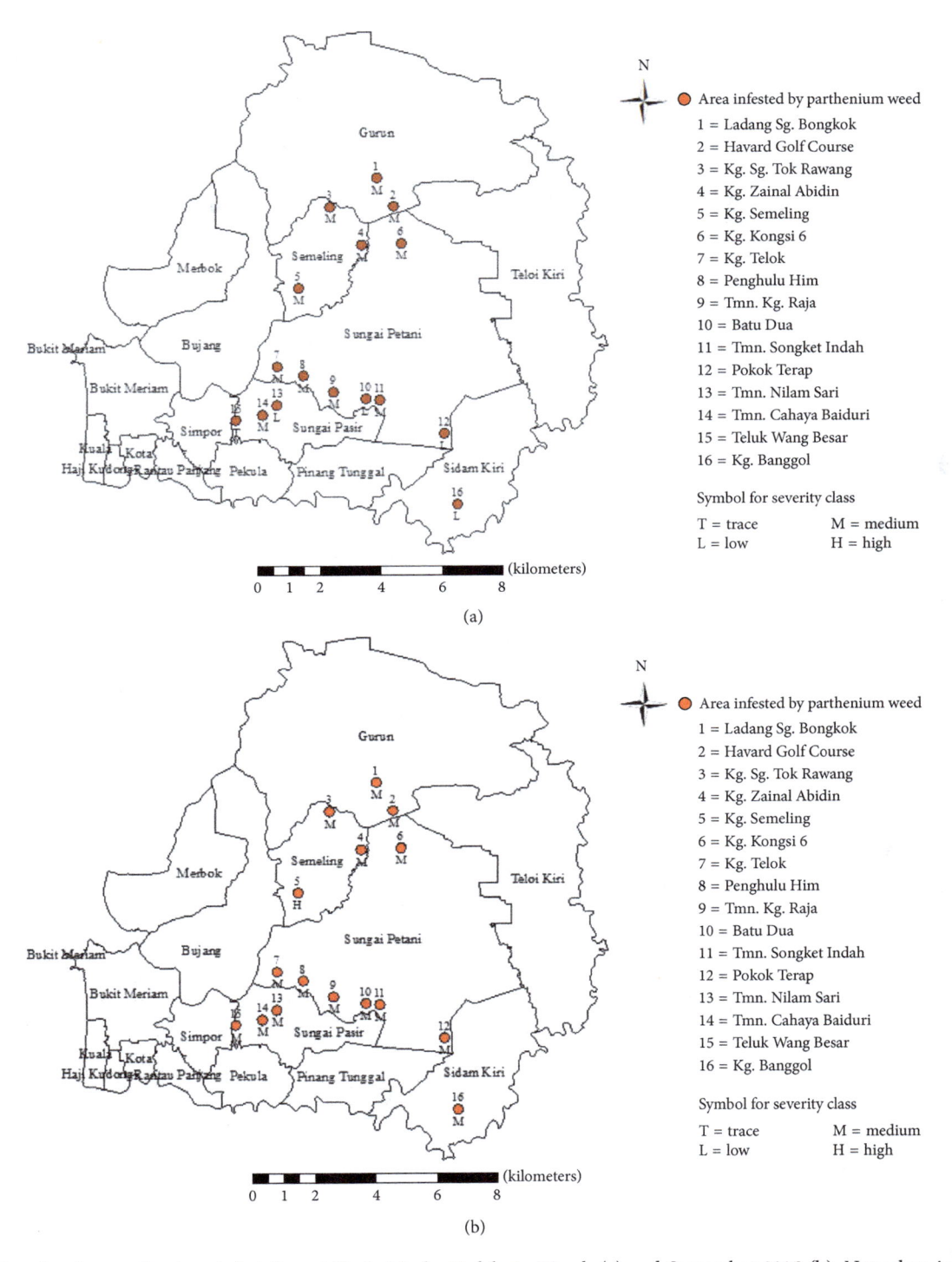

FIGURE 2: Map showing parthenium infestation at Kuala Muda, Kedah, in March (a) and September 2015 (b). Note that, in the second survey, the severity of infestation has changed from "L" to "M" in some locations and in one location from "M" to "H."

the lowest frequency was in Kg. Banggol, Kg. Teluk Wang Besar, and Pokok Terap (Table 2).

During the second survey in September 2015, a significant change occurred in weed density at Kuala Muda, Kedah. The weed density increased in most of the survey sites, ranging from 2.2 to 24.3 weeds/m^2. The highest density was recorded at Kg. Kongsi 6 (24.3 weeds/m^2), Kg. Semeling (21.2 weeds/m^2), and Kg. Zainal Abidin (15.1 weeds/m^2), and the lowest was at Kg. Banggol (2.2 weeds/m^2) (Figure 5). The average density was higher in the second survey (10.6 weeds/m^2) compared to 8.43 weeds/m^2 in the first survey, although statistically, both were similar (Figure 6). However, there was a little decrease in weed density at four sites, for example, Kg. Penghulu Him, Taman Songket Indah,

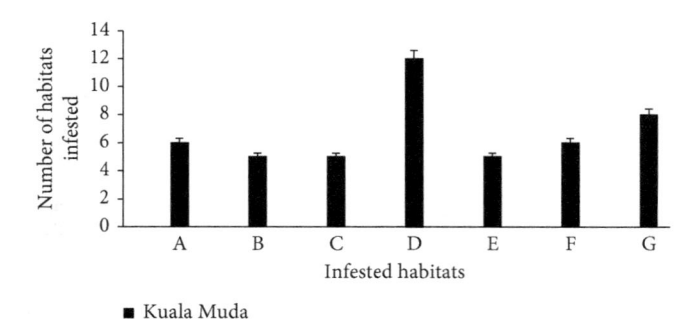

FIGURE 3: The number of habitats infested with parthenium weed at different sites of Kuala Muda. A = wasteland, B = cattle farm area, C = damping place, D = roadside, E = bank of river, F = farmlands or plantations, and G = residential area and temples/mosques. Small vertical lines above the bars indicate the standard error of means.

TABLE 2: Total number of quadrats used (NQ), other species/quadrats (OS), number of all species/quadrats (AS), number of quadrats with parthenium (QP), frequency of parthenium distribution (F), parthenium weed abundance (A), relative density (RD) (%), and symbol for severity class (SC) at Kuala Muda in March and September 2015.

Sites	NQ	OS	AS	QP	F	A	RD	SC	OS	AS	QP	F	A	RD	SC	% increase A	% increase RD
			March 2015								September 2015						
1	10	55.6	61.6	7	0.7	8.6	9.74	M	41.5	53.2	7	0.7	16.7	21.99	M	48.50	55.71
2	10	59.6	64.9	6	0.6	8.8	8.17	M	32.5	37.4	6	0.6	8.2	13.1	M	−7.32	37.62
3	10	59.9	72.6	7	0.7	18.1	17.49	M	46.4	60	8	0.8	17.0	22.67	M	−6.47	22.85
4	10	54.3	67.7	7	0.7	19.1	19.79	M	50.3	65.4	8	0.8	18.9	23.09	M	−1.06	14.22
5	10	48.2	59.4	7	0.7	16.0	18.86	M	56.6	77.8	8	0.8	26.5	27.25	H	39.62	30.77
6	10	75.9	94.7	8	0.8	23.5	19.85	M	80.7	105.0	9	0.9	27.0	23.14	M	12.96	14.22
7	10	50.2	62.5	8	0.8	15.4	19.68	M	53.8	67.4	7	0.7	19.4	20.18	M	20.62	2.48
8	10	55.4	71.2	8	0.8	19.8	22.19	M	56.7	70.0	7	0.7	19.0	19	M	−4.21	−16.79
9	10	53.0	60.9	7	0.7	11.3	12.97	M	55	63	6	0.6	13.3	12.7	M	15.04	−2.13
10	10	52.0	54.6	4	0.4	6.5	4.76	L	44.9	47.5	4	0.4	6.5	5.47	M	0	12.98
11	10	64.5	72.2	5	0.5	15.4	10.66	M	49.7	55.4	5	0.5	11.4	10.29	M	−35.09	−3.60
12	10	65.5	67.0	3	0.3	5.0	2.24	L	63	67.2	4	0.4	10.5	6.25	M	52.38	64.15
13	10	41.6	43.6	4	0.4	5.0	4.59	L	39.8	49.8	5	0.5	20.0	20.08	M	75.00	77.14
14	10	51.4	67.4	8	0.8	20.0	23.74	M	49.2	64	7	0.7	21.1	23.13	M	5.21	−2.64
15	10	83.1	83.5	3	0.3	1.3	0.48	T	51.8	56.2	5	0.5	8.8	7.83	M	85.23	93.87
16	10	41.5	42.7	3	0.3	4.0	2.81	L	39	41.2	4	0.4	5.5	5.34	M	27.27	47.28
Total	160	911.7	1046.5	95	9.5	197.8	198.02		810.9	980.5	100	10	249.8	261.51			

Note. Site 1 = Ladang Sg. Bongkok; Site 2 = Havard Golf Course; Site 3 = Kg. Sg. Tok Rawang; Site 4 = Kg. Zainal Abidin; Site 5 = Kg. Semeling; Site 6 = Kg. Kongsi 6; Site 7 = Kg. Telok; Site 8 = Kg. Penghulu Him; Site 9 = Tmn. Kg. Raja; Site 10 = Batu Dua; Site 11 = Tmn. Songket Indah; Site 12 = Pokok Terap; Site 13 = Tmn. Nilam Sari; Site 14 = Tmn. Cahaya Baiduri; Site 15 = Teluk Wang Besar; Site 16 = Kg. Banggol; Kg. = Kampung; Tmn. = Taman; Sg. = Sungai.

Taman Cahaya Baiduri, and Havard Golf Course. Probably this change is due to the greater efforts of monitoring and herbicide spraying by the Department of Agriculture, Kedah.

The relative density was in the similar trend as recorded in both the surveys. The highest relative density was in Kg. Semeling (27.25%), followed by Kg. Kongsi 6 (23.14%), Taman Cahaya Baiduri (23.13%), and Kg. Zainal Abidin (23.09%). The lowest relative density was in Kg. Banggol (5.34%) (Table 2). There was more or less similar weed frequency noticed in the second survey, and the highest was in Kg. Kongsi 6 followed by Kg. Sungai Tok Rawang, Kg. Zainal Abidin, and Kg. Semeling. The lowest weed frequency was in Batu Dua, Pokok Terap, and Kg. Banggol. Higher frequency and density of parthenium weed caused its abundance also higher in both the surveys in Kuala Muda, Kedah. From Table 2, it is obvious that the value of weed abundance increased the most in Teluk Wang Besar (85.23%) followed by Taman Nilam Sari (75.0%) and Pokok Terap (52.38%). However, there were decreases in parthenium abundance in some locations, especially in Taman Songket Indah (35.09% decrease). While an average impact of the weed distribution considered, it is clear that on an average, more than 18.0% increase in weed abundance and 28.0% increase in relative density were observed in the second survey. Nkoa et al. [5] stated that the weed abundance is related to weed number (density) or frequency, which might have influenced the abundance positively in this study.

3.2.2. Percent Weed Coverage. A significant difference in percent coverage by parthenium weed was observed in different villages ($P < 0.05$). In the first survey, parthenium weed coverage was higher at Kg. Zainal Abidin (23.5%), followed by places Kg. Semeling and Taman Cahaya Baiduri (23%) and was the lowest at Teluk Wang Besar (0.9%) (Figure 7).

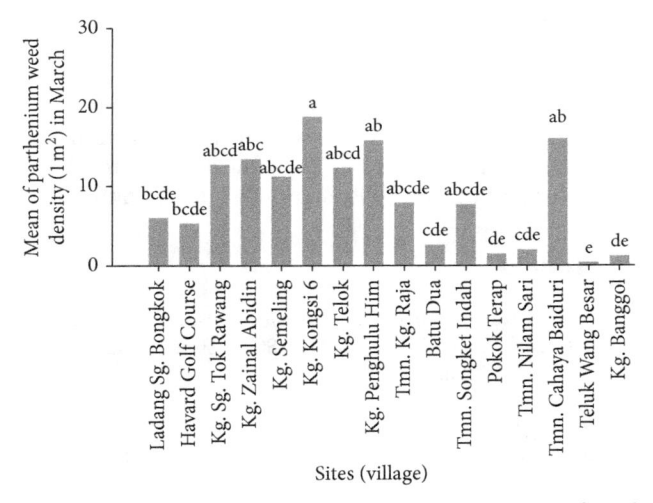

FIGURE 4: Parthenium weed density of different villages of Kuala Muda in March 2015. Kg. = Kampung, Tmn. = Taman, and Sg. = Sungai. Dissimilar small letters on bars indicate a significant difference.

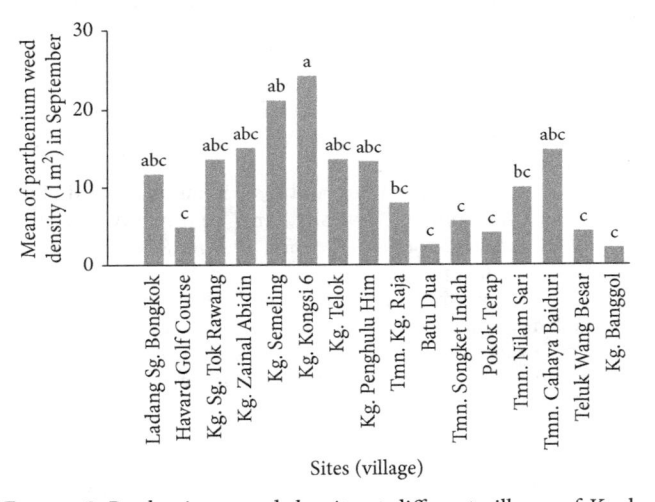

FIGURE 5: Parthenium weed density at different villages of Kuala Muda, Kedah, in September 2015. Kg. = Kampung, Tmn. = Taman, and Sg. = Sungai. Dissimilar small letters on bars indicate a significant difference.

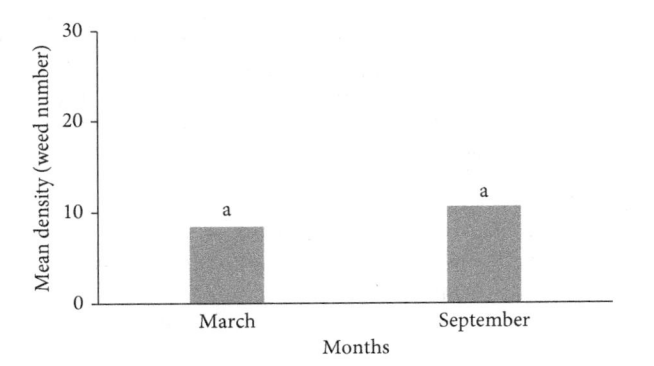

FIGURE 6: Mean density of parthenium weed during two surveys in March and September 2015. The results are the means of all villages. Similar small letters on bars indicate a nonsignificant difference.

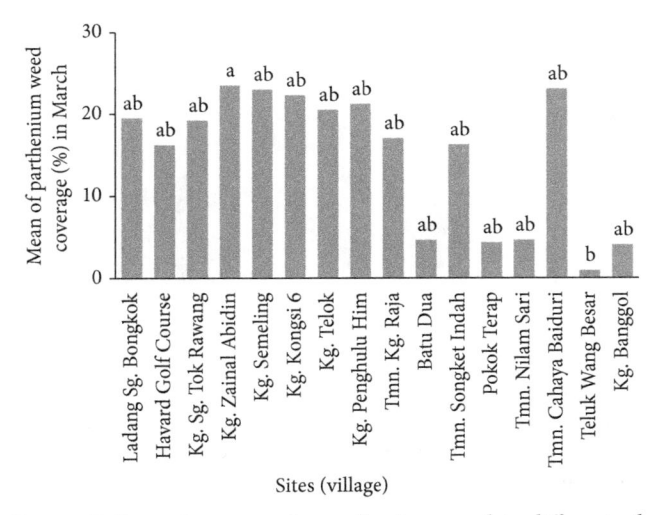

FIGURE 7: Percent coverage by parthenium weed in different villages of Kuala Muda in March 2015. Kg. = Kampung, Tmn. = Taman, and Sg. = Sungai. Dissimilar small letters on bars indicate a significant difference.

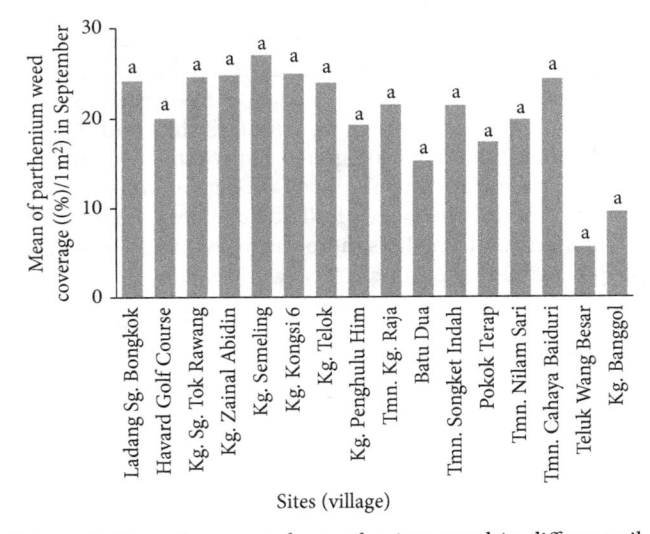

FIGURE 8: Percent coverage by parthenium weed in different villages of Kuala Muda in September 2015. Kg. = Kampung, Tmn. = Taman, and Sg. = Sungai. Dissimilar small letters on bars indicate a significant difference.

In the second survey, parthenium weed coverage (%) was higher in Kg. Semeling (27%), followed by Kg. Kongsi 6 (25%), and the lowest was at Teluk Wang Besar (5.5%) (Figure 8).

From the data of percent area coverage, the severity of infestation is regarded as "M" in most of the sites, except in Kg. Teluk Wang Besar which is regarded as trace (T) and Batu Dua, Pokok Terap, Taman Nilam Sari, and Kg. Banggol are regarded as low (L) infested areas. However, the percent coverage in the second survey increased, and the severity of infestation is regarded as "M" in those sites. However, the severity of the weed is regarded as high (H) in Kg. Semeling. The overall differences between the two surveys in March and September 2015 were statistically significant ($P < 0.05$) (Figure 9).

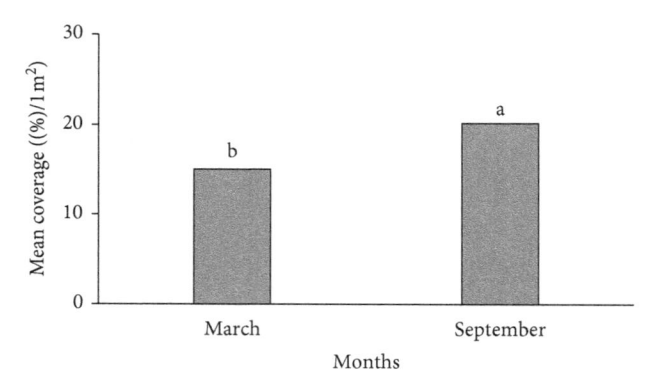

FIGURE 9: The differences of parthenium abundance as measured by weed coverage (%) for parthenium weed between March and September 2015 at Kuala Muda, Kedah. Dissimilar small letters on bars indicate significant differences.

When the comparison is made between the two surveys, it was obvious that the parthenium weed density was not different. However, the weed abundance as regarded by percent coverage was higher in September than in March 2015 (Figures 6 and 9). Parthenium weed spreads very fast just like wildfire in the forest. Therefore, the increase in weed abundance in Kuala Muda, Kedah, is a matter of great concern to the authority of the country. The control of the weed should follow an integrated approach including prevention, chemical control, physical control, and biological control. The involvement of all stakeholders, for example, researchers, policymakers, administrators, and community people, should be ensured.

4. Conclusion

The status of weed density and abundance in Kuala Muda, Kedah, Malaysia, is in a critical stage of infestation and needs quick action to tackle the problem sustainably. Timely control of the weed by adopting appropriate methods especially with an integrated weed management approach is essential.

Authors' Contributions

All the authors contributed to this research in different ways since 2014. S. M. R. Karim designed the study, and C. M. Maszura carried out the field works under the supervision of S. M. R. Karim. M. Z. Norhafizah and C. M. Maszura made the first draft. F. Kayat and M. Arifullah contributed to editing and formatting. The final draft was read and approved by all the authors.

Acknowledgments

The authors are grateful to the Ministry of Education, Malaysia, for providing the FRGS grant (FRGS/1/2014/STWN03/UMK/01/1) to carry out this research. The support from the Faculty of Agro-Based Industry, UMK, in conducting the study is acknowledged thankfully.

References

[1] S. M. R. Karim, *Parthenium Invasion in Malaysia: Weed it Out before It's Too Late*, Vol. 25, New Strait Times, Kuala Lumpur, Malaysia, October 2013.

[2] S. M. R. Karim, "Malaysia Diserang Rumpai *Parthenium hysterophorus* yang Paling Serius, *Agro Malaysia* (Malaysia invaded by the worst weed, Rumpai Miang Mexico, *Parthenium hysterophorus*)," vol. 12, pp. 40–45, 2014.

[3] S. M. R. Karim, M. Z. Norhafizah, C. M. Maszura, K. Fatimah, and M. Z. Alam, "Perception of local people about Parthenium weed (Rumpai Miang Mexico) in Kedah, Malaysia," *International Journal of Biology, Pharmacy, and Allied Sciences*, vol. 5, no. 5, pp. 1006–1015, 2016.

[4] DOA (Department of Agriculture), *Pengesahan Bancian Rumpai Miang Mexico di Semenanjung Malaysia Sehingga 10 Feb 2015, Jemputan Mesyuarat Jawatankuasa Teknikal Program Kawalan, Pembendungan dan Penghapusan Rumpai Miang Mexico (Parthenium hysterophorus)*, (Report of the meeting on parthenium weed management in Malaysia) Plant Biosecurity Division, 2015.

[5] R. Nkoa, M. D. K. Owen, and C. J. Swanton, "Weed abundance, distribution, diversity, and community analyses," *Weed Science Society of America, BioOne Research Evolved*, vol. 63, no. 1, pp. 64–90, 2015.

[6] S. M. R. Karim, "Ill impacts of parthenium weed on human and health and livestock production and environment," in *Proceedings of the Invited Seminar by Davies College of Agriculture*, West Virginia University, Morgantown, WV, USA, October 2012.

[7] S. Adkins and A. Shabbir, "Biology, ecology and management of the invasive Parthenium weed (*Parthenium hysterophorus* L.)," *Pest Management Science*, vol. 70, no. 7, pp. 1023–1102, 2014.

[8] K. V. Sankaran, "Carrot weed (*Parthenium hysterophorus*)," in Invasive Pest Fact Sheet, APFISN (Asia-acific Forest Invasive Species Network), pp. 1–3, Kerala Forest Research Institute, Peechi, Kerala, India, 2007.

[9] H. Khan, K. B. Marwat, G. Hassan, M. A. Khan, and S. Hashim, "Distribution of Parthenium weed in Peshawar valley, Khyber Pakhtunkhwa-Pakistan," *Pakistan Journal of Botany*, vol. 46, no. 1, pp. 81–90, 2014.

[10] M. Mahajan and S. Fatima, "Frequency, abundance, and density of plant species by list count quadrat method," *International Journal of Multidisciplinary Research*, vol. 3, no. 7, pp. 1–8, 2017.

[11] M. Adnan, M. Ali, A. Haider, and A. U. Khan, "Survey of District Sialkot for the infestation of *Parthenium hysterophorus*," *Journal of Biology, Agriculture and Healthcare*, vol. 5, no. 1, pp. 111–119, 2015.

[12] M. Tauseef, F. Ihsan, W. Nazir, and J. Farooq, "Weed flora and importance value index (IVI) of the weeds in cotton crop

fields in the region of Khanewal, Pakistan," *Pakistan Journal of Weed Science Research*, vol. 18, no. 3, pp. 319–330, 2012.

[13] A. I. Yakubu, J. Alhassan, A. Lado, and S. Sarkindiya, "Comparative weed density studies in irrigated carrot (*Daucus carota* L.) Potato (*Solanum tuberosum* L.) and wheat (*Triticum aestivum* L.) in Sokoto-Rima valley, Sokoto State, Nigeria," *Journal of Plant Sciences*, vol. 1, no. 1, pp. 14–21, 2006.

[14] J. Knox, D. Jaggi, and M. S. Paul, "Population dynamics of *Parthenium hysterophorus* (Asteraceae) and its biological suppression through *Cassia occidentalis* (Caesalpiniaceae)," *Turkish Journal of Botany*, vol. 35, pp. 111–119, 2011.

[15] R. Kilewa and A. Rashid, "Distribution of invasive weed *Parthenium hysterophorus* in natural and agro-ecosystems in Arusha Tanzania," *International Journal of Science and Research*, vol. 3, no. 12, pp. 1–4, 2014.

[16] D. Cooksey and R. Sheley, *Montana Noxious Weed Survey and Mapping System*, MSU Extension Service MT199613 AG 7, Bozeman, MT, USA, 2002.

[17] Z. Philippoff and E. Cox, *Measuring Abundance: Transects and Quadrats*, University of Hawaii, Honolulu, HI, USA, 2017.

[18] U. S. Department of Interior, *Guidelines for Coordinated Management of Noxious Weeds: Development of Weed Management Areas*, National Park Service, and U.S. Forest Service, Bureau of Land Management, Washington, DC, USA, 2001.

[19] M. Worku, "Prevalence and distribution survey of an invasive alien weed (*Parthenium hysterophorus* L.) in Sheka Zone, Southwestern Ethiopia," *African Journal of Agricultural Research*, vol. 5, no. 9, pp. 922–927, 2010.

[20] S. M. Masum, M. Hasanuzzaman, and M. H. Ali, "Threats of *Parthenium hysterophorus* on agroecosystems and its management: a review," *International Journal of Agriculture and Crop Science*, vol. 6, no. 11, pp. 684–697, 2013.

[21] S. Ayele, L. Nigatu, T. Tana, and S. W. Adkins, "Impact of Parthenium weed (*Parthenium hysterophorus* L.) on the above-ground and soil seed bank communities of Rangelands in Southeast Ethiopia," *International Research Journal of Agricultural Science and Soil Science*, vol. 3, no. 7, pp. 262–274, 2013.

[22] S. K. Upadhyay, M. Ahmad, and A. Singh, "Ecological impacts of weed (*Parthenium hysterophorus* L.) on saline soil," *International Journal of Scientific and Research Publications*, vol. 3, no. 4, pp. 1–4, 2013.

Evaluation of Chickpea Varieties and Fungicides for the Management of Chickpea Fusarium Wilt Disease (*Fusarium oxysporum* f.sp. *ciceris*) at Adet Sick Plot in Northwest Ethiopia

Yigrem Mengist,[1] Samuel Sahile,[2] Assefa Sintayehu,[1] and Sanjay Singh[1]

[1]*Department of Plant Sciences, College of Agriculture and Rural Transformation, University of Gondar, P.O. Box 196, Gondar, Ethiopia*
[2]*Department of Biology, College of Natural and Computational Sciences, University of Gondar, P.O. Box 196, Gondar, Ethiopia*

Correspondence should be addressed to Yigrem Mengist; yigermmengist07@gmail.com

Academic Editor: Ravindra N. Chibbar

A 2-year experiment was conducted at wilt sick plot infested with natural occurring *Fusarium oxysporum* f.sp. *ciceris* at Adet Agricultural Research Center in northwestern Ethiopia with an aim to evaluate effective chickpea varieties and fungicides for the management of chickpea fusarium wilt in order to integrate chickpea varieties and fungicides. Four varieties, namely, Shasho, Arerti, Marye, and local, two fungicides, namely, Apron Star and mancozeb, and untreated local chickpea were used as treatments. Treatments were arranged in a factorial combination in randomized complete block design in three replications. There were significant differences at $p < 0.05$ in the overall mean of fusarium wilt disease incidence, area under disease progress curve %-day, yield and yield components among varieties and fungicides treatments. Data were analyzed using SAS system version 9.2. The results indicated that the maximum disease incidence and area under disease progress curve values 65.62% and 578.5%-day, respectively, were recorded from untreated local chickpea, while the minimum disease incidence and area under disease progress curve values 23.41% and 147%-day, respectively, were recorded from Shasho variety treated with Apron Star. The maximum biomass and grain yield of 6.71 t/ha and 4.6 t/ha, respectively, were recorded from Shasho variety treated with Apron Star while the minimum biomass and grain yield of 0.62 t/ha and 0.21 t/ha, respectively, were recorded from untreated local chickpea. Thus, the experiment results suggested that the variety of Shasho treated with fungicide Apron Star caused significant reduction in chickpea fusarium wilt incidence leading to a corresponding increase in grain yield of chickpea.

1. Introduction

The average productivity of chickpea in Ethiopia is much lower than the world's average and is also lower as compared to other chickpea-growing countries such as Egypt and Turkey [1]. This low productivity is mainly due to a number of biotic and a biotic stresses. Among the biotic stresses, fusarium wilt, ascochyta blight, pod borer, and cutworm and abiotic stresses, drought, heat, soil salinity, low soil fertility, and poor crop management practices are the most important limiting factors in crop production [2]. However, fusarium wilt, caused by *Fusarium oxysporum* f.sp. *ciceris*, is one of the most important biotic stresses of chickpea (*Cicer arietinum* L.), and the average annual yield losses due to wilt have been estimated to 10 to 90% and sometimes escalate to 100% when the relative humidity is greater than 60% and temperature ranges between 10 and 25°C [3]. The disease is prevalent in the Indian subcontinent, Ethiopia, Mexico, Spain, Tunisia, Turkey, and the United States [4]. *Fusarium oxysporum* f.sp. *ciceris* is a vascular pathogen that perpetuates in seed and soil and hence is difficult to manage by the use of single control method [5].

Fusarium wilt of chickpea can be managed by using resistant varieties, healthy chickpea seed, crop rotation, biological control, and fungicides, adjusting sowing dates, and adopting integrated management practices [6]. Seid and Melkamu [7] reported that growing resistant and moderately resistant varieties treated with fungicides sown at recommended seeding rate could reduce mortality caused by chickpea fusarium wilt. In spite of the occurrence of fusarium wilt in commonly grown varieties in most parts of the country every year, limited research efforts have been made to find out sources of resistance and develop suitable methods for its management in Ethiopia. Therefore, the main objective of this study was to evaluate the effects of chickpea varieties and seed dressing with fungicides on fusarium wilt development.

2. Materials and Methods

2.1. Description of Study Area, Treatments, and Experimental Design. The field experiment was conducted during 2015 and 2016 cropping seasons on fusarium wilt sick plot at the Adet Agricultural Research Center (AARC) in the northwestern part of Ethiopia. The center is located 11° 17'N latitude, 37° 43'E longitude and lies at an altitude of about 2150 m.a.s.l. According to the meteorological data of the center, the average annual rainfall is 1250 mm ranging between 860 mm and 1771 mm with the average maximum and minimum temperature 27.5°C and 12.2°C, respectively [8].

The treatments were consisted of four chickpea varieties, namely, Shasho, Arerti, Marye, and local, two seed dressing fungicides, namely, Apron Star and mancozeb, and untreated local chickpea (control). The design was Randomized Complete Block Design (RCBD) with three replications. The gross plot size was $4\,m^2$ (1.6 m × 2.5 m), and the path plots and between blocks were 1 m and 1.5 m, respectively. Plots were prepared and fertilized with $100\,kg\cdot ha^{-1}$ diammonium phosphate at planting. The seeds were planted at spacing of 10 cm between plants and 40 cm between rows. Both the fungicides were used at the rate of 3 g for a kg of seeds, and the treated seeds were shade-dried overnight before sowing. All the recommended cultural practices were also applied in the field.

2.2. Data Collection. In the experiment, field observations of naturally occurring fusarium wilt incidence were done at 7-day interval at sick plot based on percent of wilt incidence in each plot. Initial recording data for fusarium wilt disease incidence was done when wilting symptoms were visible on the three to five basal leaves of the plants.

Disease incidence (DI) on each experimental unit was calculated by using the following formula:

$$DI(\%) = \left(\frac{\text{number of plants that shows wilt symptoms}}{\text{number of both disease infected plants and healthy plants}}\right) \times 100. \quad (1)$$

Area under disease progress curve (AUDPC) was calculated for each treatment from the assessment of disease incidence by using the following formula:

$$AUDPC = \sum_{i-1}^{n-1} 0.5\,(x_{i+1} + x_i)(t_{i+1} - t_i), \quad (2)$$

where x_i is the disease incidence in percentage at i_{th} assessment, t_i is the time of the i_{th} assessment in days from the first assessment date, and n is the total number of disease assessments [9].

Relative yield loss (RYL) was calculated using the formula of Robert and James [10]:

$$RYL = \frac{YP - YUP}{YP} \times 100, \quad (3)$$

where RYL = relative yield loss (reduction of the yield and yield component), YP = yields which were obtained from plots with maximum protection, and YUP = yields which were obtained from plots with minimum protection. Relative yield loss with different treatments was calculated with reference to the best protected plot.

2.3. Statistical Analysis. Analyses of variance (ANOVA) for the effects of varieties and fungicides on fusarium wilt disease incidence, AUDPC %-day, relative yield loss, and grain yield were used to compare the level of resistance among evaluated varieties and fungicides. All analyses were performed using statistical package SAS system, version 9.2 [11]. Least significant difference (LSD) values were used to separate differences among treatments means at 5% probability level.

3. Results and Discussion

The data of two cropping seasons were analyzed separately, but the result showed that there was no significant difference between the two season outputs in the experiments, so that the two season data were combined and analyzed together.

3.1. Disease Incidence. Significant differences at $p < 0.05$ were observed among varieties and fungicides on disease incidence percentage. Among the main effects, the maximum disease incidence of 53.45% was recorded from local chickpea while the minimum disease incidence of 29.41% was recorded from Shasho variety (Table 1). Shasho variety followed by Arerti had lower wilt incidence over the local [12]. On the contrary, the maximum disease incidence of 52.31% was recorded from untreated plots while the minimum disease incidence was ranged from 32.76% to 37.38% recorded from Apron Star and mancozeb treated plots, respectively (Table 1).

Landa et al. [13] pointed out seed dressing chemical is one of the most important factors that reduces the effects of fusarium wilt disease. The chickpea plants treated with fungicides gave 26–34% disease incidence of fusarium wilt as compared to untreated plants, which have 80% incidence. Similarly, the application of fungicides brought tremendous increment in plant growth as well as grain yield. Kamdi et al. [14] stated that seed dressing fungicide applications in chickpea plants were very effective; they have not only suppressed the pathogen activities but also increased the

TABLE 1: Main effects of chickpea varieties and fungicides on disease incidence of chickpea fusarium wilt.

Variety	Incidence (%)	Fungicides	Incidence (%)
Shasho	29.41a	Apron Star	32.76a
Arerti	34.83a	Mancozeb	37.38a
Marye	45.56b	Untreated	52.31b
Local	53.45b		
Mean	40.81		40.82
LSD	5.46		4.92
CV (%)		11.68	

LSD = least significant difference; CV = coefficient of variation; means followed by the same letter did not show significant difference at $p < 0.05$ according to least significant difference.

TABLE 2: Two-way interaction effects of chickpea varieties and fungicides on disease incidence of chickpea fusarium wilt.

Chickpea varieties	Disease incidence percentage			
	Seed dressing fungicides			
	Apron Star	Mancozeb	Untreated	Mean
Shasho	23.41a	26.68ab	40.15de	30.08a
Arerti	28.39ab	32.67bc	43.43e	34.83a
Marye	36.33cd	40.31de	60.03g	45.56b
Local	44.91ef	49.84f	65.62g	53.46c
Mean	33.26a	37.38a	52.31b	
LSD (5%)		6.125		
CV (%)		13.65		

LSD = least significant difference; CV = coefficient of variation; means followed by the same letter did not show significant difference at $p < 0.05$ according to least significant difference.

plant growth and grain yield (almost double than the control plants).

Analysis of variance showed that disease incidence was significantly affected by chickpea varieties, fungicides, and their interaction at $p < 0.05$ (Table 2). Among the interaction effects, the maximum disease incidence of 65.62% was obtained from untreated local chickpea, followed by untreated Marye variety while the minimum disease incidence of 23.41% was recorded from Shasho variety treated with Apron Star but they did not show any significant difference among Shasho variety treated with mancozeb and Arerti variety treated with Apron Star fungicides that of 26.68% and 28.39%, respectively (Table 2).

3.2. Effect of Varieties and Fungicides on the Progress of Fusarium Wilt.
The disease progress curve on chickpea varieties at two seed dressing fungicides, i.e., Apron Star and mancozeb and untreated is given in Figure 1. The fusarium wilt incidence increased from the initial to final assessment dates, and the curve showed an increasing trend of disease development for effects of four chickpea varieties in each seed dressing fungicide in the assessments. Hence, the result showed that the rate was slower on Shasho followed by Arerti variety compared to local and Marye variety (Figure 1).

On the contrary, the disease progress curve on fungicides at four chickpea varieties, i.e., Shasho, Arerti, Marye, and local is given in Figure 2. The curve showed an increasing trend of disease development for effects of seed dressing fungicides in each chickpea variety in the assessments. Hence, the result showed that the rate was slower on chickpea treated with Apron Star, followed by the ones treated with mancozeb fungicides compared to untreated plots (Figure 2).

3.3. Area under Disease Progress Curve (AUDPC).
Among the main effects, the minimum AUDPC %-day value were recorded from Shasho variety, followed by Arerti and Marye varieties while the maximum AUDPC %-day value was recorded from local chickpea (Figure 3). These mean that Shasho variety has more resistance against to the fusarium wilt disease compared to other tested varieties.

On the contrary, the minimum AUDPC %-day value was recorded from the chickpea variety treated with Apron Star, followed by mancozeb fungicides while the maximum AUDPC %-day value was recorded from untreated plots (Figure 3). This is because the chickpea variety treated with Apron Star fungicides had more resistance against to the fusarium wilt disease incidence compared to mancozeb and untreated.

Analysis of variance showed that AUDPC %-day was significantly affected by chickpea varieties, fungicides, and their interaction at $p < 0.05$. Among the interaction effects, the maximum AUDPC %-day value was recorded from untreated local chickpea, followed by local chickpea treated with mancozeb fungicides while the minimum AUDPC %-day value was recorded from Shasho variety treated with Apron Star fungicides, followed by Arerti treated with Apron Star and Shasho treated mancozeb fungicides (Figure 4).

3.4. Grain Yield of Chickpea.
Grain yield was significantly affected by chickpea varieties, fungicides, and their interaction at $p < 0.05$. Among two-way interaction effects, the maximum grain yield of 4.55 t/ha was recorded from Shasho variety treated with Apron Star, followed by Arerti variety treated with Apron Star which resulted in grain yield of 3.94 t/ha while the minimum grain yields of 0.21 t/ha were recorded from untreated local chickpea, followed by untreated Marye variety which resulted in grain yield of 0.75 t/ha (Table 3).

Yigitoglu [15] reported that the maximum grain yield was obtained from resistant variety treated with seed dressing fungicides. Apron Star seed-treatment gave minimum disease incidence of fusarium wilt and maximum grain yield [14]. Similarly, Subhani et al. [16] reported that applications of seed dressing fungicide significantly increased the plant growth and yield and also decreased chickpea fusarium wilt incidence.

3.5. Relative Losses in Grain Yield.
Among two way interaction effects, the maximum relative grain yield losses of 95.38% and 83.52% were obtained from untreated local chickpea and untreated Marye variety, respectively, while the minimum relative grain yield loss was obtained from Shasho variety treated with Apron Star fungicides which result of insignificance loss, followed by Arerti variety treated with Apron Star fungicides, which result of 13.41%

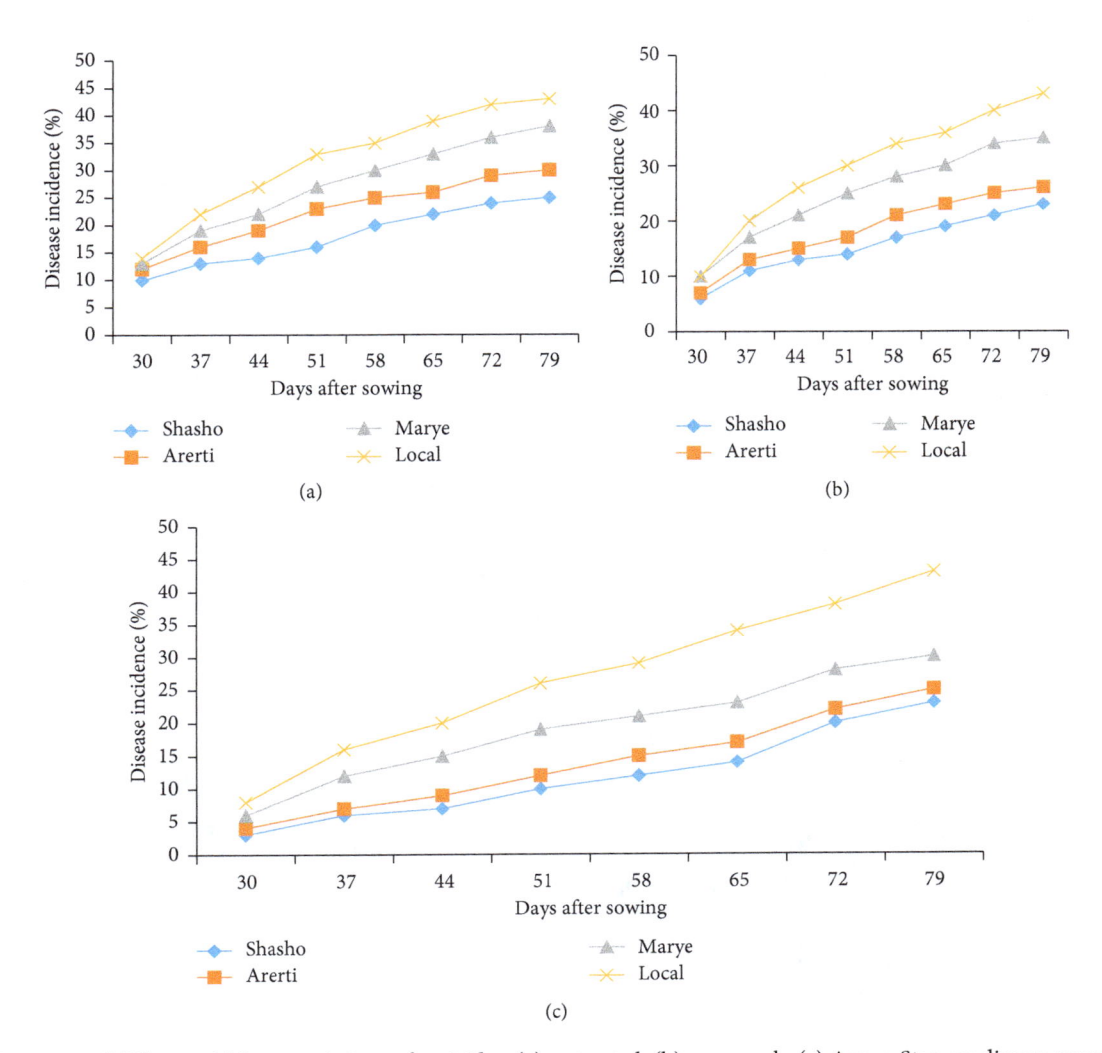

FIGURE 1: Response of different chickpea varieties to fungicides: (a) untreated; (b) mancozeb; (c) Apron Star, on disease progress curve of fusarium wilt incidence.

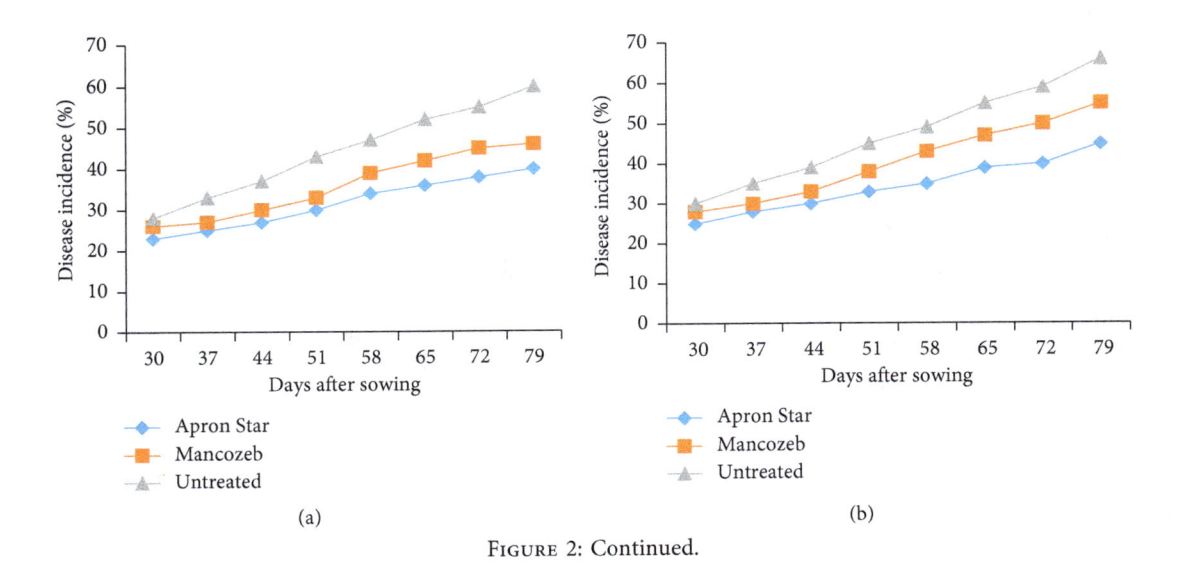

FIGURE 2: Continued.

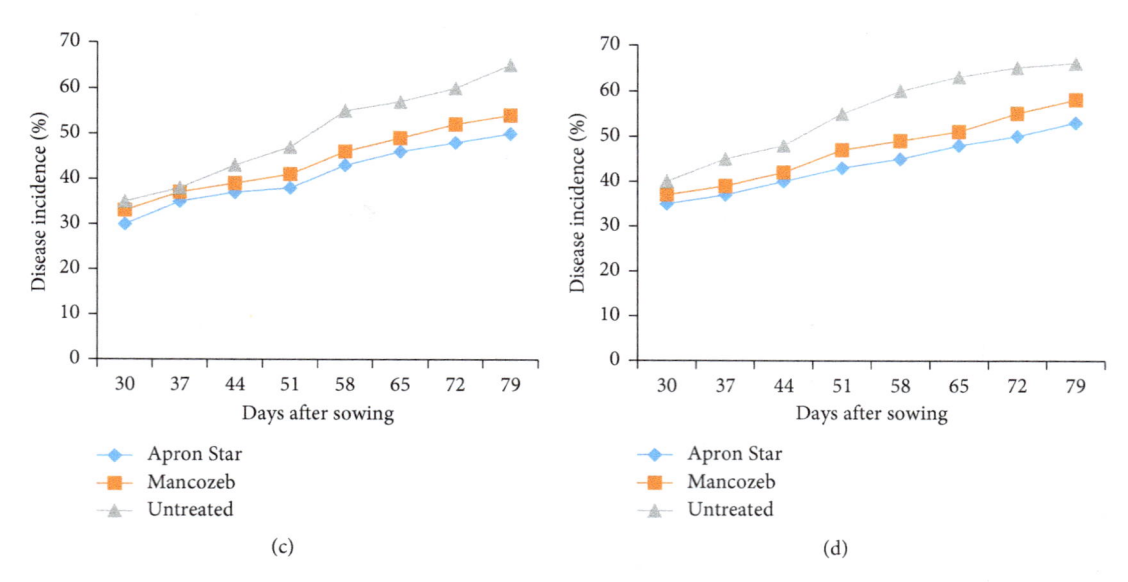

FIGURE 2: Response to different fungicides by the four varieties of chickpea: (a) Shasho, (b) Arerti, (c) Marye, and (d) local check variety, on disease progress of fusarium wilt incidence.

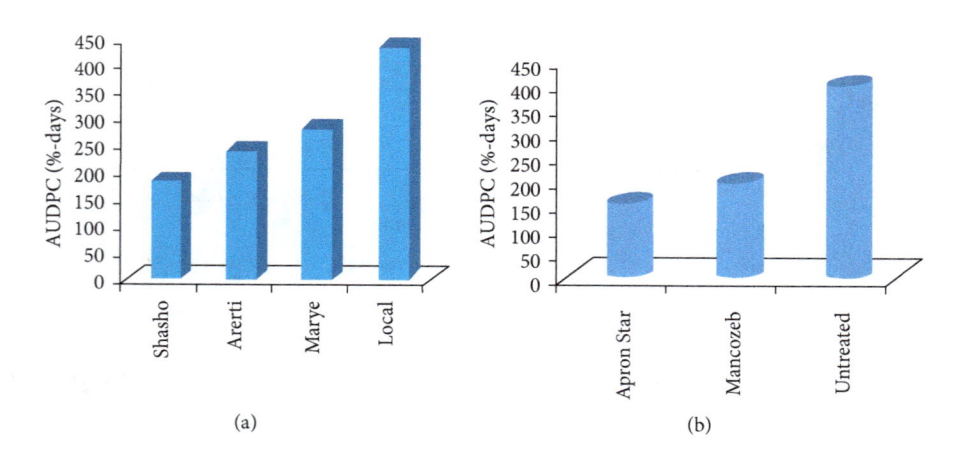

FIGURE 3: The main effects of (a) varieties and (b) fungicides on AUDPC (%-day values) of fusarium wilt of chickpea.

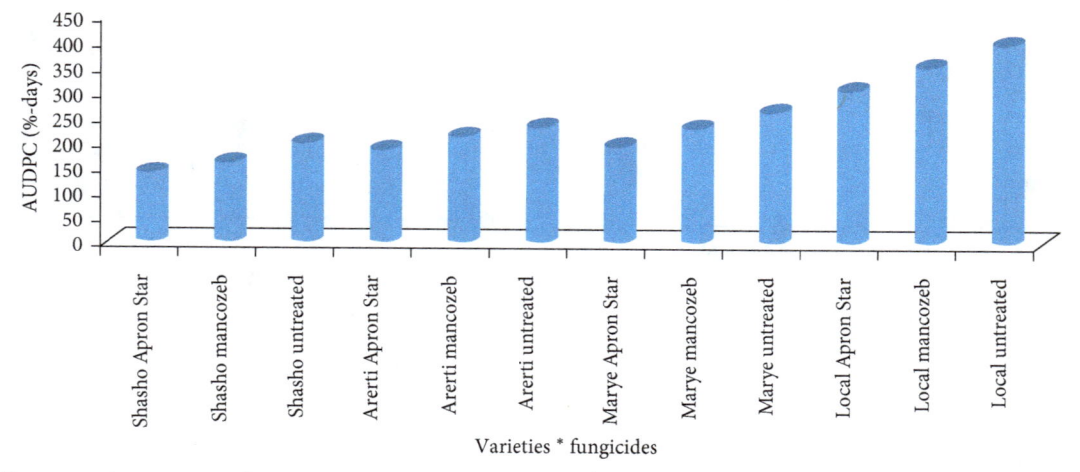

FIGURE 4: Interaction effects of varieties and fungicides on AUDPC (%-day values) of fusarium wilt disease.

TABLE 3: Two-way interaction effects of chickpea varieties and fungicides on grain yield of chickpea and their corresponding losses due to chickpea fusarium wilt.

Variety	Fungicides	Grain yield (t/ha)	Relative grain yield loss (%)
Shasho	Apron Star	4.55a	0.0
	Mancozeb	3.31bc	27.25
	Untreated	2.08de	54.29
Arerti	Apron Star	3.94ab	13.41
	Mancozeb	2.75cd	39.56
	Untreated	1.54ef	66.15
Marye	Apron Star	2.71cd	40.44
	Mancozeb	1.52ef	66.59
	Untreated	0.75fg	83.52
Local	Apron Star	2.16de	52.52
	Mancozeb	1.51ef	66.81
	Untreated	0.21g	95.38
LSD (5%)		0.812	
CV (%)		11.25	

LSD = least significant difference; CV = coefficient of variation; means followed by the same letter did not show significant difference at $p < 0.05$ according to least significant difference.

(Table 3). Grain yield losses were reduced through chickpea seeds treated with seed dressing fungicide as compared to untreated check of the respective variety [3].

4. Conclusion

The findings of the present study suggest that the adoption of resistant variety Shasho with fungicide Apron Star seed-treatment may result in reduced fusarium wilt disease progress with a corresponding increased grain yield of chickpea. Further, undoubtedly the fusarium wilt appears to be an important disease that calls for better attention in the study area in terms of economical management with fungicide seed-treatment and use of resistant varieties. The variety Shasho appears to have better resistance against the fusarium wilt disease; therefore, its genetic resistance needs to be investigated further by repeating the experiment for one more cropping season.

Acknowledgments

The authors express their profound appreciation to the University of Gondar, College of Agriculture and Rural Transformation, for granting the opportunity to the senior author to pursue this study by awarding scholarship and full financial support. Words cannot explain his appreciation for the Adet Agricultural Research Center for their material support and invaluable help in creating hospitable working environment for this research work.

Supplementary Materials

Figure 1: the data in the three tables indicate the effect of different chickpea varieties (Shasho, Arerti, Marye, and local) with the same untreated, treated with mancozeb, and treated with Apron Star on disease incidence progress on 7 days' interval. This mean that even though the chickpea varieties with the same untreated, treated with Mancozeb, and treated with Apron Star it shows difference among varieties in fusarium disease incidence progress. Hence, we have to relate these data with Figure 1. Figure 2: the data in four tables indicate the effect of the same chickpea variety with different fungicides Apron Star and mancozeb and the same untreated on disease incidence progress in 7 days' interval. This means that even though the same chickpea variety treated with different fungicides it shows difference performance among fungicides in fusarium disease incidence progress. Hence, we have to correlate these data with Figure 2. Figure 3: the data in two tables indicate the main effect of chickpea varieties and fungicides and its interaction effect on disease incidence progress. On these row data, I get good results which perform good resistance for fusarium wilt incidence per days. Hence you have to check on Figures 3 and 4. NB: even though all these agronomical parameter data were collected on the field and included in the paper, the selected data were included in the manuscript such as fusarium wilt disease incidence and grain yield of chickpea and their losses due to fusarium wilt disease. Hence, we have to check on Tables 2 and 3. (*Supplementary Materials*)

References

[1] N. S. Jodha and K. V. Subba, "Chickpea: world importance and distribution," in *The Chickpea*, M. C. Sexena and K. B. Singh, Eds., pp. 1–10, CABI, Walling ford, UK, 2012.

[2] J. B. Smithson, J. A. Thompson, and R. J. Summerfield, "Chickpea (*Cicerarietinum* L.)," in *Grain Legume Crops*, R. J. Summerfield and E. H. Roberts, Eds., pp. 312–390, Collins, London, UK, 2009.

[3] J. A. N. Cortes, B. Hav, and R. M. Jimenez-Diaz, "Yield loss in chickpea in relation to development of *Fusarium* wilt epidemics," *Phytopathology*, vol. 90, no. 11, pp. 1269–1278, 2012.

[4] Y. L. Nene, M. V. Reddy, M. P. Haware, A. M. Ghanekar, and K. S. Amin, "Field Diagnosis of Chickpea Diseases and their Control," *Information Bulletin No. 28*, International Crops Research Institute for the Semi-Arid Tropics, Patancheru, Hyderabad, 2011.

[5] D. Cook, E. Barlow, and L. Sequeira, "Genetic diversity of *Fusariumoxysporum f. sp. ciceris*: detection of restriction fragment length polymorphisms with DNA probes that specify virulence and the hypersensitive response," *Molecular Plant-Microbe Interactions*, vol. 2, pp. 113–121, 2012.

[6] R. Kakuhenzire, J. J. Hakiza, G. T. BergaLemaga, and F. Alacho, "Past, present and future srategies for integrated management of chickpea *fusarium* wilt disease in Uganda," *African Potato Association Conference Proceedings*, vol. 5, pp. 353–359, 2010.

[7] A. Seid and A. Melkamu, "Chickpea, Lentil, Grasspea, Fenugreek and Lupine Disease Research In Ethiopia," in *Food and forage legume*, K. Ali, G. Kenneni, S. Ahmed et al., Eds., p. 351, Alleppo, Syria, 2003.

[8] Y. Gelanew, "Selected chemical and physical characteristics of soils of Adet agricultural research center and its testing sites in Northwest Ethiopia," *Ethiopian J Natural Resource*, vol. 4, pp. 199–215, 2013.

[9] C. L. Campbell and V. L. Madden, *Introduction to Plant Disease Epidemiology*, John Wiley and Sons, Inc., New York, NY, USA, 1990.

[10] G. D. Robert and H. T. James, *A Biometrical Approach: Principles of Statistics*, McGraw-Hill, New York, USA, 2nd edition, 1991.

[11] SAS Institute, *SAS Technical Report, SAS system for windows V8, SAS/STAT Software Release 8.02 TS Level 02 M0*, SAS Institute Inc., Cary, NC, USA, 2001.

[12] A. Merkuz and A. Getachew, "Influence of chickpea fusarium wilt (*Fusariumoxysporum* f.sp. *ciceris*), on *Desi* and *Kabuli*-type of chickpea in integrated disease management option at wilt sick plot in North western Ethiopia," *International Journal of Current Research*, vol. 4, no. 4, pp. 46–45, 2012.

[13] B. B. Landa, J. A. Navas-Cortes, and R. M. Jimenez-Diaz, "Integrated management of *Fusarium* wilt of chickpea with host resistance, biological control and seed dressing fungicides," *Phytopathology*, vol. 94, pp. 946–960, 2014.

[14] D. R. Kamdi, M. K. Mondhe, G. Jadesha, D. N. Kshirsagar, and K. D. Thakur, "Efficacy of botanicals, bio-agents and fungicides against *Fusariumoxysporum f. sp. ciceris*, in chickpea wilt sick plot," *Annals of Biological Research*, vol. 11, pp. 5390–5392, 2012.

[15] D. Yigitoglu, *Research on the effect of different fungicides on the yield and yield components of some chickpea (CicerarietinumL.) cultivars that treated in different fungicides in kahramanmaras region*, PhD Thesis, Department of Field Crops Institute of Natural and Applied Science University of Cukurova, Adana, Turkey, 2006.

[16] M. N. Subhani, S. T. Sahi, S. Hussain, A. Ali, J. Iqbal, and K. Hameed, "Evaluation of various fungicides for the control of gram wilt caused by *Fusariumoxysporum* f. sp. *ciceris*," *African Journal of Agricultural Research*, vol. 6, no. 19, pp. 4555–4559, 2011.

Effect of Planting Material Type on Experimental Trial Quality and Performance Ranking of Sugarcane Genotypes

Michel Choairy de Moraes,[1] Ana Carolina Ribeiro Guimarães,[2] Dilermando Perecin,[2] and Manuel Benito Sainz[1,3]

[1]Syngenta Proteção de Cultivos Ltda, 18001 Avenida das Nações Unidas, São Paulo, SP, Brazil
[2]UNESP, FCAV, BR 14884-900 Jaboticabal, SP, Brazil
[3]Syngenta Crop Protection LLC, 9 Davis Drive, Research Triangle Park, NC, USA

Correspondence should be addressed to Michel Choairy de Moraes; michel.moraes@syngenta.com

Academic Editor: Isabel Marques

In recent years, the use of presprouted setts (MPB, which stands for "mudas pre-brotadas" in Portuguese) to establish commercial sugarcane nurseries has grown in Brazil. MPB and single-bud setts (SBS) have the advantage of requiring less planting material and enabling a higher multiplication rate of the source material as compared with the conventional multibud sett (MBS) planting system. Sugarcane breeding programs could also potentially benefit from the precise spacing afforded by MPB or SBS planting materials, by reducing trial variability. However, the effect of planting material type on the performance ranking and consequent selection of sugarcane clones in a breeding program has not been previously investigated. We present results on possible interactions between genotype and the type of planting material (MPB, MBS, or SBS) on key performance parameters, like sugar content, cane yield, and sugar yield, in the context of the intermediate phase of a sugarcane breeding program. Our results indicate that trial quality does not necessarily improve with the use of MPB or SBS planting materials and that type of planting material has a significant effect on the ranking of sugarcane genotypes, and this needs to be taken into consideration when considering the use of new planting technologies in breeding trials of vegetatively propagated crops such as sugarcane.

1. Introduction

Sugarcane is a vegetatively propagated crop. Manual planting of multibud setts (MBS) is the traditional planting material used in the planting of commercial nurseries and production fields. To minimize the risk of gaps in the resultant stand, manual planting rates are high (15–21 buds/meter), corresponding to 11–14 tonnes (T) of planting material/hectare (ha). With mechanized planting, the amount of planting material used is even larger, reaching levels greater than 20 T/ha. Sugarcane production costs have increased due to increased labor and agricultural input costs, with the cost of planting material accounting for almost 25% of operational production costs [1]. The large quantity of planting material required in traditional planting systems also leads to problems with logistics, storage, and loss of bud viability.

New planting systems have been developed to overcome some of the disadvantages of traditional methods. The presprouted seedling (MPB) planting system allows for a reduction in the quantity of planting material and better control of seedling vigor [2–6]. Another planting system (Plene™) developed by Syngenta uses 5 cm, single-bud setts (SBS) treated with a pesticidal slurry [7]. Bud chips are also a promising alternative for reduction of sugarcane production costs, although improvement of survival rates and plant vigor under field conditions is needed [8].

Genetic improvement of sugarcane is based on the selection and cloning of superior genotypes of segregating populations obtained through sexual crosses between different individuals [9]. Different methodologies have been used in the selection of individuals in the early stages of sugarcane breeding: mass selection [9], Australian sequential selection [10], modified sequential selection [11], and individual simulated best linear unbiased prediction (BLUP) [12, 13], among them. After the initial seedling phase, selection is based on planting clonal setts, as MBS, in plots that are assessed for key agronomic attributes. As selection progresses, the size of the plot and the number of replicates per clone increase, which allows better assessment of the materials under selection.

Sugarcane breeding trials could benefit from using planting systems that increase efficiency and data quality. Higher trial data quality could result from more precise spacing in trial plots. Planting systems with the potential to do more replications in early breeding stages due to more efficient utilization of scarce planting material would provide an additional benefit. However, planting system modifications have the potential to affect the ranking and consequent selection of genotypes, since experimental results of important performance attributes such as yield and sugar content could differ depending on the planting system used. Studies on intrarow spacing and number of buds per sett in commercial varieties [14] and optimal planting rates using whole stalks for different varieties have been reported [15]. However, to date and to our knowledge, there have been no reports in the literature about the effect of presprouted seedling (MPB) or single-bud setts planting materials on clonal selection in sugarcane breeding programs.

In the present study, we evaluate potential interactions between genotypes and the type of planting system and whether the type of planting material has an effect on trial data quality.

2. Materials and Methods

2.1. Trial Design. Three types of planting material were tested: 3-4 bud setts (MBS—conventional method), presprouted seedlings (MPB), and 5 cm, single-bud setts (SBS). All planting materials were generated from selected healthy stalks harvested approximately 9 months after planting.

All three types of planting material were planted at a single location in three different adjacent trials. All trials were in a randomized complete block design with 3 replicates. Plots consisted of two 10 m rows spaced 1.5 m apart. Adjacent plots were spaced 3 m apart along their length.

The same genotypes (cultivars and clones; see Table 1) were planted in each of the 3 different trials, adjacent to each other. Clones were part of the 3rd stage of the selection process of the Syngenta sugarcane breeding program. This stage is the first one in which replicates are used and plot weight is directly measured following mechanical harvesting.

The process for making and planting the different types of planting materials is described below and in the accompanying figures (Figures 1–3). In all cases, supplemental irrigation was applied until the crop stand was well established.

2.1.1. Presprouted Setts (MPB). Eight-month-old sugarcane stalks were harvested and 5 cm, single-bud setts were cut and planted in soil mix. The resultant sprouted seedlings were

Table 1: Varieties and clones used in the study.

Varieties	Clones		
RB86-7515	S09-0001	S09-0040	S09-0114
RB96-6928	S09-0007	S09-0046	S09-0122
SP81-3250	S09-0011	S09-0048	S09-0140
	S09-0022	S09-0052	S09-0144
	S09-0023	S09-0055	S09-0146
	S09-0031	S09-0069	S09-0148
	S09-0036	S09-0080	S09-0153
	S09-0037	S09-0081	S09-0154
	S09-0038	S09-0098	

manually planted in the field trial after 50 days at an intrarow spacing of 0.5 m.

2.1.2. Single-Bud Setts (SBS). Ten-month-old sugarcane stalks were harvested, and 5-cm, single-bud setts were cut and treated with a slurry consisting of industrial proprietary treatment. Subsequently, these were planted manually in the field at a rate of 8 single-bud setts per meter.

2.1.3. Multibud Setts (MBS): Conventional Method. Ten-month-old sugarcane stalks were harvested manually and placed in row furrows. These were cut with a machete in the furrow into 30–40 cm pieces, as per conventional manual cane planting practice.

The 3 field trials were planted in a single week in April 2014. Fertilization and cultural practices followed conventional commercial practice and were the same for all trials. Evaluations of sugarcane agronomic parameters were made over two harvest cycles (plant cane and 1st ratoon). In August 2015, a sample of 10 stalks per plot was subjected to laboratory POL analysis (a measure of sugar content). Subsequently, in the same month, the trial was mechanically harvested and whole, individual plots were weighed to estimate TCH (tonnes cane per hectare). From the POL and TCH parameters, the TPH (tonnes POL per hectare) was calculated for the plant cane harvest. It was not possible to measure POL in the 1st ratoon harvest, but in May 2016, the Brix of 5 stalks per plot was taken and averaged. Mechanized harvesting and weighing of trial plots were conducted in June 2016. Thus, Brix, TCH, and TBH parameters (tonnes Brix per hectare) were estimated for the 1st ratoon harvest.

2.2. Statistical Analysis. Analysis of variance was done by planting material (MPB, SBS, and MBS) and harvest cycle (plant cane, 1st ratoon), considering the effects of genotype (26 clones and three varieties) and blocks (3 per planting material). For each of the three traits (POL or Brix; TCH; and TPH or TBH), the ratio between the largest and smallest mean residual squares was less than three. As a result, we performed joint analyses by traditional ANOVA and mixed model restricted maximum likelihood (REML)/best linear unbiased prediction (BLUP), in which planting material, harvest cycle, and blocks nested within planting material were fixed effects, and genotype, as well as genotypic

FIGURE 1: Process for SBS (single-bud sett) production and planting.

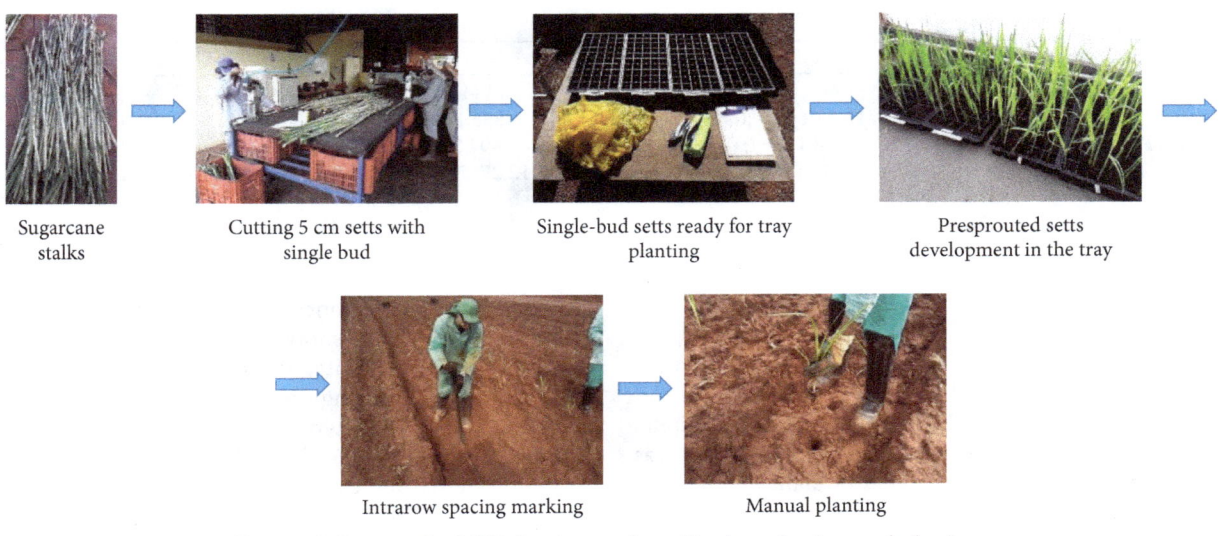

FIGURE 2: Process for MPB (presprouted seedling) production and planting.

FIGURE 3: Process for MBS (multibud sett) production and planting.

interactions with planting material and harvest cycle, were random effects. A split-plot design was not used due to possible plot border effects resulting from different growth rates of different adjacent plant material types. As stated above, mean residual squares within each trial were similar; consequently, we performed a joint analysis by ANOVA (despite the absence of randomization), much as experiments across locations are analyzed.

Within each cycle (plant and ratoon cane), statistical modeling was done as for a split-plot design in time.

TABLE 2: Average performance parameters and coefficients of variation (CV) in trials with 3 types of planting materials, 29 genotypes, and 3 blocks (replicates) per planting material.

Performance parameter	MBS average	SBS average	MPB average
POL-Brix			
Plant cane POL	12.42	13.55	13.05
1st ratoon Brix	13.59	14.71	14.71
%CV, plant cane POL	7.47	5.89	7.34
%CV, 1st ratoon Brix	8.25	7.23	7.46
TCH			
Plant cane TCH	136.11	134.78	139.10
1st ratoon TCH	104.79	97.35	106.95
%CV, plant cane TCH	12.54	15.15	13.95
%CV, 1st ratoon TCH	12.19	16.69	14.29
TPH_TBH			
Plant cane TPH	16.85	18.10	17.94
1st ratoon TBH	14.29	14.21	14.84
%CV, plant cane TPH	14.44	17.01	16.85
%CV, 1st ratoon TBH	15.81	15.53	16.81

TABLE 3: Mean squares (MS) and p values pooled joint analysis of variance. Conventional fixed model, with planting material type as test and harvest cycle (plant cane, 1st ratoon) as split-plot, was used.

Source of variation	POL_BRIX	TCH	TPH_TBH
Planting material (PM)	59.60 ($p < 0.0001$)	1484.92 ($p = 0.02$)	24.17 ($p = 0.06$)
Genotype	19.83 ($p < 0.0001$)	4855.15 ($p < 0.0001$)	104.95 ($p < 0.0001$)
Genotype × PM	1.48 ($p = 0.01$)	494.80 ($p = 0.07$)	12.48 ($p = 0.03$)
Harvest cycle	141.37 ($p < 0.0001$)	12,2717.39 ($p < 0.0001$)	1085.90 ($p < 0.0001$)
Harvest cycle × PM	1.88 ($p = 0.22$)	364.55 ($p = 0.17$)	13.61 ($p = 0.08$)
Harvest cycle × genotype	3.87 ($p < 0.0001$)	1002.86 ($p < 0.0001$)	16.51 ($p < 0.0001$)

3. Results

Mean values of TPH_TBH were similar for the three planting material types tested (Table 2). Average POL and Brix were higher for SBS and MPB than for conventional MBS planting material, whereas average TCH for SBS was lower than that of the other two types of planting material. Average TCH for MPB and the conventional MBS planting material was quite similar, as previously observed by [16], who noted that MPB required less planting material. Regarding trial coefficients of variation (CVs), these were similar across trials, planting material type, and harvest cycles, being in general lower for POL_Brix. TCH CVs for MPB and SBS were slightly higher than for the conventional MBS planting type. For TPH_TBH, the plant cane average was higher for MPB, but no differences between the 3 planting material types were observed in the 1st ratoon. The CV for TBH was slightly lower in the conventional MBS planting type than for the other two.

Conventional analysis of variance (Table 3) points to highly significant effects of genotype ($p < 0.0001$) and of the interaction of genotype with planting material type ($p \leq 0.1$) across the different parameters evaluated, indicating that rank ordering of genotypes could be affected depending on the planting material type used in the trial. However, harvest cycle and the interaction of harvest cycle with genotype also had a highly significant effect ($p < 0.0001$) on the parameters tested.

Random effect variances obtained using a mixed model (Table 4) show that genotype accounts for approximately 36–37% of the variability, which allows for selection of the best clones in a breeding program. Interactions of genotype with planting material type ranged from 2.7% to 7.1%: lower but significant ($p < 0.05$) for POL_Brix and higher for TCH and TPH_TBH, indicating that planting material type has the potential to affect the rank order of clones in the selection process. Interaction of genotype with harvest cycle was significant ($p < 0.05$), as expected, since varieties in commercial sugarcane fields will also differ in their performance depending on harvest cycle that includes the environmental effects of growing season. There is also a weak interaction of harvest cycle with planting material type ($p < 0.25$; data not shown).

Predicted genotypic differences (PGDs) obtained in the mixed model (Tables 5 to 7) are presented to show differences in genotype ranking for each performance parameter, by planting material type. For each genotype, the expected genotypic value is the mean + PGD.

For TPH_TBH (Table 5), the MPB and SBS planting materials produced similar means ($p < 0.05$) that were higher than those obtained with the MBS conventional planting material. Results with the 10 best genotypes show differences in rank ordering. For example, using conventional MBS planting material, the best clone was S09-0146, yielding 4.74 tonnes above the 15.57 TPH_TBH general average, and slightly surpassing the two RB variety checks.

TABLE 4: Components and percentage of variance as obtained in a pooled joint analysis with a mixed REML/BLUP model, considering genotype, genotype × planting material, genotype × harvest cycle, and residual as random effects.

Variable	POL_BRIX	TCH	TPH_TBH
Genotypes	0.8658 (37.08%)	232.02 (36.41%)	5.1067 (36.16%)
Genotypes × PM	0.0632 (2.71%)	37.75 (5.92%)	1.0028 (7.10%)
Genotypes × harvest cycle	0.3081 (13.20%)	93.47 (14.67%)	1.3193 (9.34%)
Residual	1.0978 (47.02%)	274.07 (43.00%)	6.6918 (47.39%)

TABLE 5: Predicted genotypic differences (PGD) for TPH_TBH across 29 genotypes (mean plant cane TPH and 1st ratoon TBH) and three planting material types (MBS, SBS, or MPB).

Rank	Genotype	MBS	p value	Genotype	SBS	p value	Genotype	MPB	p value
1	S09-0146	4.7444	<0.001	RB966928	4.4965	<0.001	RB966928	4.592	0.0002
2	RB867515	4.4646	<0.001	S09-0144	4.1737	0.0002	RB867515	4.5099	0.0003
3	RB966928	3.702	0.001	S09-0146	4.1348	0.0003	S09-0122	3.0073	0.0095
4	S09-0031	3.0899	0.006	S09-0114	3.4487	0.0021	S09-0114	2.614	0.033
5	S09-0114	2.4153	0.031	S09-0038	3.2957	0.0384	S09-0140	2.5809	0.0254
6	S09-0038	2.0233	0.07	S09-0052	2.7208	0.0325	S09-0144	2.304	0.0597
7	S09-0052	1.887	0.09	S09-0153	2.1413	0.0919	S09-0031	1.773	0.1228
8	S09-0140	1.6314	0.143	S09-0140	2.0738	0.0607	S09-0069	1.5078	0.1889
9	S09-0144	1.2664	0.254	S09-0148	1.6054	0.1454	S09-0023	1.3682	0.2614
10	S09-0148	1.2593	0.257	RB867515	0.8213	0.4548	S09-0052	1.0307	0.3682
11	S09-0037	0.7224	0.515	S09-0037	0.7444	0.555	S09-0146	0.9092	0.4548
12	S09-0122	0.3986	0.719	S09-0031	0.3428	0.7548	S09-0007	0.5837	0.6099
13	SP813250	0.3901	0.725	S09-0069	0.1064	0.9227	S09-0153	0.0999	0.9345
14	S09-0046	0.2438	0.826	S09-0055	−0.045	0.9713	S09-0046	−0.209	0.8549
15	S09-0023	0.1558	0.888	S09-0122	−0.09	0.9431	S09-0148	−0.643	0.6231
16	S09-0055	−0.394	0.722	SP813250	−0.183	0.8674	S09-0055	−0.699	0.5416
17	S09-0069	−0.435	0.695	S09-0154	−1.439	0.1916	S09-0038	−0.912	0.4259
18	S09-0153	−0.659	0.552	S09-0080	−1.678	0.1282	S09-0040	−0.943	0.4101
19	S09-0007	−0.694	0.532	S09-0022	−1.793	0.1565	S09-0001	−1.12	0.3576
20	S09-0036	−0.708	0.524	S09-0081	−1.854	0.0932	S09-0048	−1.192	0.2983
21	S09-0022	−1.703	0.126	S09-0040	−2.074	0.0607	S09-0037	−1.501	0.218
22	S09-0001	−2.118	0.058	S09-0023	−2.299	0.0379	S09-0081	−1.512	0.2148
23	S09-0154	−2.368	0.034	S09-0046	−2.327	0.1421	S09-0036	−1.605	0.1621
24	S09-0048	−2.389	0.033	S09-0007	−2.401	0.0303	S09-0022	−1.733	0.1313
25	S09-0081	−2.808	0.012	S09-0036	−2.527	0.0228	SP813250	−2.164	0.0767
26	S09-0098	−2.91	0.01	S09-0011	−2.685	0.0157	S09-0154	−2.295	0.0464
27	S09-0011	−3.422	0.002	S09-0048	−2.789	0.0285	S09-0098	−2.682	0.0203
28	S09-0080	−3.856	0.0007	S09-0001	−2.907	0.0226	S09-0080	−3.669	0.0017
29	S09-0040	−3.93	0.0005	S09-0098	−3.015	0.0068	S09-0011	−4.001	0.0013
TM[1]	—	15.57 **B**	—	—	16.16 **A**	—	—	16.26 **A**	—

[1]TM with different letters indicate significant (p value < 0.05) by T-test (LSD). TM = trial mean; p value testing PGD as null.

S09-0146 is ranked 3rd for TPH_TBH when planted as SBS, but is not among the top 10 genotypes when MPB is used as the planting material. Clone S09-0031 is ranked 4th just below the two RB checks when planted as MBS and is in the top 10 when MPB planting material is used, but not when planted as SBS. Similarly, S09-0122 is the top ranked clone when MPB planting material is used, suggesting good adaptation to this planting methodology, but is not in the top 10 when using the other planting methodologies. Among the checks, RB966928 was ranked 3rd when conventional MBS planting material is used, but 1st when MPB or SBS planting materials are used, whereas RB867515 did not perform well when SBS planting material was used.

For TCH (Table 6), the highest average yields were obtained from conventional MBS and MPB planting materials, with SBS planting material having a significantly lower ($p < 0.05$) mean. In terms of rank, MPB and SBS methods improved the performance of check variety RB966928 while SBS leads to deterioration in performance of the RB867515 check. Among the clones, S09-0146 is again at the 1st place when conventional MBS and SBS planting materials are used, but at the 4th place using MPB, whereas clone S09-114 ranks 3rd with all 3 planting material types. The clone S09-0038, at the 4th and 5th place when planted as MBS or SBS, respectively, drops to the 9th place when MPB is used, whereas the clone S0-007, at the 4th place when the MPB planting material is used, drops to 11th with MBS and 22nd with SBS.

For POL and Brix (Table 7), means were significantly different ($p < 0.05$) depending on the type of planting material, being higher (14.13) when SBS planting material was used; intermediate for MPB (13.45); and lowest for

TABLE 6: Predicted genotypic differences (PGD) for TCH across 29 genotypes (mean plant cane and 1st ratoon TCH) and three planting material types (MBS, SBS, or MPB).

Rank	Genotype	MBS	p value	Genotype	SBS	p value	Genotype	MPB	p value
1	S09-0146	39.089	**<0.001**	S09-0146	26.489	**0.0008**	RB867515	28.893	**0.0003**
2	RB867515	29.224	**<0.001**	RB966928	24.382	**0.0019**	RB966928	22.466	**0.0046**
3	S09-0114	20.817	**0.006**	S09-0114	23.463	**0.0028**	S09-0114	16.734	**0.0337**
4	S09-0038	20.557	**0.006**	S09-0153	22.079	**0.0138**	S09-0146	15.104	0.0548
5	RB966928	19.358	**0.01**	S09-0038	21.821	**0.049**	S09-0007	14.768	0.0461
6	S09-0031	14.071	0.06	S09-0144	21.342	**0.0063**	S09-0069	13.874	0.0608
7	S09-0144	12.829	0.086	S09-0052	16.24	0.0677	S09-0154	10.657	0.1486
8	S09-0052	8.2784	0.266	S09-0154	12.845	0.097	S09-0140	9.5021	0.1974
9	S09-0154	5.8661	0.43	RB867515	8.122	0.2923	S09-0038	9.2683	0.2086
10	S09-0153	5.3894	0.469	S09-0140	6.2557	0.4168	S09-0023	8.135	0.2985
11	S09-0007	4.4072	0.553	S09-0148	5.5204	0.4735	S09-0040	7.6597	0.2982
12	S09-0140	3.4683	0.641	S09-0069	3.8803	0.6142	S09-0153	6.2699	0.4226
13	S09-0148	3.4538	0.642	S09-0037	1.7567	0.8423	S09-0144	6.254	0.4238
14	SP813250	2.9916	0.687	S09-0040	0.8263	0.9145	S09-0122	4.4561	0.5445
15	S09-0069	0.5793	0.938	S09-0055	−0.536	0.9516	S09-0031	4.0024	0.5862
16	S09-0037	0.0882	0.991	SP813250	−2.002	0.7948	S09-0052	−0.356	0.9613
17	S09-0023	−2.079	0.7800	S09-0031	−2.242	0.7708	S09-0148	−1.780	0.8324
18	S09-0040	−2.310	0.756	S09-0122	−4.794	0.5873	S09-0001	−5.042	0.5189
19	S09-0036	−4.000	0.591	S09-0022	−6.746	0.4453	S09-0036	−7.231	0.326
20	S09-0122	−7.842	0.292	S09-0080	−8.124	0.2922	SP813250	−8.281	0.29
21	S09-0055	−9.301	0.212	S09-0001	−11.39	0.1986	S09-0055	−10.43	0.1572
22	S09-0022	−11.24	0.132	S09-0007	−12.61	0.1033	S09-0046	−11.11	0.1323
23	S09-0046	−11.71	0.117	S09-0081	−14.50	0.0614	S09-0037	−11.96	0.1274
24	S09-0001	−12.44	0.096	S09-0023	−20.16	0.0098	S09-0048	−14.78	0.046
25	S09-0098	−20.84	0.006	S09-0011	−21.57	0.0058	S09-0022	−15.45	0.0371
26	S09-0081	−23.99	0.002	S09-0036	−21.81	0.0053	S09-0098	−17.09	0.0214
27	S09-0048	−24.47	0.001	S09-0098	−22.26	0.0045	S09-0081	−18.45	0.0194
28	S09-0011	−28.56	2E−04	S09-0048	−22.57	0.0117	S09-0080	−25.76	0.0006
29	S09-0080	−31.69	<0.001	S09-0046	−23.71	0.0327	S09-0011	−30.31	0.0002
TM[1]	—	120.45 A	—	—	116.07 B	—	—	121.62 A	—

[1]TM with different letters indicate significant (p value<0.05) by T-test (LSD). TM = trial mean; p value testing PGD as null.

conventional MBS (13.00). Four clones showed higher POL_Brix than the RB966928 check: clones S09-0046 and S9-0122 ranked highest in POL_Brix but were not among the top 10 for TCH. On the other hand, clone S09-0146 which ranked high for TCH was not among the best clones for POL_Brix.

Results of the correlations between predicted genotypic differences (PDG; Tables 5 to 7) were used to evaluate correlations between the 3 different planting material types for the three performance parameters (Table 8). Correlations of 0.82–0.85 were observed when comparing conventional MBS and the MPB and SBS planting materials for POL_Brix, and when comparing conventional MBS with MPB for TCH. Slightly lower correlations (0.75–0.79) were observed when comparing conventional MBS with MPB and SBS for TPH_TBH, and for conventional MBS and SBS for TCH. Hence, there are relatively good but not perfect correlations between the conventional MBS and the SBS and MPB planting materials, possibly due to interactions between some genotypes and planting material type. Correlations between MPB and SBS were generally lower, ranging from 0.61 for TPH_TBH, 0.65 for TCH, and 0.76 for POL_Brix, possibly due to higher risks to stand establishment when using SBS planting materials.

4. Discussion

The observed CVs for the parameters analyzed (Table 2) were in line with those previously observed in the scientific literature. Couto et al. [17] evaluated the range of CVs observed in sugarcane experiments and concluded that the TCH and TPH parameters presented the highest CV ranges, whereas % sucrose presented the lowest. These authors also indicated that upper CV limits for good to high precision as being 10%, 15%, and 19% for % sucrose, TCH, and TPH, respectively. Using these numbers as a guide, in the present study, only the TCH in the SBS trial is higher than these limits, which indicates that trial quality was high enough to draw conclusions from this study. Although we observe lower POL and Brix CVs with the SBS and MPB planting material (Table 2), in general our data do not provide evidence that MPB or SBS planting materials decrease variability, and thus improve trial quality, in sugarcane breeding selection trials.

Conventional analysis of variance (Table 3) points to highly significant effects of genotype and harvest cycle and also of the interaction of harvest cycle with planting material. Milligan et al. [18] also found that genotype by harvest cycle interaction was important for sugar yield and its component

TABLE 7: Predicted genotypic differences (PGD) for POL_Brix across 29 genotypes (mean plant cane POL and 1st ratoon Brix) and three planting material types (MBS, SBS, or MPB). TM = trial mean; p value testing PGD as null.

Rank	Genotype	MBS	p value	Genotype	SBS	p value	Genotype	MPB	p value
1	S09-0046	1.4759	**0.0021**	S09-0046	1.7214	**<0.001**	S09-0122	1.8558	**0.0002**
2	S09-0122	1.2755	**0.0076**	S09-0140	0.9188	**0.0279**	S09-0144	1.1144	**0.0211**
3	S09-0140	0.8705	0.0667	S09-0144	0.7226	0.0828	S09-0046	1.1046	**0.0222**
4	S09-0031	0.8431	0.0757	S09-0052	0.7143	0.0864	S09-0140	0.9496	**0.0488**
5	RB966928	0.8321	0.0795	S09-0122	0.6489	0.119	S09-0081	0.9468	**0.0495**
6	S09-0055	0.7827	0.0989	S09-0036	0.6434	0.1222	RB966928	0.9301	0.0536
7	S09-0148	0.7676	0.1055	S09-0023	0.6406	0.1238	S09-0031	0.9022	0.0611
8	S09-0048	0.7429	0.1171	S09-0148	0.5182	0.2125	S09-0052	0.8589	0.0744
9	S09-0052	0.7291	0.124	RB966928	0.489	0.2392	S09-0048	0.6453	0.179
10	S09-0037	0.6221	0.1889	S09-0031	0.4834	0.2446	S09-0055	0.6145	0.2005
11	RB867515	0.3585	0.448	S09-0011	0.3832	0.3558	RB867515	0.44	0.3587
12	S09-0023	0.3516	0.4567	S09-0048	0.3053	0.4617	S09-0011	0.3297	0.4913
13	S09-0081	0.202	0.6687	S09-0037	0.2511	0.5449	S09-0037	0.2697	0.5734
14	S09-0080	0.154	0.7443	S09-0055	0.2191	0.5972	S09-0148	0.1943	0.6849
15	S09-0011	0.1306	0.782	S09-0114	0.2038	0.623	S09-0022	0.1859	0.6978
16	S09-0114	−0.1494	0.7517	SP813250	0.1482	0.7208	S09-0023	0.1105	0.8175
17	S09-0146	−0.1549	0.7429	S09-0081	0.1384	0.7384	S09-0114	0.01136	0.9811
18	SP813250	−0.1549	0.7429	S09-0146	0.137	0.741	S09-0069	−0.2428	0.6122
19	S09-0036	−0.1919	0.6843	S09-0098	−0.1356	0.7436	S09-0001	−0.3237	0.4992
20	S09-0022	−0.2455	0.6032	S09-0038	−0.1982	0.6326	S09-0080	−0.3433	0.4737
21	S09-0098	−0.2798	0.5536	RB867515	−0.2858	0.4908	S09-0098	−0.5695	0.2353
22	S09-0144	−0.3567	0.4503	S09-0069	−0.4527	0.2757	S09-0153	−0.6351	0.1859
23	S09-0069	−0.439	0.353	S09-0022	−0.5598	0.1781	S09-0146	−0.691	0.1504
24	S09-0038	−0.4733	0.3168	S09-0080	−0.5765	0.1656	S09-0036	−0.751	0.1183
25	S09-0001	−0.542	0.2519	S09-0153	−0.7323	0.0789	SP813250	−0.8515	0.0769
26	S09-0153	−0.9648	0.0424	S09-0007	−0.7852	0.0597	S09-0007	−1.1126	0.0213
27	S09-0007	−1.2119	0.0111	S09-0001	−1.272	0.0025	S09-0040	−1.4477	0.0029
28	S09-0154	−2.1933	<0.001	S09-0040	−1.7185	<0.001	S09-0038	−1.7828	0.0003
29	S09-0040	−2.7808	<0.001	S09-0154	−2.5698	<0.001	S09-0154	−2.7127	—
TM[1]	—	13.00 C	—	—	14.13 A	—	—	13.45 B	—

[1]TM with different letters indicate significant (p value<0.05) by T-test (LSD).

TABLE 8: Pearson's correlation coefficients between average performance parameters for 29 genotypes across three different planting material types, using predicted genotypic differences (PDG) from Tables 5 to 7, for each performance parameter.

Performance parameter		MBS	MPB	SBS
POL_Brix	MBS	—	0.82	0.85
	MPB	—	—	0.75
TCH	MBS	—	0.85	0.79
	MPB	—	—	0.65
TPH_TBH	MBS	—	0.77	0.78
	MPB	—	—	0.61

$N = 29$ for all correlations (Prob > (r)) < 0.05 under rho = 0.

traits. In this study, the pooled variation due to genotype (Table 4) was found to be sufficient to select the top genotypes to advance to the next stage. We found a weaker but still significant interaction ($p \leq 0.1$) of genotype by planting material, suggesting that the rank ordering of genotypes could be affected depending on the planting material type used in the trial. In contrast, planting material alone had a highly significant effect only on variation of POL_Brix. These results were verified by the differences in the rank ordering of clones we observed the TPH_TBH, TCH, and POL_Brix performance parameters (Tables 5–7).

Changes in rank order of genotypes can impact the effectiveness of clonal selection in sugarcane breeding. For example, the TPH_TBH parameter can be considered the most important one for selection in the intermediate phases of a program. Selection indices in different breeding stages vary by breeding program, as shown by [19]. In our breeding program, a 10 to 20% selection index was used in the third stage. A selection index of 20% from the trials described in this study, and considering data from 2 harvest cycles, would advance 6 clones to the next phase. If we consider conventional MBS planting material as the most relevant to the current commercial sugarcane planting practice, clones S09-0146, S09-0031, S09-0114, S09-0038, S09-0052, and S09-0140 would then be selected. However, of these 6 clones, only 4 (S09-0146, S09-0114, S09-0038, and S09-0052) would be advanced if using the SBS planting material, and only 3 (S09-0114, S09-0140, and S09-0031) would be advanced by using MPB planting material.

Orgeron et al. [15], studying whole stalks planting rating effect in 8 different genotypes, found no planting rate by genotype interaction for cane and sugar yield. Similarly, Netsanet and Tegene [14], comparing three different commercial varieties and their behavior in terms of intrarow spacing and buds per setts, found no significant interaction effect or spacing used on sugar and cane yield. On the contrary, what we found in the present study was that planting material type by genotype interaction is significant. The different conclusions between the studies possibly are due to the higher number of genotypes, and the genotypes per se, used in the present study and due to the different planting material used (MPB and SBS), compared with the other cited studies.

On an average, good correlations of performance parameter were observed between the different types of planting material. These correlations mask the significant effect on the ranking of individual clones in these trials. MPB and SBS planting methodologies have generated enormous interest in the Brazilian sugarcane industry and have undeniable advantages in terms of reduction of planting costs and material handling logistics. However, the use of these new planting methodologies in a sugarcane breeding program will result in likely selection of genotypes well adapted to the particular type of planting material used, but may not have the best agronomic performance if used in commercial plantings using other planting systems, such as the current conventional MBS planting system.

Our study indicates that trial quality does not necessarily improve with the use of MPB or SBS planting materials compared with the conventional MBS. Additionally, the type of planting material has a significant effect on the ranking of sugarcane genotypes. Because of that, when considering the use of new planting technologies in breeding trials of sugarcane, this needs to be taken into consideration for the selection of genotypes for cane yield and sugar parameters.

Acknowledgments

The authors acknowledge the support of Syngenta Proteção de Cultivos Ltda in funding this work and Marcia de Macedo Almeida for her critical reviewing of the paper.

References

[1] T. D. Chitkala, M. Bharathalakshmi, V. Gouri, M. Kumari, N. Naidu, and K. P. Rao, "Studies on the effect of sett size, seed rate and sett treatment on yield and quality of sugarcane," *Indian Journal of Sugarcane Technology*, vol. 26, pp. 4–6, 2011.

[2] M. G. A. Landell, M. P. Campana, P. Figueiredo et al., "Sistema de multiplicação de cana-de-açúcar com mudas pré-brotadas (MPB) oriundas de gemas individualizadas," *Technical Bulletin*, Instituto Agronômico, Campinas, SP, Brazil, 2012.

[3] M. A. Xavier, G. Aferri, M. A. P. Bidóia et al., "Fatores de desuniformidade e kit de pré-brotação IAC para sistema de multiplicação de cana-de-açúcar–mudas pré-brotadas (MPB)," *Technical Bulletin*, Instituto Agronômico, Campinas, SP, Brazil, 2014.

[4] M. A. Xavier, G. Aferri, M. A. P. Bidóia et al., "Mudas pré-brotadas (MPB) e produtividade de cana-de-açúcar," in *Anais do 10° Congreso Nacional da STAB*, STAB, Piracicaba, SP, Brazil, pp. 338–340, 2016.

[5] M. A. Xavier, M. G. A. Landell, M. P. Campana, G. Aferri, and D. Perecin, "Produtividade de cana-de-açúcar em fases iniciais de seleção de programa de melhoramento genético utilizando o método de multiplicação por mudas pré-brotadas," in *Anais do 10° Congreso Nacional da STAB*, pp. 334–337, STAB, Piracicaba, SP, Brazil, 2016.

[6] M. A. Xavier, M. Landell, D. N. Silva et al., *Produtividade de gemas de cana-de-açúcar para fins de abastecimento de núcleos de produção de mudas pré-brotadas-MPB*, STAB, Vol. 33 STAB, Piracicaba, SP, Brazil, 2016.

[7] L. Martinho, M. Bocchi, I. Jepson, M. Moreira, and J. C. Carvalho, "Plene, an innovative approach for sugarcane planting in Brazil," *Proceedings of the International Society of Sugar Cane Technologists*, vol. 27, 2010.

[8] R. Jain, S. Solomon, A. K. Shrivastava, and A. Chandra, "Sugarcane bud chips: a promising seed material," *Sugar Tech*, vol. 12, no. 1, pp. 67–69, 2010.

[9] S. Matsuoka, A. A. Garcia, H. Arozino et al., "Melhoramento de cana-de-açúcar," in *Melhoramentos de espécies cultivadas v.1*, A. Borém, Ed., pp. 205–252, Universidade Federal de Viçosa, Viçosa, MG, Brazil, pp. 205–252, 1999.

[10] M. C. Cox, D. M. Holgarth, and P. G. Allsopp, *Manual of cane growing*, BSES, Brisbane, Australia, 2000.

[11] J. A. Bressiani, "Seleção sequencial em cana-de-açúcar," Dissertation, ESALQ/University of São Paulo, Piracicaba, SP, Brazil, 2001.

[12] B. P. Brasileiro, T. O. De Paula Mendes, L. A. Peternelli, L. C. I. Da Silveira, M. D. V. De Resende, and M. H. P. Barbosa, "Simulated individual best linear unbiased prediction versus mass selection in sugarcane families," *Crop Science*, vol. 56, pp. 1–6, 2016.

[13] M. D. V. Resende and M. H. Barbosa, "Selection via simulated individual BLUP based on family genotypic effects," *Pesquisa Agropecuária Brasileira*, vol. 41, pp. 421–429, 2006.

[14] A. Netsanet and S. Tegene, "Effect of number of buds per setts and intra-row spacings of setts on yield and yield components of sugarcane," *International Journal of Agricultural Sciences and Natural Resources*, vol. 1, no. 5, pp. 115–121, 2014.

[15] A. J. Orgeron, K. A. Gravois, and K. P. Bischoff, "Planting rate effects on sugarcane yield trials," *Journal of Sugarcane Technologists*, vol. 27, pp. 23–34, 2007.

[16] M. P. Campana, M. A. Xavier, G. Aferri et al., "Cana-de-açúcar para a produção de material de propagação nos sistemas demudas pré-brotadas (MPB) e tradicional," *Anais do 10° Congresso Nacional da STAB*, pp. 223–224, STAB, Piracicaba, SP, Brazil, 2016.

[17] M. F. Couto, L. A. Peternelli, and M. H. P. Barbosa, "Classification of the coefficients of variation for sugarcane crops," *Ciência Rural*, vol. 43, pp. 957–961, 2013.

[18] S. B. Milligan, K. A. Gravois, K. P. Bischoff, and F. A. Martin, "Crop effects on genetic relationships among sugarcane traits," *Crop Science*, vol. 30, pp. 927–931, 1990.

[19] J. C. Skinner, D. M. Hogarth, and K. K. Wu, "Selection methods, criteria, and indices," in *Sugarcane Improvement Through Breeding*, D. J. Heinz, Ed., pp. 409–453, Elsevier, Amsterdan, Netherlands, 1987.

Genetic Variability and its Implications on Early Generation Sorghum Lines Selection for Yield, Yield Contributing Traits, and Resistance to Sorghum Midge

Massaoudou Hamidou [1,2] Abdoul Kader M. Souley,[1] Issoufou Kapran,[1] Oumarou Souleymane,[1,2] Eric Yirenkyi Danquah,[2] Kwadwo Ofori,[2] Vernon Gracen,[2,3] and Malick N. Ba[4]

[1]Institut National de la Recherche Agronomique du Niger (INRAN), BP 429, Niamey, Niger
[2]West Africa Centre for Crop Improvement, University of Ghana, PMB 30, Legon, Accra, Ghana
[3]Department of Plant Breeding and Genetics, Cornell University, 520 Bradfield Hall, Ithaca, NY 14850, USA
[4]International Crops Research Institute for the Semi-Arid Tropics, BP 12404, Niamey, Niger

Correspondence should be addressed to Massaoudou Hamidou; hmassaoud@yahoo.fr

Academic Editor: Kassim Al-Khatib

Sorghum is the second most important cereal crop in Niger. The crop is grown in a wide range of ecological environments in the country. However, sorghum grain yield in Niger is limited by both abiotic and biotic constraints. Recombinant inbred lines derived from the cross of a local variety with a midge resistant variety and two local checks were evaluated during the 2015 rainy season across two planting dates in two environments in Niger. The objective was to investigate genetic variability for yield, yield related traits, and resistance to sorghum midge. High phenotypic coefficient of variation (PCV) versus genotypic coefficient of variation (GCV) was observed in both sites and planting dates. Across planting dates at both Konni and Maradi, grain yield, plant height, panicle weight, and midge damage had high heritability coupled with high estimates of genetic advance. At Konni, high genetic advance coupled with high heritability was detected for grain yield, plant height, panicle weight, and resistance to midge. There were similar results at Maradi for grain yield, plant height, and panicle weight. Therefore, selection might be successful for the above characters in their respective environments.

1. Introduction

Sorghum (*Sorghum bicolor* L. Moench) is the second most important cereal crop in Niger after pearl millet. It is used in both human nutrition and animal nutrition. In Niger, sorghum production rarely meets the demand of the growing population. Grain sorghum yields are very low, about 0.280 tons/ha, which is far below the genetic potential of the crop compared to countries like USA (4.3 tons/ha), Argentina (4.9 tons/ha), and China (3.2 tons/ha) according to FAOSTAT [1]. The low production is attributed to abiotic and biotic stress such as sorghum midge *Stenodiplosis sorghicola*. Sorghum midge is a panicle insect found in most of the sorghum growing environments in Niger, where it causes high grain yield reduction on sorghum crop. Grain yield reduction due to midge of about 56% to 67% was documented on local sorghum varieties in Niger compared to 14% to 17% yield reduction observed on introduced, midge-resistant sorghum genotypes (Kadi Kadi et al.) [2]. These data indicate that local sorghum varieties grown in Niger are highly susceptible to midge infestation. The use of indigenous varieties with low yielding capacity also limits sorghum productivity in Niger. Farmers mostly rely on low-yielding landraces, so sorghum production fails to meet demand of increasing population and food insecurity remains a major issue (Maman et al.) [3]. Hence, it is essential that plant breeders develop and provide farmers with new improved sorghum cultivars. To stabilize sorghum production in Niger, identification of superior cultivars in terms of yield and resistance to sorghum midge is of major importance. Identification of genotypes with desirable traits and their subsequent use in breeding and establishment of suitable selection criteria can be helpful for successful varietal improvement programs. Analysis of variability among the traits and the association of a particular character with other traits contributing to yield of a crop would be of great importance in planning a successful breeding program (Mary and Gopalan) [4]. In planning sorghum improvement program, knowledge of variability of traits could be a key to the success. Comparative variability of traits is evaluated by estimating the genotypic coefficient of variation (GVC) and the phenotypic coefficient of variation (PCV) (Ahmad et al.) [5]. According to Sami et al. [6], heritable genetic effects and nonheritable environmental effects contribute to variability found in germplasm. The GCV expresses the heritable portion, while the PCV is an expression of both the genetic and environmental effects on the trait (Bello et al.) [7]. In sorghum, several studies on genetic variation have been documented. In a study of genetic variation, Warkad et al. [8] reported high genotypic and phenotypic coefficient of variation for yield and yield related traits in sorghum germplasm. Dhutmal et al. [9] also reported high genotypic and phenotypic coefficient of variation for grain yield and its components in drought-tolerant sorghum. It has been observed that the estimates of genetic variance were smaller than their respective phenotypic variances (Khan et al.) [10].

Heritability provides information on the transmissibility of traits from one generation to another. Knowledge of heritability determines the selection methods a breeder can use which could be appropriate for the improvement of traits in plants as reported [11]. Heritability is an estimating factor and indicates the consistency of a particular phenotypic observation that directs a breeding value (Falconer and Mackay) [12]. High heritability of a trait gives an indication of the progress that can be made for the improvement of that trait. However, in the absence of genetic advance, broad sense heritability may not be reliable. Therefore, broad sense heritability estimation needs to be coupled with the estimations of genetic advance for a more accurate assessment as described [13, 14].

Genetic variation evaluation that provides information on parameters like genotypic coefficient of variation, phenotypic coefficient of variation, heritability estimates, and genetic advance is absolutely necessary to start an efficient breeding program (Atta et al.) [15]. This investigation was undertaken to estimate genotypic and phenotypic coefficient of variation, heritability, and genetic advance for yield and resistance to sorghum midge in early sorghum generation.

2. Materials and Methods

2.1. Experimental Germplasm. 280 recombinant inbred lines and two local checks were evaluated. These F5 lines were obtained by crossing a local sorghum variety (MDK) and an exotic sorghum midge-resistant cultivar from ICRISAT (ICSV88032) with progeny advanced using single seed descent (SSD). The local variety has white grain with good qualities; it is widely cultivated by sorghum farmers in Niger. However, this variety is photosensitive and highly susceptible to sorghum midge.

2.2. Experimental Sites. The study was carried out during 2015 rainy season at the research stations of INRAN at Konni and Maradi. Both locations are sorghum midge hotspot in Niger. Konni has a latitude of $13°47'23''$ north and a longitude of $5°14'57''$ east and the average annual rainfall of 589.7 mm with average temperature of 29.3°C. Maradi has a latitude of $13°18'25''$ north and a longitude of $7°09'35''$ east and the average annual rainfall of 537.4 mm with average temperature of 20.5°C.

2.3. Experimental Design. The experimental design was an alpha (0.1) lattice with 2 replications in two different planting dates, giving four environments. However, blocking was not significant and the dates were reanalyzed using randomized complete block design. Environments one, two, three, and four were the first planting date at Konni, the second planting date at Konni, the first planting date at Maradi, and the second planting date at Maradi, respectively. The use of two different planting dates was to simulate early or late starting rainy season. Each genotype was grown in a single row of 3 meters; the intrarow and interrow spacing was 0.20 m × 0.80 m. The material was subjected to natural infestation of sorghum midge. In order to evaluate midge damage on the panicles, three panicles were covered at emergence using selfing bags. At harvesting, panicle and grain mass were recorded for the three covered and three uncovered panicles. The loss in grain yield in three uncovered panicles was expressed as a percentage of grain yields in covered panicles.

2.4. Data Collection and Analysis. Data collected were grain yield (GY), plant height (PH), 1000 seeds weight (TSW), days to 50% flowering (FF), midge damage (MD), and panicle weight (PW). Grain yield was measured in tons per hectare adjusted to grain moisture content at 12%. PH was measured in centimeters from the base of the plant to top of the panicle. 1000 seeds weight was measured from counted 1000 seeds in grams. Days to 50% flowering were recorded by counting the number of days from planting to the day when 50% of the plants in a plot flowered. Midge damage was calculated as loss of grain yield in three uncovered panicles expressed as a percentage of grain yield in three covered panicles. Panicles weight was recorded by weighting the entire panicle at physiological maturity. Analyses of variance were computed for all the characters evaluated using computer software system of GenStat, 12th edition. Genetic components were calculated as follows.

Heritability in the broad sense (H^2 or h^2) was estimated according to Hanson et al. [16] as follows:

$$h^2 = \frac{\sigma_g}{\sigma_{ph}}, \qquad (1)$$

where h^2 is heritability in the broad sense; σ_g is genotypic variance; and σ_{ph} is phenotypic variance.

TABLE 1: Mean square for traits across environments.

Source of variation	d.f.	GY	PH	PW	FF	TSW	MD
Replications	1	1234336	191,2	15202	193,38	120,36	0,73
Entries	281	177377**	9585,4**	4848**	219,94**	102,64**	0,14**
PD	1	3548702**	1287836,1**	339779**	49,11*	2005,55**	50,42**
Sites	1	30681580**	1640346,3**	1087245**	12805,54**	100,86^{ns}	27,80**
Entries.PD	281	129768**	1471**	2261**	70,26**	77,34^{ns}	0,11**
Entries.Sites	281	119234**	1159,3**	3165**	72,01**	80,28^{ns}	0,15**
PD.Sites	1	7723780**	16797**	258654**	379,18**	3829,83**	9,99**
Entries.PD.Sites	281	104061**	987,5^{ns}	2603**	38,59^{ns}	67,06^{ns}	0,12**
Residual	1127	62286	852,4	1540	49,91	64,47	0,06
CV		65,3	15,4	42,5	10,2	39,5	69,6

GY: grain yield; PH: plant height; PW: panicle weight; FF: days to 50% flowering; TSW: 1000 seeds weight; MD: midge damage; PD: planting dates; ns: not significant; **significant at 1%; *significant at 5%.

TABLE 2: Mean square for traits in the first planting date at Konni.

Source of variation	d.f.	GY	PH	PW	FF	TSW	MD
Replications	1	2421264	4.8	10795.5	38.49	28.5	1.56515
Entries	281	101337**	4270.8**	1451.8**	95.47*	184.9*	0.14143**
Residual	281	66375	647.9	643.6	79.77	190.6	0.06554
CV		53.8	10.5	35.4	12.5	35.5	48.1

GY: grain yield; PH: plant height; PW: panicle weight; FF: days to 50% flowering; TSW: 1000 seeds weight; MD: midge damage; ns: not significant; **significant at 1%; *significant at 5%.

Genotypic and phenotypic variances were calculated according to Burton [17] as follows:

$$\sigma^2_g = \frac{MS1 - MS2}{r},$$
$$\sigma^2_{ph} = \frac{MS1}{r}, \quad (2)$$

where σ^2_g is genotypic variance; σ^2_{ph} is phenotypic variance; MS1 is mean square for the entries; MS2 is mean square for the residuals; and r is replication.

Genotypic coefficient of variance (GCV) and phenotypic coefficient of variance (PCV) were determined according to Burton [17] as follows:

$$GCV\% = \frac{\sqrt{\sigma^2_g}}{X} * 100,$$
$$PCV\% = \frac{\sqrt{\sigma^2_{ph}}}{X} * 100, \quad (3)$$

where GCV is genotypic coefficient of variance; PCV is phenotypic coefficient of variance; σ^2_g is genotypic variance; σ^2_{ph} is phenotypic variance; and X is sample mean.

Genetic advance (GA) was calculated according to Johonson et al. [18] as follows:

$$GA = k * \sigma_{ph} * h^2, \quad (4)$$

where GA is genetic advance; K is a constant = 20.06 at 5% selection intensity; σ_{ph} is square root of phenotypic variance; H^2 is heritability in the broad sense; and

$$GA \text{ as \% of mean } (GAM) = \left(\frac{GA}{mean\ value}\right) * 100. \quad (5)$$

3. Results

3.1. Analysis of Variance. Across environments, entries and planting dates were significant ($P < 0.001$) for all traits, while sites displayed significance ($P < 0.001$) for all traits except 1000 seeds weight. The results of factors' interaction showed that entries by planting dates and entries by sites were significant ($P < 0.001$) for all traits except 1000 seeds weight. On the other hand, entries by planting dates and those by sites were significant ($P < 0.001$) for grain yield, panicles weight, and midge damage (Table 1).

In the first planting date at Konni, entries were highly significant ($P < 0.001$) for grain yield, plant height, panicles weight, and midge damage, while significant differences ($P < 0.005$) were observed for days to 50% flowering and 1000 seeds weight (Table 2).

In the second planting date at Konni, the entries were highly significant ($P < 0.001$) for all characters under study except midge damage which was significant at 5% level (Table 3).

The combined analyses of variance over the two planting dates at Konni reveal that entries were significant at 1% level for all characters except 1000 seed weight which was

TABLE 3: Mean squares for traits in the second planting date at Konni.

Source of variation	d.f.	GY	PH	PW	FF	TSW	MD
Replications	1	108631	24239	1612	456.41	4.43	0.40335
Entries	281	264307**	2706**	2336**	91.41**	85.07**	0.11681*
Residual	281	138467	1001	1138	57.52	29.6	0.07664
CV		61.8	16.7	48.8	10.5	23.8	43.2

GY: grain yield; PH: plant height; PW: panicle weight; FF: days to 50% flowering; TSW: 1000 seeds weight; MD: midge damage; ns: not significant; *significant at 1%; **significant at 5%.

TABLE 4: Mean squares for traits across planting dates at Konni.

Source of variation	d.f.	GY	PH	PW	FF	TSW	MD
Replications	1	2514949	8312.6	6006.4	298.54	8.9	1.90569
Entries	281	196938**	5738**	2476.2**	122.73**	144.7*	0.1662**
PD	1	353779*	804414.1**	3276.3*	436.87*	5832.9**	7.98005**
Entries.PD	281	166350**	1259**	1373.4**	65.26*	127.7*	0.13463**
Residual	563	92645	817.1	827	70.83	121	0.07053
CV		61.4	13.2	41.0	11.7	33.4	49.0

GY: grain yield; PH: plant height; PW: panicle weight; FF: days to 50% flowering; TSW: 1000 seeds weight; MD: midge damage; PD: planting dates; ns: not significant; *significant at 1%; **significant at 5%.

TABLE 5: Mean squares for traits in the first planting date at Maradi.

Source of variation	d.f.	GY	PH	PW	FF	TSW	MD
Replications	1	145655	20840.5	10054	26.9	346.56	0.03682
Entries	281	71434**	3253.8**	3458**	85.22**	22.98**	0.007987^{ns}
Residual	281	39197	840.2	1750	18.86	14.12	0.009737
CV		54.0	15.7	30.3	6.4	18.4	33.6

GY: grain yield; PH: plant height; PW: panicle weight; FF: days to 50% flowering; TSW: 1000 seeds weight; MD: midge damage; ns: not significant; *significant at 1%; **significant at 5%.

TABLE 6: Mean squares for traits in the second planting date at Maradi.

Source of variation	d.f.	GY	PH	PW	FF	TSW	MD
Replications	1	316788	145.3	84656	1.95	5.1	0.27202
Entries	281	44959**	2655.2**	4019**	6.61^{ns}	32.9**	0.13734*
Residual	281	21581	827.9	2532	7.81	22.19	0.09133
CV		52.6	20.3	34.5	26.7	23.8	49.2

GY: grain yield; PH: plant height; PW: panicle weight; FF: days to 50% flowering; TSW: 1000 seeds weight; MD: midge damage; ns: not significant; *significant at 1%; **significant at 5%.

significant at 5% level. Planting dates were significant at 1% level for plant height, 1000 seeds weight, and midge damage, while grain yield, panicle weight, and days to 50% flowering were significant at 5% level. The interaction of planting dates and entries was significant at 1% level for all characters except days to 50% flowering and 100 seed weight which were significant at 5% level (Table 4).

In the first planting date at Maradi, entries were highly significant ($P < 0.001$) for all characters under study except midge damage (Table 5).

In the second planting date at Maradi, entries were highly significant ($P < 0.001$) for grain yield, plant height, panicle weight, and 1000 seeds weight were significant ($P < 0.005$) for midge damage. There were no significant differences in the genotypes for days to 50% flowering (Table 6).

The combined analyses of variance over the two planting dates at Maradi show that entries were highly significant for all characters under study ($P < 0.001$). Planting dates were highly significant ($P < 0.001$) for grain yield, plant height, panicles weight, and midge damage; significant differences ($P < 0.005$) were observed for days to 50% flowering and 1000 seeds weight. The interaction of entries and planting dates was highly significant ($P < 0.001$) for grain yield and midge damage and significantly different ($P < 0.005$) for plant height, panicle weight, days to 50% flowering, and 1000 seeds weight (Table 7).

3.2. Estimates of Coefficients of Variation, Heritability, and Genetic Advance.
Across environments, the GCV values range from 6.64% for days to 50% flowering to 31.37% for

TABLE 7: Mean squares for traits across planting dates at Maradi.

Source of variation	d.f.	GY	PH	PW	FF	TSW	MD
Replications	1	1071	10381.4	9380	298.54	134.47	0.02
Entries	281	78655**	4854.7**	5051**	122.73**	37.86**	0.07**
PD	1	9165215**	506918.5**	579791**	436.87*	142.93*	52.52**
Entries.PD	281	50070**	1128.6*	3033*	65.26*	17.92*	0.07**
Residual	563	32323	854.5	2213	70.83	18.51	0.04
CV		65.5	17.9	40.9	8.3	21.4	44.3

GY: grain yield; PH: plant height; PW: panicle weight; FF: days to 50% flowering; TSW: 1000 seeds weight; MD: midge damage; ns: not significant; *significant at 1%; **significant at 5%.

TABLE 8: Genetic variation for resistance to midge, yield, and yield contributing characters for the combined environments.

Traits	GM	GV	PV	GCV (%)	PCV (%)	H_{BS} (%)	GA as % of mean
GY	382,4	14386,38	22172,13	31,37	38,94	65.58	50.63
PH	189,99	1091,63	1198,18	17,39	18,22	91.41	33.28
PW	92,44	413,50	606,00	22,00	26,63	68.38	36.41
FF	69,47	21,25	27,49	6,64	7,55	77.28	11.75
TSW	20,34	4,77	12,83	10,74	17,61	37.12	13.17
MD	0,34	0,01	0,02	30,91	39,52	61.62	49.34

GY: grain yield; PH: plant height; PW: panicle weight; FF: days to 50% flowering; TSW: 1000 seeds weight; MD: midge damage; GM: grand mean; GV: genotypic variance; PV: phenotypic variance; GCV: genotypic coefficient of variance; PCV: phenotypic coefficient of variance; $H_{(BS)}$: broad sense heritability; GA: genetic advance as a percentage of mean.

TABLE 9: Genetic variation for resistance to midge, yield, and yield contributing characters in the first planting date at Konni.

Traits	GM	GV	PV	PCV (%)	GCV (%)	$H_{(BS)}$ (%)	GA (% mean)
GY	479	17481	83856	59.09	26.98	20.84	24.90
PH	243.4	1811.45	2459.35	20.37	17.48	73.65	30.63
PW	71.7	404.1	1047.7	45.14	28.03	38.57	35.33
FF	71.25	7.85	87.62	13.13	3.93	8.95	21.65
TSW	18.3	2.85	92.45	52.54	9.22	3.08	31.25
MD	0.37	0.03	0.10	86.94	52.64	36.66	64.47

GY: grain yield; PH: plant height; PW: panicle weight; FF: days to 50% flowering; TSW: 1000 seeds weight; MD: midge damage; GM: grand mean; GV: genotypic variance; PV: phenotypic variance; GCV: genotypic coefficient of variance; PCV: phenotypic coefficient of variance; $H_{(BS)}$: broad sense heritability; GA: genetic advance as a percentage of mean.

grain yield. Grain yield, panicles weight, and midge damage recorded high GCV value. Plant height and 1000 seeds weight recorded medium GCV value, while days to 50% flowering recorded low GCV value. The PCV values range from 7.55% for days to 50% flowering to 39.52% for midge damage. Midge damage, grain yield, and panicles weight recorded high PCV value. Plant height and 1000 seeds weight recorded medium PCV value, while days to 50% flowering recorded low PCV value (Table 8). Heritability estimates ranged from 91.41% for plant height to 37.12% for 1000 seeds weight. All traits recorded high heritability estimates (Table 8). The estimates of genetic advance as percentage of mean ranged from 50.63% for grain yield to 11.75% for days to 50% flowering. Grain yield, plant height, panicles weight, and midge damage displayed high genetic advance as percentage of mean, while days to 50% flowering and 1000 seeds weight displayed medium genetic advance as a percentage of mean (Table 8).

In the first planting date at Konni, GCV ranged from 52.64% for midge damage to 3.93% for days to flowering. Midge damage, panicle weight, and grain yield showed high GCV, plant height recorded medium GCV, and days to 50% flowering had low GCV. On the other hand, PCV values ranged from 86.94% for midge damage to 13.13% for days to 50% flowering. Midge damage, grain yield, plant height, and panicle weight exhibited high PCV, while days to 50% flowering displayed medium PCV (Table 9). Heritability estimates ranged from 73.65% for plant height to 8.57% for days to 50% flowering. Days to 50% flowering exhibited low heritability, while midge damage, panicle weight, plant height, and grain yield showed high heritability estimates. The highest estimate of genetic advance was seen for midge damage (64.47%) and the lowest estimate was for days to 50% flowering (Table 9).

In the second planting date at Konni, GCV ranged from 48.42% for grain yield to 5.65% for days to flowering. Midge

TABLE 10: Genetic variation for resistance to midge, yield, and yield contributing characters in the second planting date at Konni.

Traits	GM	GV	PV	PCV (%)	GCV (%)	$H_{(BS)}$ (%)	GA (% mean)
GY	518	62920	201387	86.63	48.42	31.24	55.32
PH	189.3	852.5	1853.5	22.74	15.42	45.99	21.08
PW	69.2	599	1737	60.22	35.36	34.48	42.18
FF	72.53	16.94	74.46	11.89	5.67	22.75	5.39
TSW	22.89	27.73	57.33	33.07	23.00	48.37	32.70
MD	0.52	0.02	0.09	59.80	27.25	20.76	24.64

GY: grain yield; PH: plant height; PW: panicle weight; FF: days to 50% flowering; TSW: 1000 seeds weight; MD: midge damage; GM: grand mean; GV: genotypic variance; PV: phenotypic variance; GCV: genotypic coefficient of variance; PCV: phenotypic coefficient of variance; $H_{(BS)}$: broad sense heritability; GA: genetic advance as a percentage of mean.

TABLE 11: Genetic variability for resistance to midge, yield, and yield contributing characters across planting dates at Konni.

Traits	GM	GV	PV	PCV (%)	GCV (%)	$H_{(BS)}$ (%)	GA (% mean)
GY	495.5	52146.5	144791.5	76.79	46.08	36.01	56.95
PH	216.49	2460.45	3277.55	26.44	32.40	75.06	40.85
PW	70.11	551.4	1378.4	52.95	47.36	40.00	43.63
FF	71.88	25.95	96.78	13.68	10.02	26.81	7.33
TSW	20.59	11.85	132.85	55.97	23.64	8.91	9.22
MD	0.45	0.04	0.11	76.45	68.73	40.41	62.99

GY: grain yield; PH: plant height; PW: panicle weight; FF: days to 50% flowering; TSW: 1000 seeds weight; MD: midge damage; GM: grand mean; GV: genotypic variance; PV: phenotypic variance; GCV: genotypic coefficient of variance; PCV: phenotypic coefficient of variance; $H_{(BS)}$: broad sense heritability; GA: genetic advance as a percentage of mean.

damage, 1000 seeds weight, panicle weight, and grain yield had high GCV, plant height had medium GCV, and days to 50% flowering had low GCV. On the other hand, the PCV estimates ranged from 86.63% for grain yield to 11.89% for days to flowering. All characters exhibited high PCV except days to 50% flowering which was medium. Heritability estimates ranged from 48.37% for 1000 seeds weight to 20.76% for midge damage. All characters under study exhibited relatively high heritability estimates (Table 10). The estimates of genetic advance as a percentage of the mean ranged from 55.32% for grain yield to 5.39% for days to 50% flowering. The high estimates (>20%) were recorded by grain yield (55.32%), panicle weight (42.18%), 1000 seeds weight 32.70%, midge damage (24.64%), and plant height (21.08%). Low (<10%) GA estimate was recorded by days to 50% flowering (5.39%) (Table 10).

The estimates for GCV ranged from 68.73% for midge damage to 47.36% for panicle weight, while the PCV ranged from 76.79% for grain yield to 13.68% for days to 50% flowering. Grain yield, plant height, panicle weight, 1000 seeds weight, and midge damage recorded high PCV values, while days to 50% flowering recorded medium PCV values. On the other hand, grain yield, plant height, panicle weight, 1000 seeds weight, and midge damage showed high GCV values, whereas days to 50% flowering had a low GCV value (Table 11). The estimates of heritability ranged from 75.06% for plant height to 8.91% for 1000 seeds weight. Grain yield, plant height, panicle weight, and midge damage exhibited high heritability estimates, days to 50% flowering recorded medium heritability, and the heritability of 1000 seeds weight was low. Hence, good progress in selection could be obtained for these characters if used as selection criteria (Table 11). High GA estimates (>20%) were recorded for midge damage (62.99%), grain yield (56.95%), panicle weight (43.63%), and plant height (40.85%). Days to 50% flowering and 1000 seeds weight had low GA estimates with 7.33% and 9.22%, respectively (Table 11).

In the first planting date at Maradi, GCV ranged from 34.65% for grain yield to 8.55% for days to 50% flowering, while the PCV ranges from 64.26% for grain yield to 10.71% for days to flowering. All characters had high GCV except days to 50% flowering which had a low GCV value. All characters exhibited high PCV values except days to 50% flowering which had medium values (Table 12). Heritability ranged from 58.95% for plant height to 23.88% for 1000 seeds weight. All characters displayed high heritability in the present study. High (>20%) genetic advance as percentage of mean was detected for grain yield (38.34%), plant height (29.33%), and panicle weight (24.39%). Medium (10-20%) genetic advance was detected for days to 50% flowering (13.90%) and the low (<10%) genetic advance was for 1000 seeds weight (9.96%) (Table 12).

In the second planting date at Maradi, GCV ranged from 58.56% for grain yield to 11.71% for 1000 seeds weight, while PCV ranged from 98.80% for grain yield to 26.57% for 1000 seeds weight. All characters showed high GCV except 1000 seeds weight which showed medium GCV, while all characters displayed high PCV values (Table 13). The heritability ranged from 35.13% for grain yield to 19.85% for 1000 seeds weight. Grain yield, plant height, and panicle weight had high heritability estimates, while 1000 seeds weight and midge damage had medium heritability estimates. High (>20%)

TABLE 12: Genetic variation for resistance to midge, yield, and yield contributing characters in the first planting date at Maradi.

Traits	GM	GV	PV	PCV (%)	GCV (%)	$H_{(BS)}$ (%)	GA (% mean)
GY	366.4	16118.5	55315.5	64.26	34.65	29.13	38.34
PH	184.3	1206.8	2047	24.54	18.84	58.95	29.33
PW	137.9	854	2604	37.00	21.19	32.79	24.39
FF	67.34	33.18	52.04	10.71	8.55	63.75	13.90
TSW	20.47	4.43	18.55	21.04	10.28	23.88	9.96

GY: grain yield; PH: plant height; PW: panicle weight; FF: days to 50% flowering; TSW: 1000 seeds weight; GM: grand mean; GV: genotypic variance; PV: phenotypic variance; GCV: genotypic coefficient of variance; PCV: phenotypic coefficient of variance; $H_{(BS)}$: broad sense heritability; GA: genetic advance as a percentage of mean.

TABLE 13: Genetic variation for resistance to midge, yield, and yield contributing characters in the second planting date at Maradi.

Traits	GM	GV	PV	PCV (%)	GCV (%)	$H_{(BS)}$ (%)	GA (% mean)
GY	184.6	11689	33270	98.80	58.56	35.13	71.24
PH	141.8	913.65	1741.55	29.43	21.31	52.46	31.52
PW	92.3	743.5	3275.5	62.00	29.54	22.69	28.10
TSW	19.75	5.355	27.54	26.57	11.71	19.44	10.40
MD	0.43	0.02	0.11	78.72	35.08	19.85	30.81

GY: grain yield; PH: plant height; PW: panicle weight; FF: days to 50% flowering; TSW: 1000 seeds weight; MD: midge damage; GM: grand mean; GV: genotypic variance; PV: phenotypic variance; GCV: genotypic coefficient of variance; PCV: phenotypic coefficient of variance; $H_{(BS)}$: broad sense heritability; GA: genetic advance as a percentage of mean.

TABLE 14: Genetic variation for resistance to midge, yield, and yield contributing characters across planting dates at Maradi.

Traits	GM	GV	PV	PCV (%)	GCV (%)	$H_{(BS)}$ (%)	GA (% mean)
GY	274.7	23166	55489	47.54	30.71	41.74	72.42
PH	163.02	2000.1	2854.6	24.67	29.21	70.06	47.26
PW	115.1	1009	3222	80.96	64.07	31.31	31.49
FF	67.12	25.95	96.78	13.68	10.02	26.81	7.85
TSW	20.13	9.67	28.18	25.78	21.36	34.32	18.47
MD	0.22	0.01	0.05	111.12	79.59	25.64	57.23

GY: grain yield; PH: plant height; PW: panicle weight; FF: days to 50% flowering; TSW: 1000 seeds weight; MD: midge damage; GM: grand mean; GV: genotypic variance; PV: phenotypic variance; GCV: genotypic coefficient of variance; PCV: phenotypic coefficient of variance; $H_{(BS)}$: broad sense heritability; GA: genetic advance as a percentage of mean.

estimates of genetic advance as percentage of mean were seen for grain yield (71.24%), plant height (31.52%), midge damage (30.81%), and panicle weight (28.10%). Medium (10–20%) estimate of GA was seen for 1000 seeds weight (10.40%) (Table 13).

Across planting date, high GCV was recorded for midge damage, panicle weight, grain yield, plant height, and 1000 seeds weight across planting dates at Maradi. Days to 50% flowering recorded lower GCV. Panicle weight, grain yield, plant height, and 1000 seeds weight had high PCV, while days to 50% flowering had medium PCV (Table 14). The highest heritability was seen for plant height (70.06%) and the lowest heritability for midge damage (25.64%). All characters under study exhibited relatively high heritability. High estimates of GA (>20%) were seen for grain yield (72.42%), midge damage (57.23%), plant height (47.26%), and panicle weight (31.40%). Low (<10%) estimates of genetic advance as percentage of mean were seen for days to 50% flowering (7.85%), while medium (10–20%) GA was seen for 1000 seeds weight (Table 14).

4. Discussion

Across environments, entries, planting dates, and interaction between planting dates and sites contributed to the variations observed. However, sites and the interaction of entries by sites and entries by planting dates contributed to the variations for all except 1000 seeds weight. Entries by sites and those by planting dates contributed to the variations observed for grain yield, panicles weight, and midge damage. Therefore selection could be accomplished for grain yield, panicles weight, and resistance to midge across the study environments.

At Konni, differences due to entries, planting dates, and entries by planting dates interactions across planting dates indicate that planting date and genotypes played an important role in the genetic variation observed. Likewise, at Maradi, significant variation was observed across the two planting dates for all characters among genotypes, planting dates, and genotypes by planting dates. This shows that entries, planting dates, and entries by planting dates interactions contributed to the genetic variation observed in all the characters under study. Hence, selection for improving these characters in sorghum could be achieved in either of the two planting dates at Maradi.

In the first planting date, entries at Konni contributed to the genetic variation found in all characters, while at Maradi, in the first planting date, entries displayed differences for all characters except midge damage. Therefore, breeding for improving grain yield, plant height, panicle weight, days to 50% flowering, and 1000 seeds weight in sorghum for early planting through selection among the entries could be successful at both study sites, whereas selection for resistance to sorghum midge can only be achieved in the first planting date at Konni.

In the second planting date at Konni, entries contributed to the general genetic variability observed for all characters including resistance to sorghum midge. Selection can be attained for improving sorghum for these characters for use in late planting. In the second planting date at Maradi, genotypes were different for all characters except days to 50% flowering. Hence, selection can be achieved for grain yield, plant height, panicle weight, 1000 seeds weight, and resistance to midge in late planting sorghum at Maradi.

In general, phenotypic coefficient of variation was greater than the genotypic coefficient of variation for all the characters in the first planting and second planting dates at all locations. This suggests environment influences on the expression of these characters. Since PCV estimates the effects of genotypes and environment, higher PCV versus GCV indicates a significant contribution of environment and genotypes by environment interaction in the expression of all characters in both sites as well as planting dates. Similar results on cultivated sorghum were reported by Bello et al. [7] in Nigeria. References [19, 20] also found similar results for quantitative characters in sorghum.

The existence of high GCV in the present study suggests that selection for characters across planting dates at Konni and Maradi as well as across environments should be possible. Selection could be done for all characters except days to 50% flowering across environments and in both planting dates at Konni. At Maradi, selection could be done for grain yield, plant height, panicle weight, and 1000 seeds weight in the first planting date, while in the second planting date, selection can be successful for all characters except days to 50% flowering.

The coefficient of variation only indicates the extent of total variability present for a character and does not split the variability into heritable and nonheritable portions as reported [21, 22]. Hence, the determination of the heritability appears to be of great importance. In the present investigations, high heritability was seen for almost all the characters. Across environments, the broad sense heritability estimates were high for all traits. At Konni and Maradi, except days to 50% flowering at Konni, broad sense heritability estimates were high for all characters across planting dates. The high heritability observed for these characters indicates that genotype plays more important role than environment in determining the phenotype, suggesting the predominance of both additive and dominant gene effects in the inheritance of the characters.

In the first planting date at Konni, except for days to 50% flowering and 1000 seeds weight, all characters displayed high heritability at both sites. In the second planting date, all characters showed high broad sense heritability estimates. At Maradi, grain yield, plant height, and panicle weight displayed high heritability, while 1000 seeds weight and midge damage had medium heritability. Since high heritability estimates were observed for these characters, the contribution of the genotype was higher than that of environment to determining the phenotype. Therefore, additive and dominant gene effects probably were important in the inheritance of grain yield, plant height, panicle weight, and resistance to midge in the first planting date at Konni and for all characters at Maradi for the first planting date. Therefore, genotypes could be selected based on phenotype for improving sorghum for grain yield, plant height, panicle weight, and resistance to midge for early planting at Konni and for all the characters at Maradi.

Likewise, in the second planting date at Konni, genotype contributed to the expression of all characters, suggesting that additive and dominant gene effects are important for all characters in the second planting date at Konni. Hence, in the second planting date at Konni, entries could be selected based on phenotype for improving all characters under study.

However, in the second planting date at Maradi, genotypes contributed to the inheritance of grain yield, plant height, and panicle weight. Therefore, genotypes could be selected based on phenotype for improving grain yield, plant height, and panicle weight. High heritability for grain yield per panicle and plant height and moderate heritability for days to 50% flowering were documented by Chavan et al. [23], while Sharma et al. [24] found high heritability for number of grains per panicle and plant height.

Across all the study environments and across planting dates at both Konni and Maradi, grain yield, plant height, panicle weight, and midge damage had high heritability estimates coupled with high estimates of genetic advance. Hence, selection would be effective for these characters across planting dates at both Konni and Maradi as well as across the study environments.

At Konni, in both planting dates, all characters displayed high estimates of GA coupled with high heritability except days to 50% flowering. Hence, selection might be successful for grain yield, plant height, panicle weight, and resistance to midge for early planting materials at Konni as well as late planting date.

At Maradi, grain yield, plant height, and panicle weight showed high estimates of GA coupled with high heritability estimates in both planting dates. Therefore, selection might be successful for grain yield, plant height, and panicle weight for either early or late planting dates.

Investigation of genetic variability in sorghum by Chavan et al. [23] revealed high estimates of GA coupled with high heritability for number of grains per panicle, plant height, and grain yield per panicle, whereas high heritability and low estimates of genetic advance were identified for panicle width, panicle length, and test weight. Similar results were obtained by Bapat and Shinde [25]. High estimates of genetic advance for days to 50% flowering were documented by Kishor and

Singh [26]. High heritability coupled with high GA for plant height and grain yield was reported by Dabholkar [27]. Evidence of heritability coupled with high GA for yield and yield components in rabi sorghum was provided by Arunkumar et al. [28].

5. Conclusion

Variations were observed among landrace-derived sorghum lines for sorghum midge resistance, yield, and yield related traits in the two experimental sites and the two different planting dates within the sites. The results of this study suggest a way of improving sorghum by selecting for yield, yield contributing traits, and resistance to sorghum midge using landrace-derived germplasm in Niger. Across planting dates, at both Konni and Maradi, grain yield, plant height, panicle weight, and midge damage had high heritability estimates coupled with high estimates of genetic advance. At Konni in both planting dates, all characters displayed high heritability estimates coupled with high GA except days to 50% flowering. At Maradi, grain yield, plant height, and panicle weight had high heritability coupled with high GA in both planting dates. Therefore, there is evidence that breeding for resistance to sorghum midge can be obtained in addition to other traditional breeding objectives such as yield and yield contributing traits at Konni, whereas breeding for midge resistance may not be successful at Maradi.

Acknowledgments

The authors wish to acknowledge Alliance for a Green Revolution in Africa (AGRA) for funding this work and the West Africa Centre for Crop Improvement (WACCI) for the good training.

References

[1] FAOSTAT, *Statistical Yearbook 2013: World Food and Agriculture*, Organziation UN, Rome, Italy, 2014, http://faostat.fao.org/site/291/default.aspx.

[2] H. A. Kadi Kadi, I. Kapran, and B. B. Pendleton, "Identification of sorghum genotypes resistant to sorghum midge in Niger," *International Sorghum And Millets Newsletter*, 2005.

[3] N. Maman, S. C. Mason, D. J. Lyon, and P. Dhungana, "Yield components of pearl millet and grain sorghum across environments in the central Great Plains," *Crop Science*, vol. 44, no. 6, pp. 2138–2145, 2004.

[4] S. S. Mary and A. Gopalan, "Dissection of genetic attributes yield traits of fodder cowpea in F3 and F4," *Journal of Applied Scenences Research*, vol. 2, pp. 805–808, 2006.

[5] S. Q. Ahmad, S. Khan, M. Ghaffar, and F. Ahmad, "Genetic diversity analysis for yield and other parameters in maize (Zea mays L.) genotypes," *Asian Journal of Agricultural Sciences*, vol. 3, no. 5, pp. 385–388, 2011.

[6] R. A. Sami, M. Y. Yeye, M. F. Ishiyaku, and I. S. Usman, "Heritability studies in some sweet sorghum (Sorghum Bicolor. L. Moench) genotypes," *Journal of Biology Agriculture and Healthcare*, vol. 3, no. 17, pp. 49–51, 2013.

[7] D. Bello, A. M. Kadams, S. Y. Simon, and D. S. Mashi, "Studies on genetic variability in cultivated sorghum (Sorghum bicolor L. Moench) cultivars of Adamawa State Nigeria," *American-Euriasian Journal Agricultural Enviroment Science*, vol. 2, no. 3, pp. 297–302, 2007.

[8] Y. N. Warkad, N. R. Potdukhe, A. M. Dethe, P. A. Kahate, and R. R. Kotgire, "Genetic variability, heritability and genetic advance for quantitative traits in sorghum germplasm," *Agricultural Science Digest*, vol. 28, no. 3, pp. 165–169, 2008.

[9] R. R. Dhutmal, S. P. Mehetre, A. W. More, H. V. Kalpande, A. G. Mundhe, and A. J. Sayyad Abubakkar, "Variabity parameters in rabi sorghum (sorghum bicolor L. Moench) drought tolerent genotypes," *The Bio-Scan*, vol. 9, no. 4, pp. 1455–1458, 2014.

[10] M. Q. Khan, S. I. Awan, and M. M. Mughal, "Estimation of genetic parameters in spring wheat genotypes under rainfed conditions," *Industrial Journal of Biological Sciences*, vol. 2, no. 3, pp. 367–370, 2005.

[11] W.-U. Haq, M. F. Malik, M. Rashid, M. Munir, and Z. Akram, "Evaluation and estimation of heritability and genetic advancement for yield related attributes in wheat lines," *Pakistan Journal of Botany*, vol. 40, no. 4, pp. 1699–1702, 2008.

[12] D. S. Falconer and T. F. C. Mackay, *Introduction to Quantitative Genetics*, Benjamin Cummings, England, 4th edition, 1996.

[13] S. O. F. I. Najeeb, A. G. Rather, G. A. Parray, F. A. Sheikh, and S. M. Razvi, "Studies on genetic variability, genotypic correlation and path coefficient analysis in maize under the high altitude temperate conditions of Kashmir," *Maize Genetics Cooperation Newsletter*, vol. 83, pp. 1–8, 2009.

[14] C. M. Rafiq, M. Rafique, A. Hussain, and M. Altaf, "Studies on heritability, correlation and path analysis in maize (Zea mays L.)," *Journal of Agricultural Research*, vol. 48, no. 1, pp. 35–38, 2010.

[15] B. M. Atta, M. A. Haq, and T. M. Shah, "Variation and inter-relationships of quantitative traits in chickpea (Cicer arietinum L.)," *Pakistan Journal of Botany*, vol. 40, no. 2, pp. 637–647, 2008.

[16] C. H. Hanson, H. F. Robinson, and R. E. Comstock, "Biometrical studies in yield of segregating population of Korean lespediza," *Agronomy Journal*, vol. 48, pp. 214–318, 1956.

[17] G. W. Burton, "Quantitative inheritance in grasses," in *Proceedings of the 6th International Grassland Congress*, 1, pp. 277–283, USA, 1952.

[18] H. W. Johonson, H. F. Robinson, and R. E. Comostock, "Genotypic and phenotypic correlations in soybeans and their implication in selection," *Agronomy Journal*, vol. 47, pp. 477–483, 1955.

[19] A. K. Basu, "Variability and heritability estimate from interseason sorghum cross," *Indian Journal of Agricultural Sciences*, vol. 41, pp. 116–117, 1981.

[20] E. H. Abu-Gasim and A. E. Kambal, "Variability and interrelationship among characters in indigenous grain sorghum of the Sudan," *Crop Science*, vol. 11, pp. 308–309, 1985.

[21] D. Lakshmana, B. D. Biradar, and R. L. R. kumar, "Genetic variability studies for quantitative traits in a pool of restorers

and maintainers lines of pearl millet (Pennisetum glaucum (L.))," *Karnataka Journal of Agricultural Science*, vol. 22, pp. 881-882, 2009.

[22] M. Govindaraj, B. Selvi, S. Rajarathinam, and P. Sumathi, "Genetic variability and heritability of grain yield components and grain mineral concentration in India's pearl millet (Pennietum glaucum (L) R. Br.) accessions," *African Journal of Food, Agriculture, Nutrition and Development*, vol. 11, no. 3, 2011.

[23] S. K. Chavan, R. C. Mahajan, and S. U. Fatak, "Genetic variability studies in sorghum," *Karnataka Journal of Agricultural Sciences*, vol. 23, pp. 322-323, 2010.

[24] H. Sharma, D. K. Jain, and V. Sharma, "Genetic variability and path coefficient analysis in sorghum," *Indian Journal of Agricultural Research*, vol. 40, pp. 310–312, 2006.

[25] D. R. Bapat and M. D. Shinde, "Study of genetic variability for grain yield in sorghum," *Sorghum Newsltr*, vol. 23, pp. 27-28, 1980.

[26] N. Kishor and L. N. Singh, "Variability and association studies under irrigated and rainfed situations in the sub-montane region in forage sorghum," *Crop Research*, vol. 29, no. 2, pp. 252–258, 2005.

[27] A. R. Dabholkar, *Element of Biometrical Genetics*, Concept Publishing Company, New Delhi, India, 1992.

[28] B. Arunkumar, B. D. Biradar, and P. M. Salimath, "Genetic variability and character association studies in rabi sorghum," *Karnataka Jornal of Agricultural Science*, vol. 17, pp. 471–475, 2004.

Sesame Yield Response to Deficit Irrigation and Water Application Techniques in Irrigated Agriculture, Ethiopia

E. K. Hailu, Y. D. Urga, N. A. Sori, F. R. Borona, and K. N. Tufa

Irrigation and Drainage Research, Werer Agricultural Research Center, EIAR, P.O. Box 2003, Addis Ababa, Ethiopia

Correspondence should be addressed to E. K. Hailu; ehailu2010@gmail.com

Academic Editor: Othmane Merah

The study was conducted at Werer Agricultural Research Center, Addis Ababa, for two years, 2013 and 2015, during main seasons and for three years, 2012/13, 2013/14, and 2014/15, during the cool period cropping season (November to February) as in the local cropping calendar. The study was aimed at identifying optimum soil moisture stress for sesame and thereby determining appropriate water-saving irrigation methods and also productivity under limited water resource conditions. Nine treatments with three levels of irrigation water percentage based on evapotranspiration of the crop (ETc) (100% ETc, 75% ETc, and 50% ETc) and three types of furrow irrigation methods (alternate furrow, fixed furrow, and conventional furrow) were used. The study design was randomized complete block design (RCBD) with three replications. The yield of sesame had significant ($p < 0.05$) variation among treatments due to deficit irrigation levels and application methods for sesame planted in main seasons. The highest mean yield of 937.50 kg/ha and 2797.6 kg/ha was obtained from the treatment of 50% ETc with alternate and conventional furrow application methods in 2013 and 2015, respectively. The combined mean yield of two years (2013–2015) showed different levels of deficit irrigation, and irrigation methods had a significant effect ($p < 0.05$) on main season planted sesame. Hence, the highest mean yield of 1846.7 kg/ha was obtained from the application of 50% ETc with the conventional furrow application method. In the cool planting season, the highest mean yield of 528.55 kg/ha, 1432.3 kg/ha, and 1562.5 kg/ha was obtained from treatments of 50% ETc, 75% ETc, and 100% ETc with the conventional furrow application method in 2012/13, 2013/14, and 2014/15, respectively. Moreover, during the same period over years, combined analysis showed that the highest mean yield of 1053 kg/ha was obtained from application of 100% ETc with the conventional furrow application method. Thus, it is concluded that a deficit irrigation treatment of 50% ETc with the conventional furrow application method for main season and application of 100% ETc with the conventional furrow application method for cool planting season are best practices of water-saving strategies for irrigated agriculture system.

1. Introduction

Globally, more than 40% of annual food production comes from irrigated land, and agriculture is the largest consumer of water, at 70% of all freshwater withdrawals [1]. As water scarcity becomes more acute in many parts of the world, increasing the effectiveness with which agricultural water resources are used is a priority for enhanced food security of water [2]. In addition to this, climate change will affect the extent and productivity of both irrigated and rainfed agriculture across the globe, increasing crop water demand and decreasing crop productivity in many regions [3]. Renewable water resources for the whole of Africa amount to about 3930 km^3 or less than 9 percent of global renewable resources [4].

Water supply is often the most critical factor limiting crop growth and yield in rainfed areas and the most expensive input of irrigated crops [5]. In the dry areas, agriculture accounts for about 80% of the total consumption of water [6]. Therefore, crop production usually requires maximizing yields on limited available water resources [7]. Regardless of the irrigation potential and water availability, small area has been grown under irrigation on state farms at lower elevations [5].

In the semiarid areas of Ethiopia, water is the most limiting factor for crop production where the amount and distribution of rainfall is not sufficient to sustain crop growth and development; an alternative approach is to make use of the rivers and underground water for irrigation [6]. Based on the total irrigated area, cropping pattern, and calendar, annual agricultural water withdrawal was estimated to be in the order of 5200 million m^3 in 2002, and the number increased in 2016 so that in Ethiopia, agricultural water withdrawal is estimated at around 9000 million m^3 [4].

Studies have shown that deficit irrigation significantly increased grain yield, ET, and WUE as compared to rainfed winter wheat [7]. However, this approach requires precise knowledge of crop response to water as drought tolerance varies considerably by growth stage, species, and cultivars. The practice of limiting water applications to drought-sensitive growth stages aims at maximizing water productivity and stabilizing, rather than maximizing, yields [8]. Even a single irrigation omission during one of the sensitive growth stages caused up to 40% grain yield losses during dry years [9]. Sesame yield was reduced by up to 6.42% when the number of irrigations was reduced from seven to five [10]. According to Oweis [8], sesame crop was affected by water deficiency and was subjected to drought stress during flowering stage and grain filling stage.

Sesame seed has a long history of use for its oil as well as for other food products such as bread and bakery items. Approximately 70% of worldwide seed production is processed into oil and meal [11] as cited by USDA plant guide. Growing drought-tolerant crops as a useful strategy in many situations and efficient use of limited water are considered [12]. Sesame is a very important crop with drought-resistant characteristics and suitable for cultivation in semiarid areas than other crops. In sesame, like other crops, grain filling period is of great importance in determining productivity. Although the grain filling period is influenced by plant genetics, environmental stresses such as drought can cause yield loss [13].

This study was carried out at the Werer Research Center with the objectives to identify the level of deficit irrigation with the combination of application methods that allow achieving optimum sesame yield and its relation with WUE to develop effective water techniques for the efficient use of irrigation water in irrigated agriculture as a means of water-saving strategies under semiarid conditions.

2. Materials and Methods

2.1. Experimental Design. The experiment was laid out in RCBD with three replications. Nine treatments, with three irrigation amounts (100%, 75%, and 50% ETc) and three irrigation methods (conventional, alternative, and fixed), were tested in an experimental plot of 5 m × 10 m. Sesame Adi variety (*Sesamum indicum*) seeds were sown manually in double rows on ridges that are 80 cm apart, with row spacing of 40 cm and plant spacing of 10 cm. The size of each plot was 10 m × 5 m and was separated from adjacent plots within the replicates by 0.5 m in addition to 0.3 m bund. The field experimentation involved deficit irrigation treatments at fixed frequency (recommended amount and interval is 100 mm every twenty-one days) and different irrigation amounts and irrigation methods.

2.2. Irrigation Treatments. The treatments consisted of three irrigation methods, viz, alternate furrow irrigation (AFI), fixed furrow irrigation (FFI), and conventional furrow irrigation (CFI) and three levels of irrigation applications (50% ETc, 75% ETc, and 100% ETc) as indicated in Table 1.

2.3. Irrigation Water Application. Establishment irrigation (as preirrigation of 150 mm) was given for all plots after planting two times. Irrigation application events were monitored through soil moisture readings. Irrigation depths (amount of water applied) were calculated through cumulative ETc values in a given period, and plots were replenished with 100%, 75%, and 50% of cumulative ETc as per the treatment to be applied. Measured amount of irrigation water was applied by using a two-inch Parshall flume.

2.4. Water Use Efficiency (WUE). Sesame yield was determined by counting randomly selected plants in each plot at maturity stage from a 50 m^2 plot, which were then harvested and threshed for grain yield determination. Actual grain yield was determined on a 12.5% moisture basis. The grain weight was used for calculating the WUE. The water use efficiency (kg/ha-mm) was calculated as stated by Sinclair et al. [14]: it is ratio of the total biomass or grain yield to water supply or evapotranspiration or transpiration on a daily or seasonal basis:

$$\text{WUE} = \frac{Y}{\text{ET}}, \quad (1)$$

where WUE is the water use efficiency (kg/ha-mm), Y is the yield (kg ha^{-1}), and ET is the evapotranspiration (mm).

2.5. Statistical Analysis. Analysis of variance (ANOVA) was performed using the general linear model procedure in SAS version 9.0 with appropriate error terms. The least significant differences at a probability level of 0.05 were calculated for mean comparisons.

3. Results and Discussion

3.1. Grain Yield. The results in Table 2 indicate that sesame yield was significantly affected by irrigation levels and application method during both main cropping and cool cropping seasons. In experiment years from 2013 to 2015, the result indicated that the grain yield was significantly affected by irrigation depth applied during the main planting season (Table 2). Therefore, a high grain yield (937.50 kg/ha) was produced from T3 (irrigation amount of 50% ETc at all growth stages with the alternate furrow application method), which provided 50 mm of irrigation water every ten days, and a grain yield of 2797.6 kg/ha was obtained from T9, which was 50% ETc applied at growth stages with the conventional furrow application method. Also, in the same

Table 1: Treatment combinations.

Irrigation amount	Treatments		
	Alternative	Fixed	Conventional
100% ETc	T1	T4	T7
75% ETc	T2	T5	T8
50% ETc	T3	T6	T9

Table 2: Sesame yield response to deficit irrigation during the main cropping season in 2013 and 2015.

Irrigation depths	Irrigation methods			Mean
	Alternate furrow	Fixed furrow	Conventional furrow	
Yield (kg/ha) response to irrigation amount and methods during the main cropping season in 2013				
100% ETc	895.83^{ba}	916.67^{a}	875.00^{ba}	895.83
75% ETc	708.33^{b}	770.83^{ba}	916.67^{a}	798.61^{b}
50% ETc	937.50^{a}	833.33^{ba}	895.83^{ba}	888.88
Mean	847.22	840.28	895.83	861.11
CV%		13.22		
R-square		0.5		
LSD$_{0.05}$	Irrigation method	Irrigation depth	Method * depth	
	NS	**	**	
Yield (kg/ha) response to irrigation amount and methods during the main cropping season in 2015				
100% ETc	2142.9^{a}	2529.8^{a}	2559.5^{a}	2410.733
75% ETc	2172.6^{a}	2619.1^{a}	2648.8^{a}	2480.17^{a}
50% ETc	2202.4^{a}	2232.1^{a}	2797.6^{a}	2410.7
Mean	2172.633	2460.33	2668.63	2433.87
CV%		16.69		
R-square		0.46		
LSD$_{0.05}$	Irrigation method	Irrigation depth	Method * depth	
	NS	NS	NS	

cropping season, a minimum yield (708.33 kg/ha) was obtained from treatment T2, which was irrigated with 75% ETc at all growth stages with the alternate furrow method, and 2142.9 kg/ha was harvested from T1, which was irrigated with 100% ETc at all growth stages with the alternate application method (Table 2). Hence, the results showed that sesame yield was significantly affected by the amount of irrigation water applied and its application method with a yield advantage over 100% ETc (Table 2). Studies show that sesame yield was reduced by up to 6.42% when the number of irrigations was reduced from seven to five irrigations [9]. The yield of sesame was affected by water deficiency and the yield decreases considerably, when a crop was subjected to drought stress during flowering stage and grain filling stage [15]. This indicated that grain yield of sesame decreases with decreasing water amount.

Combined over years analysis indicated that sesame grain yield was significantly ($p < 0.05$) affected by the amount of irrigated water applied during the main cropping season. In this study, sesame plants were irrigated with 100 mm, 75 mm, and 50 mm of water every ten days for three years (2012/13–2014/15) during the main cropping season and for two years during the main planting period. From this, the maximum mean sesame grain yield (1846.7 kg/ha) was obtained from the treatment of 100% ETc with the conventional furrow application method during the main season (Table 3). This was probably associated with water availability during grain filling and establishment (pre-irrigation of 125 mm) that was applied with no stress at all growth stages. Increasing irrigation water application between irrigation frequency and amount of water to crop decreased yield and plant growth [15]. In the same experimental season, a minimum mean sesame grain yield (1440.5 kg/ha) was harvested with the treatment of 75% ETc at all growth stages with the alternative furrow application method, which is 21.99% yield difference from full irrigation applied treatment (Table 3). According to Tantawy et al. [9], sesame yield was reduced by up to 6.42% when the number of irrigations was reduced from seven to five irrigations. Fereres and Soriano [16] stated that the level of irrigation supply under deficit irrigation permits achieving 60–100% of full evapotranspiration.

The sesame yield during the cool planting season of the experiment period 2013–2015 obtained maximum yield of 1562.5 kg/ha was harvested from T7 (full irrigation at all growth stages with the conventional furrow application method), which was irrigated with 625 mm of water during the growth period including 125 mm preirrigation (Table 4). During experiment period 2012/13, 2013/14, and 2014/15, the maximum mean sesame grain yield was obtained: 528.55 kg/ha, 1276.0 kg/ha, and 1562.5 kg/ha, which were associated with 50% ETc, 100% ETc, and the conventional furrow application method, respectively (Table 5). Research studies show that the practice of limiting water applications to drought-sensitive growth stages aims at maximizing water productivity and stabilizing, rather than maximizing, yields [17]. These expressions can also be used to estimate the range of water use within which deficit irrigation would be more profitable than full irrigation [18]. This shows the potential in alleviating the adverse effects of unfavorable rain patterns, which improves and stabilizes crop yields [6]. Also, this study shows that the minimum amount of irrigation water had an advantage over full amount of irrigation with a significant difference among treatments in the sesame grain yield (Table 5).

In the experiment year 2013, the result indicated that irrigation water significantly affects sesame yield with the method of application during the growing period which saved water up to 50% with a 20% yield increase over a full irrigation (100% ETc) with the conventional method (Table 4). This result is similar to that of a previous study which stated that applying two or three irrigations (80–200 mm) to wheat increased crop grain yields by 36 to 450% and produced similar or even higher grain yields than in fully irrigated [6]. Seasonal irrigation water amounts required for nonstressed production varied by year from 390 to 575 mm [19]. Even a single irrigation omission during one of the sensitive growth stages caused up to 40% grain yield losses during dry years [18].

Over years result shown for cool planting period in Table 4 indicate that the highest sesame grain yield (1053.89 kg/ha) was obtained from treatment T7 (100% ETc with the

TABLE 3: Sesame response to deficit irrigation and combined mean yield (kg/ha) over years (from 2013 to 2015) during the main cropping season.

Irrigation depths	Irrigation methods			Mean
	Alternate furrow	Fixed furrow	Conventional furrow	
100% ETc	1519.4^{ba}	1723.2^{ba}	1717.3^{ba}	1653.27^a
75% ETc	1440.5^b	1695.0^{ba}	1782.8^{ba}	1639.43^a
50% ETc	1570.0^{ba}	1532.7^{ba}	1846.7^a	1649.8^a
Mean	1509.96	1650.3	1782.27	1647.49
CV%		18.07		
R-square		0.90		
LSD$_{0.05}$	Irrigation method	Irrigation depth	Method * depth	
	**	NS	**	

TABLE 4: Sesame yield (kg/ha) response to deficit irrigation and combined mean yield over years during the cool cropping season.

Irrigation depths	Irrigation methods			Mean
	Alternate furrow	Fixed furrow	Conventional furrow	
100% ETc	927.46	992.18	1053.89	991.17
75% ETc	995.45	946.90	1132.02	1024.79^a
50% ETc	1023.59	928.68	1018.20	990.16
Mean	982.166	955.92	1068.04	1002.04
CV%		21.08		
R-square		0.82		
LSD	Irrigation method	Irrigation depth	Method * depth	
	**	**	**	

TABLE 5: Sesame yield response to deficit irrigation in kg/ha during the cool cropping season in 2013–2015.

Irrigation depths	Irrigation methods			Mean
	Alternate furrow	Fixed furrow	Conventional furrow	
Yield (kg/ha) response to irrigation amount and methods during the cool cropping season in 2012/13				
100% ETc	438.61	476.54	427.28	447.48
75% ETc	486.35	470.90	427.30	461.52
50% ETc	492.63	442.27	528.55	487.82
Mean	472.53	463.24	461.04	465.61
CV%		14.69		
R-square		0.41		
LSD$_{0.05}$	Irrigation method	Irrigation depth	Method * depth	
	164.87	NS	NS	
Yield (kg/ha) response to irrigation amount and methods during the cool cropping season in 2013/14				
100% ETc	1224.0	1197.9	1171.9	1197.93
75% ETc	1145.8	1119.8	1432.3	1232.63
50% ETc	1119.8	1171.9	1276.0	1189.23
Mean	1163.2	1163.2	1293.4	1206.6
CV%		10.79		
R-square		0.67		
LSD$_{0.05}$	Irrigation method	Irrigation depth	Method * depth	
	NS	**	NS	
Yield (kg/ha) response to irrigation amount and methods during the cool cropping season in 2014/15				
100% ETc	1119.8	1302.1	1562.5	1328.13
75% ETc	1354.2	1250.0	1536.5	1380.23
50% ETc	1458.3	1171.9	1250.0	1293.4
Mean	1310.77	1241.33	1449.67	1333.92
CV%		26.3		
R-square		0.43		
LSD$_{0.05}$	Irrigation method	Irrigation depth	Method * depth	
	NS	NS	NS	

TABLE 6: Combined sesame yield and WUE response over years during the cool cropping season.

Treatment	Yield (kg/ha)	WUE (kg/ha-mm)
100% ETc with alternative furrow	1164.215[c]	1.114000[bc]
100% ETc with fixed furrow	1173.464[bc]	1.068000[bc]
100% ETc with conventional furrow	1242.131[abc]	1.654000[a]
50% ETc with alternative furrow	1284.603[abc]	1.166000[b]
50% ETc with fixed furrow	1246.119[abc]	1.121333[bc]
50% ETc with conventional furrow	1170.299[c]	1.230667[b]
75% ETc with alternative furrow	1319.239[abc]	1.203333[bc]
75% ETc with fixed furrow	1392.311[a]	0.994000[c]
75% ETc with conventional furrow	1349.614[ab]	1.555333[a]
CV	1.981372	1.981372
LSD$_{0.05}$	178.0989	0.2159164

conventional application method) and minimum yield (927.46 kg/ha) was attained from treatment T1 (100% ETc with the alternative application method). During this growing season, the amount of irrigation water applied was saved up to 25% with 8.15% yield advantage than full irrigation (100% ETc) as indicated in Table 6. Similar studies on sesame showed that applying two or three irrigations (80–200 mm) produced similar or even higher grain yields than in fully irrigated [20]. Even a single irrigation omission during one of the sensitive growth stages caused up to 40% grain yield losses during dry years [18]. Much greater losses of 66–93% could be expected as a result of prolonged water stress during the development stage for cereal crops like maize [19]. Partial root-zone irrigation is the most popular and effective because many field crops and some woody crops can save irrigation water up to 20 to 30% with or without a minimal impact on crop yield [21]. The application of water below the ET requirements is termed "deficit irrigation" [6, 20].

3.2. Water Use Efficiency. Water use efficiency of 1.65 kg/ha-mm was attained, which was a maximum compared with other treatments, and 39.75% of water was consumed than treatment 8 irrigated with 75% ETC applied using the fixed furrow method during the cool planting season (Table 6). These results indicated the possibilities of considerable saving of water for sesame without any decrease in grain yield, and 0.99 kg/ha-mm water use efficiency was attained by irrigating with 75% ETC using the fixed furrow method. On the other hand, 1.55 kg/ha-mm water use efficiency was attained from treatment 9 in which 75% ETC water was applied to sesame conventionally. However, as compared to 100% ETC applied with the conventional furrow method, nearly 6.06% of water was saved by irrigating 75% ETC with the fixed furrow method (Table 6). Thus, WUE increases with irrigation amount, and water-saving techniques such as deficit level have been improved water use efficiency (WUE) with minimum yield reduction.

4. Conclusion

Deficit irrigation based on growth stages affected the yield of sesame. The irrigation amount of 100% ETc applied with the conventional application method was the indicator of good relationship with the highest yield of sesame. Therefore, deficit irrigation with the conventional furrow application technique is the best practice of water saving for the irrigated agriculture system under semiarid conditions and in similar areas to produce optimum sesame yield.

Acknowledgments

The authors would like to thank Irrigation and Drainage Department, WARC, EIAR, for their financial support of the project.

References

[1] Food and Agriculture Organization (FAO), *Climate Change and Food Security: A Framework Document*, FAO, Rome, Italy, 2007.

[2] Food and Agriculture Organization (FAO), *Statistical Yearbook 2012. World Food and Agriculture*, FAO, Rome, Italy, 2012.

[3] UNEP, *The UN-Water Status Report on the Application of Integrated Approaches to Water Resources Management*, 2012.

[4] Food and Agriculture Organization (FAO), "Irrigation water requirement and water withdrawal by country," AQUASTAT Report, 2016, http://www.fao.org/nr/water/aquastat/water_use_agr/index.stm.

[5] D. Rahmato, *Water Resources Development in Ethiopia: Issues of Sustainability's and Participation, Forum for Social Studies, Addis Ababa, Ethiopia, June 1999*, 1999.

[6] T. Oweis, H. Zhang, and P. Mustafa, "Water use efficiency of rain fed and irrigated bread wheat in a Mediterranean environment," *Agronomy Journal*, vol. 92, no. 2, pp. 231–238, 2000.

[7] P. E. Abbate, J. L. Dardanelli, M. G. Cantarero, M. Maturano, R. J. M. Melchiori, and E. E. Suero, "Climatic and water availability effects on water use efficiency in wheat," *Crop Science*, vol. 4, no. 2, pp. 474–483, 2004.

[8] T. Oweis, "Supplemental irrigation: an option for improved water use efficiency," in *Proceedings of Regional Seminar on the Optimization of Irrigation in Agriculture*, pp. 21–24, Amman, Jordan, November 1994.

[9] M. M. Tantawy, S. A. Oudu, and F.A. Khalil, "Irrigation optimization for different Sesame varieties grown under water stress condition," *Journal of Applied Science Research*, vol. 3, no. 1, pp. 7–12, 2007.

[10] M. Golestani and H. Pakniyat, "Evaluation of traits related to drought stress in sesame (*Sesamum indicum* L.) Genotypes," *Journal of Asian Scientific Research*, vol. 5, no. 9, pp. 465–472, 2015.

[11] J. B. Morris, "Food, industrial, nutraceutical, and pharmaceutical uses of sesame genetic resources," in *Trends in New Crops and New Uses*, J. Janick and A. Whipkey, Eds., pp. 153–156, ASHS Press, Arlington, VA, USA, 2002.

[12] M. J. English, J. T. Musick, and V. V. N. Murty, "Deficit irrigation," in *Management of Farm Irrigation Systems. ASAE Monograph No. 9*, G. J. Hoffman, T. A. Howell, and K. H. Solomon, Eds., pp. 631–663, American Society of Agricultural Engineers, St. Joseph, MI, USA, 1990.

[13] J. R. Frederick, J. T. Wooly, J. D. Hesketh, and D. B. Peters, "Seed yield and agronomic traits of old and modern soybean cultivars under irrigation and soil water deficit," *Field Crops Research*, vol. 27, no. 1-2, pp. 71–82, 1991.

[14] T. R. Sinclair, C. B. Tanner, and J. M. Bennnet, "Water use efficiency in crop production," *BioScience*, vol. 34, no. 1, pp. 36–40, 1984.

[15] J. Sarhadi and M. Sharif, "The effect of deficit irrigation on seasam growth, yield and yield components in drought conditions on base of sustainable agriculture," *International Journal of Farming and Allied Sciences*, vol. 3, no. 10, pp. 1061–1064, 2014.

[16] E. Fereres and M. A. Soriano, "Deficit irrigation for reducing agricultural water use," *Journal of experimental botany*, vol. 58, no. 2, pp. 147–159, 2007.

[17] S. Geerts and D. Raes, "Deficit irrigation as an on-farm strategy to maximize crop water productivity in dry areas," *Agricultural water management*, vol. 96, no. 9, pp. 1275–1284, 2009.

[18] D. Molden, *Accounting for Water Use and Productivity*, Vol. 1, IWMI, Colombo, Sri Lanka, 1997.

[19] R. Cakir, "Effect of water stress at different development stages on vegetative and reproductive growth of corn," *Field Crops Research*, vol. 89, no. 1, pp. 1–16, 2004.

[20] V. Roa and C. Raju, "Effect of soil moisture stress at different development phases on growth and yield of sesame," *Journal of Oilseeds Research*, vol. 8, no. 2, pp. 240–243, 1991.

[21] Q. Chai, Y. Gan, C. Zhao et al., "Regulated deficit irrigation for crop production under drought stress," *Agronomy for Sustainable Development*, vol. 36, no. 1, 2016.

Assessment of an Invasive Weed "Maimaio" *Commelina foecunda* in the Sesame Fields of Western Zone of Tigray, Northern Ethiopia

G. Zenawi, T. Goitom, and B. Fiseha

Tigray Agricultural Research Institute, Humera Agricultural Research Center, P.O. Box 62, Humera, Ethiopia

Correspondence should be addressed to G. Zenawi; zenawigg@gmail.com

Academic Editor: Yong In Kuk

Sesame (*Sesamum indicum* L.) is probably the most ancient oilseed. It has multiple uses; it is used as a source of food and in the pharmaceutical and cosmetics industries. The average yield of sesame in western Tigray is too low (about 400 kg/ha to 500 kg/ha) due to different factors, and weed infestation takes a lion's share. More than 80 weed species were recorded and identified as weed pests for sesame in western Tigray. "Maimaio" *Commelina foecunda* is the most troublesome weed of sesame. The purpose of this study was to assess the distribution of *C. foecunda* and quantify its infestations. The survey was conducted in 24 sesame growing areas and 48 sesame farms from the three districts of western Tigray in the 2017 production season. The survey result showed that about 91.7% of the assessed sesame farms in western Tigray were found infested with *C. foecunda*. The weed frequently appeared in Kafta Humera. And, it occurred abundantly and closely in Kafta Humera, whereas it occurred poorly and irresolutely in Tsegede. Concentrated frequency, abundance, and density of the weed were recorded by large-scale sesame producers, lower growing altitudes, and early growth stage of sesame; whereas, it was limited in the small-scale farms, higher growing altitudes, and late growth stage of sesame.

1. Introduction

Sesame (*Sesamum indicum* L.) belongs to the order Tubiflorae and family Pedaliaceae cultivated for its seed. It is probably the most ancient oilseed known and used by man, so that it is almost impossible to say with any degree of accuracy where and when its domestication took place [1]. Sesame has multiple uses; it is used as a source of food: eaten raw, either roasted or parched, or as blended oil in the form of different sweets. It is also used in the pharmaceutical industry, in the manufacture of margarine and soap, as a fixative in the perfume industry, in cosmetic industry, and as synergist for insecticides. In Ethiopia, sesame is used as cash crop, export commodity, raw materials for industries, and as a source of employment opportunity. A sizable proportion of the population, therefore, generates income from oilseed farming, trade, and processing. Area allocated to sesame both at regional and national level is increasing from time to time. The average yield of sesame in western Tigray is about 400 kg/ha to 500 kg/ha (personal communication with bureau of agriculture/BoA). There are a number of constraints that limit the productivity of sesame, like pest infestations (weeds, insects, and disease), water logging (poor drainage), lack of optimum plant population, seasonal delay, low yielding varieties, postharvest loss, poor storage facility, difference in capsule maturity, and shattering.

Weed infestation is one of the major factors limiting the yield of sesame as its seedling growth is slow during the first four weeks, making it a poor competitor at earlier stages of the crop growth [2]. The early growth period of sesame is the most critical stage at which any kind of stress can affect the economic yields. Sesame is very sensitive to weeds from emergence up to 4 WAE, which can cause >80% yield loss [3]. Moreover, Upadhyay [4] has stressed that early growth stage of sesame is slow, thus suppressing the weed growth at crop establishing is important. The sesame growing areas of western Tigray are highly infested by different broad-leaved weeds and grass weeds. A weed survey conducted by Humera Agricultural Research Center/HuARC ([5], unpublished work) revealed that more than 80 weed species were recorded and identified as weed pests for sesame. Of the weed pests, *Commelina foecunda* was the most dominant, abundant, and frequently occurred weed. This species is a weed of field crops, which grows competitively and easily. The invasive plant has a native range starting from Cameroon to Ethiopia and Botswana, Arabian Peninsula, and its synonym is *Cyanotis foecunda* (http://plantsoftheworldonline.org/taxon/urn:lsid:ipni.org:names:172477-1). According to the report by Isaac et al. [6], among the Commelina weeds, *C. benghalensis*, *C. diffusa*, and *C. communis* are the most troublesome weeds in the new world, Asia and tropical Africa. But now, *C. foecunda* is becoming the most noxious weed in north Ethiopia. The problem other than abundance and density with this weed is its difficulty to eradicate from the field if it once emerged. The weed *C. foecunda* was becoming a serious weed in around 2013 (Goitom Arafaine, personal communication). In the study area; weeds are harrowed by man power using simple tool called "Mewled" (Figure 1). The weed by its nature is very complex and uprooted, and every fragments of the weed can grow overnight, if left in the field. For this reason, the sesame fields which left free of weeds by harrowing or hand weeding become highly reinvaded within few days. A survey conducted for five years in the United States revealed that *Commelina* weeds were found as troublesome weeds in cotton, maize, and wheat production areas [7]. Therefore, a sesame field once infested with *Commelina* weeds can never be free of the weed unless weeded frequently or sprayed herbicides (if available). Broken vegetative cuttings of stems of the *Commelina* weeds are capable of rooting and reestablishing after cultivation or disking [8]. In their native geographical areas, that is, tropical Asia, Africa, and the Pacific Islands, invasive *Commelina* weeds grow as a perennial, but they can survive as an annual in temperate regions, and their seeds have variable dormancy and germination features; also, they have the capacity to redevelop from stem fragments [9]. Noxious *Commelina* weeds were first observed in the continental United States in 1928 as a common weed and now listed among the world's worst weeds, affecting more than 25 crops [8–10]. It is a rainy season weed, requiring moist soil conditions for establishment and survives dry soil conditions after establishment [11]. Germination of aerial and subterranean seeds of *C. benghalensis* occurred at 25 and 30°C, but no germination occurred below 18°C [8].

Both large-scale and small-scale sesame producers from different corners of the western Tigray rise questions about the complex weed *C. foecunda* in different workshops and discussions. Although the severity of the weed is observable, its distribution and coverage in the sesame growing area is not studied yet. The purpose of this study was to assess the distribution of *C. foecunda* and quantify its infestations in different sesame growing arrears of western Tigray.

Figure 1: The simple tool "Mewled" used for harrowing and weeding.

2. Materials and Methods

The assessment was conducted in the 2017 cropping season in sesame farm lands of the sesame growing areas of the western zone of Tigray. About 24 representative sesame growing areas were selected systematically from the three districts of western Tigray (Figure 2). Two representative unharrowed farms were selected from each growing area/locality; in general, 48 sesame farms were assessed. Within the growing area, farms were about 3–5 km apart from each other. The sample was taken by throwing 0.5 × 0.5 m quadrant five times (diagonally) in each farm. The survey was conducted in the early stage of the crop before flowering and after flowering in capsule setting stages. The *C. foecunda* in the quadrant was counted separately. In each farm, altitude and farm category (both large- and small-scale farmers are there in the study area; and at the time of sampling, each farm was noted if it was for small-farm owner or large-farm owner farmer) were recorded. According to Nkoa et al. [12], prevalence, frequency, abundance, and density were also recorded and calculated as follows:

(1) *Prevalence*: it simply measures the percentage of occurrence of the pest (*C. foecunda*) in the area (location) from where the samples have been taken:

$$P = \frac{A}{N} * 100, \quad (1)$$

where P = prevalence, A = sesame farms that have *C. foecunda*, and N = total sampled sesame farms.

(2) *Frequency* (constancy): it is the percentage of sampling units on which a particular weed species is found. It explains how often a weed species occurs in the survey area:

$$F = \frac{W}{N} * 100, \quad (2)$$

where F = frequency, W = number of occurrences of a weed species, and N = sample number or quadrant number.

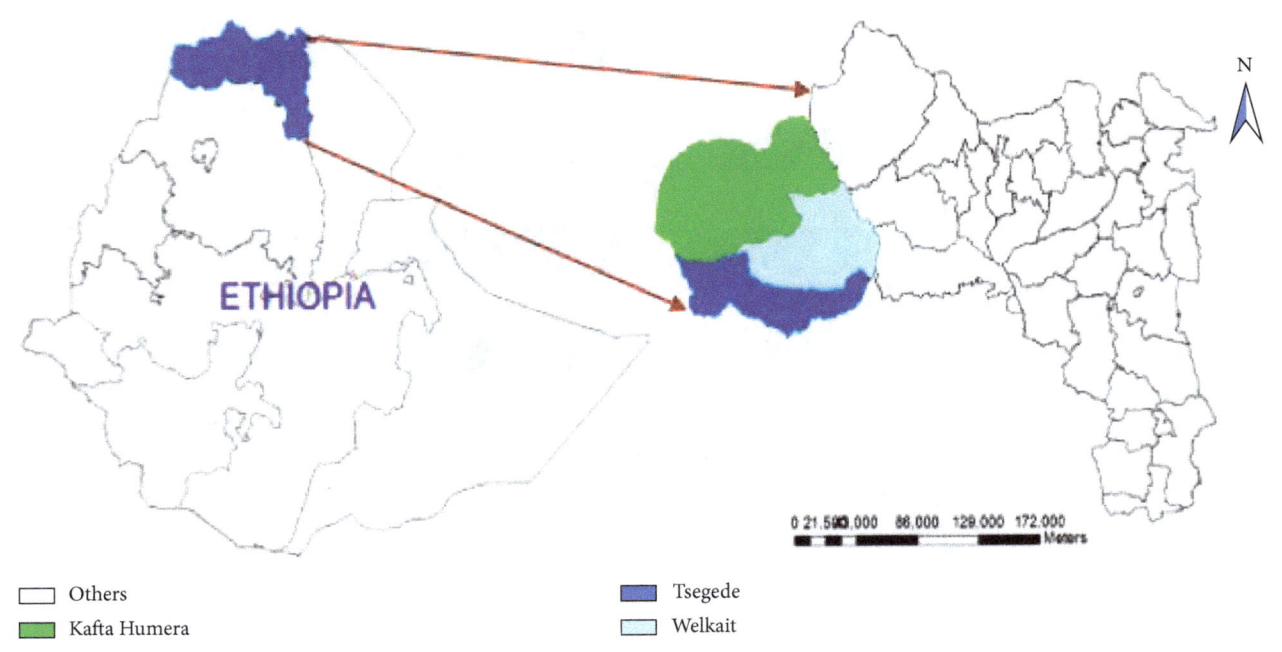

FIGURE 2: Map of the study area.

(3) *Abundance*: population density of weed species expressed as summation of the number of individuals of weed plants (*C. foecunda*) per sample number:

$$A = \sum \frac{W}{N}, \qquad (3)$$

Where A = abundance; $\sum W$ = summation of number of individual *C. foecunda* per the samples, and N = sample number.

(4) *Density*: population density of weed species expressed as the number of individuals of weed plants per unit area:

$$D = \sum \frac{W}{S_a}, \qquad (4)$$

where D = density, $\sum W$ = summation of number of individual *C. foecunda* per the samples, and S_a = sapling unit area.

3. Results and Discussion

3.1. Prevalence of C. foecunda in Western Tigray. The survey results indicated that about 91.7% (Figure 3) of the assessed sesame farms in western zone of Tigray were infested with *C. foecunda*. Sesame farms in Kafta Humera and Welkait districts were 100% positive to the weed, whereas the prevalence of the weed was about 50% in Tsegede (Figure 3). The weed *C. foecunda* did not appear in sesame fields ten years ago (personal observation) but now it has become a very problematic weed in sesame production. In 1998, *C. benghalensis* was present in Georgia but not considered a serious pest infesting cotton. However, by 2001, it had quickly become very problematic weed [13].

3.1.1. Frequency of C. foecunda. Based on the survey result, the weed frequently appeared in the districts of Kafta Humera and Welkait (>90%), whereas sparse (32.5%) occurrence was recorded in Tsegede Woreda. Particularly, *C. foecunda* was observed commonly in most of the growing areas in Kafta Humera except in Adebay and Maiweini (Table 1).

When frequency of *C. foecunda* was inspected between small- and large-scale sesame producers, more frequent observations (100%) were recorded from the large-scale sesame producers' farm, whereas about 72% was recorded from small-scale farms (Figure 4). This indicates that the weed species might be disseminated through farm equipments (tractors, disc harrow, Mewled, etc.) and natural agents (flooding, birds, etc.). Most probably, farm equipments could be the most contributors for the weed transmission among the large-scale sesame producers. For example, daily laborers use a permanent tool "Mewled" for harrowing in different farms; when weeding is completed in the first farm (in a large-scale farm), they go to another farm with their "Mewled." The same is true for tractors and disc harrows during planting time. In the other way, small-scale farmers also harrow their sesame using "Mewled" but the tool is not mobile from farm to farm, because they did it themselves using their own tools.

Shallow depth of tillage could be the other reason for higher weed frequency. Large-scale sesame producer usually use disc harrow (no more than 5 cm deep) for plowing, and the disc itself may aggravate the weeds intensity by cutting and dropping here and there, while small

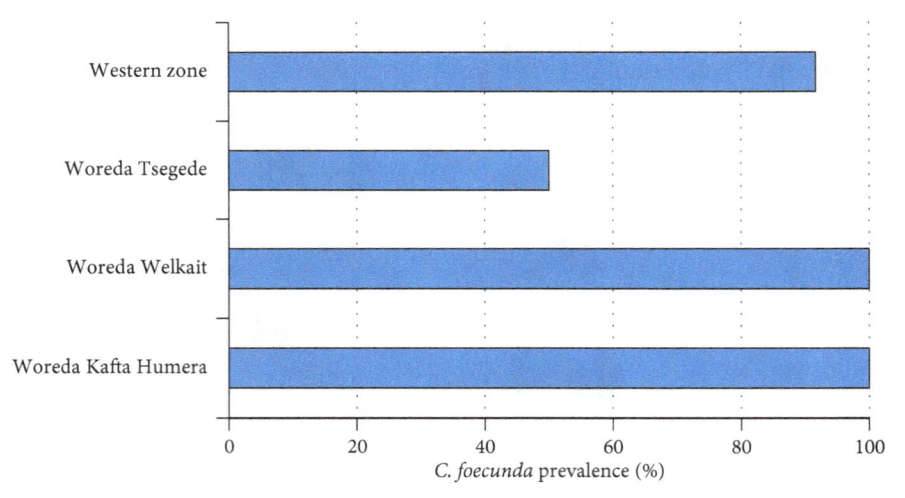

FIGURE 3: Prevalence of *C. foecunda* in sesame farms of western Tigray.

TABLE 1: Frequency, abundance, and density of *C. foecunda* in different districts and growing areas of western Tigray.

Woreda	Growing areas	Frequency (%)	Abundance	Density (m^2)
Kafta Humera	Adebay	60b	22abc	352abc
	Adigoshu	90bc	37.2a–e	595.2a–e
	Bahker	100c	46.5a–f	744a–f
	Banat	100c	71.1c–g	1137.6c–g
	Bowal	100c	45.8a–f	732.8a–f
	Central	100c	75d–g	1200d–g
	Hagre selam	90bc	19.6abc	313.6abc
	Lugdi	100c	75.3d–g	1204.8d–g
	Maikadra	90bc	21.3abc	340.8abc
	Maisegen	100c	57.4b–g	918.4b–g
	Maitemen	100c	61.3b–g	980.8b–g
	Maiweini	60b	24.1a–d	385.6a–d
	Niguara	100c	107.7gh	1723.2gh
	Rawian	100c	86.9e–h	1390.4e–h
	Shelela	100c	132.8h	2124.8h
	Sherifhamed	100c	79.6efg	1273.6efg
	Whdet	100c	102gh	1632gh
	Mean	93.5B	62.7B	1002.9B
Welkait	Maigaba	100c	95.7fgh	1531.2fgh
	Qorarit	80bc	43.2a–e	691.2a–e
	Mogue	90bc	14.6ab	233.6ab
	Mean	90B	51.2AB	818.7AB
Tsegede	Kebabo	100c	68.3c–g	1092.8c–g
	Maidelie	20a	0.3a	4.8a
	Zuriadansha	10a	0.1a	1.6a
	Rubalemin	0a	0a	0a
	Mean	32.5A	17.2A	274.8A
Grand mean		82.9	53.7	859
SE		19.33	25.02	400.2
CV (%)		23.3	46.6	46.6
LSD (0.05)		39.98	51.75	828

Means followed by the same letters are not statistically different from each other ($p < 0.01$); capital letters indicate Woreda; small letters indicate sesame growing area.

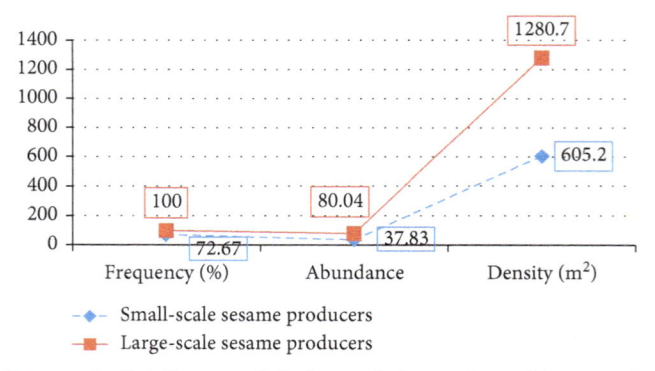

FIGURE 4: Distribution of *C. foecunda* in small- and large-scale sesame producers of western Tigray.

scale farmers most probably use oxen drawn "Mahresha" (>10 cm deep) for plowing. *Commelina* weeds can produce seeds and can regenerate from fragmented stems [9, 14]. Therefore, shallowly buried seeds and stem fragments could emerge easily and densely, whereas the deeply buried ones might not.

3.1.2. Abundance and Density of C. foecunda. Abundance and density of *C. foecunda* in western zone of Tigray were 53.7 and 859/m^2, respectively. The weed occurred richly and tightly in Kafta Humera district, whereas it was poor and irresolute in Tsegede. Growing areas with maximum abundance of the weed were Niguara (107.7), Rawian (86.9), Shelela (132.8), Sherifhamad (79.6), Banat (71.1), Lugdi (75.3), Central (75), Whdet (102), and Kebabo (68.3) (all in Kafta Humera district except Kebabo). While Rubalemin, Zuriadansha, Maidelie, Mogue, and Adebay (Welkait and Tsegede districts) have lower (<15) weed abundances. Areas with higher (>60) abundance were also with higher weed density (>1000/m^2) (Table 1). Most of the growing areas mentioned under the higher infestation of the weed are areas under semimechanized farming. And therefore, farm equipments and shallow plowing depths might have great contribution for higher weed abundance in these areas. Higher rate of stem fragments might occur in the semi-mechanized farms than the oxen plowed farms. Shallowly buried (2–4 cm) seeds and stem fragments have higher rate of germination and regrowth capacity, respectively [14], and early emerged weeds can produce seeds in about 45 days, in which they may germinate two weeks later also [8]. This means *Commelina* weeds may have two generations in a single sesame production season before the crop was going harvested. Therefore, these factors make the weed very abundant and dense in kafta Humera sesame growing areas. On the contrary, almost all small-scale farms are self-managed, and everything is done on time properly, i.e., removal of uprooted and fragmented weeds from sesame field to uncultivated land, and the deep tilling practice helps also to minimize weed population. Above ground development of the noxious *Commelina* weeds are restricted at 6 cm depth [14]. Similarly, there was no seedling emergence at soil depth of 10 cm for the noxious *Commelina* weeds [15]. In the other way, abundance and density of *C. foecunda* is higher in lower altitudes compared to higher altitudes (>1000 m) (Figure 5). Germination and development of *C. benghalensis* favor higher temperature (but below 45°C) [14]. And thus, the sesame growing areas in Kafta Humera are warmer (lower altitude) than the growing areas in Welkait and Tsegede.

3.2. Distribution of *C. foecunda* across Sesame Producers' Farm.

Concentrated frequency, abundance, and density of the weed were recorded in large-scale sesame producers, whereas it was limited in the small-scale farms (Figure 4). Therefore, *C. foecunda* was highly distributed and expanded in the farms of large-scale producers compared to the small-scale producers. This might be because of the farm equipments which are being used exchangeably among the large-scale farms and also due to the lower geographical locations of the large-scale farms which might be favorable for the weed growth. Furthermore, the farm management in general and weed management specifically in the growing areas is better in the small-scale producers' farms than the large scale producers' farms. This is because of the difficulty of managing the large-scale farms without mechanization. Most of the growing areas in Kafta Humera with higher weed infestation (Table 1) were under semimechanized production, where exchangeable farm equipment usage and shallow plowing depths are available. Germination and development of *Commelina* seeds favor higher temperature (30–35°C) [14, 16]. Moreover, the growth and reproduction rates of the weed under higher temperature were very fast [9]. And germination percentage and population stand of *C. foecunda* were very high at lower soil depth (<5 cm) [15]. Goddard et al. [17] reported that large number of mourning doves and other bird species that visit cropland help in dispersal and expansion of *C. benghalensis*.

3.3. Distribution of *C. foecunda* across Different Altitudes of the Growing Areas.

Sesame grows 500–1500 meter above the sea level in western Tigray/north Ethiopia. And the major weed of sesame *C. foecunda* was observed also to be growing in all areas in which sesame can grow (Table 1). Frequency of the weed was similar among lower, middle, and higher altitudes of sesame growing areas. But both abundance and density of

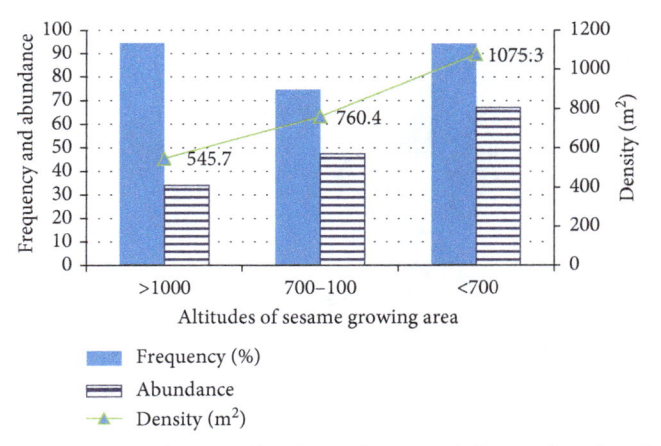

FIGURE 5: Distribution of *C. foecunda* across different altitudes of the growing areas.

C. foecunda were steeply increased from higher altitude to the lower altitudes (Figure 5). Perhaps, the weed might favor high temperature and limited moisture, since the lower altitudes areas are described with low rainfall and high temperatures. Unfortunately, most of the lower altitude growing areas are areas with large-scale or semimechanized farms. And large-scale farms were the severely infested ones (Figure 4). *Commelina* weeds (*C. benghalensis*) require higher temperature for germination, development, and fast reproduction, and it can resist moisture stress after establishment [9, 14].

3.4. Distribution of *C. foecunda* across Sesame Growth Stages.

When we examine the infestation of *C. foecunda* weed between the early and late growth stage of sesame, the early growth stage (before flower initiation) of the crop recurrently, abundantly, and densely infested compared to the late growth stage (Figure 6). Sesame is more sensitive to weeds during the first 4 weeks [3] as sesame grows very slowly in the early stages and weeds get the chance to grow vigorously. Moreover, Duary and Hazra [18] reported that sesame should be free of weeds 15 to 45 days after sowing (DAS). The authors added that 10% yield increment could be also obtained if the field is kept weed free 15 to 60 DAS. Similarly, the conventional weeding practice for sesame in the study area is weeding twice before flowering and once after flowering (optional) but not during flowering. Hence, even though the degree of severity is too different sesame could be affected by the presence of weeds up to 60 DAS. On the other way, the lower *C. foecunda* infestation in the late growth stage (capsule setting) of sesame could be because of weed removal or hand weeding. As already mentioned above, weeding practice in the study area starts two weeks after emergence (WAE) up to six WAE (sometimes up to eight WAE but optional). In addition, the weed starts dying after 45 days due to senescence. And these reasons minimized abundance and density of the weed to a certain level but could not exterminate. An experiment conducted in the same area (Humera) stated that 2-3 hand weeding is optimal

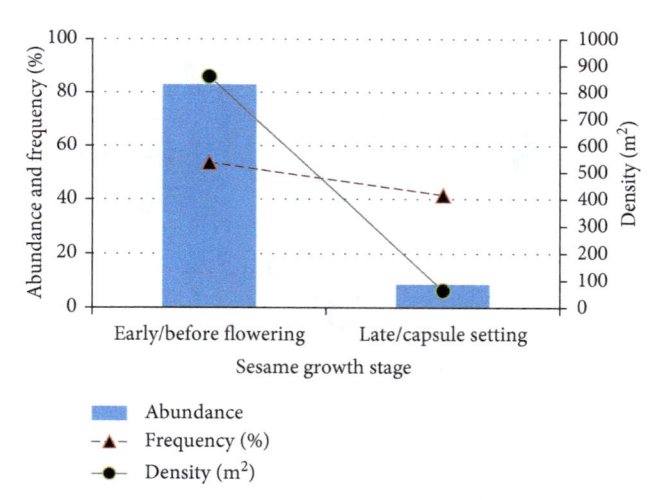

FIGURE 6: Distribution of *C. foecunda* in early and late sesame growing stages.

to keep the farm clean and can save more than 80% sesame yield losses [3]. But, occurrence of *C. foecunda* infestation still in the farm after two hand weeding was an indication of how the weed is worthy.

4. Conclusion

The survey result showed that about 91.7% of the assessed sesame farms in western Tigray were found infested with *C. foecunda*. Frequency, abundance, and density of *C. foecunda* in western Tigray were found to be 82.9%, 53.7, and 859 m^{-1}, respectively. The weed frequently appeared in Kafta Humera compared to Welkait and Tsegede. And, it occurred abundantly and closely in Kafta Humera growing areas also, whereas it occurred poorly and irresolutely in Tsegede. Concentrated frequency, abundance, and density of the weed were recorded in large-scale sesame producers' farms, in lower sesame growing altitudes, and early growth stage of sesame. And, it was limited in the small-scale farms, in higher sesame growing altitudes, and later growth stage of sesame. Since, the noxious weed has wide area coverage and distribution in western Tigray, its control method has to be developed soon.

Acknowledgments

The authors are grateful to the Tigray Agricultural Research Institute and Ethiopian Institute of Agricultural Research for the financial support, and they would like to extend their acknowledgements to all researchers of Humera Agricultural Research Center for their nice matching and expertise.

References

[1] E. A. Weiss, *Oilseed Crops*, Blackwell Science, Hoboken, NJ, USA, 2000.

[2] M. Bennett and B. Katherine, "Code, sesame recommendation for the northern territory agnote," vol. 657, no. 22, pp. 1–4, 2003.

[3] A. Mizan, "Estimation of critical period for weed control in sesame (*Sesamum indicum* L.) in northern Ethiopia," *Ethiopian Journal of Science and Technology*, vol. 2, no. 1, pp. 59–66, 2011.

[4] U. C. Upadhyay, *Weed Management in Oilseed Crops in Oilseed Production Constraints and Opportunities*, H. C. Srivastava, S. Bhasharan, B. Vatsya, and K. K. G. Menon, Eds., Oxford and IBH Publishing Company, New Delhi, India, 1985.

[5] HuARC, *Survey and Identification of the Major Weeds in Sesame Farm in the Sesame Growing Areas of Ethiopia*, Tigray Agricultural Research Institute, HuARC, Mek'ele, Ethiopia, 2017.

[6] W.-A. Isaac, Z. Gao, and M. Li, "Managing *Commelina* species: prospects and limitations," in *Herbicides-Current Research and Case Studies in Use*, IntechOpen Limited, London, UK, 2013.

[7] T. M. Webster and R. L. Nichols, "Changes in the prevalence of weed species in the major agronomic crops of the Southern United States: 1994/1995 to 2008/2009," *Weed Science*, vol. 60, no. 2, pp. 145–157, 2012.

[8] T. M. Webster, M. G. Burton, A. Stanley Culpepper, A. C. York, and E. P. Prostko, "Tropical spiderwort (*Commelina benghalensis*): a tropical invader threatens agroecosystems of the southern United States," *Weed Technology*, vol. 19, no. 3, pp. 501–508, 2005.

[9] M. K. Riar, D. S. Carley, C. Zhang et al., "Environmental influences on growth and reproduction of invasive *Commelina benghalensis*," *International Journal of Agronomy*, vol. 2016, Article ID 5679249, 9 pages, 2016.

[10] A. S. Culpepper, J. T. Flanders, A. C. York, and T. M. Webster, "Tropical spiderwort (*Commelina benghalensis*) control in glyphosate-resistant cotton," *Weed Technology*, vol. 18, no. 2, pp. 432–436, 2004.

[11] V. Kaul, N. Sharma, and A. K. Koul, "Reproductive effort and sex allocation strategy in *Commelina benghalensis* L., a common monsoon weed," *Botanical Journal of the Linnean Society*, vol. 140, no. 4, pp. 403–413, 2002.

[12] R. Nkoa, M. D. K. Owen, and C. J. Swanton, "Weed abundance, distribution, diversity, and community analyses," *Weed Science*, vol. 63, no. 1, pp. 64–90, 2015.

[13] T. M. Webster and G. E. MacDonald, "A survey of weeds in various crops in Georgia," *Weed Technology*, vol. 15, no. 4, pp. 771–790, 2001.

[14] M. K. Riar, J. F. Spears, J. C. Burns, D. L. Jordan, C. Zhang, and T. W. Rufty, "Persistence of Benghal dayflower (*Commelina benghalensis*) in sustainable agronomic systems: potential impacts of hay bale storage, animal digestion, and cultivation," *Agroecology and Sustainable Food Systems*, vol. 38, no. 3, pp. 283–298, 2014.

[15] M. Matsuo, H. Michinaga, H. Terao, and E. Tsuzuki, "Aerial seed germination and morphological characteristics of juvenile seedlings in *Commelina benghalensis* L," *Weed Biology and Management*, vol. 4, no. 3, pp. 148–153, 2004.

[16] M. H. Sabila, T. L. Grey, T. M. Webster, W. K. Vencill, and D. G. Shilling, "Evaluation of factors that influence Benghal dayflower (*Commelina benghalensis*) seed germination and emergence," *Weed Science*, vol. 60, no. 1, pp. 75–80, 2012.

[17] R. H. Goddard, T. M. Webster, R. Carter, and T. L. Grey, "Resistance of Benghal dayflower (*Commelina benghalensis*) seeds to harsh environments and the implications for dispersal by mourning doves (*Zenaida macroura*) in Georgia, USA," *Weed Science*, vol. 57, no. 6, pp. 603–612, 2009.

[18] B. Duary and D. Hazra, "Determination of critical period of crop-weed competition in sesame," *Indian Journal of Weed Science*, vol. 45, no. 4, pp. 253–256, 2013.

Screening Selected *Solanum* Plants as Potential Rootstocks for the Management of Root-Knot Nematodes (*Meloidogyne incognita*)

Benjamin A. Okorley, Charles Agyeman, Naalamle Amissah, and Seloame T. Nyaku

Department of Crop Science, University of Ghana, P.O. Box LG 44, Legon-Accra, Ghana

Correspondence should be addressed to Seloame T. Nyaku; seloame.nyaku@gmail.com

Academic Editor: Maria Serrano

Root-knot nematodes (RKNs) (*Meloidogyne* spp.) represent agricultural pest of many economic crops, including tomatoes and potatoes. They advance a complex parasitic relationship with roots of tomato plants leading to modification of host structural and physiological functions in addition to significant yield loss. Resistance in solanaceous plants to RKNs has been identified and associated with the possession of *Mi* gene. The reaction of four *Solanum* rootstocks (*S. aethiopicum* L., *S. macrocarpon* L., *S. lycopersicum* L."Mongal F1," and *S. lycopersicum* L. "Samrudhi F1") was evaluated in pots and in a natural *Meloidogyne* spp.-infested field in a two-year trial (2015–2016), to identify RKN-resistant rootstock(s), which can be utilized in tomato grafting as a management measure against these nematodes. A rootstock's reaction to RKNs was assessed using root gall scores (GSs), egg count/g of root, and reproductive factors (Rfs) at the end of 6 and 12 weeks after transplanting (wat) in infested fields, respectively. *Solanum macrocarpon*, *S. aethiopicum*, and Mongal F1 showed tolerant responses with reduced root galling and low to high reproductive factors in pot and field experimentation. Although Samrudhi F1 was resistant in both pot and field trials and consistently decreased nematode root galling (<1.00) and reproduction (Rf < 1.00), it failed to significantly increase yield, as compared with the highest yield obtained by the tolerant rootstock, Mongal F1 (870.3 and 1236.6 g/plant, respectively). Evaluation of the four rootstocks against four (0, 500, 1,000, and 5000) RKN inocula levels (Juveniles) showed no significant differences among the growth parameters (fresh and dry shoot and root weights). Root-knot nematode-susceptible tomato varieties, for example, Pectomech F1, a popular tomato variety in Ghana, can be grafted onto the RKN-resistant and RKN-tolerant rootstocks for increased yields.

1. Introduction

Tomato (*Solanum lycopersicum* L.) is an annual crop of the Solanaceae family, and the second most widely consumed vegetable after potatoes [1]. Tomato fruits are highly nutritious, and constitute a rich source of vitamins A, C, and E, and essential minerals, namely, potassium, phosphorus calcium, magnesium, and iron [2, 3]. Tomato and its products contain the antioxidant lycopene, which reduces cancers and development of atherosclerosis [4, 5]. The large diversity in the crop makes it adaptable to different climatic conditions, ranging from temperate to tropical environments, production methods, and uses [6]. Ghana has a young tomato industry (production estimated at 366, 722 tonnes from 47,000 hectares of land [7]), with the potential to increase production in the savannah and transitional zones because of the derived income and nutritional benefits. Pectomech F1, Power Rano, and "Wosowoso" are popular varieties cultivated for high yield and quality fruits. However, domestic production levels over the years have been low, mainly due to its vulnerability to pest and diseases, including plant-parasitic nematodes [8].

Plant-parasitic nematodes are economic pests of agriculture worldwide, with more than 3000 plants species as host [9–11]. Global losses associated to root-knot nematodes (RKNs) alone from 75 countries as at 2000 was valued at $121 billion [12] and cost about $500 million for their annual control [13, 14]. In Ghana, RKNs are responsible for 33% of vegetable losses per season [15], in which 73–100% are realized in tomato production alone [16]. The RKNs thus impact a key limitation to tomato production, as the crop is considered the most vulnerable in the tropics [17, 18].

Currently, crop rotation, soil solarization, nematode-free seedlings, and the use of nematicides are adopted by Ghanaian farmers for managing nematodes [16, 19]. The withdrawal of extensively used fumigants such as ethylene dibromide, because of its carcinogenic nature and its ability to contaminate groundwater [12, 20], has necessitated the adoption and worldwide use of tolerant/resistant rootstocks, as an alternative for nematode control because they are safe to both farmers and the environment [21].

Nematode-resistant plants or rootstocks have the intrinsic ability, to be unaffected significantly upon nematode attack, and will greatly contribute to reducing nematode infestations in tomato fields [22–24]. Grafting utilizing resistant rootstocks has proven to effectively manage RKNs and improve yield in tomato and eggplant cultivated in naturally infested nematode soils [25]. Resistance to RKN species (*M. incognita*, *M. javanica*, and *M. arenaria*) is conferred by a single dominant *Mi* gene, introduced from the wild relative *S. peruvianum* accession (P.I. 128657), through embryo rescue technique, into commercially cultivated tomatoes [26, 27]. Dhivya et al. [28] reported resistant and susceptible reaction, in wild *Solanum* plants after assessing their response to RKN. In another study, 33 tomato genotypes were evaluated against RKN, and variable responses in gall development and nematode reproduction existed [29]. Plants with the *Mi* gene also effectively increased yields to about ten-fold, compared with susceptible plants at high nematode inoculum levels of about 200, 000 eggs/plant [24]. Solanaceous plants, for example, *S. aethiopicum*, *S. macrocarpon*, *S. torvum*, and *S. lycopersicum* cultivars, are commercially available, and common among local farmer seed stocks in Ghana. However, the use of resistant rootstocks in tomato cultivation remains less explored, due to the unknown response of this *Solanum* spp. to RKNs. Grafting of tomato scions with superior traits onto RKN-resistant rootstocks will help manage this biotic stress in a healthy and environmentally friendly manner [30], reduce production cost, and improve yields [31].

The use of resistant *Solanum* plants in grafting experiments in Ghana is in its infancy; therefore, there is a need to identify sources of resistance in the available *Solanum* rootstocks for managing the RKN problem in tomato fields. This study was thus initiated to screen and identify potential rootstock(s) among four selected *Solanum* plants for resistance to *M. incognita*.

2. Materials and Methods

2.1. Plant Materials. Seeds of the test rootstocks, *S. aethiopicum* and *S. macrocarpon* "Gboma," were obtained from the Department of Crop Science, University of Ghana. *Solanum lycopersicum* "Mongal F1" and *S. lycopersicum* "Samrudhi F1" were supplied by Agriseed Company and East-West Seed Ltd., respectively.

2.2. Sowing of Rootstock Seeds. The seeds were sown in trays containing oven-sterilized soil. *Solanum macrocarpon* and *S. aethiopicum* were sown ten days earlier than the other tomato rootstocks because of their slow growth rate. A week after seedling emergence, N-P-K: 15-15-15 (6 g/L) nutrient solution was applied once every week by immersing in a bath for 5-6 minutes. A shade cover was placed over the seedlings to minimize the solar radiation on the leaves. Suncozeb 80 WP at 20 g/L of H_2O was sprayed, until runoff close to the crown of the seedlings, to prevent damping-off on seedlings. Two weeks before transplanting the seedling, the shade cover was completely removed to allow them harden-off.

2.3. Experimental Area (Field Trials). The field experiments were conducted at the Biotechnology and Nuclear Agricultural Research Institute (BNARI) of Ghana Atomic Energy Commission, Legon-Accra on a natural RKN-infested field. The experimental site had a sandy loam soil, humid tropical climate with average low-high day temperatures of 25–34°C (dry season) and 25–29°C (rainy season), pH ≈ 6.4, and a history of being used continuously for growing vegetables, thus making the site a hot spot for RKNs. Laboratory and nursery activities were carried out in the Department of Crop Science, University of Ghana.

2.4. Experimental Design (Field Trials). The experimental field was slashed and debris collected; then, three beds each measuring 23.2 m × 2.0 m were raised and pegged at a planting distance of 80 cm × 40 cm. A randomized complete block design (RCBD) with three replicates partitioned by two alleys of 0.5 m each was used. A block contained five plots, each measuring 4.0 m × 1.6 m. Thirty (30) plants constituted a plot, with ten plants in a row and three plants between rows. Eight middle row plants were used as record plants, from which data were taken. The experimental setup was repeated for the succeeding trial in 2016. The layout for the pot experiments was a 4 × 4 factorial, arranged in a split plot design with three replicates.

2.5. Agronomic Practices and Parameters Taken on Tomato Plants (Field Trials). Four-week-old seedlings were transplanted (one per stand), and a starter solution (N-P-K: 15-15-15 at 6 g/L of water) applied at 100 mL/plant. Two (2) kg N.P.K was mixed with 1 kg NH_3 and was also applied at 15 g/plant as a side dressing. Agricombi (fenitrothion 30% + fenvalerate 10%) were sprayed twice, at a rate of 50 mL/L of H_2O, to control whiteflies, aphids, and red spider mites. The plants were cared for by regular watering and weed control. Plant height, plant girth, and chlorophyll contents were taken 5, 7, and 9 weeks after seedling transplant.

2.6. Experimental Design (Pot Trials). The layout for the pot experiment was a 4 × 4 factorial experiment, with a control arranged in a split plot design with three replicates, each experimental plot consisting of 30 plants. Root-knot nematode inoculums (Juveniles) were prepared by extracting nematode eggs from the roots of infested plants, and hatched into juveniles [32]. These were then concentrated into the various RKN inoculum levels (0, 500, 1,000, and 5,000). The plants (rootstocks) were then inoculated with the juveniles, by making a small hole or depression in the soil, at the base plants after which the required inoculum level was gently poured into the hole and covered.

2.7. Nematode Analysis. Soil samples taken from each plot were used to determine the RKN populations, before transplanting (initial nematode population Pi), six weeks after transplanting (P6), and after harvesting (Pf) in each trial. Six random soil samples per plot were taken with a soil augur, and bulked together into a well-labeled plastic bag. A subsample of 200 cubic centimeter (cc) was used in soil extraction, through the sieving and sucrose centrifugation method [33]. Reproductive factors (Rfs) for each treatment plot, which is a ratio of the final to initial population (Pf/Pi), were used to determine the rate of nematode multiplication in the soil. Gall scoring and nematode egg extraction were carried out 6 and 12 weeks after transplanting rootstocks, to estimate RKN gall development and egg production on the host. Sampled roots were washed separately and air-dried for 5 minutes. Galls on each root system were scored on a scale of 0–10 (no damage to severe damage), using the severity rating chart by Bridge and Page [34]. Roots were then cut into pieces, and nematode eggs extracted using 10% sodium hypochlorite (NaClO) [32]. Eggs obtained were counted under the compound microscope (Exacta–OptechBiostar B5P, Germany) (magnification x100) and recorded. The mean gall scores (GSs), egg count per gram of root, and reproductive factors (Rfs) obtained from the two field trials were used as the basis to evaluate the resistance status of the rootstocks. Rootstocks were classified as resistant when their root GS < 2 and Rf < 1, tolerant when GS < 2 and Rf ≥ 1, or susceptible when GS ≥ 2 and Rf > 1 [35, 36].

2.8. Statistical Analysis. Data collected on agronomic parameters (plant height and plant girth), chlorophyll content, yield parameters, and plant biomass content were subjected to analysis of variance (ANOVA), using Genstat 12th edition software [37], and significant means separated, using least significant difference (LSD) at 5%. Where necessary, data were transformed using the equation Log $(x + 1)$, for normality.

3. Results

3.1. Root-Knot Nematode Gall Development, Egg Formation, and Reproduction among Selected Rootstocks Six and 12 Weeks after Transplanting Rootstocks. Four *Solanum* rootstocks were evaluated for resistance to RKNs in a naturally infested field, in two consecutive trials (dry season of 2015 and rainy season of 2016). Data on the RKN gall development and reproduction among the rootstocks for the two-year trial were highly comparable. Six weeks after transplanting the rootstocks, lower gall scores ranging from 0.33 to 0.67 were observed on *S. macrocarpon*, Samrudhi F1, and Mongal F1 (Table 1). In year 2, *S. macrocarpon* had the least root gall score (0.00), with a slight increase for Samrudhi F1 (0.17) compared with the others (>0.50), six weeks after transplanting the rootstocks. Significant differences were absent among the rootstocks for eggs/g of root, six weeks after transplanting in both years; however, differences were observed among the eggs numbers twelve weeks after transplanting in both years. The recovered number of eggs/g of root ranged from 2 to 14,714 in Samrudhi F1 and *S. aethiopicum*, six and twelve weeks after transplanting the rootstock seedlings for year 1. In year 2, however, eggs/g of root ranged from 20 to 11,280 in Samrudhi F1 and *S. aethiopicum*, six and twelve weeks after transplanting the rootstock seedlings. Nematode reproductive factors (Rfs) on the various rootstocks also ranged from 0.07 to 1.57 for Mongal F1 and *S. aethiopicum*, 6 and 12 weeks after rootstock transplant, respectively, in year 1. In year 2, however, Rf ranged from 0.45 to 2.37 for Samrudhi F1 and *S. aethiopicum* six and twelve weeks after rootstock transplant, respectively.

Within the pot experiments, significant differences did not exist among the rootstocks in relation to the various RKN inocula levels (500, 1000, and 5000), for mean gall scores and egg counts six weeks after treatment application (Table 2). However, significant differences were noted for mean gall scores and egg counts, twelve weeks after treatment application. The RFs for all rootstocks were below 1.

There were no significant differences among the various plant growth parameters for weeks 6 and 12, among the rootstocks for the four RKN inocula levels applied (Table 3). Generally, dry shoot and root weights were lower compared with the fresh shoot and root weights.

3.2. Mean Plant Height, Plant Girth, and Chlorophyll Contents in Rootstocks to Root-Knot Infection Five, Seven, and Nine Weeks after Transplanting Rootstocks. Growth performances varied greatly among the rootstocks and between the dry and wet season trials. Significant increases ($P \leq 0.05$) in plant height, girth, and chlorophyll contents were recorded for *S. aethiopicum* during the dry season (Figures 1(a1), 1(b1), and 1(c1)). Conversely, the vegetative growth of the three tomato rootstocks were robust in the rainy season, relative to *S. aethiopicum* and *S. macrocarpon*, but began diminishing during fruiting or after the seventh week (Figures 1(a2), 1(b2), and 1(c2)). Aside having the thickest stem, *S. macrocarpon*'s chlorophyll content was also significantly increased ($P \leq 0.05$), compared with the other rootstocks screened against RKNs (Figure 1(c2)).

3.3. RKN Infestation on the Different Rootstocks Yield Parameters. Overall, *S. macrocarpon* and *S. aethiopicum* showed significantly ($P \leq 0.05$) increased number of fruits (10 and 12/plant) and high yield performances (227.50 g and 147.61 g/plant, respectively) in the dry season (year 1). Higher fruit yields were obtained for the three tomato rootstocks in the rainy season (year 2). The number of fruits/plant increased significantly ($P \leq 0.05$), in both Mongal F1 (20) and *S. aethiopicum* (19). Fruit yield/plant for Mongal F1 (1236.60 g) greatly increased ($P \leq 0.05$) relative to other three rootstocks screened against RKNs. Furthermore, the weight of individual fruits/plant for Samrudhi F1

TABLE 1: Initial and final nematode count, mean gall scores, egg count/g of root, and reproductive factors (RFs) of RKNs after 6 and 12 weeks of transplanting rootstocks in 2015 (year 1) and 2016 (year 2) for field experiments.

Treatment	Nematode count/200 cm^3 of soil			Mean gall scores (0–10)		Egg count/gram of root (transformed)*		Reproductive factor (Pf/Pi)		^{y}Reaction
Weeks	Initial (Pi)	Six (Pf)	Twelve (Pf)	Six	Twelve	Six	Twelve	Six	Twelve	
Year 1 (dry season)										
Sam	133.00	85.33	48.67	0.50^{a*}	0.33^a	16.30 (1.06)^a	54.33 (1.23)^a	0.64	0.36	R
Mon	839.33	59.67	99.00	0.67^a	3.00^{ab}	14.30 (0.90)^a	1047.67 (2.33)^{ab}	0.07	0.12	T
S. M	197.67.00	46.67	24.00	0.33^a	0.67^a	2.00 (0.43)^a	1438.33 (3.06)^{bc}	0.24	0.12	T
S. A	255.33	112.67	400.00	1.67^a	5.17^b	17.00 (0.86)^a	14714.00 (4.07)^c	0.44	1.57	T
Year 2 (rainy season)										
Sam	397.00	178.00	227.00	0.17	0.67	20.00^a	900.00 (2.66)^a	0.45	0.57	R
Mon	220.00	176.00	331.00	0.50	1.67	50.00^a	5145.00 (3.68)^b	0.80	1.50	T
S. M	472.00	244.00	515.00	0.00	1.33	33.00^a	4267.00 (3.58)^b	0.52	1.09	T
S. A	239.00	286.00	567.00	0.50	2.17	48.00^a	11280.00 (3.91)^c	1.20	2.37	T

Gall scores: 0 = no knots on roots; 10 = all roots severely knotted, plant usually dead; rootstocks: Sam = Samrudhi F1; Mon = Mongal F1; S. M = S. macrocarpon; S. A = S. aethiopicum. ^{y}Reaction of rootstocks derived from 6 to 12 weeks after transplanting rootstocks (R = resistance and T = tolerance). *Means having different letters in a column differed significantly ($P \leq 0.05$). *Log ($x + 1$) transformed.

TABLE 2: Response of four rootstocks to four RKN inoculum levels on gall scores, egg count per gram of root, and reproductive factors 6 and 12 weeks after inoculation in pot experiments.

Rootstocks	Initial inoculum	Mean gall score (1–10)		Nematode count per 200 cc of soil (Pf)		Egg count per gram of root		Reproductive factor (Pi/Pf)		Reaction
		6 weeks	12 weeks	6 weeks	12 weeks	6 weeks	12 weeks	6 weeks	12 weeks	
Mon	500	0.51^{a*}	0.00^a	48.30	41.28	6.50^a	5.94^a	0.09	0.08	T
SA	500	0.22^a	0.20^{ab}	125.60	60.72	9.60^a	5.11^a	0.25	0.12	T
Sam	500	0.02^a	0.10^a	46.60	4.00	2.60^a	5.44^a	0.09	0.22	R
SM	500	0.38^a	0.00^a	107.90	108.56	9.60^a	3.00^a	0.22	0.01	T
Mon	1000	1.11^a	0.10^a	62.20	20.56	12.8^a	3.89^a	0.06	0.02	T
SA	1000	0.67^a	0.00^a	192.30	93.89	12.2^a	8.56^b	0.19	0.09	T
Sam	1000	0.50^a	0.00^a	124.30	2.16	7.00^a	0.15^c	0.09	0.11	R
SM	1000	0.44^a	0.00^a	86.10	108.44	15.3^a	6.33^a	0.12	0.00	T
Mon	5000	0.33^a	0.70^c	91.60	9.39	7.20^a	4.33^a	0.02	0.00	T
SA	5000	0.72^a	0.20^{ab}	158.40	90.33	11.9^a	5.00^a	0.03	0.02	T
Sam	5000	1.00^a	0.10^a	84.70	13.66	8.50^a	6.33^a	0.04	0.01	R
SM	5000	0.55^a	0.00^a	206.00	71.00	9.60^a	1.15^c	0.02	0.00	T

Sam = *Solanum lycopersicum* "Samrudhi F1," Mon = *Solanum lycopersicum* "Mongal F1," SM = *Solanum macrocarpon*, and SA = *Solanum aethiopicum* (R = resistance and T = tolerance). *Means having different letters in a column differed significantly ($P \leq 0.05$).

(66.95 g) was on the average heavier compared with the other rootstocks (Table 4).

3.4. Effects of Root-Knot Nematode Infestation on Fresh and Dry Weights of Shoot and Roots of the Rootstocks.

Twelve weeks after transplanting, *S. aethiopicum* and *S. macrocarpon* accumulated significant ($P \leq 0.05$) amounts of plant biomass, with respect to its fresh shoot (331.40 g and 159.10 g), and root weight (39.72 g and 14.05 g), obtained in the dry season trial (Table 5). In the subsequent trial, however, the tomato rootstocks picked up steadily the fresh shoot weight of *S. macrocarpon* (357.39 g) and Mongal F1 (325.28 g), which were significantly increased ($P \leq 0.05$), compared with the other rootstocks (Table 5). Significant differences also existed among the dry shoot and root weights for both year trials, and *S. aethiopicum* produced the highest dry weights (16.81 g and 19.43 g) for years 1 and 2, respectively.

4. Discussion

The reproductive factors (Rfs) of the various rootstocks together with their mean gall scores (GS) provided an estimate of host suitability, to support nematode reproduction and were used to verify host resistance [38]. A rootstock's reaction to RKN resulting to GS < 2 and Rf < 1, GS < 2 and Rf ≥ 1, and GS ≥ 2 and Rf > 1 was identified as resistant, tolerant, and susceptible, respectively [35].

There is evidence to show that fecundity increases in more vulnerable plant hosts than in resistant or tolerant hosts [39]. Sobczak et al. [40] also described a stalled response of the tomato *Hero*-gene, against invading cyst

TABLE 3: Fresh and dry shoot and root weights in four rootstocks to three RKN inocula levels (0, 500, 1000, and 5000) 6 and 12 weeks after treatment application.

Rootstocks	Initial RKN inocula levels	Fresh root weight (g) 6 weeks	Fresh root weight (g) 12 weeks	Dry root weight (g) 6 weeks	Dry root weight (g) 12 weeks	Fresh shoot weight (g) 6 weeks	Fresh shoot weight (g) 12 weeks	Dry shoot weight (g) 6 weeks	Dry shoot weight (g) 12 weeks
Mon	0	5.58^{a*}	6.32^a	1.11^a	1.69^a	50.10^a	53.78^a	5.05^a	7.91^a
SA	0	7.19^a	6.06^a	1.82^a	2.34^a	45.70^a	55.69^a	5.34^a	7.69^a
Sam	0	6.82^a	6.40^a	0.94^a	3.91^a	43.30^a	42.30^a	7.09^a	8.69^a
SM	0	13.70^a	9.40^a	4.57^a	3.68^a	40.60^a	52.16^a	11.09^a	6.21^a
Mon	500	8.12^a	4.34^a	2.13^a	2.55^a	48.80^a	57.73^a	6.16^a	7.40^a
SA	500	9.21^a	8.62^a	2.64^a	2.26^a	48.80^a	55.99^a	6.96^a	7.05^a
Sam	500	5.23^a	5.36^a	1.21^a	4.17^a	47.40^a	34.62^a	6.18^a	6.82^a
SM	500	12.20^a	12.85^a	4.38^a	3.92^a	62.00^a	51.32^a	6.73^a	6.75^a
Mon	1000	3.28^a	7.13^a	1.36^a	1.94^a	37.70^a	42.94^a	9.56^a	9.12^a
SA	1000	15.65^a	13.21^a	3.48^a	2.78^a	46.70^a	45.62^a	7.49^a	4.91^a
Sam	1000	3.08^a	10.97^a	0.73^a	3.21^a	26.20^a	34.07^a	10.29^a	7.78^a
SM	1000	15.19^a	15.24^a	5.13^a	3.10^a	36.80^a	43.68^a	7.18^a	8.79^a
Mon	5000	7.50^a	10.08^a	1.94^a	2.43^a	50.70^a	38.73^a	7.28^a	7.81^a
SA	5000	16.52^a	15.70^a	11.46^a	4.47^a	54.40^a	52.32^a	8.08^a	7.14^a
Sam	5000	8.68^a	15.12^a	1.31^a	2.72^a	49.20^a	46.15^a	4.54^a	6.75^a
SM	5000	12.60^a	15.26^a	6.56^a	3.59^a	45.70^a	41.58^a	8.71^a	7.29^a

*Means having different letters in a column differed significantly ($P \leq 0.05$).

nematode, leading to the death of nematodes at the late J2 stage, rather than the average 12 hours postinfection response [26].

Wehner et al. [41] recounted 12 weeks postinoculation nematode, counts as the best option for inferring the response of rootstocks, as compared with six weeks nematode counts. Nematodes cause severe damage depending on their population densities, host species, temperature, and the period of incubation, to allow significant differentiation between resistant and susceptible rootstocks. A remarkable increase in nematode damage, along with reproduction, was observed 12 weeks after transplanting in both trials in S. aethiopicum. However, some rootstocks (Samrudhi F1, Mongal F1, and S. macrocarpon) suppressed nematode activities, indicating the existence of genetic variability among the rootstocks tested. In a previous study, Rahman et al. [42] observed the response of S. torvum and S. sisymbriifolium as resistant to RKNs. Similarly, Dhivya et al. [28] described S. torvum and S. aethiopicum as poor hosts of RKNs, whereas the former was found tolerant by Ioannou [25]. In our study, S. aethiopicum could be classified as tolerant to RKNs, although had high root gall development (GS = 2.17–5.17), with increased reproduction (Rf = 1.58–2.37), the severest symptoms were observed during the dry season. This response could probably be due to predominant elevated soil temperatures, leading to the permanent breakdown of nematode resistance Mi gene [43, 44].

Unlike the resistance found in Samrudhi F1 (GS = 0.33–0.67 and Rf = 0.36–0.57), S. macrocarpon and Mongal F1 were found promising because of their potential in reducing nematode damage (GS = 0.67–1.33 and 1.67–3.00, respectively). Based on nematode reproduction on a host, their response could best be described as tolerant because population levels of nematodes recovered were still high (Rf > 1) in field experiments, but below 1 in pot experiments. Similarly, Lopez-Perez et al. [45] documented a tolerant response in resistant Mi gene-bearing rootstock Beaufort, for supporting RKN reproduction, with higher yields compared with a susceptible genotype (Blitz). In another test, where 33 tomato genotypes were assessed by Jaiteh et al. [29], Mongal T-11 and Mongal F1 number 5 emerged resistant (GS = 3.25 and Rf = 0.71) and susceptible (GS = 7.25 and Rf = 2.47), respectively. Tzortzakakis et al. [46] also reported S. macrocarpon's susceptibility to RKNs. However, in the current study, the observed tolerance in S. macrocarpon and Mongal F1 might be due to inactivation of the Mi gene arising from mutation, if present in the genotypes [45], or the possession of other genes (horizontal resistance) rather than Mi gene. Solanum macrocarpon's hardy roots makes penetration relatively difficult for nematodes compared with tomato roots. Alternatively, Mi gene dosage effect may account for increased nematode reproduction in tolerant Mi heterozygous individuals than in homozygous individuals [47, 48].

Apart from reducing nematode damage, resistant rootstocks are cultivated to help improve plant fitness and yield. Consequently, growth and yield performance of the test rootstocks was evaluated in the study in infected and controlled environments, to various inocula levels of RKN in pot experiments. The absence of significant differences in the rootstocks in relation to the growth parameters (fresh and dry shoot and root weights) shows the tolerance levels of the various rootstocks to the RKN, irrespective of the inocula level in the pot trials. Within the field experiments, significant differences in plant girth, height, and chlorophyll content were obtained among the tested rootstocks. These observed differences showed that although the rootstocks belonged to the same Solanum genus, these had (3) different species origins. Among the three tomato rootstocks, Mongal F1, however, presented a more robust growth after transplanting.

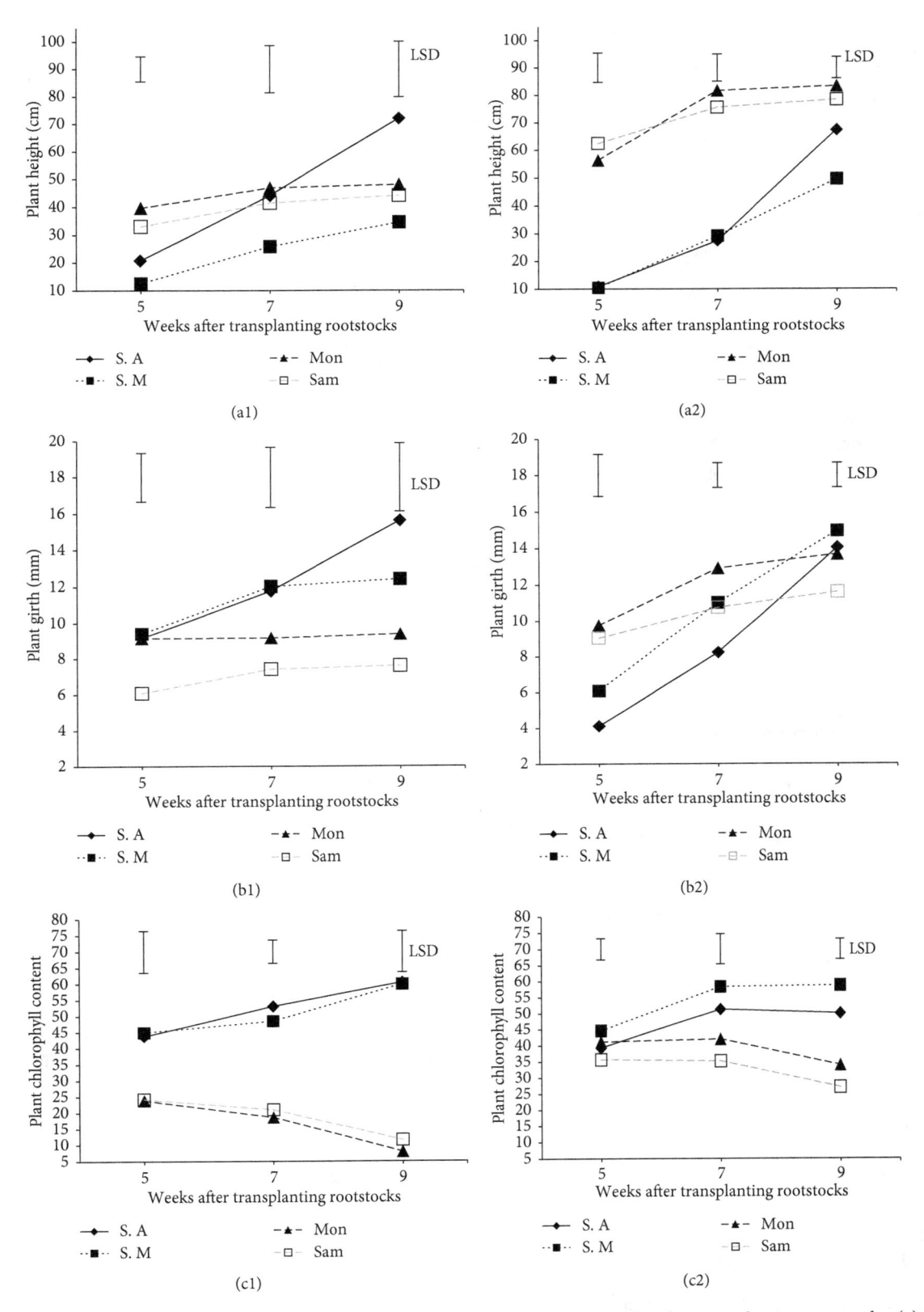

FIGURE 1: Growth parameters of rootstocks to RKN infection at five, seven, and nine weeks after transplanting rootstocks. (a) The mean plant height, (b) the mean plant girth (c) the mean chlorophyll content in 2015 (1) and 2016 (2). Rootstocks: Sam = Samrudhi F1; Mon = Mongal F1; S. M = *S. macrocarpon*; S. A = *S. aethiopicum*. Data were analyzed with two-way ANOVA and significant means separated with the least significant difference (LSD) test. LSD bars represent the least significant difference at $P \leq 0.05$.

TABLE 4: Mean yield parameters of the rootstocks after RKN infestation in 2015 (year 1) and 2016 (year 2).

Treatment	Days to 50% flowering	Number of fruits/plant	Fruit yield/plant (g)	Fruit yield (kg/ha)	Mean fruit weight/plant (g)
Year 1 (dry season)					
Sam	28.00[a]	2.33[a]	23.70[a]	2961[a]	8.89[a]
Mon	27.00[a]	5.00[ab]	44.62[a]	5581[a]	7.43[a]
S. M	39.00[b]	10.33[bc]	227.50[c]	28442[b]	22.74[b]
S. A	51.00[c]	12.00[c]	147.61[b]	18448[ab]	12.30[b]
Year 2 (rainy season)					
Sam	27.00[b]	13.00[ab]	870.30[b]	27196.87[b]	66.95[c]
Mon	29.00[b]	20.00[b]	1236.60[c]	38643.75[c]	61.82[c]
S. M	41.00[c]	9.00[a]	339.10[a]	10596.88[a]	37.68[b]
S. A	58.00[d]	19.00[b]	204.40[a]	6387.50[a]	10.75[a]

Means having different letters in a column differ significantly ($P \leq 0.05$). Rootstocks: Sam = Samrudhi F1; Mon = Mongal F1; S. M = *S. macrocarpon*; S. A = *S. aethiopicum*.

TABLE 5: Mean fresh and dry weights of shoot and roots of rootstocks after 12 weeks of root-knot nematode infestation.

Treatment	Fresh shoot weight (g)	Dry shoot weight (g)	Fresh root weight (g)	Dry root weight (g)
Year 1 (dry season)				
Sam	40.60[a]	8.74[a]	3.74[a]	0.96[a]
Mon	53.70[a]	11.13[a]	5.17[ab]	1.42[ab]
S. M	159.10[ab]	33.78[ab]	14.05[b]	5.20[b]
S. A	331.40[b]	76.73[b]	39.72[c]	16.81[c]
Year 2 (rainy season)				
Sam	257.74[a]	47.74[a]	25.13[a]	10.65[a]
Mon	325.28[ab]	65.02[a]	27.09[a]	12.09[a]
S. M	357.39[b]	62.14[a]	24.74[a]	12.67[a]
S. A	255.44[a]	63.51[a]	32.31[a]	19.43[b]

Means having different letters in a column differ significantly ($P \leq 0.05$). Rootstocks: Sam = Samrudhi F1; Mon = Mongal F1; S. M = *S. macrocarpon*; S.A = *S. aethiopicum*.

The plants were tall with thick stems and appeared dark green. Compared with Samrudhi F1, the plants stood more erect (even at flowering) with soft stem tissues that easily break upon fruit bearing (weight). The resistant response of host crops to RKNs is well known to correspond to increased yield [24, 42, 45]. The fruit yield of the tolerant rootstock (Mongal F1) was significantly increased, compared with the resistant Samrudhi F1. This suggests Mongal F1's efficiency in converting resources into fitness and in bearing of fruit rather than controlling nematodes. Restif and Koella [49] detected the distribution of tolerant plant resources towards fitness, unlike in resistant plants where this is aimed at reducing pathogen infection. Within Samrudhi F1 genotypes, appreciable yields would be obtained in addition to suppressed nematode population, making it possible for the cultivation of susceptible crops in subsequent years [50].

5. Conclusion

Our investigation showed tomato Samrudhi F1 as resistant, whereas tomato *S. macrocarpon*, *S. aethiopicum*, and Mongal F1 were tolerant to RKNs. These rootstocks, can serve as tomato rootstocks, in breeding programs targeted at nematode management. Tolerance mechanisms of plants against nematodes are less understood and should be further investigated.

Acknowledgments

The authors are grateful to the University of Ghana Research Fund (UGRF), with a grant to NA and STN in response to the 7th call for proposals.

References

[1] R. Srinivasan, *Safer Tomato Production Techniques: a Field Guide for Soil Fertility and Pest Management*, Vol. 10, AVRDC-World Vegetable Center, Tainan, Taiwan, 2010.

[2] J. O. Olaniyi, W. B. Akanbi, T. A. Adejumo, and O. G. Ak, "Growth, fruit yield and nutritional quality of tomato varieties," *African Journal of Food Science*, vol. 4, no. 6, pp. 398–402, 2010.

[3] J. Jaramillo, V. Rodriguez, M. Guzman, M. Zapata, and T. Rengifo, *Technical Manual: Good Agricultural Practices in the Production of Tomato under Protected Conditions*, FAO, Rome, Italy, 2007.

[4] B. B. Freeman and K. Reimers, "Tomato consumption and health: emerging benefits," *American Journal of Lifestyle Medicine*, vol. 5, no. 2, pp. 182–191, 2011.

[5] D. Bhowmik, K. S. Kumar, S. Paswan, and S. Srivastava, "Tomato-a natural medicine and its health benefits," *Journal of Pharmacognosy and Phytochemistry*, vol. 1, no. 1, pp. 33–43, 2012.

[6] B. van Dam, Barbara, M. D. Goffau, J. van Lidt de Jeude, and S. Naika, *Cultivation of Tomato: Production, Processing and Marketing*, Agrodok, Difrafi, Wageningen, Netherlands, 2005.

[7] Food and Agriculture Organization of the United Nations, Statistics Division (FAOSTAT), Food and Agriculture Data, July 2017, http://www.fao.org/faostat/en/#data/QC.

[8] D. Horna, M. Smale, and J. Falck-Zepeda, "Assessing the potential economic impact of genetically modified crops in Ghana: Tomato, garden egg, cabbage and cassava," Program for Biosafety Systems Report, IFPRI, International Food Policy Research Institute, Washington, D.C., USA, 2006.

[9] P. Abad, B. Favery, M. N. Rosso, and P. Castagnone-Sereno, "Root-knot nematode parasitism and host response: molecular basis of a sophisticated interaction," *Molecular Plant Pathology*, vol. 4, no. 4, pp. 217–224, 2003.

[10] P. Castagnone-Sereno, "Genetic variability and adaptive evolution in parthenogenetic root-knot nematodes," *Heredity*, vol. 96, no. 4, pp. 282–289, 2006.

[11] J. N. Sasser, *Plant-Parasitic Nematodes: the Farmer's Hidden Enemy*, North Carolina State University Graphics, Raleigh, NC, USA, 1989.

[12] D. J. Chitwood, *Nematicide*, USDA-ARS, Beltsville, MD, USA, January 2016, http://www.ars.usda.gov/SP2UserFiles/person/990/Chitwood2003NematicideReview.pdf.

[13] M. Keren-Zur, J. Antonov, A. Bercovitz et al., "Bacillus firmus formulations for the safe control of root-knot nematodes," in *Proceedings of an International BCPC Conference on Pests and Diseases*, vol. 1, pp. 47–52, British Crop Protection Council, Brighton, UK, November 2000.

[14] R. A. Sikora and E. Fernandez, "Nematode parasites of vegetables," in *Plant Parasitic Nematodes in Subtropical and Tropical Agriculture*, pp. 319–392, CABI Publishing, Wallingford, UK, 2nd edition, 2005.

[15] P. G. Addoh, "The root-knot nematode problem in Ghana: host and non-host plants of *Meloidogyne* species," *Ghana Journal of Agricultural Science*, vol. 3, no. 1, pp. 3–12, 1970.

[16] O. B. Hemeng, "Efficacy of selected nematicides for the control of root knot nematode, *Meloidogyne* spp. on tomato in Ghana," *Ghana Journal of Agricultural Science*, vol. 13, pp. 37–40, 1981.

[17] G. De Lannoy, *Vegetables: Crop Production in Tropical Africa*, DGIC, Brussels, Belgium, 2001.

[18] A. U. Charchar, J. M. Gonzaga, V. Giordano, L. D. Boiteuy, and L. S. Reis, "Reaction of tomato cultivars to infection by a mixed population of *M. incognita* race and *M. javanica* in the field," *Nematologia Brasileira*, vol. 27, pp. 49–54, 2003.

[19] S. R. Gowen, "Chemical control of nematodes: efficiency and side-effects," FAO Plant Production and Protection Paper, January 2016, http://www.fao.org/docrep/v9978e/v9978e08.htm.

[20] D. Nordmeyer, "The search for novel nematicidal compounds," in *Nematology from Molecules to Ecosystems*, pp. 281–293, European Society of Nematologists, Invergowrie, Dundee, Scotland, 1992.

[21] S. Khanzada, M. M. Jiskani, S. R. Khanzada et al., "Response of some tomato cultivars against root-knot nematode, *Meloidogyne incognita* (Kofoid & White) Chitwood," *Journal of Animal and Plant Sciences*, vol. 22, no. 4, pp. 1076–1080, 2012.

[22] I. L. P. Da Conceição, M. C. V. Dos Santos, I. M. D. O. Abrantes, and M. S. N. D. Santos, "Using RAPD markers to analyze genetic diversity in Portuguese potato cyst nematode populations," *Nematology*, vol. 5, no. 1, pp. 137–143, 2003.

[23] M. Luc, R. A. Sikora, and J. Bridge, *Plant Parasitic Nematodes in Subtropical and Tropical Agriculture*, CABI Publishing, Oxfordshire, England, 2nd edition, 2005.

[24] B. P. Corbett, L. Jia, R. J. Sayler, L. M. Arevalo-Soliz, and F. Goggin, "The effects of root-knot nematode infection and mi-mediated nematode resistance in tomato on plant fitness," *Journal of Nematology*, vol. 43, no. 2, pp. 82–89, 2011.

[25] N. Ioannou, "Integrating soil solarization with grafting on resistant rootstocks for management of soil-borne pathogens of eggplant," *Journal of Horticultural Science and Biotechnology*, vol. 76, no. 4, pp. 396–401, 2001.

[26] S. B. Milligan, J. Bodeau, J. Yaghoobi, I. Kaloshian, P. Zabel, and V.M. Williamson, "The root-knot nematode resistance gene Mi from tomato is a member of the leucine zipper, nucleotide binding, leucine-rich repeat family of plant genes," *Plant Cell*, vol. 10, no. 8, pp. 1307–1319, 1998.

[27] G. Nombela, V. M. Williamson, and M. Muniz, "The root-knot nematode resistance gene Mi-1.2 of tomato is responsible for resistance against the whitefly Bemisiatabaci," *Molecular Plant-Microbe Interactions*, vol. 16, no. 7, pp. 645–649, 2003.

[28] R. Dhivya, A. Sadasakthi, and M. Sivakumar, "Response of wild solanum rootstocks to root-knot nematode (*Meloidogyne incognita* Kofoid and White)," *International Journal of Plant Sciences*, vol. 9, no. 1, pp. 117–122, 2014.

[29] F. Jaiteh, C. Kwoseh, and R. Akromah, "Evaluation of tomato genotypes for resistance to root-knot nematodes," *African Crop Science Journal*, vol. 20, no. 1, pp. 41–49, 2012.

[30] F. J. Louws, C. L. Rivard, and C. Kubota, "Grafting fruiting vegetables to manage soil borne pathogens, foliar pathogens, arthropods and weeds," *Scientia Horticulturae*, vol. 127, no. 2, pp. 127–146, 2010.

[31] C. A. Clark and J. W. Moyer, *Compendium of Sweet Potato Diseases*, American Phyto-Pathological Society, St. Paul, MN, USA, 1988.

[32] R. S. Hussey and K. R. Barker, "A comparison of methods of collecting inocula of *Meloidogyne* spp., including a new technique," *Plant Disease Report*, vol. 57, pp. 1025–1028, 1973.

[33] W. R. Jenkins, "A rapid centrifugal-flotation technique for separating nematodes from soil," *Plant Disease Reporter*, vol. 48, no. 9, 1964.

[34] J. Bridge and S. L. J. Page, "Estimation of root-knot nematode infestation levels on roots using a rating chart," *International Journal of Pest Management*, vol. 26, no. 3, pp. 296–298, 1980.

[35] Z. Devran and I. H. Elekçioglu, "The screening of F_2 plants for the root-knot nematode resistance gene, Mi by PCR in Tomato," *Turkish Journal of Agriculture and Forestry*, vol. 28, no. 4, pp. 253–257, 2004.

[36] J. N. Sasser, C. C. Carter, and K. M. Hartman, *Standardization of Host Suitability Studies and Reporting of Resistance to Root-Knot Nematodes*, A Co-Operative Publication of North Carolina State University, Department of Plant Pathology and USAID, Raleigh, NC, USA, 1984.

[37] VSN International Ltd., *Genstat 12th Edition for Windows*, VSN Int. Ltd., Oxford, UK, 2009.

[38] G. Liébanas and P. Castillo, "Host suitability of some crucifers for root-knot nematodes in southern Spain," *Nematology*, vol. 6, no. 1, pp. 125–128, 2004.

[39] G. Karssen and M. Moens, "Root-knot nematodes," in *Plant Nematology*, R. N. Perry and M. Moens, Eds., pp. 59–90, CAB International, Wallingford, UK, 2006.

[40] M. Sobczak, A. Avrova, J. Jupowicz, M. S. Phillips, K. Ernst, and A. Kumar, "Characterization of susceptibility and resistance responses to potato cyst nematode (*Globodera* spp.) infection to tomato lines in the absence and presence of the broad-spectrum nematode resistance Hero-gene," *Molecular Plant–Microbe Interactions*, vol. 18, no. 2, pp. 158–168, 2005.

[41] T. C. Wehner, S. A. Walters, and K. R. Barker, "Resistance to root-knot nematodes in cucumber and horned cucumber," *Journal of Nematology*, vol. 23, no. 4S, p. 611, 1991.

[42] M. A. Rahman, M. A. Rashid, M. A. Salam, M. A. T. Masud, A. S. H. Masum, and M. M. Hossain, "Performance of some grafted eggplant genotypes on wild Solanum root stocks against root-knot nematode," *Journal of Biological Sciences*, vol. 2, no. 7, pp. 446–448, 2002.

[43] S. Verdejo-Lucas, M. Blanco, L. Cortada, and F. J. Sorribas, "Resistance of tomato rootstocks to *Meloidogyne arenaria* and *Meloidogyne javanica* under intermittent elevated soil temperatures above 28°C," *Crop Protection*, vol. 46, pp. 57–62, 2013.

[44] Z. Devran, M. A. Sogut, and N. Mutlu, "Response of tomato rootstocks with the Mi resistance gene to *Meloidogyne incognita* race 2 at different soil temperatures," *Phytopathologia Mediterranea*, vol. 49, no. 1, pp. 11–17, 2010.

[45] J. A. Lopez-Perez, M. L. Strange, I. Kaloshian, and A.T. Ploeg, "Differential response of Mi gene-resistant tomato rootstocks to root-knot nematodes (Meloidogyne incognita)," *Crop Protection*, vol. 25, no. 4, pp. 382–388, 2006.

[46] E. A. Tzortzakakis, F. A. Bletsos, and A. D. Avgelis, "Evaluation of Solanum rootstock accessions for control of root-knot nematodes and tobamoviruses," *Journal of Plant Diseases and Protection*, vol. 113, no. 4, pp. 188-189, 2006.

[47] E. A. Tzortzakakis, D. L. Trudgill, and M. S. Phillips, "Evidence for a dosage effect of the Mi gene on partially virulent isolates of *Meloidogyne javanica*," *Journal of Nematology*, vol. 30, no. 1, pp. 76–80, 1998.

[48] M. Jacquet, M. Bongiovanni, M. Martinez, P. Verschave, E. Wajnberg, and P. Castagnone-Sereno, "Variation in resistance to the root-knot nematode *Meloidogyne incognita* in tomato genotypes bearing the Mi gene," *Plant Pathology*, vol. 54, no. 2, pp. 93–99, 2005.

[49] O. Restif and J. C. Koella, "Concurrent evolution of resistance and tolerance to pathogens," *American Naturalist*, vol. 164, no. 4, pp. E90–E102, 2004.

[50] C. Ornat, S. Verdejo-Lucas, and F. J. Sorribas, "Effect of the previous crop on population densities of *Meloidogyne javanica* and yield of cucumber," *Nematropica*, vol. 27, no. 1, pp. 85–90, 1997.

Growth and Yield Performance of Roselle Accessions as Influenced by Intercropping with Maize in the Guinea Savannah Ecology of Ghana

Emmanuel Ayipio [ID],[1] Moomin Abu,[2] Richard Yaw Agyare [ID],[1] Dorothy Ageteba Azewongik,[2] and Samuel Kwame Bonsu[1]

[1]CSIR-Savanna Agricultural Research Institute, P.O. Box TL 52, Nyankpala, N/R, Ghana
[2]University for Development Studies, Faculty of Agriculture, Department of Horticulture, P.O. Box 1882, Nyankpala, N/R, Ghana

Correspondence should be addressed to Emmanuel Ayipio; emmayipio@gmail.com

Academic Editor: David Clay

Growing roselle with maize is common in Ghana. However, there is little information on whether the choice of accession affects maize/roselle intercrops. Also, there is little information on whether intercropping roselle with maize is beneficial or detrimental to the yields of the crops. A field experiment was conducted for two seasons to assess the growth and yield performance of three roselle accessions as influenced by intercropping with maize and to assess whether maize growth and yield would be influenced upon intercropping with roselle. The roselle accessions were *Bissap*, *Local*, and *Samadah*. Intercropped roselle accessions, their soles, and sole maize were arranged in a randomized complete block design with three replications. Intercrop performance was determined using area × time equivalence ratio (ATER) and competitive ratio (CR). The results showed significant differences in roselle accessions for growth and yield performance. *Local* outperformed the other accessions in both growth and yield attributes. Roselle calyx and shoot yields were significantly higher in intercrops than in soles. Maize growth and yield were not significantly affected by intercropping with roselle, but maize yield attributes differed between seasons. The ATERs were 1.99, 2.18, and 2.49 for *Samadah*, *Bissap*, and *Local*, respectively, indicating that there was productive use of space and time for intercropping. The CR ranged from 1.03 to 1.17, 0.96 to 1.09, and 1.12 to 1.25 for stem diameter, number of leaves per plant, and plant height, respectively. The study showed that the choice of accession did not affect the intercrop performance. It also showed that intercropping roselle with maize is beneficial to roselle without compromising the yield of maize.

1. Introduction

Roselle (*Hibiscus sabdariffa* L.) is an annual herbaceous shrub of the family *Malvaceae* [1]. Young succulent leaves of roselle are eaten in various forms as they contain nutrients such as phosphorus, calcium, magnesium, and potassium [2]. The fleshy calyces of the plant are used in various countries in Africa and the Caribbean as food or food ingredient in jellies, syrups, beverages, puddings, cakes, and wines [3]. Roselle serves as a relish that accompanies foods such as a popular northern Ghanaian food called "Touzaafi"(TZ) which is a thick or solidified porridge prepared from maize meal. A strong fiber called hemp is obtained from the stem which is used for various household purposes including making sackcloth, twine, and cord [4]. In northern Ghanaian homes, the dry stems are a source of fuel for cooking. The seed contains a substantial amount of oil that resembles cotton seed oil [5] and is found to have high antioxidant capacity and strong radical-scavenging activity [6].

Cereal cultivation predominates the agricultural landscape of Ghana. In such areas, intercropping offers an economically sustainable way of producing vegetables like roselle which is classified among underutilized crops [7]. Maize-based intercropping systems are proven to have yield advantages over sole cropping by optimizing especially the use of light [8–10] which translates into yield for both crops. Many benefits of intercropping have been reported in the literature [11, 12]. In addition to these, intercropping vegetables with cereals offers a sustainable way of cultivating vegetables [13].

Growing roselle with maize is a common practice in Ghana. However, when this is done, roselle is usually either planted around the maize farm or spot planted throughout the farm. This practice does not yield the maximum benefit of intercropping. Also, there is little information on whether planting maize with roselle affects maize yields or is beneficial to roselle in Ghana. In addition, there is no information on whether the choice of accession affects maize/roselle intercrops. The objective of the study was to test and identify roselle accession(s) whose yield would increase or be maintained when intercropped with maize without negative yield implications on the maize crop.

2. Materials and Methods

2.1. Experimental Site. A field experiment was conducted for two seasons: in 2016 at Faculty of Agriculture, University for Development Studies, and in 2017 at Savanna Agricultural Research Institute of the Council for Scientific and Industrial Research, Ghana. Both institutions are located in Nyankpala in the northern Guinea savannah ecological zone of Ghana at longitude 0°98′W and latitude 9°4′N and at an altitude of 183 m above the sea level [14]. The chemical properties of the soil for both experimental sites are presented in Table 1. The area experiences a unimodal rainfall with annual rainfall ranging between 1000 mm and 1200 mm from May to October but sporadic during other times of the year. The monthly mean temperature is 22°C in the rainy season and 34°C in the dry season [14]. The mean monthly rainfall, minimum and maximum temperatures, and relative humidity data covering the experimental periods of each year are presented in Table 2.

2.2. Plant Materials. The maize variety wang-dataa was the main crop. It matures in 90 days. It was obtained from the Savanna Agriculture Research Institute (CSIR-SARI). Three roselle accessions were obtained from WorldVeg through CSIR-SARI. The roselle accessions were *Bissap* with dark red calyces and leaves, *Local* with light green calyces and leaves, and *Samadah* with greenish red leaves and light red calyces [15]. Bissap has an average height of 132 cm, a leaf length-to-width ratio of about 1.3, and a short true stem with multiple main branches. Local can grow up to an average height of 157 cm; it has a leaf length-to-width ratio of about 1.0 and a long main stem with many primary and secondary branches. Samadah can reach an average height of 121 cm with a leaf length-to-width ratio of 0.8 [15].

2.3. Experimental Design. The experiment was laid out in a randomized complete block design (RCBD) with three replications in both seasons. A plot measured 4 m × 3 m in 2016 and 5 m × 4.5 m in 2017. Differences in plot sizes were due to seed availability in each year. Replications were separated by a distance of 1.5 m, while plots were separated by a 1 m alley. The treatments were as follows: (1) sole maize (M); (2) sole roselle *Local*, designated as A1; (3) sole roselle *Bissap*, designated as A2; (4) sole roselle *Samadah*, designated as A3; (5) A1 intercropped with maize (A1M); (6) A2 intercropped with maize (A2M); and (7) A3 intercropped with maize (A3M).

Table 1: Chemical properties of soil at experimental sites.

Location (season)	pH	% OC	% N	mg/kg·P	mg/kg·K
CSIR-SARI (2017)	5.54	0.43	0.04	10.55	52.51
UDS (2016)	5.80	0.39	0.03	4.90	46.30

Source: CSIR-SARI Plant and Soil Analytical Laboratory.

2.4. Land Preparation, Sowing, and Crop Management. Experimental fields were ploughed and harrowed after the first stable rains. Seeds of both crops were sown on 6 August 2016 and 5 July 2017 at a rate of 4 seeds per hole for both crops and later thinned to 2 plants per hill for maize and one plant per hill for roselle at 11 days after planting (DAP). In both years' trials, an interrow spacing of 0.75 m was used for both maize and roselle and an intrarow spacing of 0.4 m was used for maize and 0.5 m was used for roselle. Plant density was 4 plants·m^{-2} (48 plants/plot) for sole maize, 3 plants·m^{-2} (36 plants/plot) for sole roselle, 2 plants·m^{-2} (24 plants/plot) for intercrop maize, and 1 plant·m^{-2} (16 plants/plot) for intercrop roselle.

Fertilizer (NPK) was spot applied to both crops at a rate of 60 kg·N·ha^{-1}, 40 kg·P$_2$O$_5$·ha^{-1}, and 40 kg·K$_2$O·ha^{-1} based on the recommendation for both crops. The N was split applied. Weeds were removed as and when necessary with the aid of a hand hoe.

2.5. Data Collection. Data were collected on growth and yield attributes in both years' trials. Sampling was based on the middle rows in both intercropping and sole cropping for both seasons. Growth measurements were taken on 4 plants in 2016 and 6 plants in the 2017 cropping season. In 2017, there were more plants to sample from due to the larger plot size. Plant height (cm) was measured from the soil level to the tip of the latest apical bud. Number of leaves per plant was determined by counting all the leaves of the sampled plants and then obtaining the mean number of leaves per plant for a particular treatment. Stem diameter (cm) was measured using a GlowGeek digital caliper (Shenzhen YunYuan Technology Co., Ltd, China) at a constant height of 10 mm from the soil level for both crops. Samples were separated in various aboveground components of shoot, calyx, and seed for roselle and stalk, cob, and grain for maize. Fresh weights were taken at harvest by measuring gravimetrically using a sensitive scale (KERN & SOHN GmbH, Balingen, Germany) and then left to shade-dry and weighed till a constant weight was obtained. Yields are expressed in t/ha. The weight of thousand seeds for roselle and one hundred seeds for maize were determined gravimetrically. Attributes such as calyx and grain yield fractions were then

Table 2: Meteorological data recorded during the experimental period (2016/2017).

	Rainfall (mm)	Temperature (°C)		Rel. Humidity (%)	
		Minimum	Maximum	Minimum	Maximum
2016					
May	12.60	26.37	34.74	54.35	85.42
June	11.99	25.02	33.05	58.90	89.07
July	32.91	23.66	30.50	72.19	93.45
August	8.13	21.94	31.67	74.06	92.68
September	13.69	21.28	31.92	72.97	93.73
October	7.98	21.81	34.37	63.71	92.23
November	Nil	22.76	37.37	40.93	83.10
December	43.7	20.68	37.65	28.55	64.26
2017					
May	5.90	25.86	36.25	52.19	83.29
June	16.99	24.67	32.80	66.93	92.43
July	30.08	23.53	30.15	73.74	94.13
August	10.91	23.43	29.33	73.68	94.39
September	13.20	23.54	30.27	72.20	94.97
October	5.76	23.65	32.53	59.35	90.55
November	14.20	23.41	35.64	41.03	82.47
December	Nil	21.13	35.63	20.58	49.58

Source: CSIR-SARI Meteorological Services, Nyankpala.

estimated. Total dry biomass was a summation of the weights of the aboveground components.

2.6. Intercrop Performance.
Intercrop performance was measured using the area × time equivalence ratio (ATER), competitive ratio (CR), and percentage land saved (PLS). These are defined below.

(a) Area × time equivalence ratio (ATER): area × time equivalence ratio was calculated using the following formula:

$$\text{ATER} = \frac{([Y_{ia}/Y_{sa}] \times t_a + [Y_{ib}/Y_{sb}] \times t_b)}{T_t}, \quad (1)$$

where Y_{ia} and Y_{ib} are the intercrop yields; Y_{sa} and Y_{sb} are the yields of the sole crops; t_a and t_b are the maturity times (duration) of species a and b (a in our case is roselle accession and b is maize); and T_t is the total time occupied by the intercrop.

(b) Percentage land saved (PLS):

$$\text{PLS} = \frac{(\text{ATER} - 1)}{\text{ATER}} \times 100. \quad (2)$$

(c) Competitive ratio (CR): the concept of competitive ratio was used to estimate the competitiveness for number of leaves per plant, plant height, and stem diameter of the roselle accessions and is given as follows:

$$\text{CR}_{ab} = \frac{(Y_{ia}/Y_{sa} * F_a)}{(Y_{ib}/Y_{sb} * F_b)}, \quad (3)$$

where Y_{ia} and Y_{ib} are the number of leaves, plant height, or stem diameter of species "a" (roselle accessions) and "b" (maize) in intercropping; Y_{sa} and Y_{sb} are the values of the parameter for the species "a" and "b" in sole cropping; and F_a and F_b are the proportions of the area occupied by crops in "a" and "b" in the intercrop. Competitive ratio of 1 means the performance of species "a" and "b" is the same, greater than 1 means the performance of species "a" is better than that of species "b," while less than 1 means the performance of species "b" is better than that of species "a."

2.7. Statistical Analysis.
Analysis of variance in the RCBD model was used to determine whether statistical differences exist among the accessions, the cropping system, the different years, and the interactions in each case of the three indicated. Multiple mean comparison was used when a probability of $P \leq 0.05$ occurred using a least significant difference at 5%.

3. Results

3.1. Growth and Yield Performance of Roselle Accessions due to Intercropping.
The results showed that intercropping generally resulted in significantly higher calyx and shoot dry yields, total aboveground dry biomass yield, and thousand seed weight compared to sole cropping (Table 3). *Local* had the highest values in all growth and yield attributes assessed. *Bissap* had the lowest values for number of leaves per plant, stem diameter, calyx fresh and dry yields, shoot dry yield, and total aboveground dry biomass (Table 3).

The effect of cropping system on calyx fresh and dry yields, shoot dry yield, and thousand seed weight was influenced by the growing season. Also, variation in thousand seed weight among the accessions due to the cropping system was also dependent on the cropping season (Table 4).

TABLE 3: Main effect of accession and cropping system on roselle growth and yield performance.

	NL	PH	SD	CDY	CYF	CFY	SDY	TDB	TSW
Accession									
A1	120.00a	62.80b	0.22a	0.23a	9.32bc	1.77a	2.35a	3.02a	18.80ab
A2	185.40bc	85.40c	1.53b	0.48b	9.91b	3.53b	3.89bc	5.34	21.91b
A3	155.40b	49.90a	1.61b	0.27a	6.17a	2.28a	3.63b	4.57b	16.48a
Cropping system									
IC	150.70a	66.70a	1.44a	0.37b	8.18a	2.67a	3.64b	4.81b	22.78b
S	156.60a	65.40a	1.46a	0.28a	8.75a	2.39a	2.94a	3.81a	15.35a
LSD	ns	ns	ns	0.06	ns	ns	0.57	0.79	2.81

A1 = Bissap; A2 = Local; A3 = Samadah; IC = intercrop; S = sole; CDY = calyx dry yield (t/ha); CYF = calyx yield fraction (%); CFY = calyx fresh yield (t/ha); SDY = shoot dry yield (t/ha), NL = number of leaves per plant (n); PH = plant height (cm); SD = stem diameter (cm); TDB = total aboveground dry biomass (t/ha); TSW = thousand seed weight (g); LSD = least significant difference; ns = not significant.

The 2017 cropping season generally resulted in higher values for both intercropping and sole cropping systems except for thousand seed weight. There were significantly higher values for number of leaves per plant, stem diameter, calyx fresh and dry yields, and calyx yield fraction in 2017 than in the 2016 cropping season. Total aboveground dry biomass yield was however higher for both intercrop and sole crop systems in 2016 than in the 2017 cropping season. It was realized that thousand seed weight was highest for *Local* in the intercropping system for the 2016 cropping season and *Bissap* in the sole cropping system for the 2017 cropping season (Table 4).

The accession *Bissap* obtained lower values in number of leaves per plant, plant height, stem diameter, calyx fresh and dry yields, and calyx yield fraction but higher values in total aboveground dry biomass in 2016 than in the 2017 cropping season. The accession *Local* had higher values in number of leaves per plant, plant height, and total aboveground dry biomass but lower values in stem diameter, calyx fresh yield, and calyx yield fraction in 2016 than in the 2017 cropping season. However, *Samadah* was higher only in number of leaves per plant but lower in all the other growth and yield attributes in 2016 than in the 2017 cropping season (Table 4). There were no significant differences in the three-way interaction between accessions, cropping system, and season except for thousand seed weight.

3.2. Growth and Yield of Maize in Intercrop, Sole, and Season.

The analysis of variance results showed that maize growth and yield attributes were not significantly different between intercropping and sole cropping systems for any of the cropping seasons (Table 5). However, there was significant ($P < 0.05$) general seasonal variation in grain yield, cob yield, hundred seed weight, grain yield fraction, and stalk yield but not growth attributes (data not shown).

3.3. Intercrop Performance of Roselle.

There were no significant differences among the accessions for area × time equivalence and competitive ratios (Table 6). However, there was seasonal variation of the accessions in their ATER, whereas Local recorded the highest ATER in 2016 and Samadah recorded the highest value in 2017 (Figure 1). All roselle accessions had ATER values more than one, the lowest being 1.99 obtained by *Samadah* (A3) and the highest being 2.49 obtained by *Local* (A2). Thus, it could be seen that, in intercropping *Local* with maize, one could save up to 56% of land compared to when it is cropped as sole. The competitive ratios ranged from 1.03 to 1.17 for stem diameter, 1.12 to 1.25 for plant height, and 0.96 to 1.09 for number of leaves per plant.

TABLE 4: Seasonal variation in intercrop and accessions' performance.

	NL	CFY	CDY	SDY	CYF	TDB
Cropping system × season						
IC × '16	148.30	1.89	0.33	3.38	6.30	5.28
IC × '17	153.10	3.46	0.40	3.91	10.07	4.33
S × '16	145.70	1.50	0.19	3.34	4.44	4.73
S × '17	167.40	3.28	0.38	2.53	13.06	2.89
LSD (0.05)	36.5	0.74	0.08	0.81	2.32	1.11
Accession × season						
A1 × '16	86.80	1.50	0.25	2.93	6.37	4.03
A1 × '17	153.20	2.05	0.22	1.77	12.28	2.00
A2 × '16	189.50	2.40	0.40	4.17	6.22	6.55
A2 × '17	181.40	4.67	0.55	3.62	13.60	4.13
A3 × '16	164.70	1.18	0.13	2.98	3.53	4.43
A3 × '17	146.10	3.38	0.40	4.28	8.80	4.70
LSD (0.05)	44.7	0.9	0.10	0.99	2.85	1.36

A1 = Bissap; A2 = Local; A3 = Samadah; IC = intercrop; S = sole; '16 and '17 are the cropping seasons in 2016 and 2017, respectively; NL = number of leaves per plant; CDY = calyx dry yield (t/ha); SDY = shoot dry yield (t/ha); CYF = calyx yield fraction (%); CFY = calyx fresh yield (t/ha); TDB = total aboveground dry biomass (t/ha); LSD = least significant difference.

4. Discussion

4.1. Intercropping Effect on Growth and Yield Attributes of Roselle.

Differences in roselle growth and yield attributes observed between the two seasons could be due to differences in sowing date between both years. It has been shown that planting date affects the growth and yield of roselle [16]. Also, there were differences in precipitation between the years. However, the emphasis of this study was to observe the interspecific interaction between the two crops within each year. Genotypic variability existed among the roselle accessions assessed for both growth and yield attributes which is in conformity with previous studies [17, 18]. It was realized that *Local* which has green calyces is a potential

TABLE 5: Maize growth and yield attributes in response to intercropping with roselle accessions and season.

	NL	PH	SD	GM	GYF	GY	HY	HSW	CW	SY	TDB
Cropping system											
MA1	10.43	127.90	1.23	22.10	48.17	2.52	0.48	23.53	2.77	1.77	5.03
MA2	10.42	117.40	1.23	24.08	49.78	2.48	0.40	24.38	2.78	1.85	5.02
MA3	10.23	118.80	1.18	25.42	46.83	2.35	0.47	21.62	2.73	1.65	4.83
SM	10.80	135.10	1.29	24.75	48.33	2.58	0.53	26.02	2.97	1.78	5.27
LSD	ns	ns	ns	ns	ns	ns	ns	ns	ns	ns	ns
Cropping system × season											
MA1 × '16	11.00	129.80	1.33	18.03	42.53	1.70	0.30	16.33	1.73	1.83	3.87
MA1 × '17	9.87	125.90	1.13	26.17	53.80	3.33	0.67	30.73	3.80	1.70	6.20
MA2 × '16	10.67	113.20	1.26	22.73	44.07	2.20	0.27	16.33	2.40	2.43	5.07
MA2 × '17	10.17	121.60	1.20	25.43	55.50	2.77	0.53	32.43	3.17	1.27	4.97
MA3 × '16	10.00	109.40	1.16	24.00	38.03	1.40	0.27	15.67	1.60	1.83	3.70
MA3 × '17	10.47	128.30	1.20	26.83	55.63	3.30	0.67	27.57	3.87	1.47	5.97
SM × '16	11.33	147.80	1.31	20.83	44.40	1.90	0.33	16.33	2.03	1.90	4.27
SM × '17	10.27	122.40	1.27	28.67	52.27	3.27	0.73	35.70	3.90	1.67	6.27
LSD	ns	ns	ns	ns	ns	ns	ns	ns	ns	ns	ns

MA1 = maize intercropped with Bissap; MA2 = maize intercropped with Local; MA3 = maize intercropped with Samadah; SM = sole maize; GM = grain moisture at harvest (%); GYF = grain yield fraction (%); GY = grain yield (t/ha); HY = husk yield (t/ha); NL = number of leaves per plant; PH = plant height (cm); SD = stem diameter (cm); HSW = hundred seed weight (g); CW = cob weight (t/ha); SY = stalk yield (t/ha); TDB = total aboveground dry biomass (t/ha); LSD = least significant difference; ns = not significant.

TABLE 6: Intercrop performance of roselle yield and growth attributes.

	ATER	PLS	CRSD	CRPH	CRNL
Accession					
A1	2.18	52.2	1.06	1.12	0.96
A2	2.49	56.9	1.17	1.21	1.09
A3	1.99	48.0	1.03	1.25	1.03
LSD	ns	ns	ns	ns	ns
Accession × season					
A1 × '16			1.13	1.24	0.99
A1 × '17			0.99	0.99	0.93
A2 × '16	See Figure 1	See Figure 1	1.07	1.31	1.20
A2 × '17			1.26	1.11	0.98
A3 × '16			1.02	1.47	1.18
A3 × '17			1.04	1.03	0.88
LSD			ns	ns	ns

A1 = Bissap; A2 = Local; A3 = Samadah; '16 and '17 are the cropping seasons in 2016 and 2017; ATER = area × time equivalence ratio; PLS = percentage land saved; CRPH = competitive ratio for plant height; CRSD = competitive ratio for stem diameter; CRNL = competitive ratio for number of leaves; LSD = least significant difference; ns = not significant.

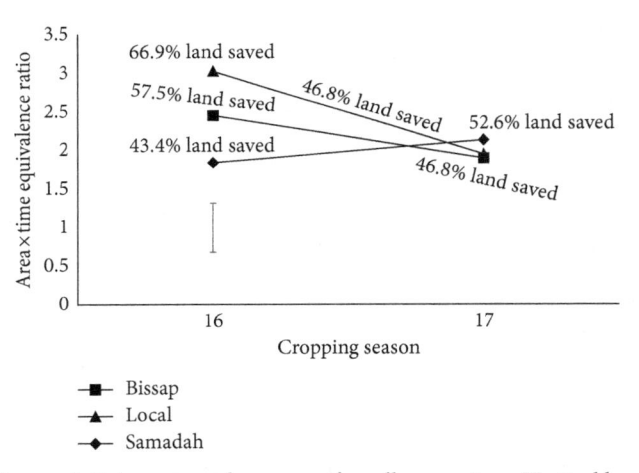

FIGURE 1: Intercrop performance of roselle accessions. Vertical bar represents the least significant difference.

good accession that could be promoted due to its performance both as sole and in intercrop. Both leaves and calyces of this accession are ideal for the preparation of soups.

Area × time equivalence ratio differed between the seasons. This could be due to the differences in sowing date and weather conditions which can affect the performance of roselle [16]. Intercropping roselle with maize has proven to result in yield benefit to roselle. This has not always been the case. It has been shown that yield losses occur in roselle when intercropped with other crops. For instance, Fbabatunde [19] revealed that roselle yields were more reduced when intercropped with cereals such as millet and sorghum than intercropping with legumes. In maize-legume intercrops, it was realized that maize reduced cowpea yields more than the effect of cowpea on maize yields because of greater competitive ability of maize when intercropped with cowpea [20]. In our study however, intercropping roselle with maize was not inimical to roselle. Intercropping did not affect growth attributes of roselle in our case and even resulted in higher yield than the sole cropping. The current finding is a revelation that intercropping roselle with maize could offer a sustainable way of producing roselle especially for calyx and shoot yield. Calyx and shoot dry yields are among the most important economic components of the roselle plant. This is because in Ghana for instance, the reddish calyx types are used in the preparation of a refreshing drink called *sobolo*, whereas the greenish calyx types are used in the preparation of soups. In traditional domestic homes of northern Ghana where fuel for cooking is scarce, dry shoots of roselle serve as a good source of fuel for domestic purposes. Thus, a significant increase in roselle calyx and shoot dry yields due to intercropping implies that intercropping could be harnessed to increase the productivity of roselle in the Guinea savannah ecology of Ghana whose cropping system is predominantly cereal based. There was no variability in the response of the accessions to intercropping. This means that intercropping was generally advantageous irrespective of accession. This conforms the conclusion drawn by Ijojah [21] who after reviewing different intercropping studies between maize and various vegetables indicated that cereal-vegetable intercrops are highly complementary.

4.2. Intercropping Effect on Growth and Yield Attributes of Maize. A nonsignificant effect of intercropping roselle with maize was realized in our study. This could be due to the fact that roselle as a C_3 crop [22] would compete less with maize which is a C_4 crop for incoming radiation. Also, roselle has a deeper root system than maize, thereby competing less with maize for water and nutrients. Therefore, it could be concluded that species complementarity exists between maize and roselle. That is, roselle can be conveniently intercropped with maize without any detrimental effects on the maize growth and yield attributes, and vice versa. This is not always the case in intercropping trials between maize and other vegetables. In some cases, maize yields were reduced upon intercropping with vegetables [23, 24]. Thus, not just the system of cultivation, but the test crop type may also affect the yield of any of the two crop species in an intercrop system. Intercropped maize was negatively affected by intercropping with wheat, especially during its early growth for a lower leaf appearance rate [25] and a lower biomass accumulation per day [26]. Depending on the vegetable, intercropping maize with vegetables can result in higher gains as reported for intercropping hybrid maize with different short-duration vegetables [24], intercropping winter maize with vegetables [23], and intercropping maize with different vegetables [27] or no effect as shown by our study (Table 5).

4.3. Intercrop Productivity of Roselle. In all intercropping systems studied, ATER values were more than one, indicating a better utilization of space and time. Competitive ability of the roselle accessions except *Bissap* was comparable with that of maize. Adhikary [28] found that productivity of vegetables reduced as a result of higher competitive ability of the baby corn. Also, the high ATER recorded by the roselle in this study could be due to the potential of intercrops to intercept higher incoming solar radiation. Kanton and Dennett [9] found that intercrops intercepted more incoming solar radiation resulting in higher biomass production of the intercrops than the sole crops. Zhu [29] found that, in all wheat-maize intercropping studies, intercropped wheat always showed yield advantage compared to sole wheat. The roselle accessions possibly portrayed characteristics such as branching habit and leaf angle that are indicative of a good intercrop.

5. Conclusion

Generally, intercropping did not affect the growth attributes of the roselle accessions studied nor did it affect maize growth and yield attributes. The yield of roselle was however greatly increased upon intercropping with maize due to the complementary effect between the two species. *Local* outperformed the other two accessions both as sole and in intercrop. Hence, when intercropping roselle with maize, a farmer should consider *Local*.

Intercropping roselle with maize gave a very productive use of land and time shown by area × time equivalence ratio of more than unity, and thus, intercropping the roselle accessions with maize is quite sustainable.

Authors' Contributions

EA conceptualized the idea and wrote the manuscript, MA and RYA supervised the work and edited the manuscript, DAA and SKB collected the data, and DAA used some of the data for her BSc thesis.

References

[1] R. S. McCaleb, *Hibiscus Production Manual*, S.l.: s.n, Herb Research Foundation, Boulder, CO, USA, 1998.

[2] S. Atta, A. B. Diallo, B. Sarr, Y. Bakasso, M. Saadou, and R. H. Glew, "Variation in macro-elements and protein contents of Roselle (*Hibiscus sabdariffa* L.) from Niger," *African Journal Food, Agriculture Nutrition and Development*, vol. 10, pp. 2707–2718, 2010.

[3] N. Mahadevan, Shivali, and K. Pradeep, "*Hibiscus sabdariffa* Linn.—an overview," *Natural Product Radiance*, vol. 8, pp. 77–83, 2009.

[4] A. Mungole and A. Chaturvedi, "*Hibiscus sabdariffa* L. a rich source of secondary metabolites," *International Journal of Pharmaceutical Sciences, Review and Research*, vol. 6, no. 1, pp. 83–87, 2011.

[5] R. Mohammed, J. Fernandez, M. Pineda, and M. Aguilar, "Roselle (*Hibiscus sabdariffa*) seed oil is a rich source of α-tocopherol," *Journal of Food Science*, vol. 72, no. 3, pp. s207–s211, 2007.

[6] N. Mohd-Esa, F. S. Hern, A. Ismail, and C. L. Yee, "Antioxidant activity in different parts of roselle (*Hibiscus sabdariffa* L.) extracts and potential exploitation of the seeds," *Food Chemistry*, vol. 122, no. 4, pp. 1055–1060, 2010.

[7] M. Modi, A. T. Modi, and S. Hendriks, "Potential role for wild vegetables in household food security: a preliminary case study in KwaZulu-Natal, South Africa," *African Journal of Food Agriculture Nutrition and Development*, vol. 6, no. 1, 2006.

[8] D. Harris, M. Natarajan, and R. W. Willey, "Physiological basis for yield advantage in a sorghum/groundnut intercrop exposed to drought. 1. Dry-matter production, production yield and light interception," *Field Crops Research*, vol. 17, no. 3-4, pp. 259–272, 1987.

[9] R. A. L. Kanton and M. D. Dennett, "Radiation capture and use as affected by morphologically contrasting maize/pea in sole and intercropping," *West African Journal of Applied Ecology*, vol. 13, no. 1, 2008.

[10] F. D. Li, P. Meng, D. L. Fu, and B. P. Wang, "Light distribution, photosynthetic rate and yield in a Paulownia–wheat intercropping system in China," *Agroforestry System*, vol. 74, no. 2, pp. 163–172, 2008.

[11] P. E. Odo and K. N. Futuless, "Millet soyabean intercropping as affected by different sowing dates of soyabean in a semi-arid environment," *Cereal Research Community*, vol. 28, pp. 153–160, 2002.

[12] B. C. Blaser, J. W. Singer, and L. R. Gibson, "Winter cereal, seeding rate and intercrop seeding rate effect on red clover yield and quality," *Agronomy Journal*, vol. 99, no. 3, pp. 723–729, 2007.

[13] M. A. Musah, I. I. Y. Baba, W. Dogbe, M. Abudulai, A. Mutari, and M. Haruna, "NERICA intercrop systems and their influence on yield components and productivity of NERICA and partner crops in the Guinea Savannah Zone of Ghana," *International Journal of AgriScience*, vol. 3, p. 12, 2013.

[14] SARI, *Annual Report of Savanna Agriculture Research Institute*, SARI, Nyankpala, Ghana, 2008.

[15] M. S. Abdulai, S. S. J. Buah, R. Y. Agyare, E. Ayipio, and S. K. Bonsu, "Enhancing productivity, competitiveness and marketing of Traditional African (Leafy) vegetables for improved income and nutrition in West and Central Africa (TAVs for Income and Nutrition)," in *CSIR-SARI Annual Report*, S. S. J. Buah, S. K. Nutsugah, M. Abudulai, W. N. Kutah, and R. K. Owusu, Eds., pp. 85-86, SARI Nyankpala, Ghana, 2015.

[16] K. Abdul Mateen, S. Muhammad, Z. S. Hamid, R. Abdur, A. Masood, and A. K. Muhammad, "Effect of Sowing Time and Plant Density on the Growth and Production of Roselle (*Hibiscus sabdariffa*)," *International Journal of Agriculture & Biology*, pp. 1814–9596, 2016.

[17] S. M. Gasim and M. O. Khidir, "Genetic variability of some characters in roselle (Hibiscus sabdariffa var sabdariffa)," *Journal of Agricultural Sciences*, vol. 6, pp. 22–34, 1998.

[18] S. Atta, H. H. Seyni, Y. Bakasso, B. Sarr, I. Lona, and M. Saadou, "Yield character variability in Roselle (*Hibiscus sabdariffa* L.)," *African Journal of Agricultural Research*, vol. 6, no. 6, pp. 1371–1377, 2011.

[19] F. E. Fbabatunde, "Intercrop productivity of roselle in Nigeria," *African Crop Science Journal*, vol. 11, no. 1, pp. 43–48, 2003.

[20] F. Ofori and W. R. Stern, "Relative sowing time and density of component crops in a maize/cowpea intercrop system," *Experimental Agriculture*, vol. 23, pp. 41–52, 1987.

[21] M. O. Ijoyah, "Review of intercropping research: Studies on cereal-vegetable based cropping system," *Scientific Journal of Crop Science*, vol. 1, no. 3, 2012.

[22] B. A. Sonar, "Physiological studies in salt tolerance of three species of Hibiscus (H. cannabinus Linn. H. sabdariffa Linn. and H. tiliaceus Linn.)," Ph. D. thesis, Shivaji University, Kolhapur, India, 2013.

[23] S. K. Choudhary, R. N. Singh, R. K. Singh, and P. K. Upadhyay, "Yield and nutrient uptake of winter maize (*Zea mays*) with vegetable intercropping," *Current Advances in Agricultural Sciences*, vol. 6, no. 1, 2014.

[24] M. A. Rahaman, M. M. Rahman, S. Roy, M. Ahmed, and M. S. Bhuyan, "On-farm study on intercropping of hybrid maize with different short duration vegetables in the Charland of Tangail," *Bangladesh Agronomy Journal*, vol. 18, no. 2, p. 65, 2015.

[25] J. Zhu, J. Vos, W. Van der Werf, P. E. Van der Putten, and J. B. Evers, "Early competition shapes maize whole-plant development in mixed stands," *Journal of Experimental Botany*, vol. 65, no. 2, pp. 641–653, 2014.

[26] L. Li, J. Sun, F. Zhang, X. Li, S. Yang, and Z. Rengel, "Wheat/maize or wheat/soybean strip intercropping. Yield advantage and interspecific interactions on nutrients," *Field Crops Research*, vol. 71, no. 2, pp. 123–137, 2001.

[27] M. R. Ali, M. S. Rahman, M. Asaduzzaman, M. M. Hossain, and M. A. Mannan, "Intercropping maize with different vegetables," *Bangladesh Agronomy Journal*, vol. 18, no. 1, p. 49, 2015.

[28] S. Adhikary, M. K. Pandit, A. V. V. Koundinya, S. Bairagi, and A. Das, "Examination of system productivity and profitability of baby corn based vegetable intercropping systems," *Journal of Crop and Weed*, vol. 11, no. 1, 2015.

[29] J. Zhu, W. Van der Werf, J. Vos, N. P. R. Anten, P. E. L. Van der Putten, and J. B. Evers, "High productivity of wheat intercropped with maize is associated with plant architectural responses," *Annals of Applied Biology*, vol. 168, no. 3, pp. 357–372, 2016.

Productivity and Water Use Efficiency of Sorghum [*Sorghum bicolor* (L.) Moench] Grown under Different Nitrogen Applications in Sudan Savanna Zone, Nigeria

Hakeem A. Ajeigbe[1], Folorunso Mathew Akinseye,[1,2] Kunihya Ayuba,[1] and Jerome Jonah[1]

[1]*International Crop Research Institute for the Semi-Arid Tropics (ICRISAT), Kano, Nigeria*
[2]*Department of Meteorology and Climate Science, Federal University of Technology Akure, Akure, Nigeria*

Correspondence should be addressed to Hakeem A. Ajeigbe; h.ajeigbe@cgiar.org

Academic Editor: Mathias N. Andersen

Nitrogen (N) is an essential nutrient for sorghum growth and development but often becomes limiting due to low availability and loss. The effects of N fertilization on water use efficiency (WUE) and physiological and yield traits of sorghum were investigated in two locations over two cropping seasons (2014 and 2015) in the Sudan Savanna zone of Nigeria. Three sorghum varieties were evaluated under six (6) N-levels (0, 20, 40, 60, 80, and 100 kg ha^{-1}) at a constant phosphorus and potassium level of 30 kg ha^{-1}. Results showed that N increased grain yield by 35–64% at the Bayero University Kano (BUK) and 23–78% at Minjibir. The highest mean grain yield in the N-fertilizer treatments (2709 kg ha^{-1} and 1852 kg ha^{-1} at BUK and Minjibir, resp.) was recorded at 80 kg N ha^{-1}. ICSV400 produced the highest mean grain yields (2677 kg ha^{-1} and 1848 kg ha^{-1} at BUK and Minjibir, resp.). Significant differences were observed among the N-levels as well as among the sorghum varieties for estimated water use efficiency (WUE). The highest mean value coincided with the highest mean grain yield at an optimum application rate of 80 kg ha^{-1}. N-fertilizer treatments increased WUE by 48–55% at BUK and increased WUE by 54–76% at Minjibir over control treatment. Maturity and physiological trait have a significant effect on WUE. The extra early maturing variety (ICSV400) recorded the highest mean WUE while late maturing variety (CSR01) recorded the lowest WUE.

1. Introduction

Nigeria is the largest sorghum (*Sorghum bicolor* L. Moench) producing country in West and Central Africa region, accounting for about 23% of the sorghum production in Africa in 2016 [1]. The country produced 8.5 million tons in 2008, 9.3 million tons in 2009, and 10.0 million tons in 2010 with a projection of being the largest sorghum grain producer in the world [2] by 2020. Sorghum was cultivated on about 10.845 million hectares in 2014 [3], representing about 50% of the total area under cereal crop production and about 13% of the total arable land in the country. Sorghum is typically grown under rain-fed conditions in regions where water and soil fertility are main factors limiting yield [4]. Suboptimal availability of water for unrestricted plant growth and transpiration, that is, drought, is a major limitation to agricultural production [5]. However, sorghum has been found to transform the available water more efficiently into dry matter than most other C_4 crops (e.g., maize) [6]; and the crop is able to utilize water from as deep as 270 cm soil depth [7].

The average grain yield of sorghum in farmers' field is estimated little below 1000 kg ha^{-1} because little or no external inputs are applied [8]. This has led to a decline in soil nitrogen resulting in low yield obtained, food insecurity, nutrient mining, and environmental degradation [8, 9]. Studies by [10–12] on soil test reported that N and P are very low for most dryland areas of Nigeria especially the sorghum growing Sudan Savanna zone. Loss of organic matter, whether by erosion or leaching, adds to the impoverishment of soil resources of several elements essential for plant growth [13]. Sorghum is known for being nutrient-use efficient and managed with lower fertilizer rates compared to maize and rice, but yields can be increased with adequate fertilizer applications [11, 14]. The use of inorganic fertilizers to boost yields of sorghum has been found to increase growth performance, as well as the chemical properties of soil [15]. In the Savanna region of Nigeria where sorghum is cultivated, nitrogen (N) is the most limiting nutrient in soil [12] and a considerable amount of soil-available N is released with the onset of rains but its uptake by crops is insignificant due to the low N requirements of plants at early growth stages [10]. As a result, much of this N is lost through leaching. Therefore, in order to maintain a positive nutrient balance, nutrient inputs from chemical fertilizers are needed to replace nutrients which are exported or lost due to leaching. Although N plays a very important role for good growth and development of sorghum, overfertilization is often harmful as it results in lower yields [16] and negative effect on soil properties. Optimum amounts of N-fertilizer combined with other input factors play a significant role in yield and overall quality of sorghum products. The optimum amount of fertilizer is related to maximum efficiency of production [17].

Consequently, N deficiency results in decreased crop photosynthetic assimilation and seed growth [18, 19]. Studies carried out by [20, 21] found that yield parameters such as plant height, tiller number, panicle number, and number of panicle per m^2 were influenced significantly by complementary application of nitrogen fertilizer on sorghum compared to application of organic fertilizer. A report by [22] also indicated that variation in leaf photosynthetic capacity for varieties, age of leaves, and growth conditions can be attributed to differences in leaf N content. In Nigeria, sorghum is traditionally a food crop grown by subsistence farmers; however, in recent years there are concerted efforts by the government as well as renewed interest by the agro-allied industry as a substitute for imported wheat and barley in the food and beverage industry [2] and call for agricultural scientists and extensions to increase its production and productivity. This has led to the release of 6 open pollinated varieties and 4 hybrids by both public and privately sponsored breeding programs [23] and the encouragement of commercial sorghum production pushing sorghum cultivation into marginal land with lower soil fertility and moisture availability. It is therefore necessary to review the nutrient needs, especially N of some of these varieties and their water use efficiency on marginal land.

2. Materials and Methods

2.1. Description of the Experimental Sites. The experiment was conducted during the 2014 and 2015 growing seasons at two locations within the Sudan Savanna region of Nigeria. The first location was in the ICRISAT research field situated within the Institute for Agricultural Research (IAR) station, Wasai Village, Minjibir (Latitude 12.17°N and Longitude 8.65°E), otherwise known as Minjibir. The second location was at the Bayero University Kano (BUK) Teaching and Research Farm (Latitude 12.98°N and Longitude 9.75°E), otherwise known as BUK. Weather information was collected from Accu Weather Stations installed in the trial sites.

Experimental Design. The experimental design was a split plot arrangement with four replications. The treatments included the six nitrogen fertilizer levels and three sorghum varieties. Six N application rates of 0, 20, 40, 60, 80, and 100 kg ha^{-1}, respectively, were applied as main plots while three sorghum varieties which include ICSV400, CSR01, and local (Kaura) varieties were considered as subplots. The fertilizer treatments were applied in the form of NPK where N-levels varied from 0 to 100 kg ha^{-1} at 20 kg interval, while P_2O_5 and K_2O were applied at a constant rate of 30 kg each per hectare. The gross size of each plot was 15 m^2 which consisted of four ridges spaced at 75 cm apart and sowing was done at 30 cm between plants that gave a total plant population of 44,444 hills ha^{-1}.

2.2. Field Management, Data Collection, and Analysis. The field was disc-harrowed and ridged at 75 cm ridges inter-row spacing before planting. The first cropping season was sown on the 19th of July at BUK and 7th of July at Minjibir in 2014; meanwhile, the second cropping season was on the 20th of July at BUK and 4th of July at Minjibir in 2015. Sowing was done after the rainfall establishment, 5–7 seeds per hole at a depth of 3–5 cm, and thinned to 2 plants per hill at 2 weeks after planting (WAP). For maintaining the optimum plant population, gap filling was done at 8–10 days after planting. The first fertilizer doses (full doses of P, K and half dose of N/plot) were applied by drilling method at sowing, while the second dose was applied at 3 WAP. Weeding was done manually to keep the field weed free. At both locations, measurements including leaf chlorophyll contents (at 3, 6, and 9 WAP) using a SPAD-502 portable chlorophyll meter (Minolta, Tokyo, Japan) were taken. All chlorophyll meter readings were taken midway between the stalk and the tip of the leaf. Leaf area index (LAI) at 6, 9, and 12 WAP (LAI) was measured with Accupar LP-80 portable canopy analyzer. Also, days to 50% flowering, days to 85% maturity, and plant height (cm) were recorded on five randomly selected plants at maturity by measuring the height from the ground to the tip of the panicle. Grain and Stover yields were measured from harvested two rows at the center of each plot [7.5 m^2 area (5 m × 1.5 m)]. Both panicle grain and Stover were sun-dried for 2 weeks before threshing. Grain yield (kg ha^{-1}) and 1000-seed weight (g) were determined while harvest index (HI) was computed as a ratio of grain yield (GY) to the total aboveground dry matter (TDM) on a sun-dried weight basis. All the data were subjected to analysis of variance (ANOVA) using GENSTAT analytical tool (14th edition). Year, N-fertilizer levels, and variety were taken as factors to determine level of significance at 5% probability. Fisher's least-significant difference (LSD) test were computed where the F values were significant at the $P = 0.05$ level of probability [24].

2.3. Water Requirement and Water Use Efficiency under Different N Application. The estimation of crop water requirements during growing seasons was determined from the crop evapotranspiration (ET_c) that was calculated by reference evapotranspiration (ET_0) and recommended crop coefficient

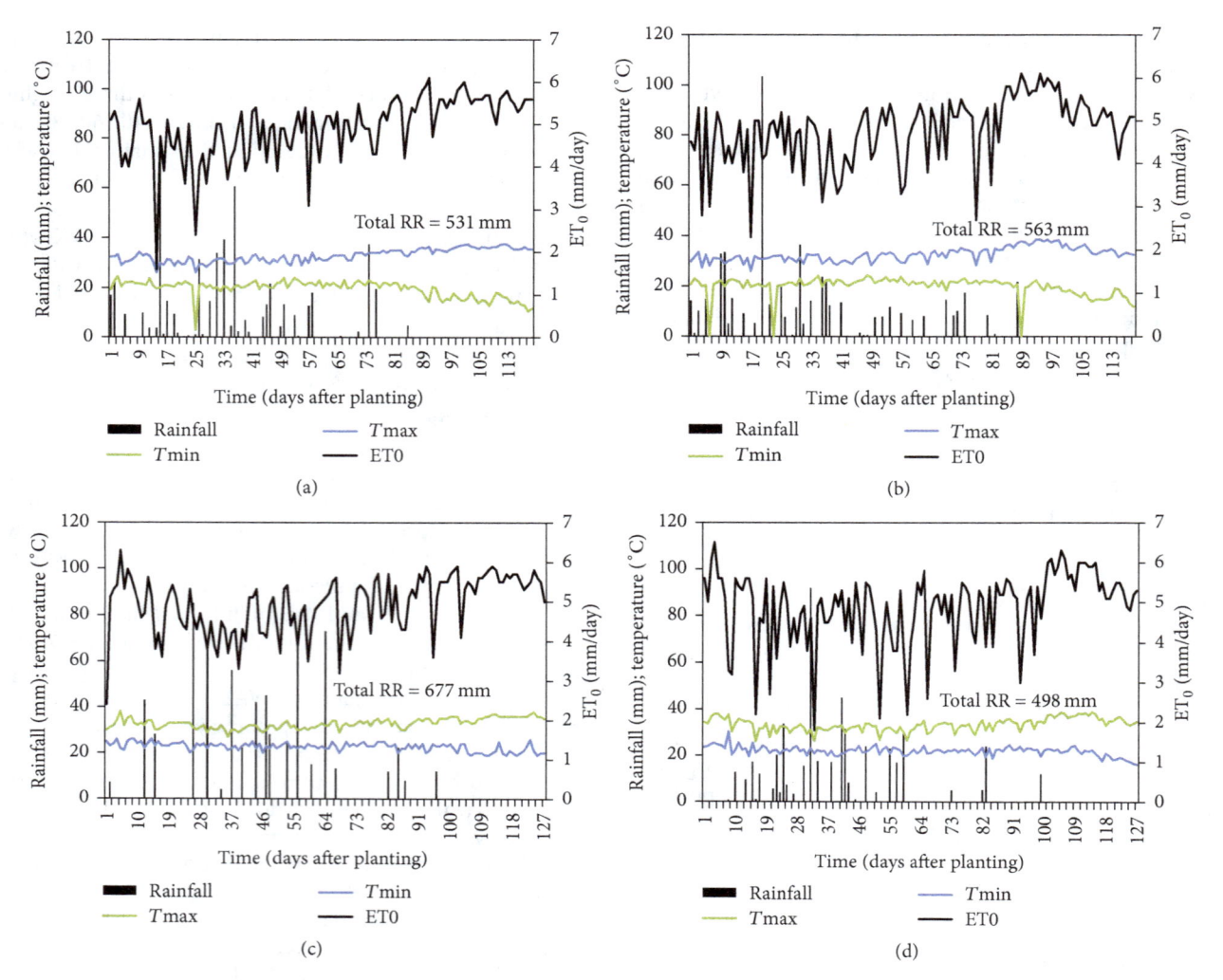

FIGURE 1: Calculated daily reference evapotranspiration (ET$_0$), minimum temperature (T_{min}), maximum temperature (T_{max}), and rainfall during the growing seasons of (a) BUK, 2014; (b) BUK, 2015; (c) Minjibir, 2014; and (d) Minjibir, 2015.

(K_c) for sorghum in (1) (Figure 1) [25]. The Penman-Monteith equation was used to calculate reference evapotranspiration (ET$_0$) in (2); the variables of this equation were described in FAO Irrigation and Drainage Paper No. 56 [26]. The method is of quite good accuracy and is usually used for calculations of evapotranspiration from farmlands.

$$ET_c = K_c ET_0, \quad (1)$$

$$ET_0 = \frac{0.408\Delta(R_n - G) + \gamma(900/(T + 273))u_2(e_s - e_a)}{\Delta + \gamma(1 + 0.34u_2)}, \quad (2)$$

where

ET$_0$ is reference evapotranspiration [mm day^{-1}],
R_n is net radiation at the crop surface [MJ m^{-2} day^{-1}],
G is soil heat flux density [MJ m^{-2} day^{-1}],
T is air temperature at 2 m height [°C],
u_2 is wind speed at 2 m height [m s^{-1}],
e_s is saturation vapour pressure [kPa],
e_a is actual vapour pressure [kPa],
$e_s - e_a$ is saturation vapour pressure deficit [kPa],
D is slope vapour pressure curve [kPa °C^{-1}],
γ is psychrometric constant [kPa °C^{-1}].

The FAO Penman-Monteith equation determines the evapotranspiration from the hypothetical grass reference surface and provides a standard to which evapotranspiration in different periods of the year or in other regions can be compared and to which the evapotranspiration from other crops can be related. However, actual field evapotranspiration (ET$_c$) obtained from (1) was used to calculate water use efficiency (WUE) in (3). Water use efficiency refers to the ratio of water used in plant metabolism to water lost by the plant through transpiration and soil evaporation (evapotranspiration). Water use efficiency was calculated for aboveground biomass at physiological maturity and grain

TABLE 1: Physical and chemical properties of the soils (0–20 cm depth) at the experimental sites.

Parameters	BUK		Minjibir	
	2014	2015	2014	2015
Soil texture	Sandy loam	Sandy clay loam	Sandy loam	Sandy clay loam
Soil pH value (1 : 2.5 soils : water)	4.86	5.98	5.01	5.75
EC (1 : 2.5 soils : water) (dS/m)	0.083	0.065	0.031	0.035
Soil organic carbon (%)	0.417	0.339	0.196	0.210
Total nitrogen (mg/kg)	311.5	700	163.3	1050
Available P (mg/kg)	2.0	2.612	4.0	3.358
Available K (cmol/kg)	0.396	0.379	0.346	0.390

yield at harvest maturity, using the following equations by [27].

$$\text{WUE} = \frac{Y}{\text{ET}_c}, \qquad (3)$$

where

Y is grain yield (kg ha^{-1}),

ET$_c$ is crop evapotranspiration (mm).

2.4. Nitrogen Use Efficiency. The nitrogen use efficiency (NUE), in terms of agronomic efficiency (AE), was calculated as per the following formula [28] and reported as kg grain/kg nutrient (NPK) applied.

$$\text{NUE} = \frac{Y_f - Y_c}{N_a}, \qquad (4)$$

where Y_f is grain yield (kg ha^{-1}) in fertilized plot; Y_c is grain yield (kg ha^{-1}) in control plot; N_a is nutrient (N + P$_2$O$_5$ + K$_2$O) applied (kg ha^{-1}).

3. Results and Discussion

3.1. Soil, Weather Variables, and Crop Evapotranspiration Effects on Yield Productivity. The locations were characterized by erratic and poorly distributed mono-modal rainfall pattern. Most of the rains came as short-duration, high-intensity storms between June and September out of which 70% of the total rainfall were received between July and August. In Minjibir, a total rainfall of 677 mm was received in 2014 and 493 mm in 2015 growing seasons while, in BUK, a total rainfall of 531 mm was received in 2014 and 563 mm in 2015 growing seasons. Mean monthly temperatures during the growing season ranged between 26 and 33°C. The soil of the experimental sites is characterized as sandy loam to sandy clay in texture with 0–20 cm depth and pH 4.86–5.98 indicating acidic, low in organic carbon, and nitrogen content varied from 183 kg ha^{-1} to 1179 kg ha^{-1} while available phosphorus varied from low to medium. A detail of the physical-chemical properties of the experimental sites is presented in Table 1.

The total rainfall recorded is within the crop water requirement for sorghum under semiarid conditions (450–650 mm) [29]. In Savanna zone of Mali, it was reported [30] that 11°C and 38°C are the lower and upper temperature threshold limit for sorghum. The present study agreed closely to both upper and lower temperature or base thresholds for the varieties used in both locations. The minimum and maximum temperatures obtained in the present study sites neither exceeded, nor went below sorghum growing temperature thresholds during the growing season. Daily crop evapotranspiration (ET$_0$) varied from 2.3 to 6.1 mm in both seasons at BUK while Minjibir varied from 1.8 to 6.5 mm. ET$_0$ recorded high value between flowering and maturity which coincided with the end of rainy season in both locations.

3.2. Effects of Nitrogen and Variety on Phenological and Morphological Traits of Sorghum

3.2.1. Days to Flowering and Maturity. There was a significant difference ($P < 0.05$) among N application rates for days to flowering and maturity at BUK while only days to maturity was significant at Minjibir (Table 2). Days to flowering and maturity were found slightly decreased with increased N application rates. Application under different N-rates attained flowering and maturity faster than control treatment (without fertilizer). Treatment with 100 kg N ha^{-1} reached maturation earlier by 3 and 5 days at BUK and Minjibir, respectively, compared to control treatment. Highly significant ($P < 0.001$) differences were observed among the sorghum varieties for flowering and maturity. ICSV400 flowered at 71 days after planting (DAP) followed by a local (85 DAP) and CSR01 (97 DAP), respectively. Also, ICSV400 variety attained physiological maturity at 91 DAP (indicating early maturing variety) while CSR01 and local varieties matured above 100 days in both locations and are therefore grouped as medium maturing varieties. Significant differences were observed between the years for sorghum days to 50% flowering in the 2 locations, though it was only significant for maturity in BUK. The differences in days to flowering could be associated with varietal response to day length (photoperiods) and soil moisture availability especially at Minjibir. The result was in agreement with earlier studied reported by [31]. ICSV400 and local varieties flowered and matured earlier in BUK than Minjibir due to higher fertility status obtained in BUK, while the plants also flowered and matured earlier in the fertilizer plots than the unfertilized plots. This implies that days to flowering and maturity are influenced by inherent

TABLE 2: Effect of fertilizer levels and variety on phenological and morphological traits of sorghum in Sudan Savanna zone Nigeria.

Treatment	BUK							Minjibir						
	50% flowering	Maturity	SPAD	LAI	SPAD	LAI	Plant height	50% flowering	Maturity	SPAD	LAI	SPAD	LAI	Plant height
	(days)		(3 WAP)		(6 WAP)		(cm)	(days)		(3 WAP)		(6 WAP)		(cm)
Year (Y)														
2014	76.32	106.31	41.90	3.78	48.83	2.84	230.6	88.4	109.5	41.72	2.50	40.18	2.78	201.1
2015	79.76	104.89	38.69	3.27	42.70	2.29	237.9	80.2	108.1	40.07	2.28	37.51	1.68	190.5
P of F	**0.005**	**0.041**	0.106	0.073	**0.024**	**0.013**	0.109	**0.002**	0.249	**0.035**	**0.036**	**0.012**	**<0.001**	0.094
SED	0.46	0.41	0.99	0.19	1.02	0.19	4.56	0.80	0.42	0.77	0.11	1.04	0.09	6.24
LSD	1.46	1.31	1.40	0.60	1.33	0.33	9.04	2.55	0.83	1.53	0.21	2.07	0.17	12.37
Fertilizer (F)														
0	79.4	107.5	38.8	3.20	40.0	2.46	232.4	84.2	111.7	39.0	1.87	33.9	1.69	180.7
20	78.7	105.5	40.7	3.58	44.6	2.65	218.3	85.0	110.2	40.0	2.23	35.8	2.12	185.0
40	77.6	104.9	41.2	3.62	47.5	2.61	258.2	84.1	108.1	38.5	2.47	37.2	2.11	186.5
60	77.8	105.3	39.8	3.57	45.0	2.46	236.8	84.2	108.8	41.1	2.50	42.0	2.31	209.7
80	77.5	105.7	40.7	3.64	46.8	2.66	233.0	84.0	107.6	44.7	2.65	42.8	2.70	209.2
100	77.3	104.7	40.7	3.53	46.9	2.53	226.8	84.2	106.2	42.1	2.61	41.4	2.46	203.6
P of F	**0.007**	**<0.001**	0.085	0.236	0.136	0.809	**<0.001**	0.792	**<0.001**	**<0.001**	**<0.001**	**<0.001**	**<0.001**	**0.015**
SED	0.64	0.66	0.88	0.20	1.56	0.19	7.90	0.84	0.72	1.33	0.18	1.81	0.15	10.81
LSD	1.28	1.30	1.75	0.39	3.10	0.37	15.66	1.67	1.44	2.65	0.36	3.58	0.30	21.43
Variety (V)														
CSR01	90.4	117.1	40.7	3.99	48.1	3.05	258.4	90.0	124.3	41.2	2.83	38.4	2.55	214.7
ICSV400	64.5	90.65	39.6	2.98	42.3	2.03	167.3	70.7	90.9	39.3	1.88	37.6	1.63	141.5
Local	79.2	109.0	40.6	3.59	46.9	2.61	277.0	85.2	111.1	42.2	2.45	40.6	2.51	231.2
P of F	**<0.001**	**<0.001**	0.156	**<0.001**	**<0.001**	**<0.001**	**<0.001**	**<0.001**	**<0.001**	**0.01**	**<001**	**0.05**	**<0.001**	**<0.001**
SED	0.46	0.46	0.63	0.14	1.10	0.13	5.58	0.60	0.51	0.94	0.13	1.28	0.11	7.64
LSD	0.90	0.92	1.24	0.27	2.19	0.26	11.07	1.18	1.01	1.87	0.25	2.53	0.21	15.15
CV (%)	2.9	2.2	7.6	17.2	11.1	25.5	11.7	3.6	2.3	11.3	26.3	16.1	23.7	19.1
Interaction														
Y × F	*	*	ns	ns	ns	ns	ns	ns	**	**	*	*	**	ns
Y × V	**	**	ns	**	**	*	**	**	**	ns	*	ns	**	**
F × V	ns	*	*	ns	ns	ns	**	ns	ns	ns	ns	ns	ns	ns
Y × F × V	*	ns	ns	ns	ns	ns	ns	ns	*	ns	ns	ns	ns	ns

SED: standard error of differences of means; LSD: least significant differences of mean (5% level); CV: coefficient of variation; ** and * mean significant different at 0.01 and 0.05 level of probability; ns: not significant.

soil fertility and drought. The sorghum plants flowered and matured significantly earlier under drought conditions as observed in Minjibir in 2015.

3.2.2. *Leaf Chlorophyll Content.* Leaf chlorophyll content is a good indicator of photosynthetic capacity of any crop [32]. Low concentrations of chlorophyll limit photosynthetic potential directly and lead to a decrease in biomass production in the plants [32, 33]. Table 2 shows that leaf chlorophyll content (SPAD) varied between years, among N application levels and sorghum varieties. While SPAD values at 3 and 6 WAP were significantly higher in 2014 than 2015 in Minjibir, it was significantly higher only at 6 WAP in BUK. This directly relate to the rainfall which was higher and better distributed in 2014 than 2015 in Minjibir. N-Fertilizer applications did not significantly affect SPAD (leaf chlorophyll content or health of the leaf) readings at 3 and 6 WAP in BUK though significant differences were observed among the fertilizer treatments in Minjibir as well as among the varieties in both locations. The soils in BUK were comparatively higher in nutrients to support the initial demand of the seedlings compared to soils in Minjibir. This was also evident in Minjibir where the leaf chlorophyll content increased with increasing N application rates. Highest SPAD value was recorded at $100\,kg\,N\,ha^{-1}$ in BUK and $80\,kg\,N\,ha^{-1}$ in Minjibir. While there was no significant differences among the varieties for SPAD readings at 3 WAP in BUK significant differences were observed at 6 WAP and in Minjibir at both 3 and 6 WAP. At BUK, CSR01 recorded the highest values (40.7 at 3 WAP and 48.1 at 6 WAP) compared to local and ICSV400. Slightly lower SPAD values recorded in Minjibir were compared to BUK confirming that BUK had higher fertility than Minjibir.

3.2.3. Leaf Area Index. There were no significant differences among N-fertilizer treatments for leaf area index (LAI) at 3 WAP and 6 WAP in BUK (Table 2) though highly significant differences were observed for LAI at 3 WAP and 6 WAP among the N-fertilizer treatments in Minjibir. The application of 80 kg N ha^{-1} produced significantly higher LAI value at 3 WAP and 6 WAP compared to the control treatment at both locations. At both locations, N-fertilizer increased early seedling vigour with higher LAI values recorded at 3 WAP but slightly decreased at 6 WAP which could be associated with increase in height of the plants compared to biomass. Similarly, reduction of LAI value was also reported in maize and forage maize by [34]. Early seedling vigour may be critical to maximizing uptake and efficient utilization of nutrients. LAI values at 3 and 6 WAP were significantly higher in 2014 than 2015 in Minjibir, though it was significantly higher only at 6 WAP in BUK. The differences in LAI between the two years in Minjibir were due to drought observed in 2015. This effect was also noticed in the plant health (SPAD values) and days to flowering.

Significant differences were also observed among the sorghum varieties in both locations for LAI at 3 and 6 WAP. At BUK, LAI value of 3.99 m^2/m^2 was recorded at highest from CSR01 and the lowest value was obtained from ICSV400 (2.98) at 3 WAP. Meanwhile, at 6 WAP, CSR01 recorded the highest of 3.05 m^2/m^2 and the lowest from ICSV400 (2.03 m^2/m^2). At Minjibir, CSR01 had highest LAI value of 2.83 m^2/m^2 and 2.55 m^2/m^2 at 3 and 6 WAP followed by a local variety, while the lowest value was recorded from ICSV400 (1.88 m^2/m^2 and 1.63 m^2/m^2 at 3 and 6 WAP).

3.2.4. Plant Height. Significant differences were observed among the different N-fertilizer rates as well as among the varieties in both locations for plant height. Increase in fertilizer application rate increased plant height up to 40 kg N/ha in BUK and up to 80 kg N/ha in Minjibir (Table 2). Nitrogen application therefore has a significant positive effect on sorghum growth in the two locations. It is interesting to note that the highest SPAD value as well as sorghum height in Minjibir was observed at 80 kg N/ha. Among the varieties, local control was significantly taller (277 cm) than CSR01 (258 cm) and ICSV400 (167 cm) in BUK. Similar observations were made in Minjibir where the local control was significantly taller (231 cm) than CSR01 (214.7 cm) and ICSV400 (141 cm).

3.3. Effect of N-Fertilizer Levels and Variety on Yield and Yield Components. The effect of different levels of N-fertilizer and varieties on the mean panicle length, 1000-seed weight, grain, and Stover yields in the two locations is presented in Table 3. N-fertilizer levels had no significant effect on panicle length in both locations, but highly significant on varieties with local variety recording the longest panicle (40.7 and 34.7 cm) over CSR-01 (30.7 and 28.1 cm) and ICSV400 (25.3 and 18.47 cm), respectively, at both locations. N-fertilizer application had no significant effect on 1000-seed weight, though there were significant differences among the sorghum varieties for 1000-seed weight. Local variety recorded highest mean value (29.2 g and 2.99 g) in both locations while the lowest value was measured from ICSV400 (22.7 g) in BUK and CSR01 (20 g) in Minjibir. The low 1000-seed weight of CSR01 in Minjibir was as a result of drought observed during the grain filling stage.

Significant differences were found among the N-fertilizer rates and varieties for both grain and Stover yields in the two locations (Table 3). Mean grain yield increased linearly with increase in N-levels across individual variety. The result showed that sorghum grain yields were higher in BUK over Minjibir by 29, 31, and 54% for local, ICSV400, and CSR01, respectively. The highest grain yield observed in BUK than Minjibir could be associated with favourable rainfall distribution, inherent higher soil micronutrients, and high water retention capacity of soil. Both the local and ICSV400 varieties had a comparative yield advantage over CSR01 in the two locations, due to earliness to flowering that coincides with the end of the rainy season for ICSV400 while the local variety possessed drought tolerance trait to complete its growth cycle even after cessation of rainfall. In BUK, N-fertilizer treatments increased yield over control treatment by 30–63% for CSR01, 38–49% for ICSV400, and 36–74% for local variety respectively. Though grain yield recorded low value in Minjibir compared to BUK, the application of N-fertilizer was highly significant. Grain yield was increased by 22–106% for CSR01, 28–96% for ICSV400, and 10–68% for local variety, respectively, over the control treatment. This result was in agreement with those reported by [4, 35] on sorghums under different N-level over Guinea Savanna, Ghana, and the semiarid region of India

N-Fertilizer treatment at 80 kg ha^{-1} produced the highest mean grain yields in BUK and Minjibir (2709 and 1881 kg ha^{-1}, resp.). This was significantly higher than control and 20 kg N/ha. The N-fertilizer rate increased grain yield in the range of 35–64% at BUK and 23–79% at Minjibir over control (no fertilizer) treatment. ICSV400 produced the highest mean grain yield (2677 kg ha^{-1}) over local (2411 kg ha^{-1}) and CSR01 (1902 kg ha^{-1}) varieties at BUK while ICSV400 recorded the highest mean grain (1848 kg ha^{-1}) over local (1693 kg ha^{-1}) and CSR01 (876 kg ha^{-1}) at Minjibir. While there were no significant interactions among the factors for grain yields in BUK, significant interactions were found in Minjibir (Table 3). Table 4 shows the interaction effects of year × fertilizer, year × variety, fertilizer × variety, and year × fertilizer × variety. Grain yield ranged from 440 kg ha^{-1} by CSR01 at 20 kg N-fertilizer application to 3096 kg ha^{-1} by local with application of 80 kgN fertilizer. In 2014, local variety produced significantly higher grain yields across the N-fertilizer levels, which ranged from 1592 kg ha^{-1} at zero fertilizer application to 3096 kg ha^{-1} at 80 kgN (indicating the optimum level) while the CSR01 produced the lowest grain yield at the optimum level application of 80 kgNha. Contrary to the result obtained in 2014, ICSV400 recorded higher grain yield across N-level ranging from 1009 kg ha^{-1}

TABLE 3: Effect of fertilizer levels and variety on panicle length, 1000-seed weight, grain and Stover yields and harvest index of sorghum in Sudan Savanna zone.

Treatment	BUK					Minjibir				
	Panicle length	1000-seed weight	Grain yield	Stover yield	Harvest Index	Panicle Length	1000-seed weight	Grain yield	Stover yield	Harvest index
	(cm)	(g)	(kg ha^{-1})		(%)	(cm)	(g)	(kg ha^{-1})		(%)
Year										
2014	34.09	29.38	2318	11435	19.92	26.69	28.53	1844	4580	28.87
2015	30.39	22.95	2342	7020	25.79	27.49	19.75	1101	4044	22.26
P of F	**<0.001**	**<0.001**	**0.873**	**<0.001**	**0.019**	**0.062**	**0.007**	**0.001**	**0.222**	**0.012**
SED	0.87	0.86	78.8	333.7	1.71	0.43	1.30	66	348	1.19
LSD	1.72	1.39	156.2	661.7	5.44	0.84	4.13	209	1108	1.84
Fertilizer (F)										
0	32.91	26.78	1649	5526	26.09	26.76	24.71	1048	3928	21.70
20	34.09	26.17	2231	6267	29.07	27.19	24.33	1288	4127	24.82
40	32.22	26.64	2452	6038	30.20	26.68	23.41	1324	4194	24.85
60	32.00	26.33	2579	6816	29.07	27.37	23.63	1534	4725	25.64
80	30.55	26.17	2709	7557	28.23	27.13	24.60	1881	4542	28.94
100	31.69	24.92	2360	6676	27.73	27.38	24.15	1757	4356	27.42
P of F	**0.286**	**0.459**	**0.001**	**0.030**	**0.148**	**0.893**	**0.790**	**<0.001**	**0.023**	**0.013**
LSD	2.97	1.92	452.1	1149	3.06	1.46	2.40	249.8	476	3.52
Variety (V)										
CSR01	30.70	26.59	1902	7162	22.6	28.11	19.99	876	4738	15.46
ICSV 400	25.33	22.70	2677	5061	36.1	18.47	22.46	1848	3750	33.32
Local	40.71	29.21	2411	7217	26.5	34.68	29.98	1693	4448	27.90
P of F	**<0.001**	**<0.001**	**<0.001**	**<0.001**	**<0.001**	**<0.001**	**<0.001**	**<0.001**	**<0.001**	**<0.001**
SED	1.06	0.68	131.2	329.7	1.17	0.52	0.86	73.8	232	1.29
LSD	2.10	1.35	266.1	668.7	2.39	1.03	1.70	149.7	463	2.63
CV (%)	16.1	3.6	28.5	17.9	32.7	9.4	17.3	32.7	26.4	23.3
Interaction										
Y × F	ns	ns	ns	ns	ns	*	ns	*	*	ns
Y × V	*	**	ns	*	ns	ns	*	**	ns	**
F × C	ns	ns	ns	ns	ns	ns	ns	*	**	ns
Y × F × V	ns	ns	ns	ns	ns	ns	ns	*	*	**

SED: standard error of differences of means; LSD: least significant differences of mean (5% level); CV: coefficient of variation; ** and * mean significant different at 0.01 and 0.05 level of probability; ns: not significant.

at zero N application to 2212 kg ha^{-1} at 60 kgN applied while the lowest grain yield was recorded from CSR01 across the N-level applied. The difference in rainfall amount received (677 mm in 2014 and 489 mm in 2015) could be directly linked to variation in yields recorded across the N-level and among varieties.

Furthermore, N-fertilizer at 80 kg ha^{-1} produced significantly higher mean Stover (7557 kg ha^{-1}) in BUK while 60 kgNha^{-1} produced significantly higher mean Stover (4725 kg ha^{-1}) in Minjibir than other treatments. N-treatments increased Stover yield by 9–35% at BUK and 5–19% at Minjibir over control treatment. The differences in N-fertilizer optimum rate at both locations could be attributed to initial soil micronutrients and rainfall distribution pattern.

Among the varieties, local produced significant higher mean Stover yield (7217 kg ha^{-1}) than CSR01 (7162 kg ha^{-1}) and ICSV400 (5061 kg ha^{-1}) at BUK; meanwhile CSR01 recorded higher mean Stover yield (4728 kg ha^{-1}) over local and ICSV-400 at Minjibir. High grain yields of sorghum obtained with increased rates of N-fertilizer can be attributed to the significant increase in the yield components, namely, a number of grains/panicle and 1000-grain weight, which corroborated the finding reported by [4]. Also, N-fertilizer application rate up to 80 kg N ha^{-1} had a profound linear effect on grain yield which implies that optimum N application rates for sorghum are 80 kg N ha^{-1}; further increased N rate did not increase grain yield proportionately. In addition, Stover yield increased with N rates as would be expected because higher

Table 4: Effects of year, N-fertilizer, and variety on sorghum grain yield Minjibir.

Treatment	2014				2015			
	CSR01	ICSV400	Local	Mean	CSR01	ICSV400	Local	Mean
0	701	1424	1592	1239	489	1009	1074	857
20	1250	1703	1965	1639	440	1428	944	937
40	897	1856	2283	1679	564	1425	919	969
60	1379	1978	2120	1826	606	2212	909	1242
80	1434	2760	3096	2430	1013	1599	1387	1333
100	1082	2627	3040	2250	657	2149	989	1265
Grand mean	1472							
LSD (Y × F)	307							
LSD (Y × V)	352							
LSD (F × V)	222							
LSD (Y × F × V)	500							

N-levels might have accelerated the conversion of rapidly synthesized carbohydrates (owing to increased N supply) into protein and developed a plant root system faster as reported by [36]. This resulted in early crop growth, which was finally expressed in taller plants, greater biomass yield, and consequently higher grain production. Reference [37] reported that the highest yield was obtained at the N-level of 125 kg ha^{-1} and likened the positive effect of increased N fertilization on dry matter yield to plant height, leaf, and stem dry weight and the amount of tillers increases which leads to the increase in total dry matter yield.

The physiological efficiency of assimilates from source into economic sinks is known as harvest index (HI). The effect N-fertilizer levels on harvest index (HI) were only significant at Minjibir but not significant differences at BUK (Table 3). Similar to observation made by [38], HI increased with increase in N-fertilizer in Minjibir; however, in BUK there was no clear cut response of HI to N-fertilizer rates. At both locations and among the varieties, ICSV400 had the highest mean HI (36.1 and 33.3%) at BUK and Minjibir followed by local variety (26.5 and 27.9%) while CSR01 recorded the lowest mean HI (22.6 and 15.5%) indicating genotypic variations in partitioning efficiency. However, higher HI may be attributed to lower mean biomass recorded while the lower HI could be linked to higher biomass as observed in ICSV400 and CSR01 accordingly. Also, lower mean HI values in this experiment among the varieties might suggest the need for the enhancement of biomass partitioning through genetic improvement.

3.4. Effect of N-Fertilizer Levels on Water Use Efficiency of Sorghums. Water use efficiency (WUE) is one of the most important indices for determining optimal crop-water management practices. Figure 2 shows the effect of N-fertilizer levels on WUE across the varieties. The estimated WUE was not significantly different among the N-treatment at BUK but highly significant effect was observed at Minjibir. In both locations, mean WUE value increased from 0 kgN to 80 kgN and dropped afterwards, though CSR01 and ICSV400 dropped after 60 kgN in Minjibir and BUK, respectively. This result agreed with similar studies reported by [39, 40] that increase in WUE with the increased N-fertilizer levels is likely to be related to the significant increase in total dry matter compared to grain yield alone. N application at 80 kg ha^{-1} estimated significantly higher mean WUE values across the N-treatments and varieties (with exception of ICSV400 at BUK and CSR-01 at Minjibir that recorded the highest WUE at 60 kg N ha^{-1}). Higher mean WUE obtained in ICSV400 implies more efficient water use than the other varieties in both locations, which could be associated with much less transpiration demand from vegetative biomass which enhanced water resources available for grain filling during the reproductive stage. In BUK, the mean WUE ranged from 4.4 to 12.9 kg ha^{-1}mm^{-1} and N-fertilizer treatments increased WUE by 48–55% over the control (no fertilizer) treatment. Meanwhile, at Minjibir, WUE varied from 1.7–11.5 kg ha^{-1}mm^{-1} and N-fertilizer treatments increased WUE by 54–76% over control treatment.

3.5. Effect of N-Fertilizer Levels on Nitrogen Use Efficiency of Sorghums. The effect of N-fertilizer levels on nitrogen use efficiency of sorghum varieties is presented in Figure 3. Nitrogen use efficiency (NUE) is calculated as a ratio of grain yield to the amount of N applied. NUE decreased as a linear function of increasing N-levels, but the estimated values varied between the two locations. This result was in agreement with several studies reported on various crops, for instance, on sorghum [9], millet [41], and maize [42]. The varieties showed significant differences for estimated NUE. In BUK, ICSV400 recorded the mean highest NUE (35.7 kg grain/kg) value, followed by local (30.4 kg grain/kg) and CSR-01 (21.2 kg grain/kg) at 20 kg N ha^{-1} and the value decreased with increased N-level. On the contrary, in Minjibir, local variety recorded the highest NUE (24.4 kg grain/kg^{-1}) with significant differences of more than 10 kg grain/kg^{-1} compared to other varieties at 20 kg N ha^{-1}. Meanwhile, CSR01 had significantly lower NUE than local and ICSV400 varieties across the different N-levels. The differences among sorghum varieties for higher NUE mechanisms could be associated with individual morphological, anatomical, and biophysical traits. This finding was similar to those reported by [4, 43]

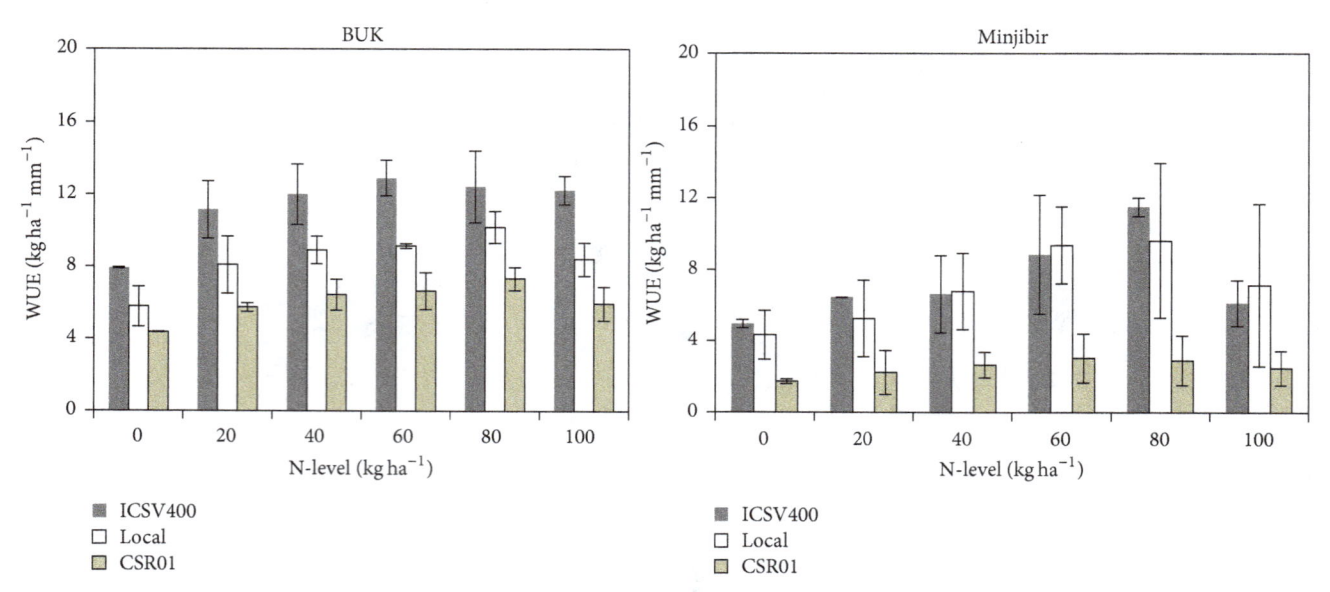

FIGURE 2: Effects of N-fertilizer rates on water use efficiency (WUE) of selected sorghum varieties in Sudan Savanna zone. *P of F equals 1.000 in BUK and 0.011 in Minjibir.*

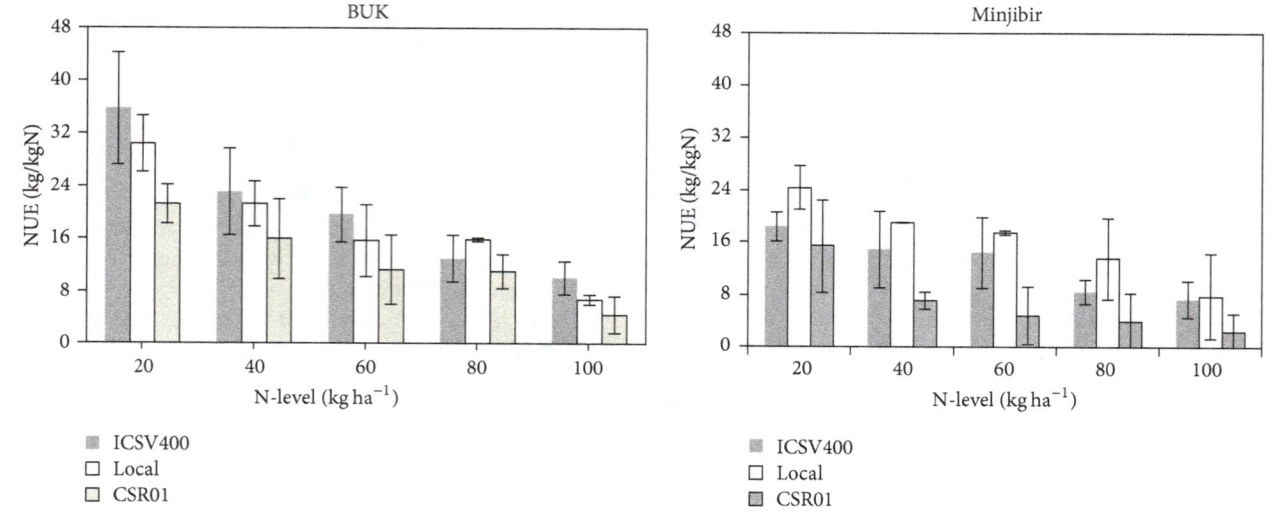

FIGURE 3: Effects of N-fertilizer rates on nitrogen use efficiency (NUE) of selected sorghum varieties in Sudan Savanna zone. *P of F equals 0.574 in BUK and 0.017 in Minjibir.*

that demonstrated genetic diversity for N use efficiency in grain sorghum. Thus, exploiting these differences in nutrient demand and efficiency is a possible alternative for reducing the cost and reliance upon fertilizer for maximizing yield productivity.

4. Conclusion

It was noted that increase in nitrogen fertilizer applications up to a point led to a significant increase in all parameters of crop growth, yield, and yield components of sorghum in the Sudan Savanna of Nigeria. Inherent soil fertility and rainfall distribution contributed significantly to the grain and Stover yields of sorghum irrespective of fertilizer rates applied. Treatment application at the rate of 80 kg N ha^{-1} was found to be the optimum application rate in both locations, above which no significant differences in grain and Stover yields. Seedling health at 3 weeks after planting was found to have significant correlation with plant height and yields implying that agronomic practices that will ensure excellent seedling vigor and health should be practiced for high grain and Stover yields. The WUE increased with increasing N-fertilizer application rates up to 80 kgN/ha and also significantly different among sorghum varieties. Grain yield was used in estimating the WUE, indicating optimum is reached at 80 kgNha^{-1} in both locations with slight differences among varieties. This implies that increased WUE between N-fertilizer levels could be more strongly associated with total dry matter produced than grain yield. The NUE decreased with increased N-fertilizer level with mean highest value recorded at $20\,kg\,N\,ha^{-1}$ at both locations. Across the three varieties, ICSV400 recorded the mean highest NUE (35.7 kg grain/kg) value at BUK and local recorded mean highest value (24.4 kg grain/kg) which decreased with increased N-fertilizer levels.

Acknowledgments

The authors would like to thank the International Crops Research Institute for The Semi-Arid Tropic (ICRISAT), Nigeria, for providing institutional support. The research was conducted under the former CG Research Program Dryland System.

References

[1] FAOSTAT 2018, http://www.fao.org/faostat/en/#data.

[2] FMARD (Federal Ministry of Agriculture and Rural Development), "Agricultural transformation agenda: we will grow Nigeria's agricultural sector," Draft for Discussion, Executive Summary Abuja, Nigeria, Federal Ministry of Agriculture and Rural Development, 2011.

[3] NAERLS (National Agricultural Extension and Research Liaison Services), *A Report on Adopted Village Concept for Agricultural Technology Transfer: NAERLS Experience*, 2014.

[4] J. S. Mishra, N. S. Thakur, P. Singh et al., "Productivity, nutrient-use efficiency and economics of rainy-season grain sorghum (Sorghum bicolor) as influenced by fertility levels and cultivars," *Indian Journal of Agronomy*, vol. 60, no. 1, pp. 76–81, 2015.

[5] D. P. Delmer, "Agriculture in the developing world: Connecting innovations in plant research to downstream applications," *Proceedings of the National Acadamy of Sciences of the United States of America*, vol. 102, no. 44, pp. 15739–15746, 2005.

[6] N. Dercas and A. Liakatas, "Water and radiation effect on sweet sorghum productivity," *Water Resources Management*, vol. 21, no. 9, pp. 1585–1600, 2007.

[7] S. Geng, F. J. Hills, S. S. Johnson, and R. N. Sah, "Potential yields and on-farm ethanol production cost of corn, sweet sorghum, fodderbeet, and sugarbeet," *Journal of Agronomy and Crop Science*, vol. 162, no. 1, pp. 21–29, 1989.

[8] H. A. Ajeigbe, B. B. Singh, and T. O. Oseni, "Cowpea-cereal intercrop productivity in the Sudan savanna Zone of Nigeria as affected by planting pattern, crop Variety and pest management," *African Crop Science Journal*, vol. 13, no. 4, pp. 269–279, 2005.

[9] S. S. J. Buah, J. M. Kombiok, and L. N. Abatania, "Grain sorghum response to NPK fertilizer in the Guinea Savanna of Ghana," *Journal of Crop Improvement*, vol. 26, no. 1, pp. 101–115, 2012.

[10] A. Y. Kamara, A. Menkir, S. O. Ajala, and I. Kureh, "Performance of diverse maize genotypes under nitrogen deficiency in the northern Guinea Savanna of Nigeria," *Experimental Agriculture*, vol. 41, no. 2, pp. 199–212, 2005.

[11] H. A. Ajeigbe, B. B. Singh, J. O. Adeosun, and I. E. Ezeaku, "Participatory on-farm evaluation of improved legume-cereals cropping systems for crop-livestock farmers: Maize-double cowpea in Northern Guinea Savanna Zone of Nigeria," *African Journal of Agricultural Research*, vol. 5, no. 16, pp. 2080–2088, 2010.

[12] F. Ekeleme, J. M. Jibrin, A. Y. Kamara et al., "Level and extent of Striga infestation of maize and cowpea crops in Bauchi and Kano States, Nigeria," Biophysical Baseline Survey, International Instititute of Tropical Agriculture, Ibadan, Nigeria, 2013.

[13] A. B. Ghosh, B. Maity, K. Chakrabarti, and D. Chattopadhyay, "Bacterial diversity of east Calcutta wet land area: possible identification of potential bacterial population for different biotechnological uses," *Microbial Ecology*, vol. 54, no. 3, pp. 452–459, 2007.

[14] J. W. Maranville, R. B. Clark, and W. M. Ross, "Nitrogen efficiency in grain sorghum," *Journal of Plant Nutrition*, vol. 2, no. 5, pp. 577–589, 1980.

[15] S. O. Ojeniyi, "Effects of goat manure on soil nutrient content and okra yield in rainforest area, Nigeria," *Applied Journal of Tropical Agriculture*, vol. 51, pp. 20–23, 2000.

[16] B. S. Parikshya Lama Tamang, *Nitrogen requirements for ethanol production from sweet and photoperiod sensitive sorghums in the Southern High Plains [M.S. thesis]*, Texas Tech University, 2010.

[17] R. Wiedenfeld, "Nutrient requirements and use efficiency by sweet sorghum," *Energy in Agriculture*, vol. 3, no. 1, pp. 49–59, 1984.

[18] T. R. Sinclair, "Nitrogen influence on the physiology of crop yield," in *Theoretical Production Ecology: Reflections and Prospects*, R. Rabbinge, J. Goudriaan, H. van Keulen, F. W. T. Penning de Vries, and H. H. van Laar, Eds., pp. 41–55, Pudoc, Wageningen, The Netherlands, 1990.

[19] R. C. Muchow and T. R. Sinclair, "Nitrogen response of leaf photosynthesis and canopy radiation use efficiency in field-grown maize and sorghum," *Crop Science*, vol. 34, no. 3, pp. 721–727, 1994.

[20] I. Dembele, Gestion des ressources organiques d'éléments minéraux dans la culture irriguée: cas des exploitations agricoles de la zone de l'Office du Niger (Mali), 2007.

[21] M. Bagayoko, "Effects of plant density, organic matter and nitrogen rates on rice yields in the system of rice intensification (SRI) in the 'office du Niger' in Mali," *ARPN Journal of Agricultural and Biological Science*, vol. 7, no. 8, pp. 620–632, 2012.

[22] T. R. Sinclair and T. Horie, "Leaf nitrogen, photosynthesis, and crop radiation use efficiency: a review," *Crop Science*, vol. 29, no. 1, pp. 90–98, 1989.

[23] NAGRAB, *Catalogue of Crop Varieties Released and Registered in Nigeria*, vol. 6, National Centre for Genetic Resources and Biotechnology (NACGRAB), Moor Plantation Ibadan, Nigeria, 2014.

[24] K. A. Gomez and A. A. Gomez, *Statistical Procedures for Agricultural Research*, John Wiley & Sons, New York, NY, USA, 2nd edition, 1984.

[25] J. Doorenbos and A. H. Kassam, "Yield response to water," FAO Irrigation and Drainage Paper 33, Food and Agriculture Organization of the United Nations, Rome, Italy, 1979.

[26] R. G. Allen, L. S. Pereira, D. Raes, and M. Smith, "Crop evapotranspiration—guidelines for computing crop water requirements," FAO Irrigation and Drainage Paper 56, Food and Agriculture Organization, Rome, Italy, 1998.

[27] Y. Kuslu, U. Sahin, T. Tunc, and F. M. Kiziloglu, "Determining water-yield relationship, water use efficiency, seasonal crop and pan coefficients for alfalfa in a semiarid region with high altitude," *Bulgarian Journal of Agricultural Science*, vol. 16, no. 4, pp. 482–492, 2010.

[28] R. Prasad, "Enhancing nutrient use efficiency: environmental benign strategies," in *Souvenir*, pp. 67–74, The Indian Society of Soil Science, New Delhi, India, 2009.

[29] G. W. P. Jewitt, H. W. Wen, R. P. Kunz, and A. M. Van Rooyen, "Scoping study on water use of crops/trees for biofuels in South Africa," WRC Report no 1772:1, Water Research Commission, Pretoria, South Africa, 2009.

[30] F. M. Akinseye, M. Adam, S. O. Agele, M. P. Hoffmann, P. C. S. Traore, and A. M. Whitbread, "Assessing crop model improvements through comparison of sorghum (Sorghum Bicolor L. Moench) simulation models: a case study of West African varieties," *Field Crops Research*, vol. 201, pp. 19–31, 2017.

[31] A. Folliard, P. C. S. Traoré, M. Vaksmann, and M. Kouressy, "Modeling of sorghum response to photoperiod: A threshold-hyperbolic approach," *Field Crops Research*, vol. 89, no. 1, pp. 59–70, 2004.

[32] A. L. Abdulai, H. Parzies, M. Kouressy, M. Vaksmann, F. Asch, and H. Brueck, "Yield stability of photoperiod sensitive sorghum (Sorghum bicolor L. Moench) accessions under diverse climatic environments," *International Journal of Agricultural Research*, vol. 7, no. 1, pp. 17–32, 2012.

[33] A. K. Van den Berg and T. D. Perkins, "Evaluation of a portable chlorophyll meter to estimate chlorophyll and nitrogen contents in sugar maple (Acer saccharum Marsh.) leaves," *Forest Ecology and Management*, vol. 200, no. 1–3, pp. 113–117, 2004.

[34] D. A. Okpara and G. P. Omaliko, "Influence of plant density and nitrogen fertilization on late-season forage maize," *Journal of Applied Chemistry and Agricultural Research*, vol. 5, pp. 54–56, 1999.

[35] S. S. J. Buah and S. Mwinkaara, "Response of sorghum to nitrogen fertilizer and plant density in the Guinea Savanna Zone," *Journal of Agronomy and Crop Science*, vol. 8, no. 4, pp. 124–130, 2009.

[36] R. K. Pandey, J. W. Maranville, and M. M. Chetima, "Deficit irrigation and nitrogen effects on maize in a Sahelian environment: II. Shoot growth, nitrogen uptake and water extraction," *Agricultural Water Management*, vol. 46, no. 1, pp. 15–27, 2000.

[37] R. P. Beyaert and R. C. Roy, "Influence of nitrogen fertilization on multi-cut forage sorghum-sudangrass yield and nitrogen use," *Agronomy Journal*, vol. 97, no. 6, pp. 1493–1501, 2005.

[38] J. R. Lawrence, Q. M. Ketterings, and J. H. Cherney, "Effect of nitrogen application on yield and quality of silage corn after forage legume-grass," *Agronomy Journal*, vol. 100, no. 1, pp. 73–79, 2008.

[39] I. de Barros, T. Gaiser, F.-M. Lange, and V. Römheld, "Mineral nutrition and water use patterns of a maize/cowpea intercrop on a highly acidic soil of the tropic semiarid," *Field Crops Research*, vol. 101, no. 1, pp. 26–36, 2007.

[40] M. E. Ahmed, M. A. M. Baldu, and M. M. B. Zaied, "Effect of tillage depth and pattern on growth and yield of grain sorghum (Sorghum bicolor L. Moench) under rain-fed," *Journal of Novel Applied Sciences*, vol. 1, no. 3, pp. 68–73, 2012.

[41] N. Gupta, A. K. Gupta, V. S. Gaur, and A. Kumar, "Relationship of nitrogen use efficiency with the activities of enzymes involved in nitrogen uptake and assimilation of finger millet genotypes grown under different nitrogen inputs," *The Scientific World Journal*, vol. 2012, Article ID 625731, 10 pages, 2012.

[42] M. M. Hefny and A. A. Aly, "Yielding ability and nitrogen use efficiency in maize inbred lines and their crosses," *International Journal of Agricultural Research*, vol. 3, no. 1, pp. 27–39, 2008.

[43] J. C. Gardner, J. W. Maranville, and E. T. Paparozzi, "Nitrogen use efficiency among diverse sorghum cultivars," *Crop Science*, vol. 34, no. 3, pp. 728–733, 1994.

Analysis of Direct and Indirect Selection and Indices in Bread Wheat (*Triticum aestivum* L.) Segregating Progeny

Zine El Abidine Fellahi [1], Abderrahmane Hannachi,[2] and Hamenna Bouzerzour[3]

[1]Department of Agronomy, Faculty of Natural, Life and Earth Sciences and the Universe, University of Mohamed El Bachir El Ibrahimi, 34034 Bordj Bou Arréridj, Algeria
[2]National Agronomic Research Institute of Algeria (INRAA), Unit of Sétif, 19000 Sétif, Algeria
[3]Department of Ecology and Plant Biology, Valorization of Natural Biological Resources Laboratory, Faculty of Natural and Life Sciences, University of Ferhat Abbas Sétif-1, 19000 Sétif, Algeria

Correspondence should be addressed to Zine El Abidine Fellahi; zinou.agro@gmail.com

Academic Editor: Iskender Tiryaki

Three selection methods including direct and indirect selection along with selection index based on the phenotypic values of eleven traits of agronomic interest were assessed for their application in F_4 bread wheat progenies. Significant genetic variation existed among parents and crosses for the traits measured. The following were the most efficient indices for simultaneous selection of superior lines for yield and its components: base index of Williams, followed by the sum of ranks index of Smith and Hazel. The selection-based index provided the highest grain yield gains as compared to the other selection criteria, except for flag leaf area, indicating that the direct and indirect monotrait selection were not appropriate in the situation analyzed in this work. PCA identified Ain Abid × Mahon-Demias, Ain Abid × Rmada, and Ain Abid × El-Wifak as the most promising populations. At 5% selection intensity, the top 30 lines selected were distinguished, in comparison with the standard check Hidhab, by significant improvements in yield and yield components.

1. Introduction

In Algeria, most of wheat producing areas are located in the High Plateaus which are characterized by cold winters, insufficient and erratic rainfall, frequent spring frosts, and late-season sirocco occurrence [1, 2]. In addition to these climatic stresses, there are some other technical constraints that essentially arise from the use of unproductive varieties and often bad agronomic practices. Selection for a better adaptation to environmental stresses is, therefore, more promising outcome in the field of wheat breeding. Breeders are continually seeking to improve the selection methods in order to develop superior wheat varieties with high grain yield, good end-use quality, and tolerance to biotic and abiotic stresses.

Direct selection based on grain yield is mainly practiced in wheat breeding programs without considering the adaptive traits that are crucial production regulators under variable environments [3–5]. In these environments, the presence of genotype × environment interactions reduces the efficiency of using grain yield as the sole selection criterion and, thus, complicates the efforts of selection [6, 7]. In addition to the environmental effects, other factors such as polygenic nature, low heritability of grain yield, linkage, and nonadditive gene action may make the selection less efficient mainly in early segregating generations.

In order to overcome these difficulties, breeders are focusing on other traits that can be used in parallel or independently of yield in a multitraits approach. Indirect selection uses some yield components that are more heritable than yield itself and more stable in relation to genetic and environmental factors affecting them. When these components are measured without error and expressed in appropriate units, their product is yield. This has created new opportunities for plant breeders to use certain morphological, physiological, and biochemical traits during selection for grain yield. In the literature, several authors have reported the use of many of these traits to improve grain yield in diverse environments [8–10].

Selection-based index is another approach, certainly complex, but can avoid the limits of the single-trait selection, particularly the undesirable between-trait relations that present an additional nuisance in breeders' work [11]. Selection-based index approach targets the simultaneous improvement of several traits at the same time, including the grain yield [12, 13]. The indices allow the use of a single value in the selection process, since the analysis is carried out by means of linear combinations of phenotypic data of different traits of agronomic interest with the genetic properties of a population [14]. The objective is to guarantee the improvement of the population genotypic values and consequently the efficiency of the selection process, maximizing the expected genetic gain. In this purpose, many selection indices have been used as an effective selection criterion in plant breeding programs on different crops [15]. However, the conditions determining the usefulness of an appropriate selection index may vary with individual plant breeder. The objective of this research paper was to evaluate the efficiency and applicability of different selection criteria based on the estimation of genetic gain in F_4 segregating populations of bread wheat evaluated under semiarid conditions.

2. Materials and Methods

2.1. Plant Material and Experimental Design. This experiment had 609 genotypes, comprising 20 F_4-derived families and their 9 parents. The history of these families is that, after the initial crosses in 2010/11 [18], 30 F_2 lines were selected in each family by using pedigree method and were evaluated during the 2012/13 cropping season [19]. A total of 600 lines were planted and harvested in bulk during two consecutive growing seasons 2013/14 and 2014/15. The F_4 lines along with their parents were planted in Sétif Research Unit (36°15′N; 5°87′E; 1081 masl) of the Algerian National Institute of Agronomic Research (INRAA) in a modified randomized complete block design with three replications. The experimental unit consisted of a single row plot of 2 m length with rows spaced 0.2 m apart. The plots were fertilized with 100 kg ha^{-1} of 46% superphosphate, before sowing, and 75 kg ha^{-1} of 35% urea, at tillering stage. Weed control was performed chemically using 12 g ha^{-1} of the Granstar *[tribenuron methyl]* herbicide. According to Chennafi et al. [20], the climate of the region is a semiarid type, continental with cold winter and hot and dry summer. Total annual rainfall is around 350 mm. The soil used was classified as a silty clay with high $CaCO_3$ content [20].

2.2. Measurements. The following traits were measured as per plot basis. Chlorophyll content (CHL, Spad) was determined at heading stage using the SPAD-502 chlorophyll meter (Minolta Camera Co., Osaka, Japan). Flag leaf area (FLA, cm^2) was obtained using the formula described by Spagnoletti Zeuli and Qualset [21]: FLA (cm^2) = L (cm) × l (cm), where L is the flag leaf length and l refers to the flag leaf width. Heading date (HD, days) was recorded as the number of calendar days from January first to the date when 50% of the spikes were half way out from flag leaf. Plant height (PH, cm) was measured at maturity from the soil surface to the top of the spike, awns excluded. Above ground biomass (BIO, g m^{-2}) was estimated from a harvested area of 0.5 m long × 0.20 m interrow spacing, which served also to obtain the grain yield (GY, g m^{-2}), number of spikes (SN, m^{-2}), and spikes weight (SW, g m^{-2}). 1000-kernel weight (TKW, g) was obtained from the count and weight of 250-kernel. The number of grains per spike (GN) was derived from estimated values of grain yield, number of spikes, and 1000-kernel weight. Harvest index (HI, %) was estimated by the ratio between grain yield and above ground biomass.

TABLE 1: The skeleton of the analysis of variance.

Source of variation	df	MS	F-test
Block	$b-1$	M1	M1/M6
Genotype	$g-1$	M2	M2/M6
Parents	$P-1$	M3	M3/M6
F_4	$n-1$	M4	M4/M6
Parents vs. F_4	1	M5	M5/M6
Residual	$(g-1)(b-1)$	M6	--
Total	$bg-1$	--	--

2.3. Data Analysis. Data collected were subjected to analysis of variance following the procedures described by McIntosh [22]; the skeleton of the analysis of variance is shown in Table 1. The statistical model used considered the complete randomized block design as $Y_{ij} = \mu + g_i + b_j + \varepsilon_{ij}$, where Y_{ij} is the observation of the ith genotype, evaluated in the jth replicate; μ is the overall mean of the experiment; g_i is the effect of the ith genotype; b_j is the effect of the jth block; and ε_{ij} is the effect of the ijth plot.

In case F-test was significant, standard error and critical differences were calculated by using the least significant difference test at 0.05 probability level (LSD0.05) according to Steel and Torrie [23]: LSD0.05 = $t_{0.05}\sqrt{2\sigma_e^2/b}$, where $t_{0.05}$ is the tabulated value of the t-test at 0.05 probability level for $(g-1)(b-1)$ residual degrees of freedom; σ_e^2 is the residual variance; and b refers to the number of replications or blocks.

The genotypic variance (σ_g^2) and residual variance (σ_e^2) were calculated and served to determine the following genetic and nongenetic parameters for each trait. The coefficient of experimental variation (CV) was calculated by CV (%) = $\sqrt{\sigma_e^2/\overline{X}}$, where σ_e^2 is the residual variance and $\overline{X}$ is the general mean of the trait. The coefficient of genetic variation (CV_g) was calculated by the following equation: CV_g (%) = $\sqrt{\sigma_g^2/\overline{X}}$, where σ_g^2 is the genotypic variance and $\overline{X}$ is the general mean of the trait. The variation index was determined as the ratio of CV_g/CV_e. Broad-sense heritability of average progenies (h_{bs}^2) was estimated by the expression [24]: h_{bs}^2 (%) = 100 × $[\sigma_g^2/(\sigma_g^2+\sigma_e^2)]$. Narrow-sense heritability of individuals within families (h_{ns}^2) was determined based on the parent-offspring regression. To do so, two methods were used: the first was by the linear regression of F_4 on the parental F_3 individual values [16], while the second approach was performed with standardized data of offspring (F_4) versus standardized of the corresponding parent (F_3) according to Frey and Horner [17].

The selection gains were estimated among families based on three selection criteria: direct selection, indirect selection, and selection-based index considering the selection intensity of 5% of top families. The expected gains by direct selection for each trait evaluated were estimated by the expression [25]: $\Delta G_i = h^2_i \times S_i = h^2_i \times (\overline{X}_{si} - \overline{X}_{0i})$, where ΔG_i is the gain with the direct selection carried for the ith trait; h^2_i is the heritability of the ith trait; S_i refers to the differential selection of the ith trait; $\overline{X}_{si}$ is the mean of the ith trait for the selected individuals; and $\overline{X}_{0i}$ is the mean of the ith trait in the base population. The expected gain of direct selection, expressed as a percentage of the population mean, is given by $\Delta G_{ii}\% = (\Delta G_i \times 100)/\overline{X}_{0i}$.

Gains from indirect response to selection were calculated using the following expression [25]: $GS_{j(i)} = h^2_j \times (\overline{X}_{ij} - \overline{X}_{0j}) = h^2_j \times DS_{j(i)}$, where $GS_{j(i)}$ is the gain on the jth trait, with selection based on the ith trait; $\overline{X}_{sj}$ is the mean of the jth trait for the selected individuals based on the ith trait; $\overline{X}_{0j}$ is the mean of the jth trait; h^2_j is the heritability of the jth trait; and $DS_{j(i)}$ refers to the differential selection of the jth trait, in which the selected lines presented the best performance for the ith trait. The expected gain of indirect selection, expressed as a percentage of the population mean, is given by $GS_{j(i)}\% = (GS_{j(i)} \times 100)/\overline{X}_{0i}$.

For the selection-based index, the following methodologies were used for gains estimation: the classic index proposed by Smith [26] and Hazel [27], the base index of Williams [28], the free weights and parameters index of Elston [29], the index of desired gains of Pesek and Baker [30], the multiplicative index of Subandi et al. [31], the sum of ranks index of Mulamba and Mock [32], and the genotype-ideotype distance index proposed by Cruz [25]. Each index displays certain particularities in its calculations and, as such, application is generally laborious due to the need to assign adequate economic weights to each trait. Based on the different analytical procedures of selection, the best populations were identified and the gains from selection were calculated. All statistical analyses were carried out using Genes software [33] and a Microsoft Excel© spreadsheet.

3. Results and Discussion

3.1. Genetic Variability and Heritability. The analysis of variance revealed significant genotype effect for almost all the traits under study except for flag leaf area and grain yield, which were not significant at 0.05 probability level (Table 2). This provides evidence of the presence of sufficient genetic variability among parents and hybrids that can be exploited in wheat breeding program through selection. Partitioning the genotype effect indicated significant differences between all parents for HD, PH, SN, TKW, SW, GN, BIO, and HI, and significant interpopulation differences for HD, PH, and GN. The contrast "Parents versus F_4," was highly significant for HD, PH, and SN, while the F_4 populations exhibited significant differences for nearly all the observed quantitative phenotypic traits, except for CHL, FLA, GN, and HI, which were not significant at the 5% probability level. The results of this study corroborate those of Abd El-Shafi [34], who reported significant and highly significant differences among genotypes (families + parents) and families for all studied traits across three segregating generations F_2, F_3, and F_4. This author, also, reported that greater response to selection can be expected from selection in cross having greater phenotypic and genotypic variances.

The coefficient of variation (CV) presented values between 0.7 and 26.8% for heading date and grain yield, respectively (Table 2). The CV above 20% is considered high, indicating high dispersion of the experimental data, which may have been caused by the genetic and phenotypic differences between the studied materials. High CV estimate obtained for grain yield can be explained by the fact that it is quantitative trait, governed by several genes and highly influenced by the environment.

The variances values, coefficient of variation, and genetic parameters estimates for wheat traits studied are presented in Table 3. Broad-sense heritability is the proportion of total phenotypic variation due to all genetic effects. The knowledge of the genotypic determination coefficient (h^2_{bs}) allows establishing an estimate of the genetic gain to be obtained and defines the best strategy to be used in the plant breeding program [35]. In this study, the estimated broad-sense heritability varied from 0.00 to 91.80%. The highest values were found for heading date (91.80%) followed by plant height (80.82%), 1000-kernel weight (72.90%), biomass (65.33%), and number of grains per spike (61.64%), indicating that these traits are highly heritable among the genotypes evaluated. These results can be confirmed with the values obtained by the CV_g/CV_e ratio that were close to or greater than 1 for these traits, suggesting satisfactory conditions for selection [36]. Moderate estimates of h^2_{bs} occurred for the number of chlorophyll contents (40.31%), number of spikes (52.00%), spikes weight (54.85%), and harvest index (40.67%). On the other hand, the lowest values of h^2_{bs} were found for the flag leaf area (0.00%) and grain yield (19.39%). These traits exhibited also low CV_g/CV_e ratio values, indicating the dominant effect of the environment on crop.

Generally, literature indicates widely varying narrow-sense heritability estimates. Mesele et al. [37] reported high heritability values for days to heading, days to maturity, and 1000-kernel weight; moderate estimates for grain filling period, spike length, number of spikelets per plant, grains per spike, and harvest index; and low values for number of tillers per plant, biomass yield and grain yield. Evaluating seven F_2 populations derived through cross combinations of five parental varieties/lines of bread wheat, Saleem et al. [38] found low to high broad-sense heritability values ranging from 4.75% to 92.6% depending on the trait and the cross. The findings of Yaqoob [39] showed that heritability estimates were low for number of tillers per plant (20%), grains per spike (26.81%), days to maturity (30.13%), spike length (36.66%), and 1000-kernel weight (38.68%), moderate for plant height (45.79%), and high for heading date (84.73%)

TABLE 2: Analysis of variance of different bread wheat traits studied.

SV	Bloc	Genotypes	Parents	F_4	Parents versus F_4	Error	CV (%)
df	2	28	8	19	1	56	
CHL	8.6	19.0**	9.5ns	23.8**	3.6ns	8.6	6.6
HD	1.5	9.0**	18.3**	2.5**	58.7**	0.7	0.7
FLA	89.4	11.1ns	6.9ns	6.6ns	0.1ns	14.9	20.4
PH	9.1	86.3**	103.0**	80.0**	72.6*	16.6	6.3
SN	1183.1	11187.5**	26092.4**	2657.8ns	53994.5**	5371.1	15.9
TKW	33.5	28.2**	39.4**	12.7ns	232.1**	7.7	7.4
SW	117034.6	44873.6**	74490.4**	18248.8ns	313746.3**	20259.8	20.9
GN	36.6	44.7**	61.3**	37.0*	41.0ns	17.1	16.8
BIO	81189.5	131077.4**	243142.1**	43601.7ns	896559.2**	45440.8	18.5
GY	34149.2	15727.9ns	26835.4ns	11096.5ns	56205.9*	12677.8	26.8
HI	122.2	62.3*	136.6**	36.1ns	21.4ns	37	16.8

CHL: chlorophyll content (Spad), HD: heading date (days), FLA: flag leaf area (cm^2), PH: plant height (cm), SN: number of spikes, TKW: 1000-kernel weight (g), SW: spikes weight (g m^{-2}), GN: number of grains per spike, BIO: above ground biomass (g m^{-2}), GY: grain yield (g m^{-2}), HI: harvest index (%), and ns, *, and **: nonsignificant and significant effect at 0.05 and 0.01 probability.

TABLE 3: Genetic and nongenetic parameters of different bread wheat traits studied.

Traits	σ^2_p	σ^2_e	σ^2_g	h^2_{bs}	h^2_{ns} (SK)	h^2_{ns} (FH)	CV_g (%)	CV_g/CV_e
Chl	3.36	2.00	1.35	40.31	28.55	30.13	2.64	0.47
HD	3.01	0.25	2.76	91.80	16.69	34.32	1.31	1.93
FLA	3.68	4.98	0.00	0.00	0.00	0.00	0.00	0.00
PH	28.78	5.52	23.26	80.82	27.27	27.27	7.43	1.19
SN	3729.17	1790.37	1938.79	52.00	0.00	0.00	9.59	0.60
TKW	9.41	2.55	6.86	72.90	31.64	25.99	6.98	0.95
SW	14957.91	6753.25	8204.67	54.85	8.66	10.70	13.19	0.64
GN	14.67	5.63	9.04	61.64	16.70	7.14	12.28	0.73
BIO	43692.45	15146.93	28545.52	65.33	3.37	5.93	14.48	0.79
GY	5242.64	4225.92	1016.71	19.39	10.13	12.31	7.58	0.28
HI	20.76	12.31	8.45	40.67	14.10	14.97	8.03	0.48

CHL: chlorophyll content (Spad), HD: heading date (days), FLA: flag leaf area (cm^2), PH: plant height (cm), SN: number of spikes, TKW: 1000-kernel weight (g), SW: spikes weight (g m^{-2}), GN: number of grains per spike, BIO: above ground biomass (g m^{-2}), GY: Grain yield (g m^{-2}), HI: harvest index (%), h^2_{bs}: broad-sense heritability, h^2_{ns} (SK): narrow-sense heritability according to Smith and Kinman [16], and h^2_{ns} (FH): narrow-sense heritability according to Frey and Horner [17].

and grain yield (99.83%). His results also indicated that most of these traits exhibited low heritability under drought stress conditions, suggesting the presence of high genotype × environment interactions that affected the crop behavior.

Narrow-sense heritability is the proportion of the total phenotypic variation that is due to the additive effects of genes. This component of variation is important because it is the only variation that natural selection can act on. Hence, h^2_{ns} determines the resemblance of offspring to their parents and the population's evolutionary response to selection. There can be considerable nonadditive genetic variance, but this does not contribute to the resemblance between parents and offspring, or the response to selection. Low to moderate narrow-sense heritability values were recorded in this study. Chlorophyll content, heading date, plant height, 1000-kernel weight, and number of grains per spike recorded the highest estimates. These traits were less influenced by the environmental factors and would respond positively to a selection pressure in the current breeding program. The low heritability values can be explained by the change occurring in the segregating lines behavior from the precedent to the current generation. This change may be due to nonadditive gene action and/or high environmental factors effects.

Means of the variables measured showed that the best values varied depending on the cross and the trait, and the few populations had the best performances for several traits at the same time (Table 4). The best grain yielding population (557.5 g) was Ain Abid × El-Wifak which had also the highest average for the number of grains per spike (31.47 grains), spikes weight (857.3 g), above ground biomass (1350.3 g), and harvest index (41.3%). Ain Abid × Hidhab cross combination had the longest vegetative cycle with an average of 127.8 days and presented the highest mean for the flag leaf area (21.0 cm^2). Acsad$_{1069}$ × El-Wifak had the highest average for the chlorophyll content (54.0 Spad). Acsad$_{1069}$ × Mahon-Demias was the tallest (75.23 cm), while Acsad$_{1135}$ × Rmada

TABLE 4: Means of the measured traits for 20 F_4 bread wheat populations.

Population	CHL	HD	FLA	PH	SN	TKW	SW	GN	BIO	GY	HI
Acsad$_{901}$ × Mahon-Demias	43.0	126.0	20.1	73.7	448.3	38.6	634.0	22.0	1135.0	379.2	33.4
Acsad$_{901}$ × Rmada	44.5	125.2	19.9	60.8	442.4	34.6	633.6	28.1	1049.7	425.5	40.5
Acsad$_{901}$ × Hidhab	42.5	125.9	19.8	65.7	426.7	33.4	616.7	23.0	1098.3	325.5	29.6
Acsad$_{901}$ × El-Wifak	45.1	124.9	20.4	70.4	467.3	33.2	684.3	30.4	1149.0	460.4	40.1
Acsad$_{899}$ × Mahon-Demias	43.7	125.7	16.9	70.8	476.0	36.4	626.0	20.5	1151.7	366.8	31.9
Acsad$_{899}$ × Rmada	43.4	125.0	16.6	59.5	479.7	37.8	625.7	20.0	1032.7	368.8	35.7
Acsad$_{899}$ × Hidhab	44.4	126.9	18.2	56.6	406.0	35.0	598.0	22.8	997.0	336.0	33.7
Acsad$_{899}$ × El-Wifak	45.4	125.0	18.1	63.1	414.7	35.5	566.7	24.2	965.7	357.7	37.0
Acsad$_{1135}$ × Mahon-Demias	41.6	126.1	17.0	68.1	456.7	39.8	604.0	21.0	1073.7	383.4	35.7
Acsad$_{1135}$ × Rmada	42.2	125.2	16.8	61.5	503.7	36.9	685.3	24.0	1113.7	440.4	39.6
Acsad$_{1135}$ × Hidhab	41.5	127.0	17.5	68.8	477.0	39.5	744.7	23.2	1268.7	450.9	35.5
Acsad$_{1135}$ × El-Wifak	45.5	125.2	17.0	64.6	397.0	34.6	489.3	22.1	879.0	309.9	35.3
Acsad$_{1069}$ × Mahon-Demias	43.8	125.8	18.2	75.2	439.0	39.2	623.3	21.7	1193.7	371.3	31.1
Acsad$_{1069}$ × Rmada	44.2	125.0	17.3	61.7	414.7	34.8	581.0	24.5	961.7	360.8	37.5
Acsad$_{1069}$ × Hidhab	46.7	125.9	18.5	60.5	411.3	35.0	617.0	26.6	976.7	395.3	40.5
Acsad$_{1069}$ × El-Wifak	54.0	125.0	18.4	66.5	423.3	34.6	611.0	27.9	994.0	403.3	40.6
Ain Abid × Mahon-Demias	42.0	127.7	20.0	72.7	454.7	38.8	747.3	27.2	1286.7	481.8	37.5
Ain Abid × Rmada	40.3	126.3	20.5	65.8	435.7	36.5	699.3	28.2	1171.7	458.1	39.1
Ain Abid × Hidhab	43.2	127.8	21.0	60.2	405.0	36.8	693.3	30.4	1130.7	462.8	40.9
Ain Abid × El-Wifak	43.6	126.9	20.3	65.1	469.3	37.7	857.3	31.5	1350.3	557.5	41.3
Parents Range	39.1	125.0	8.7	53.0	270.0	30.0	320.0	9.7	530.0	185.0	14.1
	52.2	132.0	39.1	82.0	800.0	49.0	1260.0	39.4	2330.0	809.0	63.6
F_4 Range	33.4	124.0	7.6	42.0	80.0	19.0	80.0	9.3	150.0	43.0	12.6
	53.8	136.0	36.4	100.0	850.0	51.0	1480.0	56.8	2420.0	921.0	70.0
LSD0.05	4.8	1.4	6.3	6.7	119.7	4.5	235.2	6.8	353.2	185.1	9.9

CHL: chlorophyll content (Spad), HD: heading date (days), FLA: flag leaf area (cm^2), PH: plant height (cm), SN: number of spikes, TKW: 1000-kernel weight (g), SW: spikes weight (g m^{-2}), GN: number of grains per spike, BIO: above ground biomass (g m^{-2}), GY: grain yield (g m^{-2}), and HI: harvest index (%).

exhibited the highest average for the number of spikes (503.7 spikes). For 1000-kernel weight, Acsad$_{1135}$ × Mahon-Demias was the best population with an average of 39.8 g.

As the contrast "Parents versus F$_4$" was significant for PH, SN, TKW, SW, BIO, and GY, and, compared with the least significant difference (LSD0.05), significant differences existed between the parents and hybrids values for these measured traits. These differences were 1 to 3 times higher than the LSD0.05 and were in favor of some F$_4$ lines, indicating that they perform better than the parents and suggesting the possibilities of making significant changes through effective selection. These findings were in harmony with those obtained by Abd El-Shafi [34], Löffler and Busch [40], and Alexander et al. [41].

3.2. Genetic Gain from Direct and Indirect Selection and Selection-Based Index. The gains of direct and indirect selection are shown in Table 5. The results showed a big variation in the percentages of gains among the measured traits. The total sum of gains per selection varied from −36.11% for heading date to 34.75% for flag leaf area. Generally, the direct selection based on chlorophyll content, heading date, spikes weight, grain yield, and harvest index resulted in negative total gains. On the other hand, flag leaf area, plant height, number of spikes, 1000-kernel weight, number of grains per spike, and biomass recorded positive total gains.

The gains obtained by direct selection were higher than those of indirect selection. But, sometimes the indirect selection may be more efficient, especially if the secondary trait is highly correlated with yield and is easily measurable [42]. The number of grains per spike (15.86%) followed by above ground biomass (12.30%), plant height (12.10%), and number of spikes (10.84%) exhibited the greatest gains from direct selection. Inversely, heading date (−1.13%), spikes weight (−4.99%) and harvest index (−3.81%) showed negative gain values. Negative gain for heading date is desired in the case of this study as the Algerian wheat breeding program designed for semiarid regions; precocity is a crucial criterion adopted for selection. It is related to the ability of plants to shorten their cycle, so as to decrease their exposure to the late-season sirocco weathering.

The direct selection for chlorophyll content resulted in negative indirect gains for nearly all the other traits, except for the flag leaf area, number of grains per spike, and harvest index that showed positive responses. For heading date, the indirect selection gains were only positive for chlorophyll content, number of spikes, and harvest index. The correlative effects of the selection based on the flag leaf area were desirable for grain yield, although the indirect gain for the remaining traits was practically negative.

The direct selection for plant height resulted in positive indirect gains for heading date, number of spikes, 1000-kernel weight, and spikes weight. In addition, positive responses of the selection based on yield and yield components were observed for grain yield itself, heading date, 1000-kernel weight, and spikes weight. However, negative indirect gains were exhibited for other traits including chlorophyll content, flag leaf area, and number of grains per spike, indicating that indirect selection for one variable for gain in another is unfeasible, because there will be a loss in the indirectly selected variable. In cases of negative gains, the model is considered inappropriate for selection in this plant material.

The highest correlated responses for grain yield were generated though indirect selection on the base of the flag leaf area (8.04%, i.e., 33.87 g m^{-2}), followed by the spike fertility (3.65%, i.e., 15.39 g m^{-2}), 1000-kernel weight (3.63%, i.e., 15.31 g m^{-2}), and above ground biomass (2.74, i.e., 11.54 g m^{-2}). These results indicated that the indirect selection was to be more effective in improving the primary trait than the indirect selection based on other traits and/or on the direct selection based on the grain yield itself.

Several authors have estimated the genetic gains of traits involved in yield determination. The results are often inconsistent and scarce. Our results were consistent with those of DePauw and Shebeski [43] and Inagaki et al. [44], who mentioned that direct selection on the basis of yield is ineffective in early generations. Benmahammed et al. [45] reported the same findings in barley crop. Their results showed that biomass-based direct selection appeared more discriminating than yield-based selection in their plant material. However, the results of this study do not corroborate findings reported by Lalić et al. [11], Mitchell et al. [46], Lungu et al. [47], and El-Morshidy et al. [48] and who observed the effectiveness of the direct selection on grain yield. The difference in the results may be attributed to differences of breeding material and to genotype × environments.

The gains of selection-based index are shown in Table 6. The total sum of gains from the selection-based index ranged from −9.37% to 42.86%. Five out of seven used indices showed positive total gains. The base index of Williams recorded the highest total gain (42.86%) followed by the classic index of Smith and Hazel, the index desired gains of Pesek and Baker (27.46%), and the genotype-ideotype distance index of Cruz (22.07%). The sum of ranks index of Mulamba and Mock had a very low gain of 3.74%. Subandi's and Elston's indices exhibited negative total gains.

The base index of Williams yielded positive responses for all measured traits, except for chlorophyll content, flag leaf area, and harvest index, which showed negative gains. This index also achieved the highest grain yield response (6.27%, i.e., 26.41 g m^{-2}) compared to the other indices employed in this study. The classic index of Smith and Hazel ranked second followed by the index of desired gains of Pesek and Baker, the genotype-ideotype distance index proposed by Cruz, and the sum of ranks index of Mulamba and Mock with selection responses of 36.93%, 27.46%, 22.07%, and 3.74%, respectively. The single-trait responses varied from one index to another with, a more or less, balanced distribution of positive and negative estimates among the traits. This shows that the monotrait selection was inadequate because it led to a higher final product when considering the grain yield and generated unfavorable responses in other traits. These results indicated that methods that combine favorable expected gains should be used in the evaluation of these progenies. The gains expected through indices for grain yield per se were larger than those obtained by direct and indirect monotrait selection, except for the flag leaf area (Tables 5 and 6). Mahdy [49] mentioned that selection-based index was predicted to be

TABLE 5: Selection gain estimates (GS%) obtained from direct (diagonal) and indirect (of diagonal) selection in different bread wheat traits.

Traits	X_o	X_s	GS (%) Chl	HD	FLA	PH	SN	TKW	SW	GN	BIO	GY	HI	Total
CHL	43.7	46.4	**6.15** (2.69)	−0.07 (−0.09)	1.60 (0.30)	−3.25 (−2.11)	−4.06 (−18.62)	−0.49 (−0.19)	−4.12 (−28.31)	2.14 (0.52)	−7.14 (−83.30)	−1.10 (−4.63)	5.77 (2.09)	−4.58
HD	126.5	125.0	1.43 (0.63)	**−1.13** (−1.43)	−7.74 (−1.44)	−4.69 (−3.05)	1.76 (8.09)	−2.35 (−0.88)	−10.98 (−75.43)	−4.95 (−1.21)	−12.33 (−143.86)	−3.76 (−15.84)	8.63 (3.13)	−36.11
FLA	18.6	19.4	−1.51 (−0.66)	0.54 (0.68)	**4.30** (0.80)	−0.65 (−0.42)	−4.72 (−21.69)	−0.71 (−0.27)	8.76 (60.15)	14.00 (3.43)	6.40 (74.63)	8.04 (33.87)	0.31 (0.11)	34.75
PH	64.9	72.8	−1.30 (−0.57)	0.36 (0.45)	−4.57 (−0.85)	**12.10** (7.86)	9.33 (42.85)	7.14 (2.68)	5.63 (38.69)	−15.79 (−3.86)	12.38 (144.36)	−1.23 (−5.17)	−11.08 (−4.01)	12.98
SN	459.2	508.9	−0.88 (−0.38)	0.10 (0.13)	−8.17 (−1.52)	4.58 (2.97)	**10.84** (49.76)	4.22 (1.58)	5.52 (37.90)	−12.15 (−2.98)	8.15 (95.01)	1.14 (4.80)	−4.37 (1.58)	8.97
TKW	37.5	41.3	−0.70 (−0.30)	0.96 (1.22)	−4.77 (−0.89)	3.87 (2.52)	11.19 (51.38)	**9.93** (3.73)	10.84 (74.42)	−14.46 (−3.54)	14.04 (163.76)	3.63 (15.31)	−7.41 (−2.68)	27.14
SW	686.8	652.5	−1.26 (−0.55)	0.40 (0.50)	−3.36 (−0.62)	−0.17 (−0.11)	−7.18 (−32.95)	3.59 (1.35)	**−4.99** (−34.25)	−0.13 (−0.03)	−4.07 (−47.44)	−5.05 (−21.27)	−1.50 (−0.54)	−23.71
GN	24.5	28.4	2.05 (0.89)	0.53 (0.67)	3.85 (0.72)	−5.84 (−3.79)	−10.50 (−48.20)	−0.99 (−0.37)	0.70 (4.80)	**15.86** (3.88)	−4.14 (−48.30)	3.65 (15.39)	6.87 (2.49)	12.04
BIO	1166.4	1309.9	−1.22 (−0.53)	0.63 (0.80)	−3.58 (−0.67)	7.93 (5.15)	7.44 (34.18)	6.15 (2.31)	7.36 (50.55)	−8.94 (−2.19)	**12.30** (143.50)	2.74 (11.54)	−7.44 (−2.69)	23.39
GY	421.4	422.2	−2.07 (−0.90)	0.69 (0.87)	−1.48 (−0.27)	−0.95 (−0.62)	−6.49 (−29.82)	3.40 (1.27)	−1.38 (−9.45)	4.37 (1.07)	−2.47 (−28.84)	**0.20** (0.85)	2.10 (0.76)	−4.08
HI	36.2	34.8	−0.91 (−0.40)	0.45 (0.57)	0.30 (0.06)	−2.08 (−1.35)	−5.74 (−26.35)	2.72 (1.02)	0.60 (4.09)	−0.80 (−0.19)	0.14 (1.63)	−3.54 (−14.90)	**−3.81** (−1.38)	−12.67

CHL: chlorophyll content (Spad), HD: heading date (days), FLA: flag leaf area (cm^2), PH: plant height (cm), SN: number of spikes, TKW: 1000-kernel weight (g), SW: spikes weight (g), GN: number of grains per spike, BIO: above ground biomass (g m^{-2}), GY: grain yield (g m^{-2}), HI: harvest index (%), X_0: trait mean value of the base population, and X_S: trait mean value of the selected populations; values between parentheses correspond to the genetic gain in terms of percent mean for yield and yield-related traits.

TABLE 6: Selection gain estimates (GS%) obtained from selection-based index in different bread wheat traits.

Indices	CHL	HD	FLA	PH	SN	GS (%) TKW	SW	GN	BIO	GY	HI	Total
Smith & Hazel	1.95 (0.85)	0.97 (1.23)	−0.93 (−0.17)	8.74 (5.68)	4.28 (19.65)	9.53 (3.57)	10.20 (70.02)	−8.04 (−1.97)	13.60 (158.63)	3.94 (16.59)	−7.30 (−2.64)	36.93
Mulamba & Mock	−2.71 (−1.18)	0.03 (0.04)	−0.02 (−0.00)	4.05 (2.63)	−1.60 (−7.35)	2.08 (0.78)	−2.35 (−16.11)	1.50 (0.37)	−0.58 (−6.77)	1.56 (6.57)	1.78 (0.65)	3.74
Williams	−0.43 (−0.19)	1.56 (1.98)	−0.37 (−0.07)	5.71 (3.71)	0.85 (3.91)	8.44 (3.17)	12.43 (85.35)	0.00 (0.00)	14.78 (172.43)	6.27 (26.41)	−6.40 (−2.32)	42.86
Subandi	−1.37 (−0.60)	0.78 (0.99)	0.06 (0.01)	3.27 (2.12)	−9.31 (−42.75)	4.62 (1.73)	−3.18 (−21.85)	2.53 (0.62)	−1.10 (−12.84)	−3.08 (−12.99)	−2.58 (−0.93)	−9.37
Elston	−2.07 (−0.90)	0.69 (0.87)	−1.48 (−0.27)	−0.95 (−0.62)	−6.49 (−29.82)	3.40 (1.27)	−1.38 (−9.45)	4.37 (1.07)	−2.47 (−28.84)	0.20 (0.85)	2.10 (0.76)	−4.08
Pesek & Baker	2.23 (0.97)	0.78 (0.99)	4.38 (0.81)	−5.11 (−3.32)	0.33 (1.51)	5.88 (2.21)	8.72 (59.89)	−0.31 (−0.08)	7.91 (92.23)	5.42 (22.83)	−2.76 (−1.00)	27.46
Cruz	−1.46 (−0.64)	0.95 (1.20)	2.24 (0.42)	0.15 (0.10)	−7.99 (−36.69)	3.82 (1.43)	3.63 (24.95)	10.49 (2.57)	1.59 (18.50)	5.76 (24.27)	2.88 (1.04)	22.07

CHL: chlorophyll content (Spad), HD: heading date (days), FLA: flag leaf area (cm^2), PH: plant height (cm), SN: number of spikes, TKW: 1000-kernel weight (g), SW: spikes weight (g m^{-2}), GN: number of grains per spike, BIO: above ground biomass (g m^{-2}), GY: grain yield (g m^{-2}), and HI: harvest index (%); and values between parentheses correspond to the genetic gain in terms of percent mean for yield and yield-related traits.

TABLE 7: Differences of the F_4 selected lines relatively to the standard check Hidhab for the measured traits.

Traits	F_4 selected lines mean	Hidhab	Difference	Difference in % of Hidhab	LSD0.05
Chl	42.26	43.30	−1.05	−2.41	4.80
HD	126.57	129.67	−3.10*	−2.39	1.40
FLA	19.15	19.40	−0.24	−1.26	6.31
PH	68.67	62.33	6.33	10.16	6.67
SN	615.67	463.33	152.33*	32.88	119.67
TKW	39.97	38.33	1.63	4.26	4.53
SW	1147.67	823.33	324.33*	39.39	235.17
GN	31.57	24.83	6.75*	27.18	6.75
BIO	1833.67	1376.67	457.00*	33.20	353.15
GY	758.13	441.33	316.80*	71.78	185.10
HI	41.78	32.23	9.55	29.63	9.94

CHL: chlorophyll content (Spad), HD: heading date (days), FLA: flag leaf area (cm^2), PH: plant height (cm), SN: number of spikes, TKW: 1000-kernel weight (g), SW: spikes weight ($g\,m^{-2}$), GN: number of grains per spike, BIO: above ground biomass ($g\,m^{-2}$), GY: grain yield ($g\,m^{-2}$), and HI: harvest index (%).
*Significant effect at 0.05 probability.

superior for yield improvement as compared to the monotrait selection method.

The first three axes of the Principal Component Analysis (PCA) explained more than 77.93% of the total variation available in the data subjected to analysis. Heading date (0.386*), flag leaf area (0.308*), spikes weight (0.923*), biomass (0.853*), and grain yield (0.866*) correlated significantly to PCA1. PCA2 was mainly related to the chlorophyll content (0.276*), 1000-kernel weight (0.440*), number of grains per spike (0.655*), and harvest index (0.547*) (Figure 1). The number of spikes (0.333*) was related to PCA3 (Figure 2). The PCA biplots showed that the populations Acsad$_{899}$ × Hidhab (P7), Acsad$_{899}$ × El-Wifak (P8), Acsad$_{1135}$ × Hidhab (P11), Acsad$_{1135}$ × El-Wifak (P12), Acsad$_{1069}$ × Rmada (P14), Ain Abid × Mahon-Demias (P17), Ain Abid × Rmada (P18), and Ain Abid × El-Wifak (P20) were well represented on the plane formed by the first axis PCA1 (Figure 1).

Ain Abid × Mahon-Demias (P17), Ain Abid × Rmada (P18), and Ain Abid × El-Wifak (P20) had positive coordinates with this axis. They were characterized by a long vegetative cycle and high biomass, spikes weight, and grain yield values. The last two populations Ain Abid × Rmada (P18) and Ain Abid × El-Wifak (P20) were also distinguished by high values of chlorophyll content, flag leaf area, spikes fertility, and harvest index, relatively to PCA2 but were associated with low number of ears, relatively to PCA3 (Figures 1 and 2). On the other hand, Ain Abid × Mahon-Demias (P17) was distinguished, relatively to PCA2, by high estimates of the 1000-kernel weight and plant height, and associated with high number of spikes, relatively to PCA3 (Figures 1 and 2). From these results, it could be concluded that effective selection of superior individuals within this population certainly contributes to the improvement of yield and yield components, in a semi-late genetic background.

3.3. Selection of Superior Genotypes for Grain Yield. Genotypes were first ranked according to grain yield, then the 5% highest yielding lines were selected, and their mean yield estimated. The 30 lines thus selected are derived from 15 out of 20 F_4 populations studied. Half part of these lines was equitably derived from Acsad$_{1135}$ × Hidhab (P11), Ain Abid × Hidhab (P19), and Ain Abid × El-Wifak (P20) crosses. The population Ain Abid × Mahon-Demias (P17), previously identified among the promising populations, contributed with four lines. Acsad$_{1069}$ × Rmada (P14) and Ain Abid × Rmada (P18) participated with two lines each. Seven other populations contributed by one line each. Relative to the standard check Hidhab, which is the most cultivated variety in Algeria, the top 30 lines selected were characterized by significant improvements in yield components, including the number of spikes (152.33 spikes m^{-2}), spikes weight (324.33 $g\,m^{-2}$), spikes fertility (6.75 grains spike^{-1}), above ground biomass (457.00 $g\,m^{-2}$), and grain yield (316.80 $g\,m^{-2}$) (Table 7). Besides, they were distinguished by significant reduction of the duration in the vegetative phase with 3.10 days (Table 7).

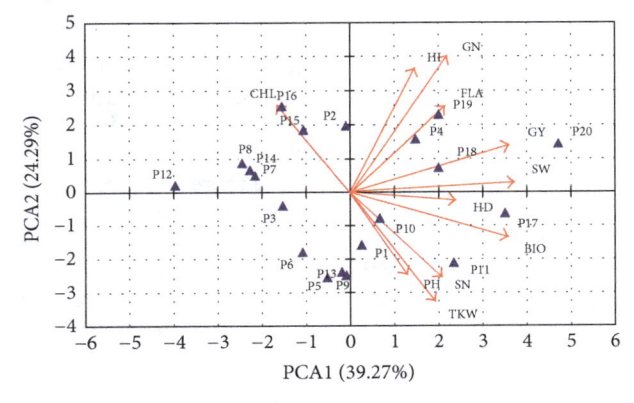

FIGURE 1: Principal Component Analysis (PCA) projections on axes 1 and 2, accounting for 63.56% of total variation, for 20 F_4 populations of bread wheat.

4. Conclusion

The results of this study indicated appreciable genetic variability among the evaluated populations. Selection based-index was more efficient to improve grain yield compared to direct

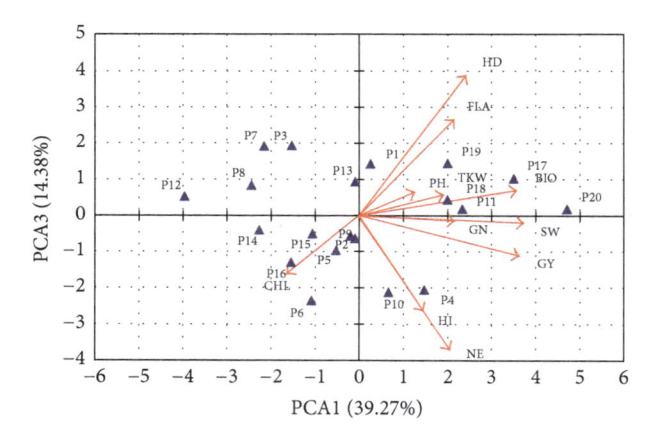

FIGURE 2: Principal Component Analysis (PCA) projections on axes 1 and 3, accounting for 53.65% of total variation, for 20 F_4 populations of bread wheat.

and indirect single-trait selection. The analytical procedures of the different selection methods showed possibilities of applications in advanced generations of breeding being superior when compared with direct and indirect selection. Williams's index was predicted to be more effective than the other selection indices for improving multiple traits at time. It brought the highest total genetic gain and the best yield gain per se associated with positive correlated responses for most of yield components. Compared to the check cultivar Hidhab, the 30 F_4 selected lines, at 5% selection intensity, were characterized by significant increase in grain yield and yield related traits.

References

[1] A. Benbelkacem, "Adaptation of cereal cultivars to extreme agroecologic environments of North Africa," *Field Crops Research*, vol. 45, no. 1-3, pp. 49–55, 1996.

[2] A. Mekhlouf, F. Dehbi, H. Bouzerzour, A. Hannchi, A. Benmahammed, and A. Adjabi, "Relationships between cold tolerance, grain yield performance and stability of durum wheat (*Triticum durum* Desf.) genotypes grown at high elevation area of Eastern Algeria," *Asian Journal of Plant Sciences*, vol. 5, no. 4, pp. 700–708, 2006.

[3] G. A. Slafer and F. H. Andrade, "Changes in physiological attributes of the dry matter economy of bread wheat (*Triticum aestivum*) through genetic improvement of grain yield potential at different regions of the world - A review," *Euphytica*, vol. 58, no. 1, pp. 37–49, 1991.

[4] S. P. Loss and K. H. M. Siddique, "Morphological and Physiological Traits Associated with Wheat Yield Increases in Mediterranean Environments," *Advances in Agronomy*, vol. 52, no. C, pp. 229–276, 1994.

[5] R. M. Trethowan, M. Van Ginkel, K. Ammar et al., "Associations among twenty years of international bread wheat yield evaluation environments," *Crop Science*, vol. 43, no. 5, pp. 1698–1711, 2003.

[6] H. Bouzerzour, A. Djekoun, A. Benmahammed, and K. L. Hassous, "Contribution de la biomasse aérienne, de l'indice de récolte et de la précocité à l'épiaison au rendement grain de l'orge (*Hordeum vulgare* L.) en zone d'altitude," *Cahiers de l'Agriculture*, vol. 8, pp. 133–137, 1998.

[7] A. Benmahammed, H. Nouar, L. Haddad, Z. Laala, A. Oulmi, and H. Bouzerzour, "Analyse de la stabilité des performances de rendement du blé dur (*Triticum durum* Desf.) sous conditions semi-arides," *Biotechnologie, Agronomie, Société et Environnement*, vol. 14, no. 1, pp. 177–186, 2010.

[8] B. R. Whan, R. Knight, and A. J. Rathjen, "Response to selection for grain yield and harvest index in F2, F3 and F4 derived lines of two wheat crosses," *Euphytica*, vol. 31, no. 1, pp. 139–150, 1982.

[9] M. P. Reynolds, R. P. Singh, A. Ibrahim, O. A. A. Ageeb, A. Larqué-Saavedra, and J. S. Quick, "Evaluating physiological traits to complement empirical selection for wheat in warm environments," *Euphytica*, vol. 100, no. 1-3, pp. 85–94, 1998.

[10] M. Balota, A. J. Green, C. A. Griffey, R. Pitman, and W. Thomason, "Genetic gains for physiological traits associated with yield in soft red winter wheat in the Eastern United States from 1919 to 2009," *European Journal of Agronomy*, vol. 84, pp. 76–83, 2017.

[11] A. Lalić, D. Novoselović, J. Kovačević et al., "Genetic gain and selection criteria effects on yield and yield components in barley (*Hordeum vulgare* L.)," *Periodicum biologorum*, vol. 112, no. 3, pp. 311–316, 2010.

[12] M. A. Babar, M. Van Ginkel, M. P. Reynolds, B. Prasad, and A. R. Klatt, "Heritability, correlated response, and indirect selection involving spectral reflectance indices and grain yield in wheat," *Australian Journal of Agricultural Research*, vol. 58, no. 5, pp. 432–442, 2007.

[13] M. M. Costa, A. O. Di Mauro, S. H. Unêda-Trevisoli et al., "Analysis of direct and indirect selection and indices in soybean segregating populations," *Crop Breeding and Applied Biotechnology*, vol. 8, no. 1, pp. 47–55, 2008.

[14] J. M. S. Viana, V. R. Faria, F. F. e Silva, and M. D. V. de Resende, "Combined selection of progeny in crop breeding using best linear unbiased prediction," *Canadian Journal of Plant Science*, vol. 92, no. 3, pp. 553–562, 2012.

[15] C. Y. Lin, "Index selection for genetic improvement of quantitative characters," *Theoretical and Applied Genetics*, vol. 52, no. 2, pp. 49–56, 1978.

[16] J. D. Smith and M. L. Kinman, "The Use of Parent-Offspring Regression as an Estimator of Heritability1," *Crop Science*, vol. 5, no. 6, p. 595, 1965.

[17] K. J. Frey and T. Horner, "Heritability in standard units," *Agronomy Journal*, vol. 49, no. 2, pp. 59–62, 1957.

[18] Z. Fellahi, A. Hannachi, H. Bouzerzour, and A. Boutekrabt, "Line × Tester Mating Design Analysis for Grain Yield and Yield Related Traits in Bread Wheat (*Triticum aestivum* L.)," *International Journal of Agronomy*, vol. 2013, pp. 1–9, 2013.

[19] Z. Fellahi, A. Hannachi, H. Bouzerzour, S. Dreisigacker, A. Yahyaoui, and D. Sehgal, "Genetic analysis of morpho-physiological traits and yield components in F2 partial diallel crosses of bread wheat (*Triticum aestivum* L.)," *Revista Facultad Nacional de Agronomia*, vol. 70, no. 3, pp. 8237–8250, 2017.

[20] H. Chennafi, A. Aïdaoui, H. Bouzerzour, and A. Saci, "Yield response of durum wheat (*Triticum durum* Desf.) cultivar Waha to deficit irrigation under semi arid growth conditions," *Asian Journal of Plant Sciences*, vol. 5, no. 5, pp. 854–860, 2006.

[21] P. L. Spagnoletti Zeuli and C. O. Qualset, "Flag Leaf Variation and the Analysis of Diversity in Durum Wheat," *Plant Breeding*, vol. 105, no. 3, pp. 189–202, 1990.

[22] M. S. McIntosh, "Analysis of combined experiments," *Agronomy Journal*, vol. 75, no. 1, pp. 153–155, 1983.

[23] R. G. D. Steel and J. H. Torrie, *Principles and procedures of statistics*, McGraw-Hill Books, New York, NY, USA, 1982.

[24] G. Acquaah, *Principles of plant genetics and breeding*, John Wiley and Sons, 2009.

[25] C. D. Cruz, *Programa GENES: Biometria*, Editora UFV, Viçosa, Brazil, 1st edition, 2006.

[26] H. F. Smith, "A discriminates function for plant selection," *Annals of Eugenics*, vol. 7, pp. 240–250, 1936.

[27] L. N. Hazel, "The genetic basis for constructing selection indexes," *Genetics*, vol. 28, no. 6, pp. 476–490, 1943.

[28] J. S. Williams, "The evaluation of a selection index," *Biometrics*, vol. 18, no. 3, pp. 375–393, 1962.

[29] R. C. Elston, "A Weight-Free Index for the Purpose of Ranking or Selection with Respect to Several Traits at a Time," *Biometrics*, vol. 19, no. 1, p. 85, 1963.

[30] J. Pesek and R. J. Baker, "Comparison of predicted and observed responses to selection for yield in wheat," *Canadian Journal of Plant Science*, vol. 51, no. 3, pp. 187–192, 1971.

[31] W. Subandi, A. Compton, and L. T. Empig, "Comparison of the efficiencies of selection indices for three traits in two variety crosses of corn," *Crop Science*, vol. 13, no. 2, pp. 184–186, 1973.

[32] N. N. Mulamba and J. J. Mock, "Improvement of yield potential of the ETO blanco maize (Zea mays L.) population by breeding for plant traits," *The Egyptian Journal of Genetics and Cytology*, vol. 7, no. 1, pp. 40–51, 1978.

[33] C. D. Cruz, "GENES - A software package for analysis in experimental statistics and quantitative genetics," *Acta Scientiarum - Agronomy*, vol. 35, no. 3, pp. 271–276, 2013.

[34] M. A. Abd El-Shafi, "Estimates of Genetic Variability and Efficiency of Selection for Grain Yield and Its Components in Two Wheat Crosses (*Triticum aestivum* L.)," *nternational Journal of Agriculture and Crop Sciences*, vol. 7, no. 2, pp. 83–90, 2014.

[35] R. W. Allard, *Principles of Plant Breeding*, John Wily and Sons, Inc, New York, NY, USA, 2nd edition, 1960.

[36] C. D. Cruz, A. J. Regazi, and P. C. S. Carneiro, *Modelos biométricos aplicados ao melhoramento genético*, UFV, Viçosa, Brazil, 4th edition, 2012.

[37] A. Mesele, W. Mohammed, and T. Dessalegn, "Estimation of Heritability and Genetic Advance of Yield and Yield Related Traits in Bread Wheat (*Triticum aestivum* L.) Genotypes at Ofla District, Northern Ethiopia," *International Journal of Plant Breeding and Genetics*, vol. 10, no. 1, pp. 31–37, 2015.

[38] B. Saleem, A. Khan, M. Shahzad, and F. Ijaz, "Estimation of heritability and genetic advance for various metric traits in seven F2 populations of bread wheat (*Triticum aestivum* L.)," *Journal of Agricultural Sciences, Belgrade*, vol. 61, no. 1, pp. 1–9, 2016.

[39] M. Yaqoob, "Estimation of genetic variability, heritability and genetic advance for yield and yield related traits in wheat under rainfed conditions," *Journal of Agricultural Research*, vol. 54, no. 1, pp. 1–14, 2016.

[40] C. M. Löffler and R. H. Busch, "Selection for Grain Protein, Grain Yield, and Nitrogen Partitioning Efficiency in Hard Red Spring Wheat," *Crop Science*, vol. 22, no. 3, p. 591, 1982.

[41] W. L. Alexander, E. L. Smith, and C. Dhanasobhan, "A comparison of yield and yield component selection in winter wheat," *Euphytica*, vol. 33, no. 3, pp. 953–961, 1984.

[42] J. Kumar and P. N. Bahl, "Direct and indirect selection for yield in chickpea," *Euphytica*, vol. 60, no. 3, pp. 197–199, 1992.

[43] R. M. DePauw and L. H. Shebeski, "An evaluation of an early generation yield testing procedure in *Triticum aestivum*," *Canadian Journal of Plant Science*, vol. 53, no. 3, pp. 465–470, 1973.

[44] M. N. Inagaki, G. Varughese, S. Rajaram, M. Van Ginkel, and A. Mujeeb-Kazi, "Comparison of bread wheat lines selected by doubled haploid, single-seed descent and pedigree selection methods," *Theoretical and Applied Genetics*, vol. 97, no. 4, pp. 550–556, 1998.

[45] A. Benmahammed, H. Bouzerzour, A. Djekou, and K. Hassous, "Efficacité de la sélection précoce de la biomasse chez l'orge (*Hordeum vulgare* L.) en zone semi-aride," *Sciences and Technologie*, vol. C, no. 22, pp. 80–85, 2004.

[46] J. W. Mitchell, R. Baker, and D. R. Knott, "Evaluation of Honeycomb Selection for Single Plant Yield in Durum Wheat," *Crop Science*, vol. 22, no. 4, pp. 840–843, 1982.

[47] D. M. Lungu, P. J. Kaltsikes, and E. N. Larter, "Honeycomb selection for yield in early generations of spring wheat," *Euphytica*, vol. 36, no. 3, pp. 831–839, 1987.

[48] M. A. El-Morshidy, K. A. Kheiralla, M. A. Ali, and A. A. S. Ahmed, "Efficiency of pedigree selection for earliness and grain yield in two wheat populations under water stress conditions," *Assiut Journal of Agricultural Sciences*, vol. 37, pp. 77–94, 2010.

[49] E. E. Mahdy, "Single and Multiple Traits Selection in a Segregating Population of Wheat, *Triticum aestivum* L," *Plant Breeding*, vol. 101, no. 3, pp. 245–249, 1988.

Insecticide Seed Treatments Reduced Crop Injury from Flumioxazin, Chlorsulfuron, Saflufenacil, Pyroxasulfone, and Flumioxazin + Pyroxasulfone + Chlorimuron in Soybean

N. R. Steppig[1], J. K. Norsworthy,[1] R. C. Scott,[2] and G. M. Lorenz[3]

[1]Department of Crop, Soil, and Environmental Sciences, University of Arkansas, Fayetteville, AR 72701, USA
[2]Department of Crop, Soil, and Environmental Sciences, Lonoke Extension Center, University of Arkansas, Lonoke, AR 72086, USA
[3]Department of Entomology, Lonoke Extension Center, University of Arkansas, Lonoke, AR 72086, USA

Correspondence should be addressed to N. R. Steppig; nsteppig@uark.edu

Academic Editor: David Clay

With increased instances of weed resistance to applications of postemergence herbicides, the use of soil-applied herbicides that offer residual activity is becoming popular. Unfortunately, under some conditions, the use of residual herbicides can result in unintentional injury to crops. However, there are a number of ways to reduce these risks, including the use of in-crop herbicide safeners. Based on previous research conducted on rice, the potential may exist for crop injury from certain soil-applied herbicides to be reduced (safened) in seeds treated with insecticides. Field trials were conducted in Marianna, Arkansas, in 2015 and 2016, and near Colt, Arkansas, in 2016, to explore this possibility in soybean. Soybean seeds were treated with the insecticide thiamethoxam and subsequently the herbicides metribuzin, saflufenacil, pyroxasulfone, sulfentrazone, chlorimuron, flumioxazin, flumioxazin + pyroxasulfone + chlorimuron, mesotrione, and chlorsulfuron were applied immediately after planting. Of the nine herbicides evaluated, the insecticide reduced crop injury for flumioxazin, chlorsulfuron, saflufenacil, pyroxasulfone, and flumioxazin + pyroxasulfone + chlorimuron. The highest degree of injury reduction was seen 1 week after emergence (WAE) at Marianna, where injury from flumioxazin + pyroxasulfone + chlorimuron was reduced from 15% to 5%. Based on the results from this study, the insecticide seed treatment thiamethoxam may have the potential to safen soybean to applications of some soil-applied herbicides.

1. Introduction

Herbicide use in the US is a vital component of agriculture production. Gianessi and Reigner [1] estimate that herbicide use provides a labor equivalent of 70 million hand laborers and increases crop yields as much as 20%. The introduction of herbicide-resistant (HR) crops has also significantly improved the efficiency of crop production, both in the US and globally [2]. Beginning with the introduction of glyphosate-resistant soybean in 1996, the widespread adoption of HR crops provided growers with the ability to effectively control a broad spectrum of weeds by utilizing just one or two postemergence (POST) applications of a herbicide with a single mode of action [3]. Unfortunately, this approach resulted in weeds that were resistant to those control strategies [4]. For example, overreliance upon glyphosate has resulted in glyphosate-resistance in 37 individual weed species since 2000 [5]. In order to effectively combat herbicide resistance, the use of herbicides with residual activity is recommended [6, 7].

Residual herbicides are applied to the soil surface, and depending on climatic, chemical, and soil properties, they can control a broad spectrum of weeds for varying lengths of time [8, 9]. The use of a residual herbicide, as a part of a sequential herbicide program, can increase crop yields as a result of increased weed control compared to programs that do not include a residual component [10, 11]. The residual activity provided by these herbicides typically allows for later applications of POST-applied herbicides and, thus, improved flexibility for crop producers [12]. Apart from being applied by themselves, residual herbicides can be tank mixed with a number of POST-applied herbicides. In these instances, the

POST herbicide controls weeds that have already emerged, whereas the residual herbicide provides lasting control of weeds that have not yet germinated at the time of application. This approach results in high levels of weed control, which can consequently improve crop yield [10].

In addition to providing the obvious benefit of successfully controlling weeds, residual herbicides are also important herbicide-resistance management tools. Because residual herbicides greatly decrease the number of weeds present early in the season, there is decreased resistance selection on POST herbicides in subsequent applications. Reduced selection results in less likelihood for herbicide resistance, which in turn increases the potential lifespan of a given herbicide [6, 13]. Including residual herbicides as part of a tank mixture with POST herbicides results in an increased number of herbicide modes of action (MOA) applied to weeds. Application of multiple effective herbicide MOA is one of the most important methods for delaying the onset of herbicide resistance [6, 14].

Unfortunately, one main drawback associated with the use of residual herbicides is crop injury following application. In some cases, herbicides that are labeled for use in crop can cause injury to young plants. Flumioxazin, sulfentrazone, chlorimuron, S-metolachlor, and pyroxasulfone are some examples in soybean production [15, 16]. Crop response to these preemergence (PRE) herbicides can be greatly variable depending upon both soil and environmental conditions, with cool, wet, and low pH conditions causing the most crop injury in soybean following applications of flumioxazin and sulfentrazone [16]. In addition to temperature, moisture, and pH, soil organic matter (SOM) and texture can impact the activity of herbicides to varying degrees, depending upon the herbicide [17, 18]. Aside from environmental effects, varietal selection can cause substantial variation in response to soil-applied herbicides [19]. Early-season injury from herbicides typically dissipates quickly with no adverse effects on crop yield, but in some cases, more severe injury symptoms and stand loss can cause reduced yields [15, 16].

Another concern with residual herbicides is injury to successive crops. Due to their relatively long half-lives, plant-back restrictions are needed for many soil-applied herbicides in order to protect crops in replant situations following crop failure, as well as crops grown the next season [20]. These plant-back restrictions can greatly limit rotational options and can drive growers' decisions on what to plant the following year. One notable example of how crop rotation is directly influenced by herbicide use in the state of Arkansas can be seen in imidazolinone-resistant (Clearfield®, BASF Corporation, Research Triangle Park, NC) rice (*Oryza sativa* L.). Imidazolinone-resistant rice is tolerant to applications of the herbicide imazethapyr, an acetolactate synthase- (ALS-) inhibiting imidazolinone. According to Renner et al. [21], imidazolinones can persist in the soil for as long as two years after their initial application. Grain sorghum, cotton, and conventional rice all have a rotational restriction of 18 months following imazethapyr applications, meaning that rice producers in Arkansas are limited to planting soybean, corn (*Zea mays* L.), or imidazolinone-resistant rice the following season [20].

A possible solution to preventing or reducing the effects of crop injury when using residual herbicides is the use of herbicide safeners. Safeners have been effectively used in crops for both PRE and POST herbicide applications and typically reduce crop injury from herbicides by increasing a plant's ability to metabolize certain herbicidal compounds [22, 23]. Through the use of safeners, crop injury can be reduced such that a herbicide can be used in crops where it would cause unacceptable levels of injury when applied without a safener [24]. One such example can be seen with the herbicide safener fluxofenin (Concep III, Syngenta Crop Protection, LLC, Greensboro, NC), which is already used extensively in grain sorghum production to prevent injury from PRE herbicides. Without a fluxofenin seed treatment, chloroacetamide herbicides such as S-metolachlor and alachlor cannot be applied in sorghum production due to high levels of injury to the crop from these herbicides [25]. While the use of safeners has generally been more successful in monocotyledonous crops [26, 27], some examples of safeners do exist in dicots [28]. As such, the potential may exist for novel herbicide safeners to be found in soybean.

Safeners can be applied to the soil, to foliage, or as a seed coating to maximize their efficacy [29]. The benefits of applying herbicide safeners as seed treatments are twofold: injury from herbicides is greatly decreased, and the safener is selectively applied to the crop [24]. Applying the safener only to the crop ensures that safening effects are not conferred to the weeds present in a field, maintaining herbicidal efficacy. This property is highly desirable, and thus, seed-applied safeners have great value. Recently, Miller et al. [30] reported that the insecticide seed treatment thiamethoxam (Cruiser 5S, Syngenta Crop Protection, LLC, Greensboro, NC), in addition to protecting seedling rice from early-season insect damage, also provided a reduction in crop injury following application of some POST herbicides. Although in-plant concentrations of insecticides decrease substantially 3 to 4 weeks after planting [31], enough insecticidal material was still present in the rice at this time to produce a safening effect. Since safening effects were seen even in the case of low thiamethoxam presence, it was hypothesized that similar effects may be seen at crop emergence, when thiamethoxam concentration is much higher in the plant. Thus, research was conducted to determine whether thiamethoxam could be used to reduce crop injury from select soil-residual herbicides in soybean.

2. Materials and Methods

An experiment was conducted at the Lon Mann Cotton Research Station (LMCRS) in Marianna, Arkansas, United States (34 43.4368N, 90 44.0390W), in 2015 to assess the potential for insecticide seed treatments to reduce crop injury following applications of residual herbicides in soybean. In 2016, experiments were repeated at LMCRS, in addition to those conducted at the Pine Tree Research Station

TABLE 1: General description of experimental sites[a].

Location	Year	Planting date	Application date	Sand	Silt	Clay	pH
					%		
LMCRS	2015	5/14/2015	5/14//2015	0.8	90.5	8.7	7.5
LMCRS	2016	5/5/2016	5/5/2016	0.8	90.5	8.7	7.5
PTRS	2016	5/19/2016	5/19/2016	0.4	78.1	21.5	7.8

[a]LMCRS: Lon Mann Cotton Research Station in Marianna, AR; PTRS: Pine Tree Research Station near Colt, AR.

(PTRS) near Colt, Arkansas, United States (35 06.3584N, 90 56.2437W). DG5067LL (Delta Grow Seed Company Inc., England, AR), a glufosinate-resistant, non-STS, maturity group 5.2 soybean, was planted at a seeding rate of 340,000 seeds ha^{-1} to an approximate 2.5-cm depth. Four-row plots were established utilizing a randomized complete block design with four replications. Row spacings were 96 cm at LMCRS and 76 cm at PTRS, with plot length at all locations of 7.2 m. Plots were managed using agronomic recommendations provided in the University of Arkansas Soybean Production Handbook [32]. The soils at LMCRS and PTRS were a Convent silt loam (fine-silty, mixed, active thermic Typic Glossaqualf) and Calhoun silt loam (coarse-silty, mixed, superactive, nonacid, thermic, Fluvaquentic Endoaquepts), respectively [33]. Prior to planting, all seeds received a fungicide seed treatment of mefenoxam + fludioxonil + sedaxane (Cruiser plus Vibrance, Syngenta Crop Protection, LLC, Greensboro, NC) at a rate of 0.075 + 0.025 + 0.025 g ai kg^{-1} seed. In addition to fungicides, seeds were treated with either no insecticide or thiamethoxam (Cruiser 5S, Syngenta Crop Protection, LLC, Greensboro, NC) at 0.5 g ai kg^{-1} seed. Both fungicide and insecticide seed treatments were made using a water-based slurry. Herbicide applications were made at planting, using a CO_2-pressurized backpack sprayer calibrated to deliver 143 L ha^{-1} at 276 kPa (Table 1). Seven herbicides that are labeled for use in soybean were applied at, or slightly above, their recommended PRE rates to encourage injurious symptomology. These herbicides included metribuzin (841 g ha^{-1}), saflufenacil (75 g ha^{-1}), pyroxasulfone (268 g ha^{-1}), sulfentrazone (533 g ha^{-1}), chlorimuron (79 g ha^{-1}), flumioxazin (107 g ha^{-1}), and chlorimuron + flumioxazin + pyroxasulfone (29 + 108 + 136 g ha^{-1}). In addition, two herbicides that commonly cause injury to soybean via carryover—mesotrione (42 g ha^{-1}) and chlorsulfuron (1.8 g ha^{-1})—were applied at reduced rates to simulate amounts that may be present following applications in the previous growing season.

Following application, visual injury ratings were collected weekly on a 0 to 100% scale, where 0% is no injury and 100% is soybean death. In addition, crop density and height measurements were made three weeks after application to allow for adequate germination across the test. Yield data were collected by harvesting the center two rows of each plot and correcting seed moisture to 13%. Data were subjected to analysis of variance, and significant means were separated using Fisher's protected LSD ($\alpha = 0.05$). Site years were analyzed separately due to considerable variation in environmental conditions at each location (Figures 1–3) and differing responses at each of the sites. For responses that did not produce a significant herbicide by insecticide seed treatment interaction, seed treatment main effects were evaluated. At evaluation timings where no measurable injury was observed for one or more herbicide treatments, the assumptions for ANOVA were not met. When either no interaction was identified, or the response did not meet the assumptions for ANOVA, individual t-tests were conducted to compare treatments with no insecticide to each insecticide seed treatment, within a herbicide.

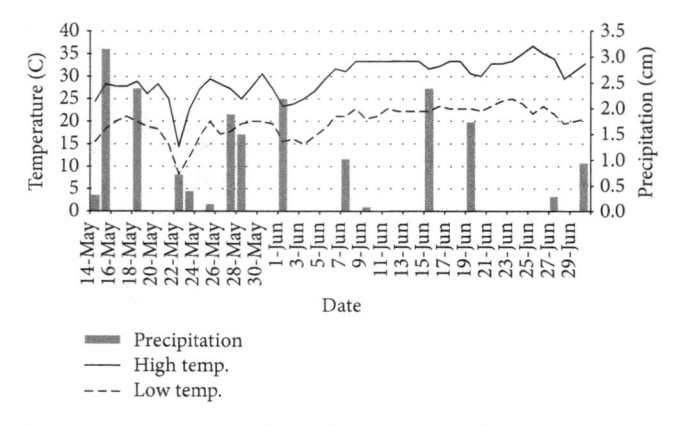

FIGURE 1: Environmental conditions at the Lon Mann Cotton Research Station in Marianna, AR, in 2015 beginning at planting date May 14.

3. Results and Discussion

Of the nine herbicides evaluated, five showed reductions in injury (safening) in at least one site year. Injury reduction was seen at two site years for flumioxazin and at one site year for chlorsulfuron, saflufenacil, pyroxasulfone, and flumioxazin + pyroxasulfone + chlorimuron. Injury from flumioxazin was reduced at LMCRS (2016) at 1 and 2 weeks after emergence (WAE), where thiamethoxam reduced injury from 13% at both evaluation timings to 8% and 5% at 1 and 2 WAE, respectively (Table 2). Additionally, at PTRS, injury caused by flumioxazin at 2 WAE was reduced from 15% to 8% (Table 3). The highest level of injury reduction occurred at LMCRS (2016), where injury was reduced 1 WAE from 15% to 5% when treated with thiamethoxam (Table 2). Chlorsulfuron

TABLE 2: Visible soybean injury, density, height, and yield at the Lon Mann Cotton Research Station in Marianna, AR in 2016[a,b].

Herbicide	Seed treatment	Injury 1 WAE (%)	Injury 2 WAE (%)	Injury 4 WAE (%)	Density (Plants m^{-1} row)	Height (cm)	Yield (kg ha^{-1})
None	None	0	0	0	27	8	2520
	Thiamethoxam	0	0	0	27	9	2550
Metribuzin	None	1	2	1	25	8	2540
	Thiamethoxam	1	0	0	26	9	2460
Saflufenacil	None	3	14	6	24	8	2400
	Thiamethoxam	1	14	1	26	8	2730
Pyroxasulfone	None	4	2	0	27	9	2630
	Thiamethoxam	1	3	0	26	8	2610
Sulfentrazone	None	1	14	4	25	8	2650
	Thiamethoxam	1	14	3	25	9	2460
Chlorimuron	None	4	5	6	27	8	2380
	Thiamethoxam	3	4	1	25	9	2540
Flumioxazin	None	13	13	5	24	8	2600
	Thiamethoxam	5*	8*	1	24	8	2480
Chl + Flu + Pyr	None	15	17	6	27	8	2620
	Thiamethoxam	5*	12*	1	25	8	2710
Mesotrione	None	10	4	3	25	8	2740
	Thiamethoxam	7	4	0	26	8	2470
Chlorsulfuron	None	5	7	8	27	8	2690
	Thiamethoxam	2	3*	4	27	8	2480
Main effect[c]	None			4	NS	NS	NS
	Thiamethoxam			1†	NS	NS	NS

[a]WAE: weeks after emergence; NS: nonsignificant; Chl + Flu + Pyr: chlorimuron + flumioxazin + pyroxasulfone; [b]means followed by an asterisk indicate a significant herbicide by insecticide interaction ($\alpha = 0.05$) or a significant injury reduction via insecticide seed treatment, within the same herbicide, compared to no insecticide. [c]Where no significant interaction is present, insecticide seed treatment's main effect is given below and marked with a cross.

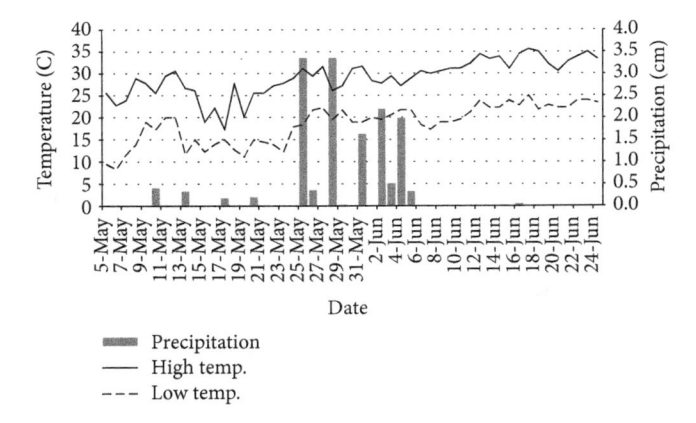

FIGURE 2: Environmental conditions at the Lon Mann Cotton Research Station in Marianna, AR, in 2016 beginning at planting date May 5.

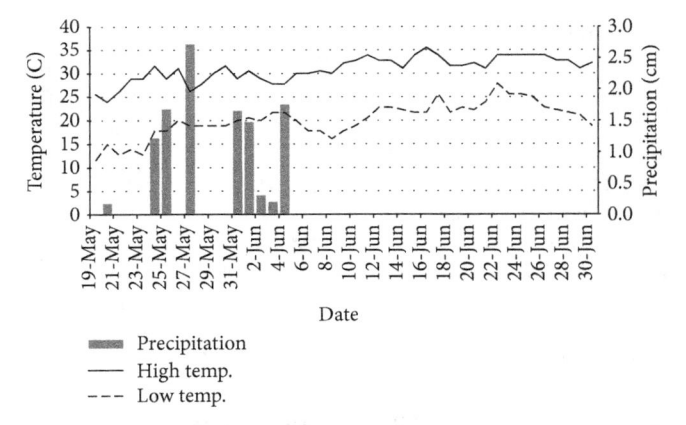

FIGURE 3: Environmental conditions at the Pine Tree Research Station near Colt, AR, in 2016 beginning at planting date May 19.

injury was reduced 1 WAE at LMCRS (2016) from 7% to 3% (Table 2). Soybean was also safened to saflufenacil at PTRS 2 WAE, where injury was reduced from 22% to 15% (Table 3). Injury from pyroxasulfone was also reduced via a thiamethoxam seed treatment, with injury being reduced at PTRS 1 and 2 WAE, from 13% to 4% and from 14% to 5%, respectively.

TABLE 3: Visible soybean injury, density, and yield at the Pine Tree Research Station near Colt, AR in 2016[a].

Herbicide	Seed treatment	Injury[b]			Density	Yield
		1 WAE	2 WAE	4 WAE		
			%		Plants m^{-1} row	kg ha^{-1}
None	None	0	0	0	16	2770
	Thiamethoxam	0	0	0	20	2930
Metribuzin	None	6	9	7	19	2700
	Thiamethoxam	0	6	0	18	3130
Saflufenacil	None	12	22	5	17	2950
	Thiamethoxam	9	15[‡]	6	18	2780
Pyroxasulfone	None	13	14	6	18	3000
	Thiamethoxam	4[‡]	5[‡]	5	17	3210
Sulfentrazone	None	8	13	0	18	3180
	Thiamethoxam	2	8	3	19	3040
Chlorimuron	None	8	10	1	17	2300
	Thiamethoxam	8	7	3	15	2810
Flumioxazin	None	9	15	10	19	3090
	Thiamethoxam	5	8[‡]	5	19	3170
Chl + Flu + Pyr	None	18	19	6	19	2850
	Thiamethoxam	15	15	5	19	2930
Mesotrione	None	9	9	5	20	2970
	Thiamethoxam	8	5	6	19	3050
Chlorsulfuron	None	3	10	8	20	2860
	Thiamethoxam	6	5	5	19	2730
Main effect	None	9	13	NS	NS	NS
	Thiamethoxam	6[†]	8[†]	NS	NS	NS

[a]WAE: weeks after emergence; NS: nonsignificant; Chl + Flu + Pyr: chlorimuron + flumioxazin + pyroxasulfone; [b]where no significant interaction ($\alpha = 0.05$) is present, insecticide seed treatment main effect is given below and marked with a cross. For responses that did not produce a herbicide by insecticide seed treatment interaction, a t-test was conducted to compare treatments with no insecticide to each insecticide seed treatment within an herbicide. Where use of an insecticide seed treatment reduced injury or increased height or yield compared to no insecticide, means are marked with a double dagger (‡).

Injury from metribuzin, sulfentrazone, chlorimuron, and mesotrione was not reduced at any evaluation timing at each of the three locations (Tables 2–4). Similar to studies by McNaughton et al. [15], soybean injury from chlorimuron, flumioxazin, or pyroxasulfone alone was less than injury seen when the three were combined. Aside from a significant seed treatment main effect at LMCRS (2016), where crop height was increased from 47 cm to 50 cm when treated with thiamethoxam, plant height was not affected by seed treatment (Tables 2–4). Additionally, while injury reduction was seen in a number of herbicide-insecticide combinations, crop yield relative to a nontreated check was not increased in these situations (Tables 2–4).

All herbicides evaluated, except for chlorsulfuron and mesotrione, are labeled for use in soybean. As a result, overall soybean injury was low in many cases. Additionally, based on the low levels of injury following application of both metribuzin and sulfentrazone, it is likely that the variety chosen for these studies was tolerant to these herbicides. Choosing a susceptible variety would likely increase crop injury response to these herbicides, which may make the safening benefits associated with insecticide seed treatments more obvious than the ones in this study. In future research, variety selection should be heavily scrutinized in order to select crops that will exhibit high levels of injury.

4. Conclusions

In these experiments, insecticide seed treatments caused significant reductions in soybean injury following applications of flumioxazin, chlorsulfuron, saflufenacil, pyroxasulfone, and flumioxazin + pyroxasulfone + chlorimuron. Because thiamethoxam is a commonly used insecticide seed treatment in soybean production, and all of these herbicides except chlorsulfuron are frequently applied PRE in soybean, it is likely that some growers who use these insecticide/herbicide combinations will likely see reduced early-season injury from these herbicides. Although yield increases were not seen as a result of decreased crop injury in the trials in this study, reduced injury to seedling crops has been shown to result in increased yield in some cases. Future research examining injury reduction from other insecticide seed treatments (aside from thiamethoxam) and PRE herbicide combinations, under a variety of environmental conditions, may show that these yield increases are possible in soybean.

TABLE 4: Visible soybean injury, density, height, and yield at the Lon Mann Cotton Research Station in Marianna, AR in 2015[a].

Herbicide	Seed treatment	Injury[b]			Density	Height[c]	Yield
		1 WAE	2 WAE	4 WAE			
		%			Plants m^{-1} row	cm	kg ha^{-1}
None	None	0	0	0	21	57	3890
	Thiamethoxam	0	0	0	23	58	3740
Metribuzin	None	2	5	3	19	59	3900
	Thiamethoxam	0	6	3	22	62	3720
Saflufenacil	None	15	29	13	17	51	4040
	Thiamethoxam	14	24	15	18	57	3640
Pyroxasulfone	None	14	24	14	19	53	3650
	Thiamethoxam	10	25	11	22	53	3800
Sulfentrazone	None	24	43	24	15	47	3740
	Thiamethoxam	21	40	21	17	48	3500
Chlorimuron	None	3	6	4	20	38	3790
	Thiamethoxam	1	4	3	23	45	3820
Flumioxazin	None	2	1	3	21	57	3700
	Thiamethoxam	1	0	1	22	61	3770
Chl + Flu + Pyr	None	28	49	39	15	39	3250
	Thiamethoxam	26	48	41	13	41	3290
Mesotrione	None	1	13	3	20	57	3880
	Thiamethoxam	1	9	3	20	58	4050
Chlorsulfuron	None	18	53	83	21	12	1870
	Thiamethoxam	13	51	81	21	12	1350
Main effect	None	12	NS	NS	NS	47	NS
	Thiamethoxam	10[†]	NS	NS	NS	50[†]	NS

[a]WAE: weeks after emergence; NS: nonsignificant; Chl + Flu + Pyr: chlorimuron + flumioxazin + pyroxasulfone; [b]where no significant interaction ($\alpha = 0.05$) is present, insecticide seed treatment main effect is given below and a significant main effect is denoted with a cross (†).

References

[1] L. P. Gianessi and N. P. Reigner, "The value of herbicides in U.S. crop production," *Weed Technology*, vol. 21, no. 2, pp. 559–566, 2007.

[2] G. Brookes and P. Barfoot, "The income and production effects of biotech crops globally 1996-2010," *GM crops & food*, vol. 3, no. 4, pp. 265–272, 2012.

[3] B. G. Young, "Changes in herbicide use patterns and production practices resulting from glyphosate-resistant crops," *Weed Technology*, vol. 20, no. 2, pp. 301–307, 2006.

[4] W. K. Vencill, R. L. Nichols, T. M. Webster et al., "Herbicide resistance: toward an understanding of resistance development and the impact of herbicide-resistant crops," *Weed Science*, vol. 60, no. 1, pp. 2–30, 2012.

[5] I. Heap, "International Survey of Herbicide Resistant Weeds," 2017, http://www.weedscience.com/summary/home.aspx.

[6] J. K. Norsworthy, S. M. Ward, D. R. Shaw et al., "Reducing the risks of herbicide resistance: best management practices and recommendations," *Weed Science*, vol. 60, no. 1, pp. 31–62, 2012.

[7] M. D. Owen, B. G. Young, D. R. Shaw et al., "Benchmark study on glyphosate-resistant crop systems in the United States. Part 2: Perspectives," *Pest Management Science*, vol. 67, no. 7, pp. 747–757, 2011.

[8] R. P. Dewerff, S. P. Conley, J. B. Colquhoun, and V. M. Davis, "Weed control in soybean as influenced by residual herbicide use and glyphosate-application timing following different planting dates," *Weed Technology*, vol. 29, no. 1, pp. 71–81, 2015.

[9] C. J. Meyer, J. K. Norsworthy, B. G. Young et al., "Early-season palmer amaranth and waterhemp control from preemergence ptograms utilizing 4-hydroxyphenylpyruvate dioxygenase-inhibiting and auxinic herbicides in soybean," *Weed Technology*, vol. 30, no. 1, pp. 67–75, 2016.

[10] J. S. Aulakh and A. J. Jhala, "Comparison of glufosinate-based herbicide programs for broad-spectrum weed control in glufosinate-resistant soybean," *Weed Technology*, vol. 29, no. 3, pp. 419–430, 2015.

[11] M. M. Loux, A. F. Dobbels, W. G. Johnson, and B. G. Young, "Effect of residual herbicide and postemergence application timing on weed control and yield in glyphosate-resistant corn," *Weed Technology*, vol. 25, no. 1, pp. 19–24, 2011.

[12] J. M. Ellis and J. L. Griffin, "Benefits of Soil-Applied Herbicides in Glyphosate-Resistant Soybean (Glycine max)," *Weed Technology*, vol. 16, no. 3, pp. 541–547, 2002.

[13] H. J. Beckie, "Herbicide-resistant weeds: management tactics and practices," *Weed Technology*, vol. 20, no. 3, pp. 793–814, 2006.

[14] A. J. Diggle, P. B. Neve, and F. P. Smith, "Herbicides used in combination can reduce the probability of herbicide resistance

in finite weed populations," *Weed Research*, vol. 43, no. 5, pp. 371–382, 2003.

[15] K. E. McNaughton, C. Shropshire, D. E. Robinson, and P. H. Sikkema, "Soybean (*Glycine max*) Tolerance to Timing Applications of Pyroxasulfone, Flumioxazin, and Pyroxasulfone + Flumioxazin, and pyroxasulfone+flumioxazin," *Weed Technology*, vol. 28, no. 3, pp. 494–500, 2014.

[16] S. Taylor-Lovell, L. M. Wax, and R. Nelson, "Phytotoxic Response and Yield of Soybean (Glycine max) Varieties Treated with Sulfentrazone or Flumioxazin," *Weed Technology*, vol. 15, no. 1, pp. 95–102, 2001.

[17] C. V. Eberlein, A. G. Dexter, J. D. Nalewaja, and W. C. Dahnke, *Soil organic matter, texture, and pH as herbicide use guides*, vol. 14, North Dakota State University Cooperative Extension Service 14 AGR-8, 1984.

[18] T. W. Gannon, A. C. Hixson, K. E. Keller, J. B. Weber, S. Z. Knezevic, and F. H. Yelverton, "Soil properties influence saflufenacil phytotoxicity," *Weed Science*, vol. 62, no. 4, pp. 657–663, 2014.

[19] J. M. Swantek, C. H. Sneller, and L. R. Oliver, "Evaluation of soybean injury from sulfentrazone and inheritance of tolerance," *Weed Science*, vol. 46, no. 2, pp. 271–277, 1998.

[20] T. Barber, J. Norsworthy, and B. Scott, *Row-Crop Plant-Back Intervals for Common Herbicides*, vol. MP519 6, The Arkansas Cooperative Extension Service Publications MP519, Little Rock, AR, USA, 2014.

[21] K. A. Renner, O. Schabenberger, and J. J. Kells, "Effect of tillage and application method on corn (Zea mays) response to imidazolinone residues in soil," *Weed Technology*, vol. 12, no. 2, pp. 281–285, 1998.

[22] M. Barrett, "Protection of corn (Zea mays) and sorghum (*Sorghum bicolor*) from imazethapyr toxicity with antidotes," *Weed Sci*, vol. 37, pp. 296–301, 1989.

[23] R. F. Spotanski and O. C. Burnside, "Reducing herbicide injury to sorghum with crop protectants," *Weed Sci*, vol. 21, pp. 531–536, 1973.

[24] J. Davies and J. C. Caseley, "Herbicide safeners: a review," *Journal of Pesticide Science*, vol. 55, no. 11, pp. 1043–1058, 1999.

[25] L. Espinosa and J. Kelley, *Arkansas Grain Sorghum Production Handbook. Arkansas Cooperative Extension Service Miscellaneous Publications 297*, University of Arkansas, Little Rock, AR, USA, 2004.

[26] K. K. Hatzios, "Mechanisms of action of herbicide safeners: an overview," in *Crop Safeners for Herbicides: Development, Uses, and Mechanisms of Action*, K. K. Hatzios and R. E. Hoagland, Eds., pp. 65–101, Academic Press, San Francisco, Calif, USA, 1989.

[27] D. E. Riechers, K. Kreuz, and Q. Zhang, "Detoxification without intoxication: herbicide safeners activate plant defense gene expression," *Plant Physiology*, vol. 153, no. 1, pp. 3–13, 2010.

[28] Y. Ferhatoglu, S. Avdiushko, and M. Barrett, "The basis for the safening of clomazone by phorate insecticide in cotton and inhibitors of cytochrome P450s," *Pesticide Biochemistry and Physiology*, vol. 81, no. 1, pp. 59–70, 2005.

[29] A. W. Abu-Qare and H. J. Duncan, "Herbicide safeners: uses, limitations, metabolism, and mechanisms of action," *Chemosphere*, vol. 48, no. 9, pp. 965–974, 2002.

[30] M. R. Miller, R. C. Scott, G. Lorenz, J. Hardke, and J. K. Norsworthy, "Effect of insecticide seed treatment on safening rice from reduced rates of glyphosate and imazethapyr," *International Journal of Agronomy*, vol. 2016, Article ID 7623743, 2016.

[31] W. Bailey, C. DiFonzo, E. Hodgson et al., "The effectiveness of neonicotinoid seed treatments in soybean," 2015, https://www.extension.umn.edu/agriculture/soybean/pest/docs/effectiveness-of-neonicotinoid-seed-treatments-in-soybean.pdf.

[32] L. Purcell, M. Salmeron, and L. Ashlock, *Soybean Growth and Development. Arkansas Soybean Production Handbook. Arkansas Cooperative Extension Service Miscellaneous Publications 197*, University of Arkansas, Little Rock, AR, USA, 2014.

[33] Anonymous (2016), "USDA web soil survey," 2017, https://websoilsurvey.sc.egov.usda.gov/App/WebSoilSurvey.aspx

Transforming Triple Cropping System to Four Crops Pattern: An Approach of Enhancing System Productivity through Intensifying Land Use System in Bangladesh

Md. Aminul Islam,[1] Md. Jahedul Islam,[2] M. Akkas Ali,[3] A. S. M. Mahbubur Rahman Khan,[3] Md. Faruque Hossain,[3] and Md. Moniruzzaman[4]

[1]On-Farm Research Division, BARI, Masterpara, Gaibandha, Bangladesh
[2]Regional Wheat Research Center, BARI, Shyampur, Rajshahi, Bangladesh
[3]On-Farm Research Division, BARI, Joydebpur, Gazipur, Bangladesh
[4]On-Farm Research Division, BARI, Pabna, Bangladesh

Correspondence should be addressed to Md. Aminul Islam; amin.agron@gmail.com

Academic Editor: David Clay

Changing three crops pattern to four crops can play a potential role for achieving countries food security. With this view to increase crop productivity, production efficiency, land use efficiency, and economic return through intensifying cropping intensity as well as crop diversity by transforming three crops pattern to four crops, the experiment was conducted in High Ganges River Floodplain Soils under the Agro-Ecological Zone (AEZ) 11 at Pali, Durgapur, under the Multilocation Testing Site, Puthia, Rajshahi, for two consecutive years 2014-15 and 2015-16. Four crops pattern mustard-onion/maize-T. *Aman* rice was tested at on-farm condition over the existing three crops pattern mustard-onion-T. *Aman* rice. Maize was introduced here as a relay crop with onion to fit it in the four crops pattern. The experiment was laid out in RCB design with six dispersed replications. Two-year crop cycles were completed, and data regarding component crops yield were considered for assessing the performance of the two cropping patterns for making a sense of comparing productivity. Although there was no significant difference in component crops yield between four crops and three crops pattern, as an additional crop, maize tremendously increased the system productivity and economic return of the four crops pattern. Higher rice equivalent yield 28.96 t·ha^{-1} in 2013-14 and 30.95 t·ha^{-1} in 2014-15 was recorded from the four crops pattern with a mean rice equivalent yield (REY) 29.95 t·ha^{-1} over the existing pattern with a mean value 21.76 t·ha^{-1}. However, four crops pattern resulted in higher cultivation cost due to growing maize as an additional crop; nevertheless, it gave the higher gross return, marginal return, marginal benefit cost ratio, and production efficiency. The four crops pattern resulted averagely 37.63% higher production (REY) compared to the existing three crops pattern. Production as well as land use efficiency were increased by 9.33% and 19.18%, respectively, from the intensified alternate pattern.

1. Introduction

Trend of available agricultural land over the time is decreasing with an amount to 1.33 lac ha in 1976 (91.83% of total land area) that is decreased to 1.27 ha in 2000 with yearly loss of 23,391 ha. The area is further dropped to 1.21 lac ha in 2010 with yearly loss of 56,537 ha. A sum of 1.13 lac ha land has been lost during the past 34 years from 1976 to 2010. Rate of cropland shifting to nonagricultural land (housing, industry, etc.) is alarming as it is associated with the food security of the country. Total cropland was estimated to be 9,761,450 ha, 9,439,541 ha, and 8,751,937 ha in 1976, 2000, and 2010 with an average decrease of 0.14% during 1976 to 2000 and 0.73% during 2000–2010, respectively. Hasan et al. reported the rate of change over the 34 years is 0.30% which is still declining [1].

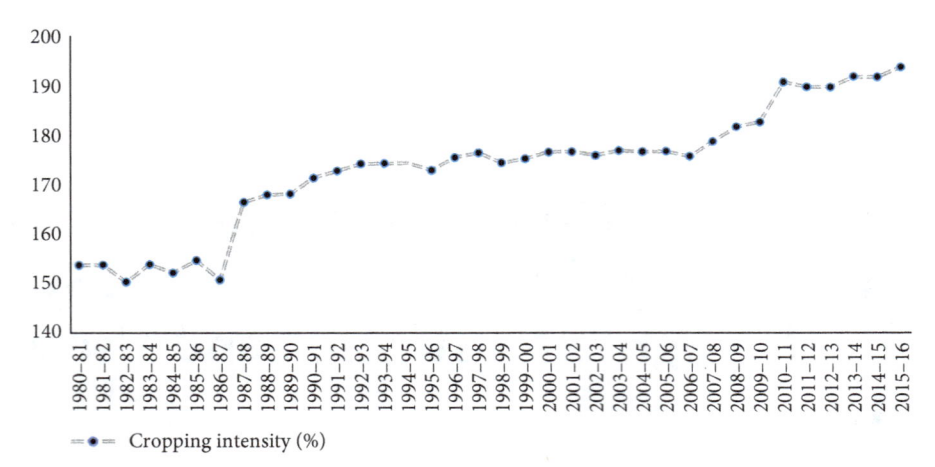

FIGURE 1: Trend of cropping intensity (%) in Bangladesh since 1980-81 to 2015-16 (BBS).

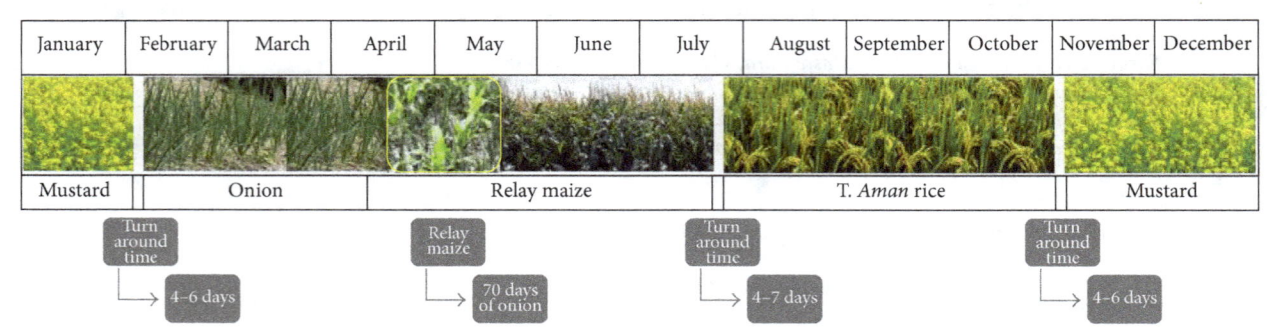

FIGURE 2: Calendar year of four crops pattern.

As per Bangladesh Bureau of Statistics (BBS) [2], around 3795 and 1688 thousand ha of land remain under double and triple cropped area, respectively, which means that 48.41% and 21.53% of the country's net cropped area has avenues partly or a major portion to be brought under quadruple cropping system. The area of cropland is decreasing, that is why there is no option of horizontal expansion but intensifying land use system through multiple cropping or by growing more and more crops on the same piece of land in a calendar year. However, the agriculture is heading towards a new paradigm to address the country's food security a concerned issue of Bangladesh. Alauddin and Tisdell [3] investigated that food production has been increased to 3.67 folds over the period of time, from an estimated amount of 1.01 m ton in 1971 during the independence of Bangladesh to 3.71 m ton in 2016 as recorded in BBS [2]. Cultivation of modern crop varieties, improvising cultural operations, and crop protection measures as well as increasing crop intensity (Figure 1) collectively contributed to such achievement.

Mustard-onion-T. *Aman* rice is a popular cropping pattern in Durgapur Upazilla of Rajshahi district under High Ganges River Floodplain. Upazila agricultural office of DAE claims that there are 2000 hectare of land in the upazila in which the pattern mustard-onion-T. *Aman* rice is being practiced during 2012 cropping year. Mustard is usually grown in Rabi season. Sowing is started in the last week of October to 1st or 2nd week of November, and the crop is harvested in the end of January to mid-February followed by cultivation of onion planted at 2nd week to end of February. After harvest of onion in April, farmers go for T. *Aman* rice cultivation in the end of July to 2nd or 3rd week of August. Thus, after harvest of onion, the land remains fallow from last week of April to July until T. *Aman* rice being transplanted. Some commercial maize hybrids grown in *Kharif* season mature in 95 to 105 days. Therefore, there is a chance of developing four crops pattern as maize can easily be grown as relay crop with onion in between the fallow period as shown in Figure 2. Tanveer et al. [4] reviewed the benefits of relay cropping that enhanced sustainable system productivity through efficient use of available resources like microclimate, nutrients dynamics [5–8]. Inclusion of fourth crop in the sequence will increase system productivity as well as improve farmer's economic condition. Keeping that in the point of view, the study was undertaken to increase diversification and intensification of existing Mustard-onion-T. *Aman* rice cropping pattern in terms of productivity, production efficiency, land use efficiency, and economic return.

2. Materials and Methods

2.1. Site Selection. The experiment was conducted at the farmers' field of Pali, Durgapur, under the Multilocation Testing Site (MLT), Shibpur, Puthia, Rajshahi, during 2014-15 to 2015-16 to assess the performance of alternate four crops pattern against existing three crops pattern (mustard-onion-T.

Table 1: Details of cultural practices adopted for different crops in field experiments.

Crop	Cultivar	Seed rate (kg·ha^{-1})	Spacing (cm^2)	Date of sowing	Date of harvesting	Rate of fertilizer application (kg·ha^{-1})					
						N	P	K	S	Zn	B
Mustard	BARI Sarisha-14	07	30 cm solid row	01–05 Nov	01–05 Feb	78	15	42	10	2	1.5
Onion	BARI Piaz-2	08	15 × 10	05–10 Feb	10–15 May	93	29	85	19	2	—
Maize	NK-40	21	60 × 20	10–15 Apr	15–20 Jul	250	53	100	40	—	—
T. *Aman* rice	BRRI dhan33	30	20 × 15	20–25 Jul	25–31 Oct	62	6	19	7	2	—

Aman rice) in High Ganges River Floodplain soils. Geographically, the experimental field is located at 20°024.49/N latitude and 88°47.20/E longitude with the elevation of 20 m above the sea level. The land was selected based on the discussion with local farmers, DAE personnel, and available secondary information.

2.2. Information Related to AEZ-11.
There is an overall pattern of olive-brown silt loams and silty clay loams on the upper parts of floodplain ridges and dark grey, mottled brown, mainly clay soils on ridge sites and in basins. Most ridge soils are calcareous throughout. General soil types predominantly include calcareous dark grey floodplain soils and calcareous brown floodplain soils. Organic matter content in brown ridge soils is low and higher in dark grey soils. Soils are slightly alkaline in reaction. General fertility level is low.

2.3. Nutrient Status.
The experimental soil was slightly alkaline (8.4) having low organic matter (1.78%). The soil contained 0.08% total N (very low), 25.85 ppm available P (optimum), 0.16 me% exchangeable K (low), 15.85 ppm available S (medium), 0.50 ppm available B (high), and 0.20 ppm available Zn (very low). The rates of the fertilizers for different crops were calculated using soil test values based (STB) on high yield goal as per Fertilizer Recommendation Guide' 2012 [9].

2.4. Experimental Procedure.
The experiment was designed with two cropping patterns following randomized complete block design with six dispersed replications. The unit plot size was 8 m × 4 m. The treatments were C_1: mustard-onion/maize-T. *Aman* rice and C_2: mustard-onion-T. *Aman* rice. The sources of nutrients were urea for N, TSP for P, M_OP for K, gypsum for S, zinc sulphate for Zn, and Boric acid for B. The details of the varieties used and cultural operations adopted in different crop sequences are given in Table 1. For mustard, half of urea and all other inorganic fertilizers were applied according to individual plot and mixed with soil at the time of final land preparation. The rest of urea was top dressed before flowering just one day after first irrigation. In case of onion, half of urea and all other inorganic fertilizers were applied to individual plot and mixed with soil at the time of final land preparation. The rest of urea was top dressed 30 days after planting. Maize was planted 70 days after planting of onion. For maize, half of urea and all other inorganic fertilizers were applied according to individual plot after harvesting of onion at 6-leaf stage of maize as side dressing followed by earthing up. The rest urea was top dressed at 10-leaf stage. In case of T. *Aman* rice, all the fertilizers except urea were applied as basal. Urea was applied as top dress in three equal splits at 15, 30, and 45 days after transplanting. All the crops were harvested at maturity from an area of central 5.0 m^2. Data on yield of component crops in sequences were recorded and converted to ton per hectare. Total system productivity was calculated as summation of individual crop yield of each cropping cycle. The productivity of different crop sequences was compared by calculating their economic rice equivalent yield (REY) using formula given by Ahlawat and Sharma [10], where

$$\text{REY} = \frac{\text{Yield of each crop } (t \cdot ha^{-1}) \times \text{economic value of respective crop } (Tk \cdot t^{-1})}{\text{Price of rice grain } (Tk \cdot t^{-1})}. \quad (1)$$

2.4.1. Land Use Efficiency.
Land use efficiency (LUE) was estimated by the total duration of crops in the sequence divided by 365 days and expressed in % as outlined by Jamwal [11]:

$$\text{LUE} = \frac{\sum \text{Dc}}{365} \times 100, \quad (2)$$

where Dc = duration of crops in the sequence.

2.4.2. Production Efficiency.
Production efficiency (PE) was calculated by taking total economic yield of the sequence on wheat equivalent basis divided duration of crops using per formula by Jamwal [11].

$$\text{PE} = \frac{\text{REY}}{\sum \text{Dc}}, \quad (3)$$

where REY = rice equivalent yield in a sequence and Dc = duration of crops in that sequence.

2.4.3. Marginal Benefit Cost Ratio (MBCR).
The economic analysis was done following the method suggested by CIMMYT [12]. The MBCR can be computed as the marginal

TABLE 2: Productivity of component crops (mustard and onion) in different cropping patterns.

Cropping pattern	Cropping year 2013-14		Cropping year 2014-15	
	Seed yield of mustard (t·ha^{-1})	Bulb yield of onion (t·ha^{-1})	Seed yield of mustard (t·ha^{-1})	Bulb yield of onion (t·ha^{-1})
Mustard-onion/maize-T. *Aman* rice	1.56	14.60	1.61	15.68
Mustard-onion-T. *Aman* rice	1.48	14.82	1.55	15.10
Level of significance	NS	NS	NS	NS
SE (±)	0.35	0.51	0.26	0.57
CV (%)	4.23	5.47	5.21	4.37

TABLE 3: Productivity of component crops (maize and T. *Aman* rice) in different cropping patterns.

Cropping pattern	Cropping year 2013-14				Cropping year 2014-15			
	Yield of maize (t·ha^{-1})		Yield of T. *Aman* rice (t·ha^{-1})		Yield of maize (t·ha^{-1})		Yield of T. *Aman* rice (t·ha^{-1})	
	Grain	Stover	Grain	Straw	Grain	Stover	Grain	Straw
Mustard-onion/maize-T. *Aman* rice	7.65	8.43	4.54	5.88	8.39	9.13	4.77	6.06
Mustard-onion-T. *Aman* rice	—	—	4.44	5.46	—	—	4.48	5.75
Level of significance	—	—	NS	NS	—	—	NS	NS
SE (±)	—	—	0.36	1.23	—	—	0.19	0.16
CV (%)	—	—	4.59	5.74	—	—	10.32	8.50

value product (MVP) over the marginal value cost (MVC). It can be computed as

$$\text{MBCR} = \frac{\text{MVP (over control)}}{\text{MVP (over control)}}. \quad (4)$$

All the data were statistically analyzed following the F-test and the mean comparisons were made by DMRT at 5% level as per the outline by Gomez and Gomez [13].

3. Results and Discussion

3.1. Yield of Mustard. Mustard was the first crop in both the cropping patterns. There was no significant difference of mustard yield in between the cropping patterns. However, numerically higher yield (1.56 t·ha^{-1} in 2013-14 and 1.61 t·ha^{-1} in 2014-15) was observed in the improved pattern followed the existing one (Table 2).

3.2. Yield of Onion. Onion was the second crop in both the sequences and planted in the late *Rabi* season (February). There was no significant difference between the cropping patterns on the bulb yield of onion. However, from the 1st year observation, numerically higher bulb yield was recorded in the existing cropping pattern (mustard-onion-T. *Aman* rice) than four crops pattern. The 2nd year trend was contrary to the first year result though the difference was statistically nonsignificant (Table 2).

3.3. Yield of Maize. Maize was the included crop in the improved pattern and grown as a relay crop with onion in *Kharif-I* season (April), whereas, in the existing pattern, this period remained fallow. From the two years' observation, it was clear that maize can effectively be grown as relay with onion to fit it in the four crops pattern by saving time. Inclusion of maize in the fallow period characteristically

TABLE 4: System productivity (REY*) of component crops in different cropping patterns.

Cropping pattern	System REY (t·ha^{-1})	
	2013-14	2014-15
Mustard-onion/maize-T. *Aman* rice	28.96	30.95
Mustard-onion-T. *Aman* rice	21.51	22.01
Level of significance	0.05	0.05
SE (±)	2.65	2.96
CV (%)	6.21	8.62

*Rice equivalent yield.

increased the system's equivalent yield. Generally, farmers use excess amount of fertilizers for preceding crop onion. Therefore, maize can be grown with a minimum support regarding input cost. However, maize produced 7.65 ton grain and 8.43 ton stover ha^{-1} in 2013-14 and 8.39 ton grain and 9.13 ton stover ha^{-1} in 2014-15 which contributed to a higher REY of 29.95 t·ha^{-1} in four crop-based pattern (Tables 3–5). Correia et al. [14] observed 209.57% higher yield from maize as intercropped/relayed with mucuna.

3.4. Yield of T. Aman Rice. T. *Aman* rice is the common crop in the *Kharif-II* (July–October) season in both cropping patterns. Grain yield of T. *Aman* rice is a complex character depending on a large number of environmental, morphological, and physiological characters. Grain yields also depend upon their yield components. However, there was no significant difference on grain yield of rice in between the cropping patterns but numerically higher (2.25% in 2013-14 and 6.47% in 2014-15) grain yield observed in the imposed pattern might be due to the residual effect of fertilizers applied to the preceding maize crop (Table 3). Jabbar et al. [15] observed the positive impact on residual soil fertility when crops are grown as inter/relay cropping system.

TABLE 5: Field crop duration, production efficiency, and land use efficiency as influenced by cropping patterns.

Cropping pattern	Mean REY (t·ha^{-1})	Field duration of crop sequence (day)	Production efficiency (kg^{-1}·ha^{-1}·day^{-1})	Land use efficiency (%)
Mustard-onion/maize-T. *Aman* rice	29.95	340	88.10	93.15
Mustard-onion-T. *Aman* rice	21.76	270	80.58	73.97

TABLE 6: Profitability of the four crops pattern over the existing pattern.

Cropping pattern	Gross return (Tk.·ha^{-1})	Cultivation cost (TVC) (Tk.·ha^{-1})	Marginal value (MVP) (Tk.·ha^{-1})	Marginal cost (MVC) (Tk.·ha^{-1})	MBCR*
Mustard-onion/maize-T. *Aman* rice	539180	285542	147540	46894	3.14
Mustard-onion-T. *Aman* rice	391640	238648	—	—	—

*Marginal benefit cost ratio; input: urea: 16 Tk·kg^{-1}, TSP: 22 Tk·kg^{-1}, MoP: 15 Tk·kg^{-1}, gypsum: 6 Tk·kg^{-1}, zinc sulphate: 120 Tk·kg^{-1}, boric acid: 150 Tk·kg^{-1}, furadan: 150 Tk·kg^{-1}, mustard seed: 100 Tk·kg^{-1}, tillage cost: 10000 Tk·ha^{-1}, irrigation (1 time): 1000 Tk·ha^{-1}, and labour: 200 Tk·day^{-1} (8 hours); output: rice grain: 18 Tk·kg^{-1}, mustard seed: 50 Tk·kg^{-1}, onion bulb: 15 Tk·kg^{-1}, maize grain: 16 Tk·kg^{-1}, maize stover: 1 Tk·kg^{-1}, and rice straw: 2 Tk·kg^{-1}.

3.5. System Productivity. System productivity was considered as rice equivalent yield (REY). The system REY significantly differed between the cropping patterns. However, data are presented in Table 2. In general, the pattern involving four crops produced significantly greater REY than that having three crops pattern. However, mustard-onion/maize-T. *Aman* rice showed higher productivity in terms of REY (28.96 t·ha^{-1} in 2013-14 and 30.95 t·ha^{-1} in 2014-15) with a mean REY 29.95 t·ha^{-1} (Table 5) than mustard-onion-T. *Aman* rice cropping pattern (21.76 t·ha^{-1}). Total productivity increased by 37.63% in the maize-included four crops pattern. Mondal et al. [16] also claimed of having 49 to 67% higher productivity from the intensified land use system. Total field duration of crops by cropping pattern is presented in Table 5. Total field duration was the higher in mustard-onion/maize-T. *Aman* rice cropping pattern (340 days) due to additional maize crop cultivation and found lower (270 days) in the existing mustard-onion-T. *Aman* rice cropping pattern because of not cultivating maize in *kharif* season.

3.6. Production Efficiency (PE). The cropping patterns showed variation on production efficiency (PE) (Table 5). The pattern having four crops generated the higher PE (88.10 kg^{-1}·ha^{-1}·day^{-1}). This is due to the higher productivity of this sequence in which the contribution of maize is quite obvious. However, the mustard-onion-T. *Aman* rice gave the 9.33% lower PE (80.58 kg^{-1}·ha^{-1}·day^{-1}) in the sequences. Higher PE associated with improved cropping pattern coupled with modern management practices were noted by Nazrul et al. [17], Khan et al. [18, 19], and Krrishna and Reddy [20]. Though not studied here, inter/relay crops could increase system production by suppressing the weed growth [21].

3.7. Land Use Efficiency (LUE). Land use efficiency (LUE) varied according to the cropping patterns (Table 5). In general, patterns intensified by four crops resulted in 19.18% higher LUE than the triple cropping system. The higher LUE (93.15%) was recorded in mustard-onion/maize-T. *Aman* rice whereas the lower LUE (73.97%) was recorded in mustard-onion-T. *Aman* rice cropping pattern. The results are in agreement with Kamrozzaman et al. [22].

3.8. Economic Performance. Based on two years' observation, economic performance of the patterns is presented in Table 6. Cropping pattern attributed a remarkable impact on variable cost, marginal return, and marginal benefit cost ratio (MBCR). The annual gross return, cultivation cost, marginal return, and marginal cost were considered for assessing the suitability of the cropping pattern. In general, inclusion of the fourth crop (maize) markedly enhanced both the return and cultivation cost. Though maize accounted for 19.64% higher cultivation cost, consequently it contributed to 37.63% higher return in the four crops pattern. The mustard-onion/maize-T. *Aman* rice had a maximum gross return (Tk. 539180 ha^{-1}) along with a higher cultivation cost (Tk. 285542 ha^{-1}), which also contributed to higher marginal return (Tk. 147540 ha^{-1}) and MBCR (3.14) than mustard-onion/maize-T. *Aman* rice-based three crops pattern.

4. Conclusion

Relaying maize with onion is a simple but an effective time space saving technology for shifting the thee crops pattern to four crops (mustard-onion/maize-T. *Aman* rice) one. Considering systems REY, LUE, PE, and economic performances of the two-year crop cycle, it is revealed that four crops pattern is the best option for greater productivity and profitability over the triple cropping system.

Acknowledgments

The authors are thankful to the On-Farm Research Division, Bangladesh Agricultural Research Institute, Gazipur, for providing financial help and logistic support, respectively.

References

[1] M. N. Hasan, M. S. Hossain, M. A. Bari, and M. R. Islam, *Agricultural Land Availability in Bangladesh*, SRDI, Dhaka, Bangladesh, 2013.

[2] BBS (Bangladesh Bureau of Statistics), *Yearbook of Agricultural Statistics*, Statistics and Informatics Division (SID) Ministry of Planning, Dhaka, Bangladesh, 2016.

[3] M. Alauddin and C. Tisdell, "Trends and projections for Bangladeshi food production: an alternative viewpoint," *Food Policy*, vol. 12, no. 4, pp. 318–331, 1987.

[4] M. Tanveer, S. A. Anjum, S. Hussain, A. Cerdà, and U. Ashraf, "Relay cropping as a sustainable approach: problems and opportunities for sustainable crop production," *Environmental Science and Pollution Research*, vol. 24, no. 8, pp. 6973–6988, 2017.

[5] H. Knörzer, H. Grözinger, S. Graeff-Hönninger, K. Hartung, H. P. Piepho, and W. Claupein, "Integrating a simple shading algorithm into CERES-wheat and CERES-maize with particular regard to a changing microclimate within a relay-intercropping system," *Field Crops Research*, vol. 121, no. 2, pp. 274–285, 2011.

[6] A. C. Gaudin, K. Janovicek, R. C. Martin, and W. Deen, "Approaches to optimizing nitrogen fertilization in a winter wheat–red clover (*Trifolium pratense* L.) relay cropping system," *Field Crops Research*, vol. 155, pp. 192–201, 2014.

[7] C. Amossé, M. H. Jeuffroy, B. Mary, and C. David, "Contribution of relay intercropping with legume cover crops on nitrogen dynamics in organic grain systems," *Nutrient Cycling in Agroecosystems*, vol. 98, no. 1, pp. 1–14, 2014.

[8] J. S. Schepers, D. D. Francis, and J. F. Shanahan, "Relay cropping for improved air and water quality," *Journal of Biosciences*, vol. 60, pp. 186–189, 2005.

[9] BARC (Bangladesh Agricultural Research Council), *Fertilizer Recommendation Guide*, BARC, Farmgate, Dhaka, Bangladesh, 2005.

[10] I. P. S. Ahlawat and R. P. Sharma, *Agronomid Terminology*, Indian Society of Agronomy, New Delhi, India, 3rd edition, 1993.

[11] J. S. Jamwal, "Productivity and economics of different maize (*Zea mays*) based crop sequences under dryland conditions," *Indian Journal of Agronomy*, vol. 46, no. 4, pp. 601–604, 2001.

[12] CIMMYT, *From Agronomic Data to Farmer Recommendations: An Economic Training Manual*, International Maize and Wheat Improvement Center, El Batan, MEX, Mexico, 1988.

[13] K. A. Gomez and A. A. Gomez, *Statistical Procedures for Agricultural Research, An International Rice Research Institute Book*, John Wiley and Sons, New York, NY, USA, 2nd edition, 1984.

[14] M. V. Correia, L. C. R. Pereira, L. D. Almeida et al., "Maize-mucuna (*Mucuna pruriens* (L.) DC) relay intercropping in the lowland tropics of Timor-Leste," *Field Crops Research*, vol. 156, pp. 272–280, 2014.

[15] A. Jabbar, R. Ahmad, I. H. Bhatti, T. Aziz, M. Nadeem, and R. A. Wasi-u-Din, "Residual soil fertility as influenced by diverse rice-based inter/relay cropping systems," *International Journal of Agriculture and Biology*, vol. 13, pp. 477–483, 2011.

[16] R. I. Mondal, F. Begum, A. Aziz, and S. H. Sharif, "Crop sequences for increasing cropping intensity and productivity," *SAARC Journal of Agriculture*, vol. 13, no. 1, pp. 135–147, 2015.

[17] M. I. Nazrul, M. R. Shaheb, M. A. H. Khan, and A. S. M. M. R. Khan, "On-Farm evaluation of production potential and economic returns of potato-rice based improved cropping system," *Bangladesh Agronomy Journal*, vol. 16, no. 2, pp. 41–50, 2013.

[18] M. A. Khan, S. M. A. Hossain, and M. A. H. Khan, "A study on some selected jute-based cropping patterns at Kishoregonj," *Bangladesh Journal of Agricultural Research*, vol. 31, no. 1, pp. 85–95, 2006.

[19] M. A. H. Khan, M. A. Quayyum, M. I. Nazrul, N. Sultana, and M. R. A. Mollah, "On-farm evaluation of production potential and economics of mustard-rice based improved cropping system," *Journal of Social and Economic Development*, vol. 2, no. 1, pp. 37–42, 2005.

[20] A. Krishna and A. K. Reddy, "Production potential and economics of different rice (*Oryza sativa*) based cropping systems in Andhra Pradesh," *Indian Journal of Agricultural Sciences*, vol. 67, no. 12, pp. 551–553, 1997.

[21] C. Amossé, M. H. Jeuffroy, F. Celette, and C. David, "Relay-intercropped forage legumes help to control weeds in organic grain production," *European Journal of Agronomy*, vol. 49, pp. 158–167, 2013.

[22] M. M. Kamrozzaman, M. A. H. Khan, S. Ahmed, and A. F. M. Ruhul Quddus, "On-farm evaluation of production potential and economics of Wheat Jute-T.aman rice-based cropping system," *Journal of the Bangladesh Agricultural University*, vol. 13, no. 1, pp. 93–100, 2015.

PERMISSIONS

All chapters in this book were first published in IJA, by Hindawi Publishing Corporation; hereby published with permission under the Creative Commons Attribution License or equivalent. Every chapter published in this book has been scrutinized by our experts. Their significance has been extensively debated. The topics covered herein carry significant findings which will fuel the growth of the discipline. They may even be implemented as practical applications or may be referred to as a beginning point for another development.

The contributors of this book come from diverse backgrounds, making this book a truly international effort. This book will bring forth new frontiers with its revolutionizing research information and detailed analysis of the nascent developments around the world.

We would like to thank all the contributing authors for lending their expertise to make the book truly unique. They have played a crucial role in the development of this book. Without their invaluable contributions this book wouldn't have been possible. They have made vital efforts to compile up to date information on the varied aspects of this subject to make this book a valuable addition to the collection of many professionals and students.

This book was conceptualized with the vision of imparting up-to-date information and advanced data in this field. To ensure the same, a matchless editorial board was set up. Every individual on the board went through rigorous rounds of assessment to prove their worth. After which they invested a large part of their time researching and compiling the most relevant data for our readers.

The editorial board has been involved in producing this book since its inception. They have spent rigorous hours researching and exploring the diverse topics which have resulted in the successful publishing of this book. They have passed on their knowledge of decades through this book. To expedite this challenging task, the publisher supported the team at every step. A small team of assistant editors was also appointed to further simplify the editing procedure and attain best results for the readers.

Apart from the editorial board, the designing team has also invested a significant amount of their time in understanding the subject and creating the most relevant covers. They scrutinized every image to scout for the most suitable representation of the subject and create an appropriate cover for the book.

The publishing team has been an ardent support to the editorial, designing and production team. Their endless efforts to recruit the best for this project, has resulted in the accomplishment of this book. They are a veteran in the field of academics and their pool of knowledge is as vast as their experience in printing. Their expertise and guidance has proved useful at every step. Their uncompromising quality standards have made this book an exceptional effort. Their encouragement from time to time has been an inspiration for everyone.

The publisher and the editorial board hope that this book will prove to be a valuable piece of knowledge for researchers, students, practitioners and scholars across the globe.

LIST OF CONTRIBUTORS

Atef M. K. Nassar and Danielle J. Donnelly
Plant Science Department, Macdonald Campus of McGill University, 21111 Lakeshore Road, Sainte Anne de Bellevue, QC, Canada H9X 3V9

Atef M. K. Nassa and Stan Kubow
School of Dietetics and Human Nutrition, Macdonald Campus of McGill University, 21111 Lakeshore Road, Sainte Anne de Bellevue, QC, Canada H9X 3V9

Atef M. K. Nassar
Plant Protection Department, Faculty of Agriculture, Damanhour University, Damanhour, Albeheira 22516, Egypt

Carlos Vilcatoma-Medina and Flávio Zanette
Programa de Pós-Graduação em Agronomia-Produção Vegetal, Departamento de Fitotecnia, Federal University of Paraná, Curitiba, PR, Brazil

Glaciela Kaschuk
Programa de Pós-Graduação em Ciência do Solo, Departamento de Solos e Engenharia Agrícola, Federal University of Paraná, Curitiba, PR, Brazil

Rahena Parvin Rannu, Razu Ahmed and Alamgir Siddiky
Soil & Water Management Section, Horticulture Research Centre, Bangladesh Agricultural Research Institute (BARI), Gazipur 1701, Bangladesh

Abu Saleh Md. Yousuf Ali
Regional Agricultural Research Station, Bangladesh Agricultural Research Institute (BARI), Gazipur 1701, Bangladesh

Khandakar Faisal Ibn Murad and Pijush Kanti Sarkar
Irrigation & Water Management Division, Bangladesh Agricultural Research Institute (BARI), Gazipur 1701, Bangladesh

Y. S. Chen
Malaysian Pepper Board, Jalan Utama, Pending Industrial Area, Sarawak 93916, Malaysia

M. Dayod
Agriculture Research Centre, Semongok, Kuching, Sarawak 93720, Malaysia

C. S. Tawan
Faculty of Resource Science and Technology, Universiti Malaysia Sarawak, Jalan Datuk Mohd Musa, Kota Samarahan, Sarawak 94300, Malaysia

Carlos H. Galeano Mendoza
Colombian Corporation for Agricultural Research (Corpoica), CI Palmira, Palmira, Colombia

Edison F. Baquero Cubillos, José A. Molina Varón and María del Socorro Cerón Lasso
Colombian Corporation for Agricultural Research (Corpoica), CI Tibaitatá, Mosquera, Colombia

M. Azalia Lozano-Grande and Alma Leticia Martínez-Ayala
Centro de Desarrollo de Productos Bióticos, Instituto Politécnico Nacional, San Isidro, 62731 Yautepec, MOR, Mexico

Shela Gorinstein
Institute for Drug Research, School of Pharmacy, Hadassah Medical School, The Hebrew University, Jerusalem 91120, Israel

Eduardo Espitia-Rangel
Instituto Nacional de Investigaciones Forestales, Agrícolas y Pecuarias, Campo Experimental Valle de México, Coatlinchan, 56250 Texcoco, MEX, Mexico

Gloria Dávila-Ortiz
Instituto Politécnico Nacional, Escuela Nacional de Ciencias Biológicas, Delegación Miguel Hidalgo, 11340 Ciudad de México, Mexico

Licet Paola Molina-Guzmán, Paula Andrea Henao-Jaramillo and Lina Andrea Gutiérrez-Builes
Grupo Biología de Sistemas, Facultad de Medicina, Universidad Pontificia Bolivariana, Calle 78B No. 72A-109, Medellín, Colombia

Licet Paola Molina-Guzmán and Leonardo Alberto Ríos-Osorio
Grupo de Investigación Salud y Sostenibilidad, Escuela de Microbiología, Universidad de Antioquia, Calle 67 No. 53-108, Medellín, Colombia

Krishna Bhattarai and Dilip R. Panthee
Department of Horticultural Science, North Carolina State University, Mountain Horticultural Crops Research and Extension Center, Mills River, NC 28759, USA

List of Contributors

Sadikshya Sharma
Department of Environmental Horticulture, University of Florida, 2550 Hull Rd, Gainesville, FL 32611, USA

Cristina Monteiro and Helena Oliveira
Laboratory of Biotechnology and Cytomics, University of Aveiro, Campo Santiago, 3810-193 Aveiro, Portugal

Sara Sario, Rafael Mendes, Nuno Mariz-Ponte, Verónica Bastos, Conceição Santos and Maria Celeste Dias
Department of Biology, LAQV/REQUIMTE, Faculty of Sciences, University of Porto, Rua Campo Alegre, 4169-007 Porto, Portugal

Sónia Silva
QOPNA, Department of Chemistry, University of Aveiro, 3810-193 Aveiro, Portugal

Maria Celeste Dias
Center for Functional Ecology (CEF), Department of Life Science, University of Coimbra, Calçada Martim de Freitas, 3000-456 Coimbra, Portugal

Tolera Abera, Tolcha Tufa, Tesfaye Midega, Haji Kumbi and Buzuayehu Tola
Natural Resources Management Research Process, Ambo Agricultural Research Center, Ethiopia Institute of Agricultural Research Institute, Ambo, West Showa, Oromia, Ethiopia

C. M. Maszura, S. M. R. Karim, M. Z. Norhafizah, F. Kayat and M. Arifullah
Faculty of Agro-Based Industry, Universiti Malaysia Kelantan, Jeli Campus, Kota Bharu, Malaysia

Yigrem Mengist, Assefa Sintayehu and Sanjay Singh
Department of Plant Sciences, College of Agriculture and Rural Transformation, University of Gondar, Gondar, Ethiopia

Samuel Sahile
Department of Biology, College of Natural and Computational Sciences, University of Gondar, Gondar, Ethiopia

Michel Choairy de Moraes and Manuel Benito Sainz
Syngenta Proteção de Cultivos Ltda, 18001 Avenida das Nações Unidas, São Paulo, SP, Brazil

Ana Carolina Ribeiro Guimarães and Dilermando Perecin
UNESP, FCAV, BR 14884-900 Jaboticabal, SP, Brazil

Manuel Benito Sainz
Syngenta Crop Protection LLC, 9 Davis Drive, Research Triangle Park, NC, USA

Massaoudou Hamidou, Abdoul Kader M. Souley, Issoufou Kapran and Oumarou Souleymane
Institut National de la Recherche Agronomique du Niger (INRAN), BP 429, Niamey, Niger

Massaoudou Hamidou, Oumarou Souleymane, Eric Yirenkyi Danquah, Kwadwo Ofori and Vernon Gracen
West Africa Centre for Crop Improvement, University of Ghana, PMB 30, Legon, Accra, Ghana

Vernon Gracen
Department of Plant Breeding and Genetics, Cornell University, 520 Bradfield Hall, Ithaca, NY 14850, USA

Malick N. Ba
International Crops Research Institute for the Semi-Arid Tropics, BP 12404, Niamey, Niger

E. K. Hailu, Y. D. Urga, N. A. Sori, F. R. Borona and K. N. Tufa
Irrigation and Drainage Research, Werer Agricultural Research Center, EIAR, Addis Ababa, Ethiopia

G. Zenawi, T. Goitom and B. Fiseha
Tigray Agricultural Research Institute, Humera Agricultural Research Center, Humera, Ethiopia

Benjamin A. Okorley, Charles Agyeman, Naalamle Amissah and Seloame T. Nyaku
Department of Crop Science, University of Ghana, Legon-Accra, Ghana

Emmanuel Ayipio, Richard Yaw Agyare and Samuel Kwame Bonsu
CSIR-Savanna Agricultural Research Institute, Nyankpala, N/R, Ghana

Moomin Abu and Dorothy Ageteba Azewongik
University for Development Studies, Faculty of Agriculture, Department of Horticulture, Nyankpala, N/R, Ghana

Hakeem A. Ajeigbe, Folorunso Mathew Akinseye, Kunihya Ayuba and Jerome Jonah
International Crop Research Institute for the Semi-Arid Tropics (ICRISAT), Kano, Nigeria

Folorunso Mathew Akinseye
Department of Meteorology and Climate Science, Federal University of Technology Akure, Akure, Nigeria

Zine El Abidine Fellahi
Department of Agronomy, Faculty of Natural, Life and Earth Sciences and the Universe, University of Mohamed El Bachir El Ibrahimi, 34034 Bordj Bou Arréridj, Algeria

Abderrahmane Hannachi
National Agronomic Research Institute of Algeria (INRAA), Unit of Sétif, 19000 Sétif, Algeria

Hamenna Bouzerzour
Department of Ecology and Plant Biology, Valorization of Natural Biological Resources Laboratory, Faculty of Natural and Life Sciences, University of Ferhat Abbas Sétif-1, 19000 Sétif, Algeria

N. R. Steppig and J. K. Norsworthy
Department of Crop, Soil, and Environmental Sciences, University of Arkansas, Fayetteville, AR 72701, USA

R. C. Scott
Department of Crop, Soil, and Environmental Sciences, Lonoke Extension Center, University of Arkansas, Lonoke, AR 72086, USA

G. M. Lorenz
Department of Entomology, Lonoke Extension Center, University of Arkansas, Lonoke, AR 72086, USA

Md. Aminul Islam
On-Farm Research Division, BARI, Masterpara, Gaibandha, Bangladesh

Md. Jahedul Islam
Regional Wheat Research Center, BARI, Shyampur, Rajshahi, Bangladesh

M. Akkas Ali, A. S. M. Mahbubur Rahman Khan and Md. Faruque Hossain
On-Farm Research Division, BARI, Joydebpur, Gazipur, Bangladesh

Md. Moniruzzaman
On-Farm Research Division, BARI, Pabna, Bangladesh

Index

A
Acaulospora Scrobiculata, 29, 31-32
Alternate Furrow Irrigation, 150
Arbuscular Mycorrhizal Fungi, 29-30, 32-34, 77-79, 84-85

B
Barley, 108-112, 114-115, 179, 194, 198
Black Pepper, 45-47, 51-55
Black Polythene Mulch, 35-38, 40-42
Bovine Production Systems, 77-78
Bulb Onion, 56, 59
Bunching Onion, 56-57, 59-63
Butylated Hydroxyl Toluene, 4

C
Carotenoids, 1-2, 4, 14-15, 19-24, 26, 28, 61
Chromium, 19, 101, 107-108
Commelina Foecunda, 155-156
Conventional Furrow Application Method, 149-150
Crop Injury, 200-201, 204

D
Dentiscutata Heterogama, 29, 31-32
Disease Progress Curve, 124-127

F
Field Capacity, 36, 38, 40-42
Flag Leaf Area, 189-197
Flumioxazin, 200-206
Fruit Spike Length, 47, 50, 52-53
Fungicides, 61, 124-130, 202
Fusarium Wilt Disease, 124-126, 128-129

G
Genotoxicity, 101-102, 107-108
Genotypic Coefficient of Variation, 139-140, 146
Grains Per Spike, 190-197
Greenhouse, 1-3, 6-8, 10-11, 18, 22-24, 26, 33, 44, 62-63, 77, 102, 115
Guaiacol, 102

H
Harvest Index, 110, 112, 179, 185, 190-198
Herbicide, 118, 120, 190, 200-206

I
Insecticide Seed Treatment, 200-201, 203-206
Integrated Weed Management, 116, 122
Intercropping, 114, 171-174, 176-177, 212
Internode Length, 47, 52-53
Irrigation, 35-38, 40-44, 57, 79, 87, 101, 149-150, 152-154, 187-188, 198
Isopentenyl Diphosphate, 66, 74

L
Lactuca Sativa, 101-102
Land Use System, 207-208

M
Microorganisms, 64, 67, 77-79, 81-85, 102
Midge Damage, 139-147
Mitotic Index, 103, 105-106
Mulch, 35-38, 40-44
Multibud Sett, 131, 133

N
N Fertilization, 85, 178
Nitrogen, 2-3, 34, 43-44, 77-85, 111, 114-115, 178-179, 181, 183, 185-188, 199, 212

O
Organic Acids, 1, 5, 15, 17-18, 20, 26
Organic Fertilizer Rate, 110
Organic Fertilizers, 109-111
Oxidative Stress, 2, 101-102, 106-108

P
Panicle Weight, 139, 141-147
Parthenium Hysterophorus, 116, 122-123
Parthenium Infestation, 117-119
Pathogens, 77, 84, 95, 100, 169-170
Percent Fruit Set, 47, 51-53
Percentage Land Saved, 173
Phenotypic Coefficient of Variation, 139-140, 146
Phytoremediation, 101, 107
Plant Height, 38, 40, 43, 57, 59, 110-111, 139-147, 163-164, 167, 171-174, 179, 182-183, 185-186, 190-197, 204
Plant-parasitic Nematodes, 162, 169
Presprouted Seedling, 131, 133
Presprouted Setts, 131-133
Principal Component Analysis, 2, 5-6, 55, 77, 82, 86, 197-198

R
Reactive Oxygen Species, 101
Rice Equivalent Yield, 207, 210
Rice Straw Mulch, 35, 37-38, 40-42

Root Gall Scores, 162
Root-knot Nematodes, 162, 169-170
Roselle, 171-174, 176-177

S
Selection-based Index, 189-191, 194, 196
Sesamum Indicum, 150, 153, 155, 160
Single-bud Setts, 131-133
Soil Temperature, 37-38
Sorghum Midge, 139-140, 146-147
Squalene, 64-76

T
Tomato Genotypes, 86-91, 93, 95-96, 100, 163, 166, 170
Total Anthocyanin Content, 4, 12
Triple Cropping System, 207
Triterpene, 64-65, 67, 73

W
Water Use Efficiency, 36, 43, 150, 153-154, 178, 185-187
Weed Density, 116-119, 121-123
Weed Infestation, 116, 118, 155
Weed Monitoring, 117
Wilting Point, 36

CPSIA information can be obtained
at www.ICGtesting.com
Printed in the USA
BVHW011751200820
586899BV00003B/25